Introduction to Canadian Business

Introduction to Canadian Business

BRIAN E. OWEN
University of Manitoba

FREDERICK A. STARKE
University of Manitoba

JOHN A. REINECKE
University of New Orleans

WILLIAM F. SCHOELL
University of Southern Mississippi

Allyn and Bacon Canada Ltd.
Toronto

Copyright © 1981 by Allyn and Bacon Canada Ltd., Toronto. All rights reserved. No part of the material protected by this copyright notice may be reproduced or utilized in any form or by any means, electronic or mechanical, including photocopying, recording, or by any information storage and retrieval system, without written permission from the copyright owners.

Canadian Cataloguing in Publication Data

    Main entry under title:

    Introduction to Canadian business
    Includes index.
    ISBN 0-205-06998-3
    1. Business. 2. Canada – Commerce.
    I. Owen, Brian E., 1943-
    HF5351.I57   380.1'0971   C80-094308-2

Chapter opening photographs: pages 2, 250, 339, 459, 508 courtesy of Montreal Star-Canada Wide; pages 27, 189 courtesy of Miller Services Ltd.; page 49 by Fred Herzog/Image Finders; page 74, Saskatchewan Government Photograph; pages 106, 425 by Ted Grant/NFB Photothèque ONF ©; page 136 by George Hunter/NFB Photothèque ONF ©; pages 166, 534 courtesy of Image Finders; page 224 by Talbot Lovering; page 284 by David Portigal/NFB Photothèque ONF ©; page 309 by Karl Sommerer/NFB Photothèque ONF ©; page 371 by Neville Bell/NFB Photothèque ONF ©; page 402 by Crombie McNeill/NFB Photothèque ONF ©; page 480 courtesy of the Ontario Ministry of Industry and Tourism.

Portions of this book previously appeared in *Introduction to Business: A Contemporary View*, second edition, by John A. Reinecke and William F. Schoell, Copyright © 1977, 1974 by Allyn and Bacon, Inc.

10 9 8 7 6 5 4   85 84 83 82
Cover Designer: Christy Rosso
Series Editor: Jack B. Rochester
Managing Editor: Michael E. Meehan

For Gladys, Rosie, Wendy, and Aurelia

Contents

Preface	xv
SECTION ONE INTRODUCTION	1
1 Economic Ideas for Business	2
The Economic Problem	4
Needs and Wants	5
The Factors of Production	7
The Purpose of an Economic System	11
Collectivism	13
Capitalism	13
Modern Collectivism	19
Modern Capitalism	20
Summary and Look Ahead	22
Key Concepts	22
Questions for Discussion	23
2 The Business Firm	27
The Mixed Market Economy	29
The Business Firm	34
Summary and Look Ahead	44
Key Concepts	45
Questions for Discussion	46
SECTION TWO BUSINESS BASICS	48
3 Ownership	49
Public vs. Private Ownership	51
The Sole Proprietorship	51
The Partnership	52
The Corporation	54
Co-operatives	63

CONTENTS

    The Effects of Size   65
    Summary and Look Ahead   68
    Key Concepts   69
    Questions for Discussion   70

4 Organization   74
    What Is an Organization?   76
    Organization as Structure   77
    Grouping and Assigning Activities   80
    The Span of Management   81
    Types of Organization Structures   86
    The Organization Chart   90
    Individuals and Groups   94
    The Informal Organization   97
    Summary and Look Ahead   102
    Key Concepts   102
    Questions for Discussion   104

5 Management   106
    The Nature of Management   108
    The Manager as a Person   110
    The Functions of Management   113
    The Decision-making Process   124
    Summary and Look Ahead   128
    Key Concepts   129
    Questions for Discussion   130

SECTION THREE   BUSINESS DECISIONS   135

6 Producing Goods and Services   136
    What Is Production?   138
    What Is Production Management?   141
    Production and Ecology   155
    Materials Management   156
    The Purchasing Task   156
    Operations Management—A Broader View   159
    Summary and Look Ahead   161
    Key Concepts   162
    Questions for Discussion   163

7 Marketing   166
    What Is Marketing?   168
    Marketing and Utility   169
    The Marketing Concept and the Managerial Approach   170
    Product Differentiation and Market Segmentation   177

	Knowing the Consumer	178
	Patterns of Consumption	179
	Consumerism	180
	Marketing of Services	183
	Summary and Look Ahead	185
	Key Concepts	186
	Questions for Discussion	187
8	**Marketing Decisions**	189
	Product	191
	Distribution	196
	Promotion	204
	Price	210
	Summary and Look Ahead	215
	Key Concepts	216
	Questions for Discussion	221
9	**Accounting**	224
	What Is Accounting?	226
	Financial Accounting	228
	Financial Statements	231
	Managerial Accounting	242
	Summary and Look Ahead	246
	Key Concepts	247
	Questions for Discussion	248
10	**Financial Institutions**	250
	The Financial System	252
	Financial Intermediaries	254
	Credit Instruments	264
	The Public Market	266
	Stocks and Bonds	267
	Investment Dealers	269
	Securities Exchanges	270
	The Over-the-Counter Market (OTC)	272
	Stock and Bond Prices	272
	Speculating and Investing	274
	Securities Regulation	276
	Summary and Look Ahead	278
	Key Concepts	279
	Questions for Discussion	281
11	**Financial Decisions and Insurance**	284
	The Financial Manager and Financial Planning	286
	Protecting the Firm's Resources	299

	Extraordinary Financing Arrangements	302
	Summary and Look Ahead	304
	Key Concepts	304
	Questions for Discussion	306
12	**Personnel Management**	309
	Human Resource Management	311
	Personnel Administration	311
	Personnel Management and the Personnel Department	312
	Determining Human Resource Needs	313
	Searching for and Recruiting Applicants	314
	Selecting Applicants for Employment	316
	Training and Developing Employees	320
	Appraising Employee Performance	322
	Compensating Employees	325
	Promoting Employees	328
	Terminating Employees	329
	Providing Personnel Services	330
	Performing Other Personnel Activities	332
	Summary and Look Ahead	333
	Key Concepts	333
	Questions for Discussion	338
13	**Labour Relations**	339
	Why Unions Began	341
	Federal Legislation—The Canada Labour Code	346
	Provincial Labour Legislation	347
	The Organizing Drive	348
	The Collective-Bargaining Process	349
	Types of Unions	349
	Union Objectives	350
	Why Do Workers Join Unions?	354
	Sources of Labour-Management Conflict	355
	Developing the Collective-Bargaining Agreement	358
	Living with the Collective-Bargaining Agreement	361
	Factors Affecting Unionization	362
	The Future of Unionism	362
	Summary and Look Ahead	364
	Key Concepts	364
	Questions for Discussion	368
14	**The Computer and Other Tools**	371
	What Is a Computer?	373
	Basic Computer-related Ideas	375
	How Are Computer Systems Created?	384

Some Quantitative Tools for Management Decision Making	390
Summary and Look Ahead	396
Key Concepts	397
Questions for Discussion	399

SECTION FOUR   SPECIAL TYPES OF BUSINESS   401

15   Small Business   402

What Is a Small Business?	404
Becoming a Small-Business Owner	404
Opportunities in Small Business	406
Should You Go into Business for Yourself?	407
Starting a Small Business—The Preliminaries	409
Franchising	411
Government and Small Business	415
Survival of the Small Firm	418
Growth of the Small Firm	420
Summary and Look Ahead	422
Key Concepts	422
Questions for Discussion	423

16   International Business   425

Why Nations Trade	427
Barriers to International Trade	429
Removing Trade Barriers	437
State Trading	439
Exporting and Importing	440
Types of International Business	441
The Multinational Company	444
The "Why" of Multinational Business	444
Foreign Investment in Canada	445
Summary and Look Ahead	448
Key Concepts	450
Questions for Discussion	454

SECTION FIVE   BUSINESS ENVIRONMENTS AND YOUR CAREER   455

What Is an Environment?	455

17   The Physical, Ecological, and Technological Environments   459

The Physical Environment	461
The Ecological Environment	463
The Technological Environment	466
Summary and Look Ahead	476
Key Concepts	476
Questions for Discussion	477

18	**The Economic and Social Environment**	480
	The Economic Environment	482
	Microeconomic Environmental Factors	483
	Macroeconomic Environmental Factors	485
	The Social Environment	491
	The Social Responsibility of Business	500
	Summary and Look Ahead	503
	Key Concepts	504
	Questions for Discussion	505
19	**The Political Environment**	508
	The Government of Canada	510
	The Role of Government in the Economic System	516
	Government Influence in Business Decisions	519
	Administration of Government Programs	527
	Change in Government Policy	528
	Summary and Look Ahead	531
	Key Concepts	531
	Questions for Discussion	532
20	**Your Career in Tomorrow's Business**	534
	Environmental Trends Affecting Business and Your Career in It	536
	Choosing a Career	540
	A Final Word	551
	Key Concepts	552
	Questions for Discussion	552
Appendix	**Student Data Sheets, Resumes, Letters of Inquiry, and Job Interviews**	555
	The Student Data Sheet	555
	The Resume	556
	Preparing Your Letters of Inquiry	559
	Preparing for Job Interviews	561
Index		567

Rocketoy Company episodes appear on the following pages:

Rocketoy Company I	25
Rocketoy Company II	72
Rocketoy Company III	132
Rocketoy Company IV	164
Rocketoy Company V	222

Rocketoy Company VI   307
Rocketoy Company VII   368
Rocketoy Company VIII   453
Rocketoy Company IX   553

Careers in Business commentaries appear on the following pages:
21, 39, 43, 70, 99, 128, 153, 184, 207, 245, 277, 301, 331, 360, 389, 417, 449, 549

Preface

In developing *Introduction to Canadian Business* we began with an excellent basic text written by John Reinecke and William Schoell and revised it for use by Canadian students. In some chapters dealing with universally applicable concepts such as those of management, marketing, finance, production, and personnel, this meant relatively minor changes. In others, particularly those dealing with the environment of Canadian business, the material was completely rewritten. Between these two extremes are a number of chapters including those on ownership, financial institutions, labour relations, and international business which have been extensively revised.

Our objective was to provide an improved text for students of Canadian business while retaining the outstanding features of the two prior editions by Reinecke and Schoell.

The language of the book is concrete and within the grasp of the average reader. Complicated, abstract terminology is avoided. The interest level is kept high by providing many real-world examples drawn from familiar experiences.

Many concepts will be new to you. New concepts, especially those important to the understanding of business, are given special treatment. These key concepts are introduced at the beginning of each chapter in the order of their appearance and appear in boldface print in the text discussion. They are highlighted in the margins of the text and listed alphabetically, with definitions, at the end of each chapter. This is to help you become thoroughly acquainted with common business terms.

Review and discussion are aided by setting clear learning objectives at the start of each chapter and by providing discussion questions and brief incidents or cases at the end of each chapter. Throughout the book there are discussion-stimulating materials that have been boxed for easy reference. They present two opposing points of view or ask you to reach a conclusion on a controversial contemporary issue. These questions, incidents, and cases have been developed for your interest in and involvement in applying the concepts you learn in the text.

In this book we use two means of linking the chapters together. First, the chapters are arranged in logical groups and sequence. Five sections or chapter groups are set apart. Each section has a goal and each one builds on the previous one. Secondly, the continuing case history of the Rocketoy Company is used to tie the learning together.

The ongoing case traces the birth, development, and growing pains of a toy manufacturer. It shows how many of the concepts in the preceding chapter or two fit into the life of the firm.

The book starts with the basic economic ideas that explain business activity. Next, you will learn about the legal forms of business ownership, the organization of a firm, and basic management. Then the book "tours" the functional areas—production, marketing, and finance—and looks at accounting, personnel activities, labour relations, and computers. Here you will learn about the kinds of decisions managers face. The last two sections introduce the areas of small and international business; the factors of human values, environments, and expectations; and career opportunities in business.

Look for the Careers in Business symbol throughout the text. Every chapter, particularly the last one, provides special insight into business careers. The effect of this career orientation is to help you develop a taste for business and clarify a career choice.

We are deeply indebted to the following people for their helpful suggestions during the development of this manuscript: G. Edward Bissell, British Columbia Institute of Technology; Philip Bomeisl, Bergen County Community College; Sonya Brett, Macomb County Community College; Don R. Brown, Northern Arizona University; Harold Buck, Mohawk College; Frank Collom, Queen's University; Kathy Hegar, Mountain View College; Murray Hilton, University of Manitoba; Robert Litro, Mattatuck Community College; James C. Manning, Dawson College; John Marts, University of Southwest Louisiana; Donald Mask, Dawson College; John McCallum, University of Manitoba; Jack Miller, St. Petersburg Junior College; Robert Nelson, University of Illinois at Urbana-Champaign; Robert Ristau, Eastern Michigan University; Donald Sedik, Harper College; Douglas R. Sherk, Seneca College of Applied Arts and Technology; James Wallace, Loyola College; and Ian A. Wilson, Saint Lawrence College.

We are very grateful to Maryann Urban and Suzanne Frenette for their efforts in typing the manuscript.

We also owe special thanks to our families for their patience and understanding.

Brian E. Owen
Frederick A. Starke

Section One

Introduction

Our study of business begins with a discussion of its economic setting and the reasons why business firms exist.

In Chapter 1 we examine what an economic system is and why it exists. The main purpose of any economic system is to enable its people to cope with the economic problem—how to satisfy their unlimited wants with limited resources. This is the overall economic problem which confronts all people. Vastly different types of economic systems can be utilized to address the economic problem. The two ends of the spectrum are collectivism and capitalism. In Canada we have a "mixed" economic system.

In Chapter 2 we take a look at reasons business firms exist and at the top management of business firms. The business firm is the basic building block for organizing production activity in our "mixed" economic system.

1
Economic Ideas for Business

1 ECONOMIC IDEAS FOR BUSINESS

OBJECTIVES: After reading this chapter, you should be able to:

1. Define and discuss the economic problem and how we cope with it.
2. Distinguish between needs and wants.
3. Discuss how an economic system's performance is measured.
4. Identify and explain two basic concepts of value.
5. List and define the factors of production.
6. State the basic purpose of an economic system.
7. Compare and contrast the collectivist and capitalist economic systems.
8. List and discuss the chief characteristics of a capitalist economic system.
9. Explain the characteristics of the mixed Canadian economy.

KEY CONCEPTS: In reading the chapter, look for and understand these terms:

THE ECONOMIC PROBLEM
SPECIALIZATION
EXCHANGE
UTILITY
VALUE IN EXCHANGE
STANDARD OF LIVING
GROSS NATIONAL
   PRODUCT (GNP)
DISPOSABLE PERSONAL
   INCOME (DPI)
FACTORS OF PRODUCTION
LAND
LABOUR

CAPITAL
ENTREPRENEURSHIP
MERCANTILISM
LAISSEZ FAIRE
CAPITALISM
MIXED ECONOMIC SYSTEMS
COLLECTIVISM
CENTRAL PLANNING
INDIVIDUALISM
PROTESTANT ETHIC
CAPITAL FORMATION
CONSUMER SOVEREIGNTY

Human beings have always been "wanting" animals. From earliest times peoples' unlimited needs and wants had to be satisfied with the limited resources available to them. This is what the economic problem is all about—satisfying unlimited wants with limited resources.

Earliest societies lived in isolation. But modern people live with others in social and economic systems. In such systems, we learn to want more things than what we really need in order to survive. Our natural resources, however, have not increased in the same manner as our wants. Thus, the economic problem has been brought into sharper focus.

By specializing and exchanging, we can satisfy more wants with our limited resources. This raises our standard of living. We have more goods and services to consume. But some people question whether this really means that we are "better off".

An economic system is a framework for satisfying human wants. We will discuss two vastly different types, collectivist and capitalist, in their pure forms. Then we will discuss the "mixed" economic system in Canada. Let's begin with a discussion of the economic problem which faces all economic systems.

THE ECONOMIC PROBLEM

The economic problem

Because our wants are unlimited and our resources are limited, we face a problem. It is the economic problem. **The economic problem is concerned with how we can satisfy our unlimited wants with our limited resources.**

We must deal with the questions of: What goods and services to produce and in what quantities? How to produce goods and services? How to allocate the goods and services among members of society?

We have always had to face the economic problem. Our material progress depends on our ability to cope with it. Today, in Canada, for example, we have to make decisions on how our scarce natural resources are to be used to satisfy our wants.

Specialization

Specialization

Sharing in the task to provide basic needs means that people do not have to provide for all their needs alone. In earliest times each person had to hunt and cook alone. Later, each person had more freedom to do only one task—either to hunt or to cook. This is an example of specialization in a simple economic system. **Specialization means giving more effort to a specific task instead of giving less effort to a greater number of tasks.** By specializing, better use is made of each person's limited time. For the cave dwellers, for example, this means one person would specialize in hunting, and the other would specialize in cooking.

A modern example of specialization is the assembly line in an automobile plant. Each worker on the assembly line performs a highly specialized task.

Exchange

Specialization is pointless, however, unless specialists can exchange. **Exchange means trade, or giving up one thing to get another thing.**

Let's assume that the man specializes in hunting and the woman specializes in cooking. The man exchanges part of his hunt for part of the meal prepared by the woman. Specialization and exchange, of course, can be extended beyond the family—and have been.

Over time we have come to live in larger groups. From the simple family, we have progressed to more complex groups such as tribes, villages, towns, cities, and nations. Production has become organized in shops, stores, and factories. Thus, the specialization and exchange process now includes a great many people. Each person is dependent on more people to satisfy his or her wants. The process of exchange organizes people into groups. An economic system is the result. Although there are many different types of economic systems, all have one element in common—they exist to satisfy human wants.

Exchange

NEEDS AND WANTS

An isolated person has basic needs like food, clothing, and shelter. These needs must be satisfied if the person is to survive. However, once the person relates to other people, specialization and exchange begin and result in new needs which are learned. These new needs are not needs in the sense that a person must satisfy them in order to survive. They are wants.

A person needs food to survive. He or she can satisfy this need by eating wild berries. But a person may also learn, as a result of coming into contact with other people, to want fancy cuts of meat and pastries. Many other wants are also learned. No one really needs a television set, a piano, an automatic dishwasher, or a car in order to survive. But a lot of us want them and many other products and services as well.

Satisfaction of Wants

A want is satisfied by consuming an object or a service which is useful in relation to the want. If a good can satisfy a certain want, then it has utility. **Utility means usefulness.**

Wants are very specific to individuals. A list of wants made by one person would probably be somewhat similar to that made by another. Basic wants such as food would appear in both lists. Beyond these, however, there would be many differences in the types of wants.

In a modern economy, people have different wants. Thus, different goods have different degrees of utility for different persons. Suppose two people have the same want. Even then, we could not assume that a certain good that satisfies the want would have the same utility to both people.

Utility

SECTION ONE INTRODUCTION

Value in exchange

If something has utility for someone, it is valuable to that person. The concept of value is important to any economic system. All of us must breathe. Thus, air has value to all of us. Although it has value, people will not ordinarily pay anything for air. It is too plentiful. It has value in use, but it does not have value in exchange. **Something has value in exchange when it can command something else in return for it.**

An example will help to illustrate the meaning of value in exchange. Mr. App has some apples. Mr. Oran has some oranges. Suppose both men want apples and oranges. Mr. App can exchange some apples for some oranges. Mr. Oran can exchange some oranges for some apples. Exchange would occur if both men thought they would benefit from it. Mr. App, since he already has apples, values them less than Mr. Oran, who has none. Mr. Oran, since he already has oranges, values them less than Mr. App, who has none.

The exchange does not involve things of equal value. Mr. App's apples are worth less to him than the oranges he gets in return. Mr. Oran's oranges are worth less to him than the apples he gets in return. How much anything is worth to a person depends partly on how much of it he or she already has. The more a person has of it, the less that person wants (and is willing to pay) to get more units. This is the principle of diminishing marginal utility.

One of the important functions of an economic system is to determine what goods and services to produce to satisfy people's wants.

Standard of Living

Standard of living

The purpose of an economic system is to satisfy human wants. It is, therefore, desirable to measure the system's performance. Comparisons of different systems may be made by a measure called the standard of living. **The standard of living is a measure of economic well-being.** It indicates how much income is available to allocate to all members of society.

The standard of living helps us compare the well-being of one society with that of another society. It also helps us to observe change in well-being over time. The standard of living estimates the value of all goods and services produced by a country during a period of time. This figure is then divided by the population. The result is a measure of the economic well-being of an average person in that country.

Gross National Product (GNP)

A popular measure of the standard of living is Gross National Product (GNP). **The Gross National Product is the sum of the values of all goods and services produced in a nation during a given year.** (See Figure 1-1.)

Disposable Personal Income (DPI)

Some people think that Disposable Personal Income (DPI) is a better measure of the people's welfare. **Disposable Personal Income is a smaller amount than GNP. It indicates the total amount of buying power from current sources available to the nation's people. It is equal to their incomes minus the taxes they pay.** (See Table 1-1.)

Figure 1-2 compares the GNP of Canada with the GNP of several other countries. It shows per capita GNP for each of the countries in U.S. dollars. (Per capita GNP is a country's total GNP divided by its population.)

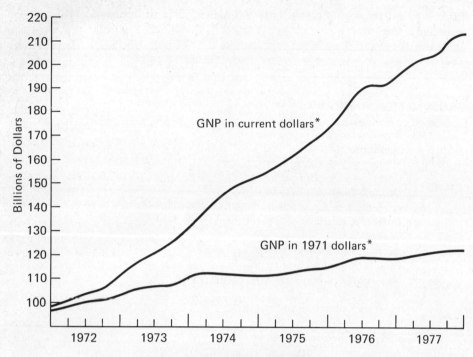

*GNP in current dollars refers to the current dollar value of GNP at a point in time, whereas GNP in 1971 dollars adjusts the GNP for subsequent years to the value of the dollar in 1971. The difference between the two values reflects the impact of inflation.

Figure 1-1. Gross National Product of Canada. (Source: *Bank of Canada Review,* April, 1978, pp. S99-101.)

Measures such as GNP and DPI, however, do not indicate the quality of life in a nation. There are side effects of high productivity which are not measured by GNP. These include pollution, congested cities, and rapid depletion of basic natural resources. They may have a strong negative effect on our enjoyment of our productive efforts. In other words, the value of things is subjective—it depends on the judgment of the people who possess or consume them.

THE FACTORS OF PRODUCTION

Limited resources must be wisely used to achieve a high standard of living. **These limited resources are the factors of production: (1) land, (2) labour, and (3) capital. The factors of production are the inputs of the productive system.** One of the issues addressed in any economic system is how the factors of production are combined to produce outputs. The outputs are the goods and services produced to satisfy human wants.

Factors of production

SECTION ONE INTRODUCTION

Table 1-1. Relationship of Gross National Product, national income, and personal income and saving

	Item	1978	1970	1950
		....millions of dollars....		
	GROSS NATIONAL PRODUCT	230,407	85,685	18,491
	Composed of National Income, i.e.:	179,426	64,235	14,553
i.e.:	a) Wages, salaries, & supplemental labour income	129,885	46,706	8,998
	b) Military pay & allowances	1,609	914	154
	c) Corporation profits before taxes	26,069	7,699	2,608
	d) Deduct: Dividends paid to non-residents	−2,355	−952	−412
	e) Interest & miscellaneous investment income	15,174	3,428	396
	f) Accrued Net Income of farm operators from farm production	3,740	1,211	1,301
	g) Net income of non-farm & incorporated businesses (incl. rent)	9,612	5,424	1,882
	h) Inventory valuation adjustment	−4,308	−195	−374
Plus:	Indirect taxes, less subsidies	25,423	11,299	2,065
	Capital consumption allowances and miscellaneous valuation adjustments	25,146	9,806	1,876
	Residual error of estimate	412	345	−3
	Net national income, adjusted for transfer payments and earnings not paid to persons yields:			
	Personal Income	136,205	66,633	14,262
Less:	a) income taxes	18,019	8,811	
	b) succession duties and estate taxes	156	266	915
	c) contributions to social insurance and government pension funds	5,895	2,470	
	d) other	1,139	1,077	62
Equals:	Disposable Personal Income	110,996	54,009	13,285
Less:	a) personal expenditures on consumer goods and services	96,995	50,327	12,482
	b) transfers to corporations (interest)	1,616	641	29
	c) transfers to non-residents	246	169	36
Equals:	Personal Savings	12,139	2,872	738

Source: Statistics Canada 13-001 & 13-531

Land

Land, as a factor of production, means natural resources. Examples are petroleum, iron ore, and farmland. In the long run, all natural resources can run out. With proper management, some will last for a very long time (air, water, forests). Others, however, are non-renewable (petroleum, uranium, other metals). Physical land is limited, too. We have a choice between thoughtless and rapid use of these resources and carefully planned resource use.

Labour

Labour means human mental and physical effort. Much of our economic progress results from substituting mental effort for physical effort. Mental effort leads to technological "breakthroughs" which result in greater output from the same physical effort. This enables us to get more satisfaction from a given quantity of natural resources. For example, if such progress enables us to get 10 per cent more energy from a ton of coal, the effect is the same as adding 10 per cent to the coal supply.

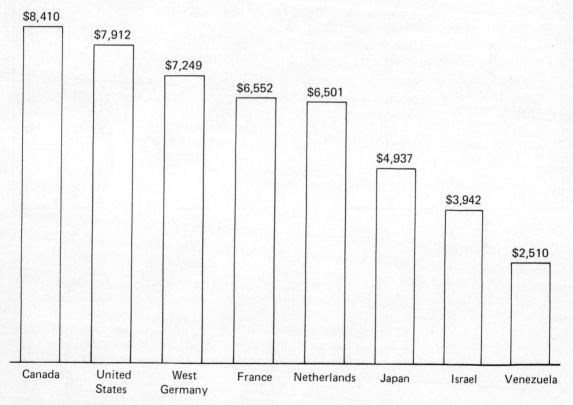

Note: GNP per capita is a country's total GNP divided by its population.

Figure 1-2. Gross National Product per capita for selected countries (in U.S. dollars).
(Source: Prepared from *Statistical Yearbook*, 1974. Copyright © 1975 by the United Nations. By permission.)

> ## WHAT DO YOU THINK?
>
> ### How Do You Measure the Quality of Life?
>
> A 1974 study for the Economic Council of Canada by D.W. Henderson on Social Indicators indicated three basic goals related to the quality of life. These were (1) material well-being, (2) sociocultural well-being, and (3) equity of opportunity with respect to the first two.
>
> These three were further sub-divided into nine different major areas of socioeconomic concern. These were:
>
> 1. Individual rights and responsibilities (legal rights and participation in public decision making)
> 2. Social rights and national identity (domestic social rights and international relations)
> 3. Command over knowledge and skills (basic and higher education, etc.)
> 4. Health (mortality, morbidity, and positive health)
> 5. Natural environment (soil, water, and air)
> 6. Human-made environment (social and political)
> 7. Production and consumption of final goods and services (inputs, outputs, and efficiency)
> 8. Employment (labour-management relations, job security and satisfaction, etc.)
> 9. Financial status (income and assets)
>
> Is this listing of concerns useful in measuring the "quality of life"? WHAT DO YOU THINK?
>
> Reproduced by permission of the Minister of Supply and Services Canada.

Capital

Capital

Capital, as a factor of production, means tools and machinery or anything made by humans that aids in producing and distributing goods. It is human-made productive capacity.

To add to its capital equipment, an economy must produce more than it currently consumes. But how much of its output can a country devote to its productive facilities? This depends largely on the willingness and ability of its people to postpone consumption.

Entrepreneurship

Entrepreneurship

Many people would add a fourth factor of production in a free-enterprise system—entrepreneurship. **Entrepreneurship occurs when a person or firm assumes risk in the hope of making a profit.** People who go into business have no guarantee of earning a

---

A CONTEMPORARY ISSUE

The Energy Crunch

In recent years, we have become painfully aware that the days of cheap energy are over.

As one alternative, many want to use coal to generate electricity. Many also want to move ahead with the construction of nuclear power plants.

Some people, however, are against these moves. They fear sulphur pollution from burning coal. They fear nuclear plants because of their alleged safety hazard. But if our economy is to continue growing, the controversy must be resolved. HOW WOULD YOU RESOLVE IT?

---

profit. They assume the risk of losing what they invest in their firms. Entrepreneurship is a major part of productive effort in a free-enterprise economy.

The factors of production can be combined and used in many ways to produce many different things. The amount and variety of goods produced depends mainly on the way economic life is organized.

THE PURPOSE OF AN ECONOMIC SYSTEM

The purpose of an economic system is to provide a framework for satisfying human wants. The functions of an economic system are to determine what products to produce, how to produce them, and how to allocate income among members of society. In Canada, most productive economic activities (producing and selling) are channelled through business firms. They operate through the market system. Historically, there have been different types of economic systems.

In ancient Greece, for example, producing goods for sale was not common. Agriculture was the main economic activity. Business, economic activity, and profit seeking were considered acceptable but lowly activities.

During the Middle Ages, the Church taught that people should not seek economic betterment but should concentrate on salvation. Although there was a great deal of economic activity, it was considered worldly and often sinful.

The next major "age", as far as economic philosophy is concerned, was the age of mercantilism. **Mercantilism is an economic philosophy which advocates building strong national states (nations) from warring feudal kingdoms. A major goal of mercantilism is to increase the government's holdings of precious metals.** Foreign trade between two countries was viewed as involving a gain for one and a loss for the other. The nation with the greatest supply of precious metals was considered the strongest.

Mercantilism

Laissez faire

The citizens could be poor, but as long as gold was in the state treasury, the state was considered wealthy.

Mercantilism was followed by a period of "laissez faire" economics in Europe and, later, in the United States. **Laissez faire means "let businessmen compete".** This new economic philosophy began in France, but it was first presented in a complete form in 1776 by Adam Smith in his book, *The Wealth of Nations.* Smith believed in free competition and capitalism.

Capitalism

Capitalism is an economic system based on private ownership of the factors of production. The major features of capitalism are individualism, private property, profit incentive, consumer sovereignty (or consumer power), freedom to compete, occupational freedom, freedom of contract, and limited role of government.

Collectivism, or socialism, is another type of economic means of production and distribution by society (the state) rather than by private individuals. Collectivism is characterized by features such as government ownership.

Our Canadian economic system has its roots in the philosophy of laissez faire capitalism. However, at present we are far from a pure capitalist economic system. **In Canada, we have a "mixed" economic system, characterized by an active market economy with significant direct and indirect involvement by government in economic decision making.**

Mixed economic system

Some portions of the Canadian economy can be accurately described in terms of the features which characterize capitalism. There is also, however, a substantial portion of the economic activity in this country owned and operated by governments.

There are also many significant government involvements in the Canadian economy which stop short of direct ownership and control. There are a wide range of

---

WHAT DO YOU THINK?

How Do We Deal with Limited Natural Resources?

Two ways of dealing with the growing scarcity of natural resources (land as a factor of production) are to limit wants and/or to increase the capacity of natural resources to satisfy our wants.

Government can adopt policies to limit consumption. For example, to discourage consumption of gasoline, it could be rationed or prices could be raised.

Another approach is to use the other factors of production to help us to expand the want-satisfying capacity of our natural resource base. Human mental effort (labour) leads to new technology. Entrepreneurs put this technology to work by building new plants and equipment (capital) based on that technology. To get more miles per gallon from gasoline, for example, technological advances are made in engineering and producing cars. Which approach is better? WHAT DO YOU THINK?

regulations, regulatory boards, incentive programs, tariffs, taxes, and other government influences.

In the sections below we discuss the characteristics of pure collectivist and pure capitalist economic systems. The mixed Canadian economic system is also discussed in greater depth.

COLLECTIVISM

In any economic system decisions are made about the alternative uses of resources and how the goods produced from those resources are to be distributed. In a purely collectivist system, the government controls social and economic decision making. **Collectivism means government ownership of the factors of production and government control of all economic activities.** There is little or no private property. The government determines such things as the economy's rate of growth, the amount of investment, the actual allocation of resources, and the division of output. Direct government means are used to achieve the desired results. Wage rates, production volumes, and prices of goods would be set by the government as well.

Collectivism

The government likely also practices central planning. **Central planning means that the government drafts a master plan of what it wants to accomplish and directly manages the economy to achieve the plan's objectives.** The total supply of goods available for household consumption is fixed. This supply is distributed to households in limited amounts and at fixed prices. There is no guarantee, however, that the goods produced are what consumers want. Thus, consumers spend their fixed incomes on fixed amounts of goods at fixed prices.

Central planning

A collectivist system seeks to achieve what it alleges to be the "greatest good for the greatest number". People contribute to it on the basis of their abilities. They receive from it on the basis of their needs. The individual is of less importance than the system, and the government largely determines each person's role in the system.

Collectivist systems, however, stress social and economic equality among their citizens. They seek to eliminate differences in economic welfare among people of different occupations, races, and backgrounds. A uniform standard of living is sought in order to eliminate friction among the various classes of people.

CAPITALISM

Individualism

The basic idea underlying capitalism is individualism. **Individualism is the idea that the group, the society, and the government are necessary but are of less importance than the individual's self-determination.** The ancient Greeks valued the dignity and uniqueness of the individual. Although this value was lost sight of during some of the darker periods of history since the Greeks, the idea of individualism remains a cornerstone of most of the Western democracies.

Individualism

Capitalism is based upon the strength of individual self-interest. The basic belief is that a person will, if left alone, seek his or her economic betterment. Further, this will lead to the economic betterment of the whole society and a more economical use of resources.

Protestant ethic

The full development of individual initiative and some other basic ideas of capitalism required a shift in Christian philosophy from medieval notions of "other-worldliness" and distrust for business to the more practical Protestant ethic. **The Protestant ethic is a tradition which emphasizes the value of hard work, accumulation of property, and self-reliance.** Its most extreme form existed in the United States during the nineteenth century. It became a belief in the survival of the economically fittest. It gave the mark of Christian respectability to the rugged individualists who dominated that country's economic growth in the latter half of the nineteenth century.

Private Property

Related to individualism is the right of private property. This is an individual's right to acquire, to use, to accumulate, and to dispose of things of value. The right to own and accumulate property is felt to be a contributing factor to individual initiative.

Private property has existed in most cultures. In any society, certain limitations must be placed on the right of ownership. For example, a person who owns a house cannot use it to conduct illegal activities. Nor can the owner set fire to it. In Canada one is restricted from selling certain properties to foreigners subject to approval by the Foreign Investment Review Agency. A balance is struck between the individual's private property rights and the society's "common good".

TWO POINTS OF VIEW

Egalitarianism

Collectivist systems are egalitarian in that they seek to equalize incomes and achieve a uniform standard of living for their people.

Egalitarians in Canada want to do the same. They see government's social programs as the equalizing and levelling force in society. Through those programs, they believe, the "have-nots" will have more and the "haves" will have less. This redistribution of income and wealth would continue until there is total equality. The egalitarian movement has a lot of appeal to some groups in our society.

Critics of egalitarianism argue that an egalitarian system cannot exist in a capitalist-based economy because equal wealth would destroy the profit incentive.

Can those who favour egalitarianism and those who oppose it resolve their differences? WHAT DO YOU THINK?

An important aspect of private property is that a person can accumulate it and use it as he or she pleases within the very broad limits set by society.

Profit Incentive

The success of a capitalistic system in contributing to a high standard of living depends on having many people use their private property to generate more property. They invest it to make a profit. The incentive to invest lies in the chance of getting a return—making a profit—from it. If people could not profit, they would have little reason to use their property in such a way. Instead, they would use up or consume their property. If people did not make a profit from investing or going into business, they would not bother to accumulate property.

The right to pursue profits involves risk. The degree of risk depends on the use to which property is put. When people willingly risk their private property, such as their savings, by going into business, they do so to try to make a profit. This process of putting money into business firms in order to try to make more money is called investment.

Investment is necessary for a nation's economic growth. The more complex forms of productive business activity require large amounts of equipment, tools, land, and other things. Capital formation is another way to describe the process of investment. **Capital formation is the process of adding to an economy's productive capacity.** The profit incentive encourages investment in a capitalistic economic system.

Capital formation

Consumer Sovereignty

Individualism also affects the demand for goods. Unless a firm can cultivate a group of customers, it will not survive. In a capitalistic system, the consumer enjoys a position

Table 1-2. Government vs. consumer control over economic decision making

Collectivism	*Capitalism*
1. Government practises central planning.	1. The individual makes independent choices.
2. Central planning determines how resources will be allocated.	2. These choices determine how resources will be allocated.
3. Resources are allocated so as to achieve the goals set by government.	3. Resources are allocated in response to decisions made by consumers.
4. Government determines: (a) what will be produced. (b) how it will be produced. (c) how much will be produced. (d) how the goods will be distributed.	4. Through the market consumers contribute to decisions about: (a) what will be produced. (b) how it will be produced. (c) how much will be produced. (d) how the goods will be distributed.

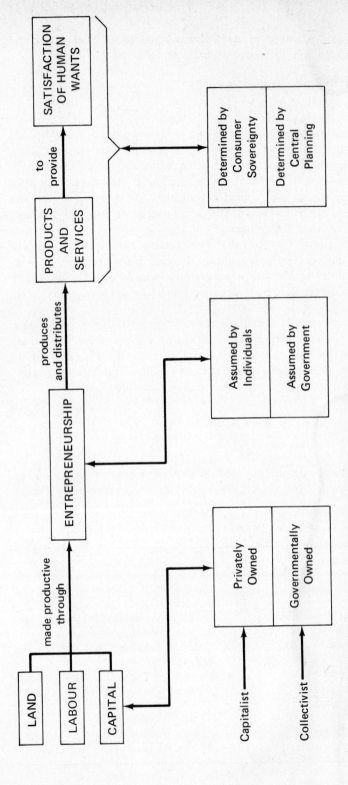

Figure 1-3. Characteristics of capitalist and collective economic systems

of sovereignty. Through individual decisions to patronize one firm or another, consumers decide the economic fate of a firm.

No one "guarantees" success to a firm. **Consumer sovereignty means that the consumer is free to do business with whomever he or she chooses.** A firm, of course, does what it can to influence consumer decisions through, for example, its advertising. The consumer, however, has the final say as to what will be bought. (See Table 1-2.)

Consumer sovereignty

Freedom to Compete

A person who is free to compete can risk his or her private property in the hope of earning a profit. Within very broad limits, that person can go into any business. This is true no matter how much those already in the same type of business would like to keep the field to themselves.

Competition among firms benefits both consumers and firms. When competitors stand ready to take sales away from a firm, that firm has a strong incentive to remain efficient and to please its customers. Thus, both the firm and its customers benefit when rivals seek to earn the customer's favour. (See Figure 1-3.)

Occupational Freedom

Still another example of individual freedom is the freedom of occupational choice. You are free to start up a new firm—go into business for yourself and become your own boss. You are also free to work for someone else. The choice of occupation is a highly personal one. No one forces a person to be a plumber, a teacher, or a lawyer. The choice is made by each person. Individuals make this choice guided by their own best economic interests and within the limits of their talents and education.

In some economic systems, however, central planning determines the need for persons to fill various job categories. People are trained for, and assigned to, those jobs. The choice of a job is not made by the person. It is made by the government.

Freedom of Contract

One of the most important freedoms is freedom of contract. This enables a person to enter into contracts with one or more other persons. The contract may involve relationships like that between an employer and an employee or that between a seller and a buyer. As long as a contract is legal, it is protected by law and is legally enforceable.

Limited Role of Government

A basic characteristic of a capitalist economic system is the limited role of governments. In general, the government is permitted to step in only when the welfare of citizens is threatened. The manner in which government regulates can be hotly debated.

Table 1-3. Basic ideas underlying capitalism and collectivism

Capitalism

1. The individual is of primary importance
2. Private property
3. Profit incentive
4. Consumer sovereignty (consumer power)
5. Freedom to compete
6. Occupational freedom
7. Freedom of contract
8. Limited role of government

Collectivism

1. The individual is less important than the system, which seeks equality for all of its citizens
2. No private property
3. No recognition of profit
4. Central control over social and economic decision making
5. Competition results in economic waste
6. Central planning determines the need for various types of occupations
7. Little freedom of contract
8. The government is the primary decision maker

Closely related to the concept of limited role of government is a belief that price is the basic regulator of the economic system.

Table 1-3 compares some of the basic ideas which underlie capitalism and collectivism, the two ends of the spectrum of economic systems.

The Canadian Mixed Economy

In Canada, as mentioned previously, we have neither a purely capitalist nor a collectivist economy. We have a "mixed" economy. In a speech given in January, 1976, to the Canadian Club in Ottawa, Prime Minister Trudeau said, "The fact is that for over 100 years, since the government stimulated the building of the Canadian Pacific Railway by giving it Crown land, we have not had a free market economy in Canada, but a mixed economy—a mixture of private enterprise and public enterprise."

Sectors of our economy are characterized best as capitalist. Examples of these include the retail and service industries, the printing industry, newspapers, sporting goods, and small business in general.

We have a significant number of business-related activities which are government-owned. According to the 1976 Public Accounts of Canada, the total assets controlled by federal Crown corporations in 1975 amounted to $35.2 billion for 43 Crown corporations. In Canada, the federal government has been responsible for a large number of Crown corporations for many years. They operate, for example, in transportation (Canadian National Railway [CNR], Air Canada, Eldorado Aviation Ltd., Northern Transportation Company Ltd., St. Lawrence Seaway, and more recently VIA Rail Canada, Inc.), communications (CBC and Telsat), housing (Central Mortgage and Housing Corporation), and many others such as the Farm Credit Corporation and National Arts Centre Corporation. In addition, many provincial governments have established Crown corporations. The most prominent of these are involved in the generation and distribution of hydro-electricity.

Between the sector of the economy which is government-owned and which is best characterized as free enterprise or capitalist, there are varying degrees and types of government intervention. In its November, 1976, issue the magazine, *Canadian Insurance/Agent and Broker,* records Professor D. H. Thain of the University of Western Ontario as referring to a medium amount of government involvement and a public utility type of system as different degrees of government involvement between the two ends of the spectrum. The medium amount of government involvement was referred to as a "welfare state based on an industrial system with a high degree of free enterprise"; he said, "this is what has been going on in Canada for the last twenty or thirty years and involves a continual expansion of government involvement, regulation and planning to achieve a higher level of employment and welfare services for the redistribution of wealth."

Thain refers to the next stage as the public utility stage. At this stage companies must report some intended decisions to regulatory agencies for prior approval. This takes away the autonomy of companies to make independent decisions about those issues for which prior approval is required. It also puts a company in a position of being subject to an arbitrary decision of a regulatory agency. There are numerous examples of individual company decisions which must receive prior approval. One very general example was the wage-and-price-controls program initiated in November, 1975, and phased out beginning in 1978. Under this program many companies could not change prices or finalize a wage settlement with employees before receiving approval from the Anti-Inflation Board. This act turned many companies into quasi-public utilities. The relationships between business and government are expanded upon in Chapter 19.

The economic systems of many other countries also cannot be categorized as either purely capitalist or collectivist.

MODERN COLLECTIVISM

Just as our economic system is mixed and not purely capitalist, neither are the economic systems of all communist countries purely collectivist. Pure collectivism involves setting exact rates of growth, exact control of investment, exact prices and wages, and so on.

Some forms of collectivism now recognize how hard it is to set exact prices and objectives through strict central planning. Increasingly, they settle for approximations.

In some collectivist countries, there is a degree of self-management of business

Figure 1-4. Some socioeconomic systems in the world today

firms. Managers are expected to earn a profit. They, not the government, determine what consumer goods will be produced. Competition among firms in these nations is a fact of economic life. Wage rates, and even the prices of the goods the firms sell, are essentially determined in the market. The firms pay taxes, and part of their profits are distributed to the workers and managers. Since the consumers buy those goods and services they desire, firms have an incentive to produce them efficiently. This involves making decisions about how the factors of production will be combined and employed. Those firms which anticipate consumer demand for various products and services and offer them for sale are the ones which profit most.

There are, of course, still examples of strict collectivism, such as that found in the People's Republic of China, where economic freedom is severely limited. In the Soviet Union itself, the degree of freedom is much smaller than it is in some of the "satellite" countries (Romania, Hungary, Czechoslovakia) or the "independent" communist nation of Yugoslavia. (See Figure 1-4.)

MODERN CAPITALISM

The socioeconomic systems of some Western European nations are not easy to categorize as capitalist or collectivist. These systems might be lumped under the heading of "democratic socialism". As a result of peaceful, democratic means, these nations have adopted some rather severe restrictions of the classic capitalist freedoms. In many cases, the practice of medicine and ownership of certain basic industries are in the hands of the government. Other economic activities are highly regulated by government. Advertising is restricted in some of these nations. Consumption of goods is sometimes controlled by rationing. Taxation is, in many of these nations, much heavier than it is in Canada.

Even the United States is far from the classical ideal of capitalism. Public utilities, such as telephone and power companies, are strictly regulated.

Many forms of regulation and taxation further distinguish the United States today from the capitalist ideal. Progressive income taxes, laws against price discrimination, the Small Business Administration (through which the federal government gives aid to small firms), and many other governmental activities represent a retreat (or an advance?) from pure capitalism.

CAREERS IN BUSINESS

This symbol is something we will be using throughout the book to tie together comments about your career. The symbol stands for "careers in business". This concept will link together a series of commentaries on career choice—one in each chapter.

It is never too early to start planning a career. The dizzy pace at which the world is changing makes nearly all choices of occupations somewhat risky. This underlines the importance of choosing carefully. It also brings home the idea that the more hectic the competition for jobs becomes, the more important your preparation for each becomes. You face the double challenge of being well-prepared for a specific job and being prepared to change jobs.

Your education at this time, then, must teach you how to learn and how to adjust to change. This book can help, especially if your career is in the business world. It will give you some ideas about the whys of business. It will provide you with a survey of the major business career areas—jobs like selling, banking, accounting, and computer operation.

We wish you good luck in the greatest adventure of a lifetime—the search for a career. We hope that your choice will be a career in the business world!

WHAT DO YOU THINK?

Are Capitalist Freedoms Declining?

According to some people, many of the characteristics of capitalism were more in evidence during our early history as a nation than they are now. Individualism and individual self-determination have taken a back seat to the collective well-being of society. The right of private property has been diminished by government programs such as the Foreign Investment Review Act. The profit incentive has been dealt a large blow by progressive income taxes and, from time to time, governmentally enforced price-and-wage controls. The government does not, it is argued, play a limited role in the economy.

Some believe that the only characteristic of capitalism which has been strengthened is consumer sovereignty. Our governments are more committed to protecting the consumer in the marketplace. But, they argue, this has been accomplished at the expense of some of the other freedoms.

On the other hand, some people argue that these freedoms have never really been characteristic of our mixed economic system. Are capitalist freedoms declining in Canada? WHAT DO YOU THINK?

SUMMARY AND LOOK AHEAD

Economic systems help us to satisfy our unlimited wants with our limited resources. Because our wants are so varied and because our resources, in many instances, have become more scarce, the economic problem today is more complex than it was in the past.

Several elementary concepts (utility, value, specialization, exchange, standard of living, diminishing marginal utility, and the factors of production) were introduced. These concepts are relevant in any type of economic system.

The major ideas underlying a capitalist private-enterprise system are individualism, private property, profit incentive, consumer sovereignty, freedom to compete, occupational freedom, freedom of contract, and limited role of government.

The type of economic system most unlike the capitalist system is the collectivist system. In this type of system, private property and the other institutions and ideas of the private-enterprise system are largely absent. Central planning by government replaces individual freedoms and initiative.

Most economic systems in the world today fall somewhere between the two extremes of pure capitalism and pure collectivism. In Canada we have a "mixed" economic system.

In the next chapter we will see how people use the freedoms accorded them under our economic system to form business firms. We will discuss why business firms exist.

KEY CONCEPTS

Capital A factor of production. Tools and machinery or anything human-made that aids in producing and distributing goods and services. Human-made productive capacity.

Capital formation The process of adding to an economy's productive capacity.

Capitalism An economic system in which the factors of production are privately owned. The bulk of economic decision making is in the hands of individuals and privately owned business firms.

Central planning Practised in collectivist economic systems. Government decides how productive resources will be used and how the system's output will be distributed.

Collectivism An economic system in which factors of production are owned by the government. It controls social and economic decision making.

Consumer sovereignty Concept that the consumer dictates quality, style, etc., of products produced by business firms and determines the success of those firms. Consumer power.

Disposable Personal Income The total amount of buying power from current sources available to the people of a nation. Basically, the incomes of people minus taxes paid by them.

(The) economic problem The problem we face in trying to satisfy our unlimited wants with limited resources.

Entrepreneurship A factor of production. Occurs when a person or firm assumes risk in the hope of making a profit.

Exchange Makes specialization or division of labour possible. A specialist trades part or his or her output for part of the output of other specialists. Trading one thing for another thing.

Factors of production The elements needed for producing goods and services. The inputs of the productive system: land, labour, capital, and entrepreneurship.

Gross National Product (GNP) The market value of all goods and services produced in a country during a year.

Individualism The idea that the group, the society, and the government are necessary but are of less importance than the individual's self-determination. A characteristic of a free-enterprise or capitalist economic system.

Labour A factor of production. Human mental and physical effort needed to produce goods and services.

Laissez Faire "Hands off" or "let businesspeople compete". Economic philosophy which advocates government playing a very limited role in business and economic affairs.

Land A factor of production. Includes all natural resources.

Mercantilism An economic philosophy which advocates building strong national states and which views the strength of a nation to be in its supply of precious metals.

Mixed economic system An economic system characterized by an active market economy with significant direct and indirect involvement by government in economic decision making.

Private enterprise Private ownership of business firms.

Protestant ethic A tradition emphasizing the values of hard work, accumulation of property, and self-reliance.

Specialization Dividing work into several tasks and having one person perform only a limited number of those tasks. An assembly-line worker is a specialist in performing a very limited number of tasks. Also called division of labour.

Standard of living A measure of economic well-being. Often expressed as GNP per capita.

Utility The ability of a good or service to satisfy a human want. Usefulness.

Value in exchange The ability of one good or service to command another good or service in an exchange.

QUESTIONS FOR DISCUSSION

1. Do you think that "the economic problem" has become more complex in modern times than it was in earlier times? Explain.
2. How do you, as an individual, cope with "the economic problem"?
3. Should future generations of people be considered when we cope with "the economic problem"? Why or why not?
4. If all countries specialized in producing what they could produce best, would that help the cause of world peace? Discuss.

5. Economic systems exist to satisfy human wants. Thus, those systems should change as the underlying wants change. How has our economic system changed over the last century?
6. Should comparisons of the standards of living in different countries be based on per capita GNP? Why or why not? Discuss other approaches that might be used.
7. Do some present-day Canadians still believe in mercantilism? Discuss.
8. What are the implications of the tendency of modern capitalism and modern collectivism to "borrow" from each other?
9. Individual initiative is important in our system. What factors determine a person's initiative?
10. Capital formation is necessary for fuller economic development. How could it occur in the absence of private property and profit incentive?
11. Does consumer sovereignty guarantee that consumer welfare is maximized? Why or why not?
12. Give some examples of government limitations on free enterprise. Why have they become necessary, or are they?
13. Does collectivism mean the same thing in the People's Republic of China as it does in Yugoslavia? Discuss.
14. Many services such as pollution control, health services, education, transportation, and communication are provided by the various levels of government in Canada. Does the profit incentive have any meaning here? Explain.
15. One of the features of capitalism is freedom to compete. Should limitations be placed on this freedom? If so, how should they be determined and carried out?

INCIDENT:

Jane Patterson

Jane Patterson, a first-year student in business administration at a community college, was recently asked by a friend why she chose business as her course of study. This friend, Charles Smith, was also a student, majoring in political science.

Actually, Charles' question was preceded by a friendly debate with Jane. It seems that Charles is really down on "the system". He told Jane that there is no excuse for our system's failure to solve all our social problems. "Canada is one of the richest nations in the world, but still diseases remain unconquered, pollution is a problem, poverty remains in the midst of plenty, and our country even threatens to break up."

Questions:
1. Do you think Charles' negative sentiments about "the system" are justified? Why or why not?
2. Suppose you were Jane. How would you answer Charles' question as to why you are studying business?

INCIDENT:

Anthony B. Mark

Anthony B. Mark, a syndicated columnist whose column appears in newspapers across Canada, has for years been a leading social critic of contemporary life in North America. On one of his numerous speaking engagements, Mr. Mark commented that too many of our wants are frivolous and result in a misallocation of our resources. Millions and millions of dollars are spent every year by Canadians buying the latest gadgets, such as new automobiles, when the previous year's models are still capable of transporting people. Even home appliances, such as dishwashers, refrigerators, and washing machines, now go through annual model changes in an attempt to induce the buying public to become dissatisfied with older models which still perform the job they were bought to perform. Millions of dollars are spent every year by men and women on cosmetics which, in some cases, are actually dangerous to their health.

    Mr. Mark believes that our wants have to be channelled in new directions. "As our resources become increasingly scarce, it is idiotic to waste what remains on frivolous forms of consumption."

Questions:
1. How do you think Mr. Mark would define "frivolous wants"?
2. Do you think that the "frivolous wants" to which he refers are learned wants? How do we learn them?
3. How do you think Mr. Mark would have us go about "channelling our wants in new directions"? What "directions" do you think he has in mind?

ROCKETOY COMPANY I

In this chapter and in eight others we will use the Rocketoy Company to illustrate how the ideas in the text fit into the life of a firm. In each of the nine episodes the history of Rocketoy will unfold, revealing the variety of problems faced by its management. This will give the concepts you learn in the book greater meaning.

    The Rocketoy Company episodes illustrate the use of the case method. The case method can be exciting. It shows that "real-world" solutions to problems require more from a manager than can be found in textbooks. Your judgment of the personalities which will appear in the Rocketoy episodes will affect the responses you make to the questions we raise. It is important to "hash it all out" with the others in your class. Let's get on with the facts!

While Terry Phillips was growing up in Toronto, he spent a lot of time "tinkering" in his father's workshop. As a result, he developed great skill in wood-working, carpentry, and whittling. At age 10, he sold his first wooden toy. By the time he was 16, Terry was making and selling 50 items per month for sale to friends and relatives.

After he graduated from high school, Terry borrowed $2,000 from his uncle, Joe Phillips, who had faith in Terry's creativity and good business sense. This interest-free loan enabled Terry to rent a garage in his neighbourhood. It also enabled him to buy the basic equipment he needed to make up to 500 toys per month in five basic models. He sold these to stores in Toronto—three variety stores, a novelty shop, and a large toy store.

Terry's sister, Pam, had recently graduated from a community college where she studied accounting. She agreed to keep Terry's financial records on a part-time basis for a wage of $4 per hour. She became his first employee.

Terry also got help from a lawyer he hired to check sales contracts. A local banker helped Terry to set up a chequing account for the business and arranged for a short-term loan, at a reasonable interest rate. The bank loan made it possible for Terry to buy wood and paint in larger quantities and to hire a shop assistant.

In the first year, after expanding the shop, sales grew a little and, despite the added wages and interest cost, Terry was able to pay his uncle back one-fourth of his original loan. By the end of the year, Terry got a year-long trial contract with a national toy distributor (Toyco) to supply them with 1,000 units of his "Rocketoy". The Rocketoy is a simple, durable toy space rocket that appeals to children 8 to 10 years old. At this stage, the Rocketoy Company seemed to be a fantastic success.

Questions:
1. How does the Rocketoy Company illustrate the principles of specialization and exchange?
2. Describe the factors of production used by Terry.
3. Could the Rocketoy Company have been established and operated profitably under a collectivist economic system? Why or why not?
4. Why are Uncle Joe and the bank willing to lend Terry money? Do their motivations for lending money differ?

2

The Business Firm

OBJECTIVES: After reading this chapter, you should be able to:

1. Explain what a mixed market economy is and how it works.
2. Define the "law of demand" and the "law of supply".
3. Discuss the factors which influence the overall levels of supply and demand.
4. Explain how the concepts of supply, demand, and price influence the allocation of resources for production of goods and services in a free-market economy.
5. Draw supply and demand curves.
6. Explain the significance of the intersection of a supply and a demand curve on a graph.
7. Discuss reasons why governments become involved in decisions about what goods and services to produce.
8. Discuss how governments become involved in decisions about how to produce goods and services.
9. Discuss the reasons why people form business firms.
10. Define profit and give examples of ways firms try to increase their profits.
11. Discuss the role of top management of a business firm.
12. Understand the concept of business strategy.
13. Differentiate between strategy formulation and strategy implementation.
14. Explain why business owners assume risk.

KEY CONCEPTS: In reading the chapter, look for and understand these terms:

PRICE	BUSINESS FIRM
DEMAND	PROFIT
EFFECTIVE DEMAND	OPPORTUNITY
LAW OF DEMAND	RISK
SUPPLY	BUSINESS POLICY
LAW OF SUPPLY	STRATEGY
DEMAND CURVE	STRATEGY FORMULATION
SUPPLY CURVE	STRATEGY IMPLEMENTATION

There are not enough resources to satisfy our unlimited wants. Choices, therefore, must be made about what will and what will not be produced. This determines how resources will be used. Under our economic system, the allocation of resources results in part from independent decisions made by consumers and producers and in part from decisions of governments.

The business firm is the basic building block for the production of goods and services in our system. Most economic activity is channelled through business firms, which gather and organize resources for production. They do so in the hope of making a profit. Significant economic activity is also undertaken by governments in Canada. They are undertaken for a variety of social and economic reasons.

THE MIXED MARKET ECONOMY

As indicated in the previous chapter, we have a mixed economic system in Canada. Government and business are both involved in economic activity.

We do not have a strictly free-market economy in the sense that prices alone determine how resources will be allocated and how the goods and services produced will be allocated to members of society. Nor is it strictly a free-market economy in the sense that only market supply and demand forces determine what all prices are to be.

In the Canadian economy, some prices *are* established solely by supply and demand. Also, many resource allocation decisions are made strictly on the basis of their anticipated impact upon profitability, and much income is allocated to individuals on the basis of the perceived relationship of their contribution to economic activity. All of these are typical of a free-market economy. However, there are also prices which are not established strictly by supply and demand, and resource and income allocation decisions which are not made strictly on the basis of prices.

Prices

The price of an item is the quantity of money (or other goods and services) which is paid in exchange for it. The demand for a good results from the wants of people and their belief in the ability of different goods to satisfy these wants. The quantity demanded is the number of units of a good people will buy at a certain price. There is an inverse relationship between price and quantity demanded known as the "law of demand"—as one (price) goes up, the other (quantity demanded) goes down. (See Figure 2-1.)

The other half of the price system is supply. **The supply of a good results from the effort of producers.** The quantity supplied is the number of units of a good producers will offer for sale at a certain price. There is a direct relationship between price and supply. **This is the "law of supply"—as price goes up, the quantity supplied also goes up; alternatively, as price goes down, the quantity supplied decreases.** (See Figure 2-2.)

The supply and demand curves can be depicted graphically. (See Figure 2-3.) Price is established at the point where the two curves intersect. The supply and

Margin notes: Price / Demand / Quantity demanded / Law of demand / Supply / Law of supply

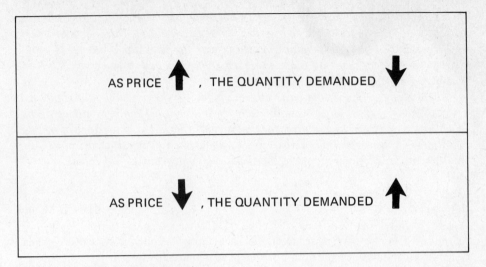

Figure 2-1. The law of demand

demand curves cross at a price of 50 cents per unit. Only at this price is the quantity suppliers are willing to offer exactly equal to the quantity buyers are willing to buy. The market is cleared at this price. At higher prices, suppliers would be willing to supply more units than buyers would be willing to buy. At lower prices, buyers would be willing to buy more units than suppliers would be willing to supply.

At any rate, Figure 2-3 gives us a basic insight into the nature of price. Whether we are talking about the price we pay for hamburgers or cars, the price we get for our labour (wage), or the price we pay to borrow money (interest), the forces of supply

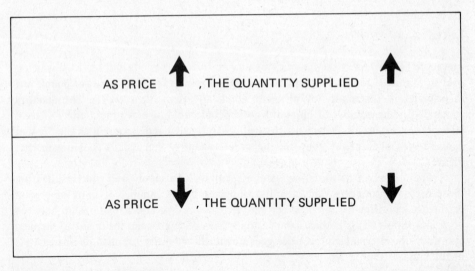

Figure 2-2. The law of supply

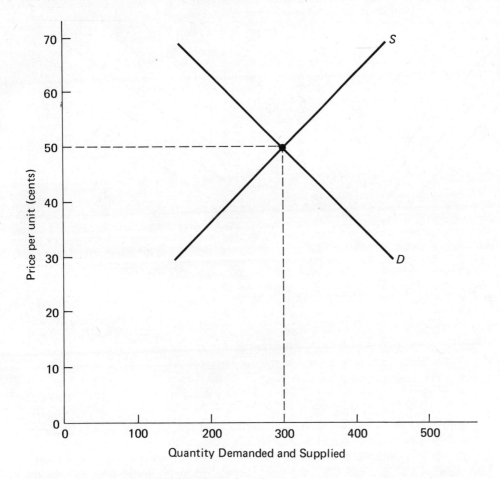

Figure 2-3. Determination of market price

and demand are at work. For some items, for example, primary products such as grains and minerals, the market forces of supply and demand are the primary determinants of price. Even in instances, such as public utility price regulation, where prices are administered, the concepts of supply and demand underlie the final decisions made.

Figure 2-4 shows shifts in the demand and supply curves. The shifts are due to changes in the underlying forces of supply and demand. The shift from D_0 to D_1 means that, at any given price, demand is greater than it was before the shift. A firm advertises its product in the hope that it will shift its demand curve up and to the right. More units are demanded at any given price.

The shift from D_0 to D_2 means that, at any given price, demand is smaller than it was before the shift. One real example is men's hats. Over the years, the demand for men's hats has shifted down and to the left. Fewer units are demanded at any given price.

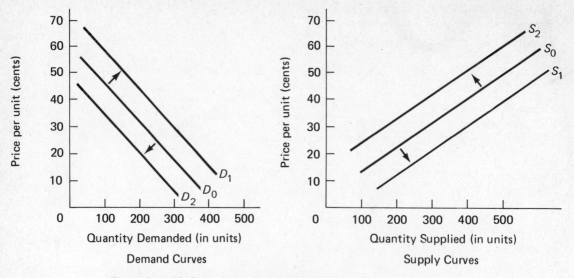

Figure 2-4. Shifts in demand and supply curves

The shift from S_0 to S_1 means that, at any given price, supply is greater than it was before the shift. More units are supplied at any given price.

The shift from S_0 to S_2 means that, at any given price, supply is smaller than it was before the shift. Fewer units are supplied at any given price.

Table 2-1 explains several concepts of supply and demand.

However, supply and demand are not always the only factors influencing the establishment of prices. One example of such a situation is rail freight rates for grain. The Crow's Nest rates established in federal legislation have kept the returns to Canada's railway companies for handling grain at a level below what would be established by supply and demand. After the energy crisis, when the world price of oil rose dramatically, Canadian federal and provincial governments followed policies which artificially kept the price of oil in this country below the world level. The intention was to bring the Canadian oil price up to the world level gradually rather than allow it to move up all at once with the resulting pressures on industrial and personal customers.

It is also argued that large companies, large unions, and some large marketing boards have the power to set prices independently of the forces of supply and demand. It was in response to these types of market power that a wage-and-price controls program was introduced in Canada in late 1975. The wage-and-price controls program introduced another non-market factor into the establishment of prices.

Resource Allocation

In a strictly free-market economy resources are allocated to the production of certain goods and services by the price system. For example, if the market price of a product

Table 2-1. Demand and supply concepts

Change in demand. Means that a greater or lesser number of units is bought without changing price. This means a shift in the demand curve. If it shifts up and to the right, a greater number of units is demanded at any given price. If it shifts down and to the left, a lesser number of units is demanded at any given price.

Change in supply. Means that a greater or lesser number of units is supplied without changing price. This means a shift in the supply curve. If it shifts up and to the left, a lesser number of units is supplied at any given price. If it shifts down and to the right, a greater number of units is supplied at any given price.

Change in quantity demanded. Means that a greater or lesser number of units is bought because of a change in price. An increase in quantity demanded means that a greater number of units is bought because the price has been lowered. We are moving down a particular demand curve. A decrease in quantity demanded means that a lesser number of units is bought because the price has been raised. We are moving up a particular demand curve.

Change in quantity supplied. Means that a greater or lesser number of units is supplied because of a change in price. An increase in quantity supplied means that a greater number of units is supplied because the price has been raised. We are moving up a particular supply curve. A decrease in quantity supplied means that a lesser number of units is supplied because the price has been lowered. We are moving down a particular supply curve.

---

is high in relation to the cost of producing it, as determined by the prices of inputs, a substantial profit can be made on it. This will motivate people to allocate resources to the production of this product. An example is the number of companies which allocated resources to the production of electronic desk calculators when their profit potential initially became known. Another example is the increased allocation of resources to uranium exploration in Canada in the mid- and late-1970s in conjunction with the energy crisis and an increase in the price of uranium. Conversely, during the same time period there was a cutback in the resources (money, labour, and equipment) allocated to exploration for minerals such as copper, zinc, and nickel as their world prices weakened and their profit potential declined.

The price system in a free-market economy also influences the way the various factors of production are combined. For example, if the cost of producing an item can be reduced by substituting machinery (a capital input) for labour, this action will be taken. In this sense the price system, which establishes the prices of the factors of production, would determine how goods and services are to be produced. Chapters 8 and 18 contain further discussion of the various types of market structures which can exist. For example, oligopoly and monopoly markets are discussed in addition to perfect competition.

Many decisions about what goods and services to produce and how to produce them are made after consideration of social, ecological, and/or political factors as well as economic considerations. The Canadian Broadcasting Corporation is an example of a service which would not be offered if the only consideration were economic. The

---

GOVERNMENT ECONOMIC DECISION MAKING

In June, 1978, the Cabinet of Canada's federal government was debating a $200 million proposal for 4,000 new (grain) hopper cars and rehabilitation of 5,000 grain cars. This decision was to be made by the government rather than the railway companies (Canadian National and Canadian Pacific) because, as a company spokesman said, "the government controls the purse strings" in determining what the railways can charge for moving grain. "And it's up to it to determine what happens to the grain fleet."

Another railway company spokesman said the company isn't "making a nickel moving grain now so it just isn't logical that it would put money into grain car repairs." Another spokesman said that his company, although continuing to make minor repairs to grain cars, hadn't undertaken major repairs of grain cars for at least two years because there's no money to be made in moving grain.

Officials of both CN and CP had indicated that their companies would be willing to undertake grain car repair programs almost immediately if the government agreed to underwrite the cost of repairs.

The issue was heightened because there had been recent criticism of the rail companies by grain companies. The grain companies had questioned the railways about withholding cars when all available resources are needed to move export grain to waiting ships at Thunder Bay and Vancouver.

WHAT DO YOU THINK? Should the federal government invest this money? Should the government be involved in making this type of decision or should it be left to the market?

---

corporation annually records significant deficits. A government decision has been made that the service offered by the corporation is valuable to Canadians for social reasons.

Actions like tax incentives and the formulation of regulations also influence the manner in which resources are combined to produce goods and services. These actions are taken by the federal, provincial, and municipal governments.

THE BUSINESS FIRM

Business firm

Business firms play an integral role in our economic system. **A business firm is an entity (thing) which seeks to make a profit by gathering and allocating productive resources to satisfy demand.** Both public (government) and private organizations can be defined as business firms. Most private organizations, with the exception of those which are specifically non-profit, are business firms. Government organizations which are intended to be self-supporting (i.e., make a profit or break even) also qualify as business firms. This means that firms such as Air Canada, Canadian National Railways, The Saskatchewan Government Insurance Office and provincially owned telephone and hydro-utilities will be viewed as business firms even though they are government-owned.

On the other hand, the Canadian Broadcasting Corporation, which is not viewed as self-supporting, is not a business firm. Services offered to the public which are not intended to break even are public services. In addition to the CBC illustration above, public education (elementary, secondary, and post-secondary), police and fire services, hospitals, garbage collection, roads, parks, the Mint, the postal service, and so on, are examples of public services. Many people are employed by organizations offering public services, and they account for a major share of governments' budgets and of Gross National Expenditure. Many of the concepts of accounting, personnel, labour relations, production, finance, and marketing are as applicable to public services as they are to business firms. Also, there is considerable scope for innovation and improvement in these services. However, they are not the focus of this text.

The business firm is the basic building block for economic decision making in our system. Through it, resources are organized for production. Land, labour, and capital are gathered and converted into products or services which can be sold. This activity is directed and guided by managers in business firms.

Business activity, regardless of whether the firm conducting it is publicly or privately owned, requires decision making to produce and sell goods and services at a profit. It requires buying as well as selling. Thus, the market plays a role. How resources are used depends mainly on choices made by firms and consumers. Both are primarily guided by market prices. The firm is the key to the market's operation. It guides the flow of resources through the marketplace. The firm is an input-output system. The inputs are productive resources which the firm buys in the market. The outputs are the goods it makes and sells in the market. Both input and output depend on market prices. In some instances, decisions will be guided by social and political considerations in addition to economic ones. (See Figure 2-5.)

Resources and the goods made from them are both scarce. Thus, they command prices. The firm's costs of doing business (converting resources from one form to another) must be less than its returns if it is to earn a profit. To determine how profitable it is, a firm must keep records of its costs and sales. Accounting traces the effects of resource flows on its profits. **Profit is the difference between the cost of inputs and the revenue from outputs.** Profit

It is hard to define some of the terms used in modern business. The following statements, however, summarize some key points:

1. A business firm is an entity (thing) which seeks to make a profit by gathering and allocating productive resources to satisfy demand.
2. Most economic activity in our system is channelled through privately owned business firms, although some is conducted by publicly owned firms.
3. Business activity involves gathering and allocating resources to make a profit.
4. Since resources are scarce and choices must be made, business activity requires that costs be recognized.
5. Producer and consumer choices are primarily guided by prices.

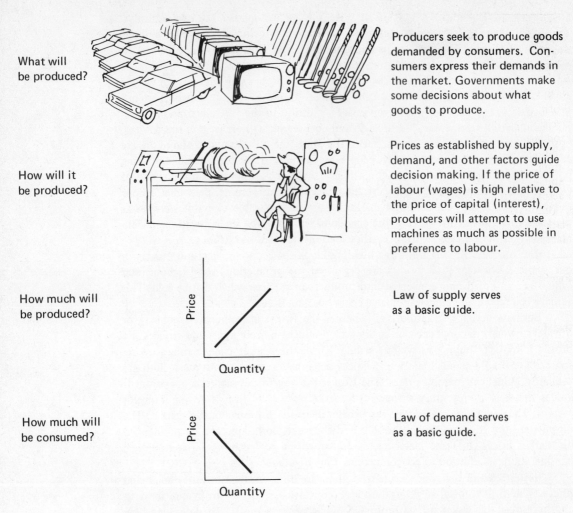

Figure 2-5. Economic decision making by business firms is guided primarily by price.

6. Most prices are determined by the price system.
7. Business activity requires that demand is present and saleable goods or services can be supplied to satisfy buyers.

You might find it helpful to study Figure 2-6.

The Motivation of the Firm

People form business firms to produce and sell want-satisfying goods and services so that they can make a profit. Profit, however, does not appear until it is earned. In our system there is no guarantee that a firm will make a profit. It is the hope for profit

2 THE BUSINESS FIRM

Effort	is exerted to exploit	Opportunity
Supply	is created to satisfy	Demand
Most prices	are determined in	Markets
Entrepreneurship	is concerned with	Risk assumption
Risk	is assumed in the hope of	Profit
Profit	involves a recognition of	Costs
Decision making	is necessary because of the	Choice process
Production activity	is primarily guided by the	Price system
Land ⎫ Labour ⎬ Capital ⎭	are made productive through	Entrepreneurship
Factors of production (inputs)	are converted to	Marketable goods and services (outputs)
Marketing activity	is primarily guided by the	Price system
Markets	exist because of	Specialization and exchange

Figure 2-6. The nature of business activity

that leads people to start and operate businesses. By providing products and services that satisfy customers, a firm may make a profit. (See Figure 2-7.)

Suppose Tommy Fields, age 7, opens a lemonade stand because he hopes to

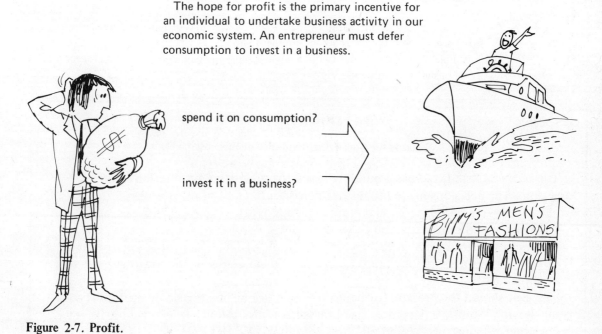

The hope for profit is the primary incentive for an individual to undertake business activity in our economic system. An entrepreneur must defer consumption to invest in a business.

spend it on consumption?

invest it in a business?

Figure 2-7. Profit.

make a profit. During the first week he sells 20 cups of lemonade at 5 cents per cup. Is his profit for the week $1?

The answer depends on many things. If the cost of sugar, lemons, and cups is 2 cents per cup, Tommy's profit is not $1 but 60 cents—the difference between $1 in revenue and 40 cents in costs (2 cents per cup times 20 cups). If his mother wanted to be paid for the things she supplied him, Tommy would realize that they are scarce resources. His profit is sales revenue minus the cost of doing business.

If revenues are greater than costs, a profit is earned. Profit may be increased by raising prices, lowering costs, or selling more units. But most firms cannot raise prices very much without reducing sales. Most of them try to increase profit by cutting costs and/or increasing the number of units sold.

If a firm's revenues and costs are equal, it earns no profit. It only breaks even. Very few people, however, go into business to break even. This is especially true if there is no "owner's salary" included in the firm's list of expenses. Firms that just break even give their owners no economic return from being in business.

Any after-tax profit earned by a firm is reinvested in the firm and/or is distributed to its owners. For many firms, reinvested profit is the major source of funds to finance their growth. The owners, in effect, are willing to reinvest their profit in the firm rather than taking it out and spending it on consumption.

Governments also have profitability in mind when they form business firms. However, they also take into account other social and political factors.

Profit Opportunity and Organized Effort

It is an accepted fact in capitalist-based economies that all types and sizes of business firms seek profit. There is some argument, however, whether profit should be their only objective. Later in our book, we will discuss the concept of the social responsibility of business. At any rate, it's one thing to seek profit and another to make it.

WHAT DO YOU THINK?

What Is the Social Responsibility of Business?

The business firm is the basic building block for organizing production in our system. The decisions made by businesspersons affect all of us. Business touches our lives in our various roles as consumers, employees, taxpayers, citizens, and so on. Thus, we all have a stake in those decisions.

A private individual goes into business hoping to make a profit. That person can stay in business only by continuing to make a profit over the long haul. This is a simple fact of business life in our economic system.

But should the "reason for being" for one of our most basic institutions be so self-serving? Should a person engage in business activity only to make a profit or does that person "owe something" to society? WHAT DO YOU THINK?

A HEALTHY ATTITUDE TOWARDS BUSINESS

Most people who have attended college work for businesses. Some are there because they couldn't find anything else and needed to earn a living. Some are very unhappy, because their only reason for staying in their jobs is to earn a living. This situation is unfortunate, because these people are not nearly as productive or happy as they would be if they were doing work that really gave them satisfaction.

If you're going to work in business, it helps to start with the right attitude towards business itself and towards your particular job and employer. Most people who are taking this course in business administration already have a generally favourable attitude towards business. They know that it is a worthwhile activity. They know that: (1) business is the chief provider of goods and services to our nation; (2) business is working hard now to improve opportunities for all and to eliminate discrimination among workers and managers; (3) business has to be flexible, inventive, and open to new ideas if it is to grow; and (4) business holds the key to economic growth and technological development. In short, business can be exciting and satisfying for most people.

If you are going to be motivated in the specific firm and job you choose, however, you must examine your own values and attitudes. In some cases, values and attitudes are only loosely held; that is, you don't have a very strong basis for holding them. If this is true about some of your presently held reasons for accepting or rejecting a career, you'd better get informed. There is nothing like knowing the facts to repair distorted or biased attitudes and values. A good example is the female graduate of a western Canadian university. She nearly turned down a job offer in Toronto because she had some very bad impressions of the city. This was due to her exposure to some people who had previously lived and worked in Toronto and did not like the city. It was also due in part to the fact that none of her friends and classmates was going there. After flying down for one interview, she discovered her previous perceptions were wrong. She took the job and eventually found a rewarding career as a director of personnel for a large electronics firm.

You may need to get more facts, too, so that the values you employ in choosing a job are reasonable ones. Don't assume that all advertising people are phonies or that all accountants are dull or any of the clichés about occupations. Get the facts. They will strengthen your system for evaluation.

A firm must find a way to use its scarce resources to its advantage by converting them into saleable goods. It must both identify an opportunity for profit and use its resources to try to make that profit.

Before an opportunity can be exploited, it must exist and be recognized. **An opportunity exists where there is a "set of circumstances" which may enable a firm to make a profit.** This set of circumstances can be the result of decisions made by persons inside or outside the firm or it can be mostly good luck.

Opportunity

> ## THINK ABOUT IT!
>
> ### The Importance of Profit to You
>
> The profits earned by business are important to you in many ways. Profits reinvested in firms for growth create jobs. Business profits are an important source of federal and provincial tax revenues. These tax payments by businesses help to pay for schools, hospitals, and other social services.
>
> Not only does profit reward a firm's current owners, but it also attracts new investors. This stimulates investment, creates more jobs, and raises our standard of living.
>
> Profitable firms can afford to set up programs to train people who want to work but cannot find jobs because they lack job skills. Profitable firms can afford to invest in costly pollution control devices. Unprofitable firms cannot.
>
> In other words, business profits are important to all of us. THINK ABOUT IT!

An announcement in a town's newspaper that a large factory will soon be located there can represent opportunity for many firms. The expected inflow of workers may lead True Realty Company to build a new apartment complex. This is an externally created opportunity for True Realty. The decision to locate the factory in the town was not made by True Realty but by the firm which is moving into town.

For an example of internally created opportunity, consider the case of Gem Chemical Company. It recently discovered a new cleaning compound. This set of circumstances was the result of decisions made by Gem's management. Their own research has made profit more likely.

The framework for evaluating opportunity is the market. A firm searches the market for unsatisfied wants which could be satisfied at a profit to the firm. General Motors introduced the Chevette automobile after recognizing a market opportunity and deciding to respond to it with an organized effort. The Chevette satisfied the wants of a specific group of customers who wanted a car which was cheap to operate.

Exploiting opportunity, however, need not involve developing new products. A grocer is doing it simply by staying open an hour longer than the competitors in order to satisfy late-night shoppers. In any case, exploiting opportunity requires an ability to meet it with organized effort and productive resources. Unless the firm can do this, that opportunity will be lost.

The Role of Risk

Risk

It is the hope of profit which motivates people to go into business. Since hopes are not always realized, risk is present. **Risk is the chance of loss.** The hope for profit explains risk assumption. The person thinks the expected profit is worth the risk involved. The greater the reward a business owner expects, the more risk he or she is likely to take.

> **WHAT DO YOU THINK?**
>
> **Risks of New Product Development**
>
> In his monograph, *Winning the New Product Game,* Professor G. Cooper, of the Faculty of Management at McGill University, reported that "huge sums could be lost" if efforts to develop and introduce new products failed. In the course of a survey of 150 companies in Ontario and Quebec he reported that "in more than one case a new product failure had actually resulted in corporate bankruptcy, while several firms had been forced to refinance with new owners because of substantial losses from a new product failure."
>
> Only 6 per cent of the firms studied thought their new product development programs were "extremely successful". Almost 60 per cent of the 150 firms rated their product development activities as only "moderately successful" or worse. Only 49.4 per cent of new products were considered a commercial success in another study conducted by Professor Cooper.
>
> On the other hand, successful new product development had very attractive rewards associated with it. For successful new products the "median ratio of profits returned to dollars risked was over five to one"! WHAT DO YOU THINK? Are the risks of investment in new product development worth taking?

People see risk differently. What one person sees as a very risky investment, another may see as "quite safe". This perception of risk is important in understanding why people are willing to risk their money in the hope of profit. Each year, thousands of new firms are started in Canada. Their owners invest their money in them in the hope of making a profit from serving customer wants.

Much of our material progress is due to our stable political and economic systems. This helps to reduce the amount of risk seen in a possible investment. Thus, a person is more likely to invest in Canada than in a country where frequent and violent revolutions occur. (See Figure 2-8.)

Top Management of Business Firms

The study of top management of business firms is the subject of business policy. The major decisions of a business firm as discussed in this chapter are made by a firm's top management. These include, among others, the decisions about what business to be in (what products to produce); what markets to serve (geographical area to sell in); what methods of production to employ (how to produce); where to locate production facilities; what profit goals to aspire to; what level of risk to accept; and how to distribute the firm's revenues among employees, dividends to owners, and reinvestment in the firm (allocation of income).

Business policy

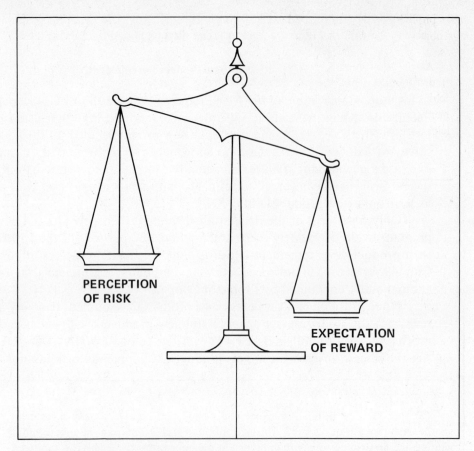

Figure 2-8. The businessperson weighs perceived risk against the expected reward—profit. When the expected profit outweighs the perceived risk, the business opportunity probably will be undertaken.

All firms, large and small, have top management which is responsible for these types of major corporate decisions. The job of top management will differ between smaller and larger businesses. In small businesses the top manager, who will usually also be an owner of the firm, will tend to be involved in all aspects of the firm's operation.

In large, functionally organized, single-product firms, the top management job tends to be one of co-ordination of the various functional activities of the firm. In large, multi-product (diversified) firms, which have a large number of operating units, top management will be primarily involved in finance, control, and provision of advice to the operating divisions.

Strategy

For all types of firms the key concept of business policy is strategy. **A firm's strategy is what it is currently doing, what it intends to do in the future (its goals), and how it intends to achieve these goals.** A firm's strategy embodies its decisions about what products or services to produce, how to produce and distribute them, the goals and objectives the firm aspires to, and the intended plan of action for achieving these goals.

As the term implies, **strategy formulation is the activities (formal and/or informal) determining what a firm's strategy is to be. Formulating a strategy involves matching company strengths with environmental opportunity**. The corporate top management strategist must have a good understanding of the environment. He or she must anticipate and cope with market conditions and trends and with political, social, and

Strategy formulation

KNOW YOUR STRENGTHS

Few of you will be responsible for general management decisions in the near future. The concepts of strategy and strategy formulation can, however, be immediately useful to you in making career decisions.

In making a decision about a career it is important, as it is in establishing a new business venture, to be aware of both the opportunities available and the strengths you possess to exploit these opportunities.

A good understanding of a wide range of available opportunities can be obtained from information gleaned from a number of sources. These might include discussions with professors, recent graduates, businesspersons, civil servants, and others. Other information can be obtained from libraries and from contacting the various industry associations and organizations indicated in this book.

It is sometimes harder to assess your own strengths than it is to obtain information about job opportunities. You start from the premise that every individual has things he or she is better at than other things.

Your grades and intellectual ability should be one thing you consider. However, you must not neglect your interpersonal and other types of skills. Also, assess your motivation and desire when you are considering your strengths.

You must, however, be objective and neither underrate or overrate yourself. Self-analysis can be a difficult thing. One way to approach the task is to list in two columns your strengths and your weaknesses. Another way is to sit down and draft a hypothetical letter of reference for yourself for a variety of different types of jobs. You will find these tasks difficult but useful.

After you have completed your analysis of strengths and opportunities, you will find that you have more information on which to base a career choice. Remember, the potential is greater when you attempt to apply strength to opportunity than when you attempt either to apply strength to an area where there is no opportunity or to take advantage of an opportunity even though you are weak in the area and are unwilling to develop the appropriate strength.

technological developments. The top manager must also have a good understanding of the strengths and limitations of the resources at his or her disposal. A strength is present when a firm has a competitive advantage over other firms in an industry. A strength can be any of a large number of things. For example, a firm could have a strength by being superior to others in terms of lower production costs, faster delivery, higher quality, better financial structure, more skilled work force, good corporate image, better marketing force, or many others.

The skill of the strategist lies in the capacity to identify strengths, identify opportunities, match the two, and take decisive and appropriately timed action. In matching strength to opportunity, top managers must also take into account their personal desires and aspirations. There is great scope for innovation and creativity in formulating strategies. The various aspects of the firm and the environment will be discussed throughout the remainder of this book.

Implementation of strategy

After a strategy is formulated, top management must ensure that it is implemented. The tools available to top management for strategy implementation are discussed in detail in later chapters. Basically, however, **top management can influence implementation of strategy by design of organizational structure; by design and management of various organizational systems such as accounting and control, planning, reward/punishment, staffing, training, and the like; and by the personal leadership style used in dealings with others in the firm.**

The essence of the top management job is to relate the firm to the environment in which it operates. The firm must be positioned so that it exploits market opportunities which arise in the environment while at the same time coping with other economic as well as political, social, technological, and ecological environmental factors. The remainder of this text will provide you with a framework for understanding the various functional aspects of a business firm. It will also indicate the different environmental factors a Canadian business manager must take into account.

SUMMARY AND LOOK AHEAD

Ours is a market-based economy. In many instances, relative prices determine how limited resources will be used to satisfy unlimited wants. Because resources are limited, choices must be made about what will be produced. Many of these choices are made by consumers and producers working through the price system. In other instances, governments determine what goods and services will be provided.

Prices induce or limit production and consumption. The price system helps firms decide which goods will be produced, how much will be produced, how they will be produced, and how they will be distributed.

The law of demand means that more of a given good is demanded at lower prices than at higher prices. The law of supply means that more of a given good is offered for sale at higher prices than at lower prices. These forces of supply and demand work to determine prices.

The firm is the basic building block for organizing production in our economy. It gathers productive resources (inputs) and converts them into saleable products or services (outputs). Both input and output depend on the price system.

The hope of profit motivates people to go into business. They earn profit from identifying opportunity and exploiting it through organized effort. But risk is always present. One role of top management is to identify opportunities and use the resources of the firm to take advantage of these opportunities.

Now that we know what a firm is and why it exists, we can go on to discuss the different forms of legal ownership of business firms.

KEY CONCEPTS

Business firm An entity (thing) which seeks to make profit by gathering and allocating productive resources to satisfy demand.

Business policy The study of top management of a business firm.

Demand The demand for a good results from the wants of people and their belief in the ability of different goods to satisfy those wants. The quantity demanded is the number of units of a good which people will buy at a certain price.

Demand curve A curve showing the number of units which will be demanded (bought) at various prices. Usually slopes down and to the right.

Law of demand More of a good is demanded (will be bought) at a lower price than at a higher price. Graphically depicted as a demand curve.

Law of supply More of a good is supplied (offered for sale) at a higher price than at a lower price. Graphically depicted as a supply curve.

Market economy An economic system in which relative prices determine how productive resources (the factors of production) will be allocated and how the goods and services produced will be distributed. These prices are determined in markets through the interaction of supply and demand.

Opportunity Exists when a set of circumstances may enable a person or firm who exploits it to reap some benefit or profit.

Price The quantity of money (or other goods and services) which is paid in exchange for something else.

Profit A firm's sales revenues minus its costs of doing business.

Risk The chance of loss.

Strategy What a firm is currently doing, what it intends to do in the future (its goals), and how it intends to achieve these goals.

Strategy formulation The activities involved in determining what a firm's strategy is to be. Includes assessing corporate strengths and environmental opportunities and matching them.

Strategy implementation The activities involved in putting a firm's strategy into effect. Includes activities associated with design of organization structure, reward/punishment system, and control system of the firm.

Supply The supply of a good results from the effort of producers. Price induces supply. The quantity supplied is the number of units of a good sellers are willing to offer for sale at a certain price.

Supply curve A curve which shows the number of units which will be supplied (offered for sale) at various prices. Usually slopes up and to the right.

QUESTIONS FOR DISCUSSION

1. "Since prices determine how much of a given product will be produced (supplied) and sold (demanded), there should never be 'too much' or 'too little' produced." Do you agree with that statement? Why or why not?
2. What is a market?
3. Can products or services really be "overpriced"? Explain.
4. What are the "underlying forces" of supply and demand?
5. What is profit? What determines how much profit a business firm will make?
6. What is a "favourable business climate"? How does such a climate affect the willingness of individuals to undertake business activity?
7. In our chapter, we said that the hope of profit is the main reason why business people undertake risk. What other reasons for going into business can you identify?
8. Are there any types of business opportunity in your community which are not being exploited? If there are, why are they not being exploited?
9. Explain the increase in the price of gasoline during recent years by using the laws of supply and demand. Discuss the important consequences this rise in price has

---

INCIDENT:

Sue Martin

Sue Martin is the owner of a small hamburger stand in a Maritime province college town. Several years ago, she inherited $10,000 and immediately quit her job as a delivery-driver for a restaurant supply firm. As a delivery-driver, she had become rather closely acquainted with several small-business owners who had, in Sue's words, "made it big" in the pizza, fried chicken, and hamburger business. Sue believed that if they could do it, so could she.

While many of these successful business owners had franchise operations such as Kentucky Fried Chicken, some were not affiliated with any franchises. Several of these business owners had boasted to Sue that their firms were so profitable that they got back their original investments in less than three years. This played a large part in Sue's decision to "get in on a good thing". She figured that after three years, everything was gravy.

However, Sue now finds her business in less than good shape. After several years of long hours and hard work, she can hardly pay her bills.

Questions:
1. Why do you think Sue decided to go into this business?
2. What do you think of Sue's analysis of opportunity?
3. What do you think of Sue's concept of profit?

had on consumers, oil companies, automobile manufacturers, and any others you think are important.
10. Define the strategy of a firm you are familiar with.
11. What factors must be taken into account when formulating a strategy?
12. What tools are available to the top manager of a business firm to implement its strategy?

INCIDENT:

Kristine Hoffman

Kristine Hoffman recently bought a new refrigerator from a local appliance store. She's been having problems with it ever since she bought it. The service department has been unable to repair it to her satisfaction.

Kristine has been very unhappy about the entire situation. She's written several letters to the manufacturer but has never received a reply. As a result, she's decided to never again buy an appliance manufactured by that company.

Her frustration runs even deeper. In one of her letters to the manufacturer, Kristine stated that she believed "all you companies are concerned about is profit—you don't really care if the product works or not as long as you make a sale at your inflated prices."

Questions:
1. What is the real source of Kristine's bitterness?
2. Is "making a sale" the same thing as "making a profit"? Explain.
3. Do you believe that Kristine's experience is too often shared by other persons? Discuss your answer in terms of exploiting opportunity to make a profit.

Section Two

Business Basics

In this section we place the business firm under a microscope in order to get a sharper focus on its nature and workings.

Our attention in Chapter 3 is on the forms of business ownership. Although many people think mainly of corporations when they think of business firms, there are other legal forms of ownership. In fact, the majority of Canadian firms are not corporations. But corporations do conduct most of the business activity in Canada.

In Chapter 4 we study the firm as an organization. Its goals can be achieved only if its human, financial, and physical resources are meaningfully related to each other. This is why organization is so important. We discuss the formal and informal organizational aspects of the business firm.

It is within the organization that managers perform functions which enable a firm to reach its goals. In Chapter 5, therefore, we look at the nature and functions of management.

3

Ownership

SECTION TWO   BUSINESS BASICS

OBJECTIVES: After reading this chapter, you should be able to:

1. Identify the reasons for the growth in public ownership of the factors of production in Canada.
2. List and define the four major forms of legal ownership of business firms.
3. Discuss the relative advantages and disadvantages of each major form of legal ownership of business firms.
4. List and define the different types of partnerships and partners.
5. List and define the different types of corporations.
6. Discuss the relative importance of the four major forms of legal ownership in terms of the number of firms.
7. Draw up a partnership agreement.
8. Explain how a corporation is formed and who controls it.
9. Identify and discuss other business structures in addition to the four major forms.
10. Understand the co-operative form of business ownership.
11. Discuss the advantages and disadvantages of large-scale operations.

KEY CONCEPTS: In reading the chapter, look for and understand these terms:

SOLE PROPRIETORSHIP
UNLIMITED LIABILITY
PARTNERSHIP
CORPORATION
BOARD OF DIRECTORS
STOCKHOLDERS
PROXY
PROFESSIONAL MANAGERS
CO-OPERATIVE
COUNTERVAILING POWER

Businesses vary in size from single-owner firms to giant corporations owned by thousands of people. The type of ownership is important. For example, some types of opportunity can be exploited only by large corporations. It would be hard to imagine a single-owner firm undertaking a project as vast as building a ship for the Canadian navy. Furthermore, a firm's growth is affected by its type of ownership. There are limits to expansion of a single-owner firm.

We begin by briefly comparing private and public ownership. The majority of the chapter is devoted to an analysis of the four common forms of business ownership in Canada—sole proprietorships, partnerships, corporations, and co-operatives.

PUBLIC VS. PRIVATE OWNERSHIP

In our capitalist system a person has the right to save and to invest money to make more money. The same right holds for a group of people. Persons, alone or in groups, can risk their money by going into business to try to make a profit. Private ownership of the means of production is a basic part of our economic system. Private ownership, however, is not the only form. Public or government ownership has become important in recent years.

Public ownership of the means of production may be undertaken for many reasons. When private investors are unwilling to assume the risk in some types of investment, the government may do it. In other cases, the investment needed may be too great for private investors and/or the potential payoff may be too intangible. Or, the government may wish to provide employment in an area and will therefore set up a government-owned firm to accomplish this goal. The government might also believe that businesses in the private sector are not providing a product or service properly, or are charging too high a price for it, so they will become competitors to firms in the private sector.

If we compared the growth rates of the private and the public sectors of our economy, we would find that the public sector has grown faster. Because the founders of our country feared too much government interference and control, they resisted public ownership. However, the Great Depression, which began in 1929, caused us to re-examine the basic reasons for distrusting government. During the depression, many people questioned the ability of the capitalist system to survive.

Today most Canadians accept the idea of some form of government ownership. Private ownership, however, dominates the society. The four most common legal forms of private ownership are the sole proprietorship, the partnership, the corporation, and the co-operative.

THE SOLE PROPRIETORSHIP

The sole proprietorship is the oldest and is still the most common form of legal ownership in Canada. **A sole proprietorship is a business owned and managed by one person. That person, however, may have help from others in running the business.** The sole proprietor is the classic case of the entrepreneur. Only a sole proprietor can say, "I am the company" or "This is my business."

Sole proprietorship

Advantages of the Sole Proprietorship

Suppose that Alice Stone wanted to go into the florist business. She might find that the sole proprietorship is the easiest way for her to start. There are no general laws which regulate the setting up of a sole proprietorship. Of course, the business activity must be legal. Furthermore, there may be municipal and provincial laws which require licences and permits. These are discussed in Chapter 15. Otherwise, Alice can go into business anytime she pleases. Thus, simplicity in starting the business is a definite advantage.

As the sole owner, Alice owns the firm outright. She is the sole owner of any profits (or losses). Alice may also get a lot of personal satisfaction out of seeing her firm grow under her direct guidance. She does what she believes is best for her firm. She makes decisions without required approval from anyone else.

Because Alice is the firm, she pays only personal income taxes on the firm's profits. If she wants to go out of business, she simply sells her inventory and equipment. She needs permission from no one. A sole proprietorship is easy to dissolve.

Disadvantages of the Sole Proprietorship

Unlimited liability

Since Alice is the firm, she is legally liable for all its debts. She has unlimited liability. **Unlimited liability means that a proprietor is liable for claims against the business which go beyond his or her ownership in the firm. The liability extends to his or her personal property and, in some cases, real property.**

Suppose Alice goes out of business and still owes her creditors $10,000 after selling her inventory and equipment. They can legally lay claim to her personal property (car, furniture, clothing, etc.) and even her real property (house and land).

The amount of money Alice is able to invest in her firm is limited to what she has and what she can borrow. In many cases, the difficulty of borrowing more money discourages this type of ownership.

As the firm grows, Alice may find that she is spreading herself too thin. A sole proprietor usually takes on the entire task of running the business. Thus, the entire burden of management is borne by the owner.

If Alice were to die, go to prison, or go insane, the business would cease to exist. There is a built-in lack of permanence. This makes it hard to attract employees who want a permanent job. It also makes it hard for the firm to grow.

THE PARTNERSHIP

Partnership

A partnership comes into being when two or more individuals agree to combine their financial, managerial, and technical abilities for the purpose of operating a company for profit. The partnership came about to overcome some of the more serious disadvantages of the sole proprietorship. It also dates back to ancient times.

There are several different types of partnerships. (See Table 3-1.) Our discussion, however, focuses on the most common type—the general partnership.

Table 3-1. Types of partnerships and partners

Types of partnerships

General partnership	All partners have unlimited liability for the firm's debts.
Limited partnership	This partnership has at least one general partner and one or more limited partners. The latter's liability is limited to their financial investment in the firm.

Types of partners

General partner	Actively involved in managing the firm and has unlimited liability.
Secret partner	Actively participates in managing the firm and has unlimited liability. A secret partner's identity is not disclosed to the public.
Dormant partner	Does not actively participate in managing the firm. A dormant partner's identity is not disclosed to the public. Has unlimited liability.
Ostensible partner	Not an actual partner but his or her name is identified with the firm. Usually an ostensible partner is a well-known personality. Promotional benefits accrue from using that name for which the person is usually paid a fee. Has unlimited liability.
Limited partner	Liability is limited to the amount invested in the partnership.

Advantages of the Partnership

Instead of a proprietorship, suppose Alice decided to form a partnership with Joe Gunn. Getting started requires that the partners agree on their duties, distribution of profits, and other features of the proposed association. While the agreement may be written or oral, a written agreement is superior, since it helps to avoid future disagreement between partners. Like a sole proprietorship, a partnership is easy to set up.

Since Alice and Joe are in business together, they can pool their funds and invest more than either one could invest alone. They have a greater ability to borrow money, since their combined personal and real property are available to creditors. They can also pool their talents and divide the tasks of the business. This brings the advantages of specialization to the firm.

Like a proprietorship, a partnership is not taxed as a business separate from its owners. The owners, not the firm, are taxed.

Disadvantages of the Partnership

Partners, like sole proprietors, have unlimited liability for the partnership's debts. It is a joint liability. This means that Alice is responsible for business debts incurred by Joe and vice-versa. In a limited partnership, however, only one partner must have unlimited liability. The other(s) may have limited liability.

In forming a partnership, one must choose a partner(s) with great care. Personal disagreements have caused many failures. This is why the partnership agreement should be written. Disagreement can stem from such things as how long the partners intend to be in business; the amount of money each is to invest; their salaries; how profits or losses will be shared; the duties of each; what procedure will be followed in admitting new partners; and a procedure for dissolving the partnership. Figure 3-1 shows the partnership agreement between Alice and Joe.

The death, withdrawal, or insanity of a partner automatically ends the business. Since more persons are involved, this lack of permanence is more serious in a partnership than in a proprietorship.

THE CORPORATION

Corporation

A corporation has been defined as "an artificial being, invisible, intangible, and existing only in contemplation of law." Unlike a sole proprietorship or partnership, the corporation has a legal existence apart from its owners. It can buy, hold, and sell property in its own name, and it can sue and be sued.

Originally, the corporate form of ownership was used most frequently for charitable, educational, or public purposes. In order to incorporate, a charter was required from the federal government. That is why the corporation is legally separate from its owners. It is a creation of governmental authority.

WHAT DO YOU THINK?

How Should Partners Deal with Personal Disagreements?

Personal disagreements between partners are a major cause of partnership failures. Even with a written partnership agreement, conflict will arise from time to time. But disagreement can be healthy if the partners can resolve their differences to the benefit of the firm.

Deep-seated personality differences which lead to endless petty arguments, however, can wreck a partnership. Unfortunately, these differences usually do not surface until after the partners are in business. The saying, "You really don't know a person until you live with that person," applied to a partnership, becomes, "You really don't know a person until you go into a partnership with that person." Do you agree? How should partners deal with personal disagreements? WHAT DO YOU THINK?

PARTNERSHIP AGREEMENT

THIS PARTNERSHIP AGREEMENT made and entered into this first day of January, 1980, and between Alice Stone of Victoria, B.C., and Joseph Gunn of Victoria, B.C.

WITNESSETH:

1. The parties hereby agree to form a partnership.
2. The name of the partnership shall be S & G Florists.
3. The business to be conducted shall be a florist business.
4. The principal place of business of the partnership shall be at 807 East Main Avenue.
5. The capital of the partnership is to consist of the sum of $30,000.00.
   Alice Stone is to contribute $15,000 in cash and Joseph Gunn is to contribute $15,000 in cash. No interest shall be paid to the partners on any contributions to capital.
6. Whenever required, additional capital shall be contributed by the partners in the proportion of the initial capital contribution.
7. The net profits of the partnership shall be divided equally and the partners shall equally bear the net losses.
8. Each partner shall be entitled to a drawing account as may be mutually agreed upon.
9. Neither partner shall receive a salary.
10. Each partner shall have an equal right in the management of the partnership.
11. Alice Stone shall devote her entire time and attention to the business. Joseph Gunn shall devote his entire time and attention to the business.
12. Either partner may retire from the partnership after giving the other partner at least 90 days' written notice of his or her intention so to do. The remaining partner shall have the option of purchasing the retiring partner's interest or to terminate and liquidate the business. The purchase price shall be the balance in the retiring partner's capital account based upon an audit by an independent public accountant to the date of retirement. The purchase price shall be payable 50 per cent in cash and the balance in 36 equal monthly instalments and shall not bear interest.
13. Upon the death of a partner, the surviving partner shall have the option to either purchase the interest of the decedent or to terminate and liquidate the business. The purchase price and payment shall be the same as above set forth.
14. The partnership shall begin the tenth day of January, 1980, and shall continue until dissolved by retirement or death of a partner or by mutual agreement of the partners.

IN WITNESS WHEREOF, the parties have signed this agreement.

Witnesses:

Roger Allen   *Alice Stone* (SEAL)

Vernon Collins   *Joseph Gunn* (SEAL)

Figure 3-1. A partnership agreement

Table 3-2. Types of corporations

	Definition	Example
1. Crown Corporations		
a. Departmental	Responsible for administrative, supervisory, and/or regulatory government services	Atomic Energy Control Board, Economic Council of Canada, Unemployment Insurance Commission
b. Agency	Management of trading or service operations on a quasi-commercial basis	Atomic Energy of Canada Ltd., Loto Canada, Royal Canadian Mint
c. Proprietary	Management of lending or financial operations; management of commercial or industrial operations	Air Canada, CBC, Central Mortgage & Housing Corp., St. Lawrence Seaway Authority
2. Business Corporations		
a. Private	Formed to carry on production of goods and/or services at a profit; number of shareholders limited to 50; board of directors must approve transfer of shares; stock cannot be traded on the open market	Eaton's Many other less well-known companies
b. Public	Formed to carry on production of goods and/or services at a profit; none of the restrictions on private corporations applies	Neonex Massey-Ferguson Steel Co. of Canada Imperial Oil Many Others

Formation of the Corporation

Although there are a number of different ways to form a corporation, the two most widely used avenues are federal incorporation under the Canada Business Corporations Act and provincial incorporation under any of the provincial Incorporations Acts. The former is used if the company is going to operate in more than one province; the latter is used if the founders intend to carry on business in only one province.

Except for banks and certain insurance and loan companies, any company can be federally incorporated under the Canada Business Corporations Act. To do so, Articles of Incorporation must be drawn up. These articles include such information as the name of the corporation (which must not already be in use), the type and number of shares to be issued, the number of directors the corporation will have, and the location of the company's operations. All companies must attach the word "Limited"

(Ltd.) or "Incorporated" (Inc.) to the company name to clearly indicate to customers and suppliers that the owners have limited liability for corporate debts.

Provincial incorporation takes one of two forms. In certain provinces (British Columbia, Alberta, Saskatchewan, Manitoba, Ontario, Newfoundland, Nova Scotia, and the two territories), the registration system or its equivalent is used. Under this system, individuals wishing to form a corporation are required to file a memorandum of association. This document contains the same type of information as required under the Canada Business Corporations Act discussed above. In the remaining provinces, the equivalent document is called the letters patent. The specific procedures and information required varies from province to province. The basic differences between these incorporation systems is that the registration system forms corporations by authority of parliament, while the letters-patent system forms corporations by royal prerogative.

Types of Corporations

Corporations can be found in both the private and the public sector in Canada, although our emphasis is on the private sector. The different types of public and private corporations and examples of each kind are indicated in Table 3-2.

Control of the Corporation

Two groups in the corporation typically exercise control—the board of directors and the shareholders. A corporation's board of directors is elected by the corporation's stockholders. (See Figure 3-2.) **The board of directors is the group which is given the power to manage the corporation. This power comes from the corporate charter and the corporation's stockholders.** In some corporations, the board of directors runs the corporation with an iron hand. In others, the board provides only minimal guidance.

Some people believe the board has one main task—to select a president. After that, it should keep hands off corporate management and leave it to the president. Others believe the board should play an active role in managing the corporation. Such a board meets more often and is more involved in managing the corporation. The iron-hand and hands-off approaches are the extremes. There are, however, many positions between these.

In many corporations, the board concentrates on developing long-range goals and long-range plans and does not get involved in day-to-day management. The board sets the goals, and the officers manage the corporation to achieve the goals.

Board members have certain legal obligations. They must function in the best interest of the owners and be reasonable and prudent in making decisions which affect the corporation. They must be as careful in managing the corporation's affairs as they are in managing their personal affairs. The board also must oversee and evaluate the job being done by the corporate officers. In Canada, it is generally difficult to successfully prosecute a board member for a breach of general duty. Directors are, however, liable for employee wages for six months after a corporation has become insolvent.

Board of directors

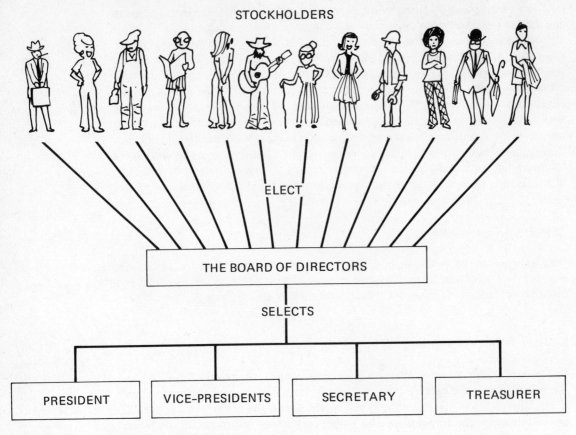

Figure 3-2. The corporate structure

Stockholders

The other group which is important in controlling the corporation is the stockholders. **Stockholders are persons who own the common and preferred stock of a corporation.** The number of votes a stockholder has depends on the number of shares he or she owns. It is not a case of "one person, one vote." This is why many small stockholders do not vote their shares in large corporations. For example, a person who owns 10 shares of INCO may choose not to vote. Control of such a corporation can be effective if a group of stockholders pools their votes to get a voting majority. In some cases, a person with 10 per cent or even less of a corporation's stock can exercise much control over its affairs.

Proxy

The corporate secretary must notify stockholders of the date, time, and place of stockholder meetings. Since many stockholders are unable or unwilling to attend, a proxy form is usually included along with the meeting notice. **A proxy is a person who is appointed to represent another person. By signing a proxy form, a stockholder transfers his or her right to vote at a stockholders' meeting to someone else.** A stockholder who does not attend the meeting or does not return the proxy form loses voting rights in that meeting.

By gaining enough proxy forms, present directors can keep their seats on the board. The job of soliciting proxies is easy for present board members. They can get the names and addresses of stockholders, and the corporation pays the cost of solicitation. Thus, it is easy for a board to keep itself in power as long as it does a satisfactory job in the opinion of voting stockholders. Unseating a board member can be tough.

Institutional investors, such as mutual funds, pension funds, investment clubs, and insurance companies, are professionally managed investors. They usually do not seek to exercise control of the corporations whose stock they hold. They buy and sell those shares mainly to make profits. But they may enjoy some degree of control because of their tremendous capacity to invest. A stock certificate is shown in Figure 3-3.

A corporation has many "publics"—customers, employees, labour unions, suppliers, creditors, and government. Each of these publics has a different interest in its affairs. In some cases, these interests are in conflict. For example, when union and management are in conflict, they often try to resolve their differences by bargaining. As we will see in several later chapters, there is growing consideration of the obligations of corporations to society as a whole.

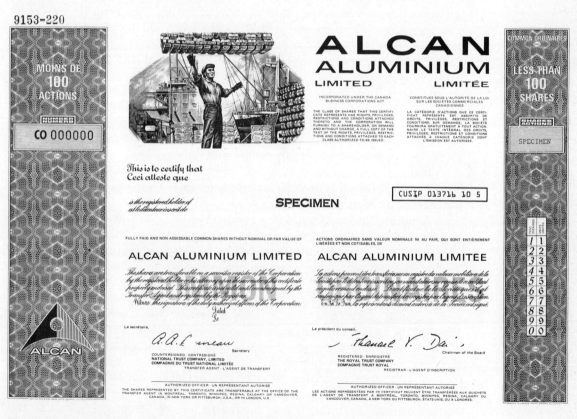

Figure 3-3. Ownership in Alcan Aluminum is shown by stock certificate

Advantages of the Corporation

A corporation's stockholders are not the corporation. They have only limited financial liability. A corporation's debts are separate and distinct from those of its owners. If you purchased stock in a corporation, the most you could lose is what you paid for the stock.

Let's return to Alice and Joe's business discussed previously. Suppose the partnership is doing well and the owners want to expand. They have little money to finance it, and they have borrowed up to their credit limit. A logical step would be for them to incorporate. This would give them access to people who might want to buy shares in the new corporation. Suppose Alice and Joe need $500,000 to expand. They might find it easier to get 1,000 persons to invest $500 each than to borrow a lump sum of $500,000 or to find another partner willing to put up $500,000. Thus, corporations usually find it easier to get money for expansion.

A corporation, for all practical purposes, can exist forever. The death, insanity, or imprisonment of a stockholder or a corporate officer has no direct effect on the corporation's existence.

A proprietor may have a lot of trouble selling his or her business. Selling one's interest in a partnership requires the approval of the other partners. But transfer of ownership is simple in a corporation. Stockholders simply sell their shares of stock. No permission from anyone else is needed. All that is needed is a buyer and a seller. Organized stock exchanges make it easy for them to get together. Millions of shares in hundreds of corporations are traded each day. In most cases, buyers and sellers never see each other. They deal through their stockbrokers. This is discussed in Chapter 10.

The benefits of specialization are apparent in a corporation. Corporations are more easily able to hire specialists to do specific types of work because of their typically larger size. This applies to workers and managers. Of course, the same is true for the few proprietorships and partnerships which are very large.

Most corporations have thousands of shares and hundreds of stockholders. Some, however, remain quite small in terms of number of owners and the number and market value of shares outstanding.

A stockholder need not, and generally does not, participate in managing the corporation. Ownership and management are separate. Most of the stockholders in Canada do not participate in managing the corporations of which they are part owners. The stockholders come from many backgrounds—schoolteachers, plumbers, salaried executives, and so forth.

Stockholders are the direct owners of a corporation. There are, however, millions of other people who have an indirect ownership in many corporations. For example, the members of many labour unions make payments to their union pension funds. Some of this money is used to buy stock in corporations. Thus, the union members are indirect owners of stock.

A stockholder owns a partial interest in the whole corporation. Suppose you own one share of stock in Norcen Energy Resources. You are not entitled to walk into the corporation's headquarters and demand to see the "property" you own. The property is owned by the corporation. What you own is a small part of the entire corporation. The value of that part varies with changes in the value of the shares of stock which you own. This is determined by the supply of, and demand for, the shares on the market.

Most corporations are managed by professional managers. **Professional managers are people whose profession or career is management. Such a person participates in managing a firm in which he or she is not a major owner.** Whereas sole proprietors and partners are owner-managers, most managers in most corporations are professional managers. As long as they produce profits, which show up as dividends or increased market value of the owners' shares, the majority owners are content to let the professional managers run the corporation. Although many managers own some shares of stock in the corporations they run, most of those managers do not own enough to control the corporations.

Professional managers

Disadvantages of the Corporation

A major disadvantage is that a corporation is subject to double taxation. As an entity separate from its owners, a corporation pays federal and provincial taxes on its profits. When the after-tax profits are paid to stockholders as dividends, they pay income taxes on them. Thus, corporate profits are taxed twice.

Another disadvantage is that corporations must conform to precise legal requirements to be granted a charter. Thus, they are more complicated and expensive to form than proprietorships and partnerships. To sell its shares of stock nationally, a corporation must get prior approval from a Securities Commission. In addition, federal and provincial regulation of corporations has increased over the past several decades.

Another disadvantage relates to the separation of ownership and management. Some people believe this makes a corporation's management too conservative. The proprietor who assumes risk stands to gain all the rewards. The hired manager of a corporation stands to gain less of the rewards from risk assumption. A hired manager does not own the business but tends to be blamed for all that goes wrong. He or she may avoid going out on a limb even when that may be best for the firm. This tendency towards conservatism in management, where it exists, may be a disadvantage of the corporation. Of course, this is less likely in a very small corporation where the president owns a large percentage of the outstanding stock.

The separation of ownership and management also reduces personal contact between owners and managers and makes managers even more conservative. Lack of personal contact between managers and workers also tends to make workers feel "outside" the business. How serious a problem this is depends mainly on the corporation's size.

Finally, there is less privacy in a corporation in terms of its financial performance. Whereas proprietorships and partnerships need not publish yearly financial statements for public consumption, corporations must do so. Since knowledge of how well or poorly the firm is doing can be useful to competitors, this publishing of financial data is considered to be a disadvantage.

Table 3-3 is a summary of the advantages and disadvantages of the three forms

Table 3-3. Relative advantages and disadvantages of the sole proprietorship, partnership, and corporate forms of ownership

Sole proprietorship	Partnership	Corporation
Advantages:		
Simplest to start	Few restrictions on starting	Limited financial liability
Sole ownership of all profits	Pooling of capital	Expansion easier due to greater capital resource
Greatest degree of personal involvement and satisfaction	High degree of personal involvement and satisfaction	Long life
No tax on the business as distinct from owner	No tax on the business as distinct from owner	Easy to transfer ownership
Maximum ease of dissolution	Greater borrowing power	Maximum borrowing power
	Greater managerial talent	Maximum availability of managerial talent
	More opportunity for specialization	Maximum opportunity for specialization
		Separate legal entity
Disadvantages:		
Unlimited financial liability	Unlimited and joint financial liability	Special and double taxation
Most difficulty in raising capital	Frozen investment	Complicated to form
Entire burden of management on one person	Personal disagreements	Much governmental regulation
Impermanence	Impermanence	Tendency towards overconservatism in management
Difficult to attract and retain employees		Impersonal
Difficult to expand		
Little opportunity for specialization		

of ownership discussed to this point. Table 3-4 indicates the relative importance of different forms of ownership in certain sectors of the economy, and Table 3-5 discusses other business structures.

Table 3-4. Form of ownership in selected sectors of the Canadian economy, 1971

	Number of establishments in				
Form of ownership	Mfg.	Service trade	Retail trade	Wholesale trade	Agriculture
Sole Proprietorship	7,043	71,173	89,757	11,243	336,175
Partnership	1,751	16,905	15,662	1,940	21,019
Corporation	22,622	19,486	28,329	18,110	7,992
Co-operative	492	39	629	595	–
	31,908	107,996	134,470	31,899	365,186

Source: Statistics Canada data for each industry listed.

CO-OPERATIVES

Co-operative

A co-operative is an organization which is formed to benefit its owners in the form of reduced prices and/or the disbursement of surpluses at year-end. The process works like this: suppose a group of farmers felt they could get cheaper fertilizer prices if they formed their own company and purchased in large volume. They might then form a co-operative (either federally or provincially chartered) and operate it. Prices are generally lower to buyers, and at the end of the fiscal year any surpluses are distributed to members on the basis of how much they have purchased. If Farmer Jones bought 5 per cent of all co-op sales, he would receive 5 per cent of the surplus.

Voting rights are markedly different than in a corporation. In the co-operative, each member is entitled to one vote, irrespective of how many shares he or she holds. This voting system is entirely consistent with the egalitarian view generally held by co-op members, but it may be a disadvantage, since others who do not hold that view may not become members in the co-op. This system may also be plagued with poor management. It does, however, prevent voting and financial control of the business by a few wealthy individuals.

Types of Co-operatives

There are hundreds of different co-operatives, but they generally function in one of five main areas of business:

1. Consumer co-operatives—these organizations sell goods to both members and the general public (e.g., co-op gasoline stations, agricultural implements dealers).
2. Financial co-operatives—these organizations operate much like banks, accepting deposits from members, giving loans, and providing chequing services (e.g., credit unions).

Table 3-5. Other business structures

Form	Nature	Example
Joint venture	A special type of temporary partnership. Set up for a specific purpose and ends when that purpose is accomplished. The death or withdrawal of a partner does not end the joint venture. Usually one partner manages the venture and has unlimited liability. The other partners have limited liability. Sometimes called a syndicate.	Several brokerage firms get together to sell a new stock issue for a client. These firms make up an underwriting syndicate.
Business trust	A trustee (or trustees) is created by an agreement. The trustees hold property, run the business, and accept funds from investors. Investors receive trust shares but they do not vote for trustees. Investors have limited liability.	A mutual fund accepts funds from investors. It pools their investment dollars and uses them to buy stock in other companies.
Mutual company	A firm which is owned by its user-members.	A life insurance company that is owned by its policyholders.
Holding company	A firm which holds enough of the stock of another firm(s) to control it or them. The controlled firm is called a holding company subsidiary.	Firm A buys controlling interest in Firms 1 and 2. Firm B buys controlling interest in Firms 5 and 6. A is a holding company and 1 and 2 are its subsidiaries. B is a holding company and 5 and 6 are its subsidiaries. If Firm C buys out controlling interest in A and B, C becomes the holding company.
Conglomerate	A firm which controls a number of firms in unrelated fields of business activity. A temporary decline in business in one industry will not have a large effect on the conglomerate's performance because its operations are spread out into several different industries. It is highly diversified.	International Telephone and Telegraph (ITT) Corporation's business activities include space, defence, industrial products, consumer products, natural resources, and telecommunications.

3. Insurance co-operatives—these organizations provide many types of insurance coverage such as life, fire, liability, etc. (e.g., the Co-operative Hail Insurance Company of Manitoba).
4. Marketing co-operatives—these organizations sell the produce of their (farm) members and also purchase inputs for the production process (e.g., seed and fertilizer).
5. Service co-operatives—these organizations provide members with services, such as recreation.

In terms of numbers of establishments, co-operatives are the least important form of ownership. (See Table 3-4.) However, they are important to society and to their members, since they may provide services which are not readily available or cost more than the members would otherwise be willing to pay.

THE EFFECTS OF SIZE

The relative advantages of the four forms of ownership we have discussed relate basically to the *form of ownership*. While we usually think of sole proprietorships as very small firms and corporations as very large firms, this is not always accurate. Some sole proprietorships, for example, are as large or larger than some small corporations. In other words, the *size* of a business firm has no necessary relationship to its *form of ownership*.

It is also true, however, that when we talk about "big business," we are talking about big corporations. Some of our biggest corporations, however, have very little in common with small corporations. Small, locally owned corporations have more in common with sole proprietorships and partnerships than they have with Canadian Pacific, Imperial Oil, and other huge corporations. In fact, in discussing "business interests" we often find vast differences of opinion between big corporations, medium-size corporations, and small corporations, sole proprietorships, and partnerships regarding what is "good" for business. Table 3-6 shows the 40 largest industrial corporations in Canada.

Advantages of Large-scale Operations

The larger a firm's size, the more likely it can afford to set up research and development, product testing, marketing research, and advertising departments and hire specialists such as engineers, chemists, market researchers, and advertising copy writers to staff them. A small firm usually cannot afford this degree of specialization. It may have one manager in charge of all production operations instead of separate managers for warehousing, traffic, production scheduling, and product quality control.

Large firms also are able to borrow more money and get favourable interest rates. The prime interest rate which often is reported in the news is the rate banks charge their most creditworthy customers—the big corporations. It is lower than the rate most smaller firms pay to borrow money.

Table 3-6. The forty largest industrial corporations in Canada

Rank 76	Rank 75	Company	Sales[1] $000	% change	Assets[2] $000	Rank	Net income[3] $000	Rank	% sales	Employees	Head office
1	2	General Motors of Canada	5,189,764*	19.5	1,296,733	17	159,808	6	3.1	31,639	Oshawa
2	1	Ford Motor Co. of Canada	4,768,900	7.5	1,666,900	14	126,200	8	2.6	36,682	Oakville
3	3	Imperial Oil	4,367,000⁰	6.3	3,139,000	4	264,000	2	6.0	14,753	Toronto
4	4	Canadian Pacific	4,016,671[4]	9.9	6,820,507	1	190,424	4	4.7	90,100	Montreal
5	5	Bell Canada	3,158,558[5]	12.3	6,657,038	2	287,524	1	9.1	77,522	Montreal
6	7	Chrysler Canada	2,941,053*	19.8	721.161	38	42,525	30	1.4	14,958	Windsor
7	6	Massey-Ferguson[6]	2,771,696*†	10.3	2,305,145	7	117,914	9	4.3	68,200	Toronto
8	8	Alcan Aluminum	2,585,426†	15.9	3,090,239	5	44,007	28	1.7	60,000	Montreal
9	10	Shell Canada	2,111,185	13.0	1,753,098	13	135,828	7	6.4	6,812	Toronto
10	9	The Seagram Company[7]	2,048,970†[12]	6.1	2,161,193	8	80,523	13	3.9	17,000	Montreal
11	12	Inco[8]	2,040,282*†	20.4	3,628,311	3	196,758	3	9.6	38,696	Toronto
12	11	Gulf Oil Canada	1,924,100[9]	13.1	2,045,200	10	165,900	5	8.6	11,100	Toronto
13	13	Canada Packers[10]	1,701,081*	4.0	321,059	73	20,142	69	1.2	15,000	Toronto
14	14	MacMillan Bloedel	1,520,207	17.2	1,280,060	18	22,842	62	1.5	23,601	Vancouver
15	21	TransCanada Pipe Lines	1,499,137	62.9	1,614,996	15	76,779	16	5.1	1,541	Toronto
16	15	Steel Co. of Canada	1,359,755	13.1	1,840,843	11	90,605	11	6.7	22,691	Toronto
17	17	Brascan	1,284,088†	11.4	2,583,661	6	102,523	10	8.0	31,000	Toronto
18	16	Noranda Mines	1,234,754	6.5	2,092,669	9	46,735	27	3.8	42,100	Toronto
19	18	Northern Telecom	1,112,009	9.2	705,039	39	73,936	17	6.6	—	Montreal
20	19	Moore Corporation	1,053,241†	4.7	764,262	34	61,633	21	5.9	25,964	Toronto
21	20	Imasco Corporation	1,031,642[12]	9.6	455,278	61	34,921	41	3.4	8,200	Montreal
22	22	Texaco Canada	989,308	14.4	940,602	25	28,982	48	2.9	3,841	Toronto
23	30	Dominion Foundries & Steel	903,874	22.5	1,033,151	20	66,699	19	7.4	11,500	Hamilton
24	26	Rothmans of Pall Mall Canada[13]	897,489[12]	10.7	424,272	65	21,084	65	2.3	7,000	Toronto
25	27	The Molson Companies[11]	889,012[12]	9.6	437,810	63	26,031	51	2.9	10,758	Toronto
26	25	Domtar	886,769	8.8	767,904	33	10,563	106	1.2	17,520	Montreal
27	33	Genstar	883,349*	22.4	1,232,843	19	55,071	24	6.2	10,695	Montreal
28	28	Abitibi Paper	880,351	15.2	899,364	28	13,024	89	1.5	17,000	Toronto
29	24	Canadian General Electric	879,427	7.0	571,187	47	32,699	43	3.7	17,512	Toronto
30	23	Hiram Walker-Gooderham & Worts[14]	874,955†	1.3	912,388	27	55,283	22	6.3	7,500	Walkerville
31	32	John Labatt[15]	837,218[12]	15.1	437,183	64	24,327	54	2.9	12,150	London
32	34	IBM Canada	837,137	16.4	490,173	57	79,067	14	9.4	10,680	Toronto
33	—	Alberta & Southern Gas	792,091	35.3	160,972	116	67	186	—	70	Calgary
34	—	Canadian International Paper	750,000	n.a.	n.a.	—	n.a.	—	—	12,500	Montreal
35	36	Consolidated-Bathurst	745,193*	15.8	742,667	36	18,240	72	2.4	17,557	Montreal
36	29	Cominco	725,005	2.5	973,205	21	47,673	26	6.6	10,696	Vancouver
37	37	Burns Foods	721,820[12]	16.0	163,950	113	5,691	137	0.8	6,971	Calgary
38	31	International Harvester Canada[6]	714,988	0.1	519,929	53	20,824	66	2.9	6,556	Hamilton
39	35	Mitsubishi Canada	704,801	5.6	91,622	156	1,589	170	0.2	92	Vancouver
40	44	Norcen Energy Resources	627,897	31.2	949,969	23	36,796	39	5.9	3,250	Toronto

* Net Sales
† U.S. Funds
0 Gross Revenue
n.a. Figures not available
1 Sales on operating revenue, unless otherwise indicated.
2 Total Assets
3 Net income before extraordinary items.
4 Revenue from CP Rail, CP Trucks, CP Telecommunications, CP Air, CP Ships, miscellaneous and CP Investments. Includes cross charging representing less than 5% of revenue.
5 All figures consolidate subsidiary Northern Telecom.
6 Fiscal year ended October 31, 1976.
7 Fiscal year ended July 31, 1976.
8 Formerly International Nickel.
9 Net Sales and Operating Revenue.
10 Fiscal year ended March 26, 1977.
11 Fiscal year ended March 31, 1977.
12 Excise and sales taxes not subtracted.
13 Fiscal year ended March 31, 1976. Percentage change in sales based on fiscal year ended June 30, 1975.
14 Fiscal year ended August 31, 1976.
15 Fiscal year ended April 30, 1976.

Source: *Canadian Business,* July, 1977. (Used with permission.)

Large firms tend to be more permanent, and this helps them in hiring professional managers who value permanence in employment. These managers participate in managing a firm in which they are not a major owner. Sole proprietors, on the other hand, are usually owner-managers. Only the larger ones can afford to hire professional managers. Partners in successful partnerships, however, can attract high-quality personnel who might expect to be offered a chance to buy into the firm. This is common in law and accounting firms.

As noted previously, while stockholders are the direct owners of a corporation, there are millions of other people who have ownership in large corporations through mutual funds and union pension funds. Investors in these funds, therefore, are *indirect owners* of stock. This, of course, gives large corporations access to hundreds of millions of investment dollars.

The advantages of large-scale operation, regardless of the form of ownership, are:

- greater opportunity for specialization by workers and managers
- greater borrowing power
- greater availability of managerial talent
- greater access to investment money

Disadvantages of Large-scale Operations

Some economists, government regulators, small-business owners, politicians, and labour leaders believe that businesses can be too big and result in reduced competition and greater concentration of economic power in the economy. Some of them would like to see giant corporations like Gulf Oil and GM broken up.

In order to counter the power of big business, big labour and big government also are part of our business system. The overall balancing of power between them is called countervailing power. Hopefully, the size and power of each of these three will prevent any one of them from becoming too powerful and dominant in our society.

Countervailing power

In any large organization, there is some tendency towards impersonality. In some of our largest corporations, for example, the lack of personal contact between workers and managers may lower the morale of the workers. They may have trouble identifying with the corporation, its owners, and its managers. There is a lot more personal contact in small firms.

In smaller corporations, manager's accomplishments can be observed more easily than in large corporations, where there are many levels of managers. In large corporations, therefore, hired managers may get less of the rewards from assuming risk. Some may become overly conservative and avoid taking risks, even when that is best for the firm.

The disadvantages of large-scale operation, regardless of the form of ownership, are:

- the potential for too much concentration of economic power
- the potential for reduced management efficiency
- some tendency towards impersonality
- the potential for overconservatism in management

> ## WHAT DO YOU THINK?
>
> ### When Is a Corporation Too Big?
>
> The corporation is one of our most basic business institutions. Without it, business activity as we know it would be impossible.
>
> Large corporations can afford to invest in costly research and development. This leads to new manufacturing techniques, improvements to established products, and the introduction of new products. Large-scale operations are also economical. The classic example is the modern assembly line in manufacturing. As a greater number of units are produced, the manufacturing cost per unit tends to decline. This can lead to lower prices for consumers.
>
> However, some people believe that corporations can be too big. They argue that the tremendous size of some corporations contributes to reduced management efficiency, which, in turn, leads to lower profits and higher prices to consumers.
>
> Other people are critical of corporations from a different point of view. Bigness, to them, is "badness". Many of these critics believe that the huge size and economic power of some corporations lead to reduced competition among firms and greater concentration of economic power. When is a corporation too big? WHAT DO YOU THINK?

SUMMARY AND LOOK AHEAD

The four major legal forms of business ownership are the sole proprietorship, the partnership, the corporation, and the co-operative. These are all forms of private ownership in contrast to public ownership. Private ownership is the most common form of ownership in Canada, but there is more public ownership today than there was fifty years ago.

Most privately owned firms are sole proprietorships. Most of these are small and employ only a handful of people. In many, the owner is the only employee.

A partnership is a firm owned by two or more persons who voluntarily go into business together. Like the proprietorship, it also dates back to ancient times.

A corporation is something separate and distinct from its owners. It is a creation of governmental authority. It comes into existence when its owners are granted a corporate charter by the provincial or federal government. Ownership is shown by shares of stock.

A co-operative is formed expressly for the benefit of its owners. It is premised on the belief that members should directly share in the benefits of the co-operative's existence.

What form is "best" depends on the circumstance in each situation. In no sense is one form of ownership always better than another.

In the next chapter, we view the firm as an organization. As we will see, a firm has both formal and informal dimensions.

YOU BE THE JUDGE!

Do We Need a Co-Determination Act in Canada?

In July, 1976, West Germany passed a "co-determination law" which gives workers in about 600 to 650 major companies a nearly equal voice with stockholders in running the companies. The idea is to extend political democracy to economic life. Other terms for this type of legislation are worker participation and industrial democracy.

Worker participation in company decision making exists at two levels in West Germany. At the top, workers are represented on supervisory boards which are similar to boards of directors in Canadian firms. At the bottom, there is worker representation on the plant level in works councils. These councils have an equal voice with management in decisions about hiring, firing, and working conditions. The basic idea of worker participation has existed for a number of years. The West German coal and steel industries have operated under worker participation since 1951 and big firms in other industries have been required since 1952 to allot one-third of supervisory seats to labour. The 1976 law raises the percentage to half for the largest firms.

In the United States in 1976, the United Auto Workers (UAW) asked Chrysler Corporation to give the union two seats on Chrysler's board but dropped the idea when Chrysler resisted. Except for the UAW, most American unions oppose the idea for a number of reasons. Some union people see worker participation as a threat to union strength because board members representing labour might be independent of *unionized* labour. Some unionists also contend that union workers already enjoy enough participation through the collective-bargaining process. Do we need a co-determination act in Canada? WHAT DO YOU THINK?

Source: James Furlong, *Labor in the Boardroom: The Peaceful Revolution* (Princeton, N.J.: Dow Jones & Company, Inc., 1977).

KEY CONCEPTS

Board of directors The group of persons, elected by a corporation's stockholders, which is ultimately responsible for the management of that corporation.

Co-operative An organization formed to benefit its owners in the form of reduced prices and/or the disbursement of surpluses at year-end.

Corporation The legal entity (thing) created by law which is granted rights set out in the corporate charter. A separate and legal entity apart from its owners.

Countervailing power The balancing of power between big labour, big government, and business.

WOULD YOU LIKE TO BE YOUR OWN BOSS?

To listen to a lot of people these days—and that includes some business teachers and textbook writers—you'd think everybody works for a big corporation. Of course, a lot of people do. But a lot find it much better working for themselves— running their own businesses.

Why not ask yourself this question: "Am I better equipped to work for somebody else or to be my own boss?" Weigh the points on each side. Can you handle a lot of uncertainty? Can you come up with enough money to start a business? Do you have something to sell? If the answer is yes to all of these, then maybe you should start out on your own. If you value security, if you have a small bank account and no one to back you financially, if you don't have a strong, clearly defined service or product to offer, then go to work for an established firm.

Maybe you have only some of the things you need. In that case it might be a good idea to go to work for a small business. In this way you may learn a lot and earn enough to overcome your handicaps. Later you can start your own business— older, wiser, and richer. As somebody with a lot of sense once said, "Making mistakes with somebody else's money is much easier to take than making mistakes with your own money!"

Partnership An association of two or more persons to carry on as co-owners of a business for profit.

Professional managers People whose career is management. Used when there is a separation of ownership and management in a firm.

Proxy A person appointed to represent another person. A stockholder transfers his or her right to vote at a stockholders' meeting to a proxy by signing a proxy form.

Sole proprietorship A business firm owned by one person. The firm and the proprietorship are the same.

Stockholders Also called shareholders. A corporation's stockholders are the persons who own it. The owners of the entity which is the corporation.

Unlimited liability The proprietor in a sole proprietorship and general partners in a partnership are liable for claims against their firms which are not limited to their ownership in those firms. Their liability extends to their personal property and, in some cases, to their real property.

QUESTIONS FOR DISCUSSION

1. What are some of the reasons for public ownership of the means of production?
2. List the business firms with which you did business in the past week. Which were proprietorships, which were partnerships, which were corporations, and

which were co-operatives? Why do you think that form of ownership existed in each case?
3. Do you think the president of Canada Packers gets less personal satisfaction from his job than the proprietor of a small business? Why or why not?
4. What, in your opinion, is the most serious disadvantage of the partnership form of legal ownership? Why?
5. If you decided to form a business organization and had already made up your mind to form a partnership, what factors would you consider in the selection of a partner?
6. Suppose that you own one share of stock in MacMillan Bloedel. Exactly what do you "own"?
7. What problems, if any, do you think go along with the separation of ownership and management in the typical corporation?
8. Which form of legal ownership is "best"? Why?
9. What is the function of the board of directors?
10. Do you think that the hired managers of a corporation are likely to assume more social responsibility than the sole proprietor or partner? Explain.

INCIDENT:

Ted Adkins

Ted Adkins, a sole proprietor, has been in the business of installing carpets for the past five years. He gets a lot of jobs from two local carpet retailers who also sell other home and building supplies. These retailers do not want to employ their own installers. When they sell a carpet, they recommend Ted to the buyer. Because he has a fine reputation for quality work in the community, Ted also gets business from word-of-mouth advertising. Ted has only two employees who help him install the carpets.

Business is doing so well that Ted finds himself having to turn away customers. Being a perfectionist about his work, he won't let anybody else do a job for him unless he personally oversees the installation. He refuses to hire more employees because of the experience of some other installers who had thriving businesses until they hired more employees to do the actual installing while the owners concentrated on drumming up new business. Their reputations as quality installers suffered, and they lost customers as a result.

Questions:
1. Do you think that Ted is really a businessman or more like a craftsman who takes great pride in the quality of his work? Explain.
2. Why do you think the other installers ran into trouble?
3. Do you think Ted's business is typical of many small proprietorships? Why or why not?

11. Who really "controls" the corporation?
12. Do you think the profit incentive is as important to the manager in a corporation as it is to the sole proprietor? Why or why not?
13. What do you think is the primary concern of a stockholder in a "widely held" corporation?
14. Why are most big businesses corporations?
15. Why do people join co-operatives?
16. What are the advantages of large-scale operations? The disadvantages?

INCIDENT:

The Paluzzi Brothers

Georgo and Tony Paluzzi are brothers studying business administration at a community college in Ontario. They are energetic, industrious, and want to put to good use what they are learning in school. They have a hobby of developing new recipes for Italian foods. Not too long ago they came up with a dish which proved to be very popular among some of their friends at college. In fact, word of the "new food" spread so fast that the brothers find that their hobby has grown into something of a business. No other Italian food on the market is quite like theirs.

Thus, they are considering going into business on a somewhat larger scale. They are convinced that there are many people in the area who would buy their product if it could be canned and distributed to grocery stores. They also believe that it's a product which would appeal to almost all Canadians.

They are very enthusiastic about going ahead with plans to can and sell the new product. The only obstacles now seem to be money and "know-how"

Questions:
1. What would you advise the brothers to do at this point in their new venture?
2. What are the relative advantages and disadvantages of the four forms of legal ownership as far as this business is concerned?

ROCKETOY COMPANY II

The large toy distributor (Toyco) which signed a trial contract for 1,000 Rocketoys negotiated with Terry for a longer-term contract for 5,000 units each of three of Terry's most popular toys. Terry recognized the opportunity here, and he wanted to close the deal.

Terry's only problem was his shortage of funds. This is common in many sole proprietorships like Terry's. Terry could not afford to buy the additional equipment and materials needed to meet the production requirements called for in the proposed contract. Although he reinvested all his profit in his business, Terry still needed more money for expansion.

Furthermore, Terry found that he needed more employees. Although he had three full-time production workers, he knew that the new contract would require the hiring of at least two more workers.

Toyco offered Terry a $15,000 loan if he would agree to sign the proposed contract. Terry figured the loan would be adequate to enable him to "tool up" for the new order. Because Toyco was convinced that it could make a large profit from sales of Rocketoy's products, Toyco was willing to lend the money at a very low rate of interest. Terry is seriously considering the offer.

Meanwhile, Joe Phillips approached Terry with a proposal to take him in as a general partner. Since Joe is quite wealthy, the new partnership would have no financial problems. Besides what Joe could contribute in cash, his being a partner would make Rocketoy a much better credit risk for any potential creditors.

Finally, Terry was also approached by three Toronto investors who wanted to make Rocketoy a corporation. They assured Terry that they were seriously interested investors who would gladly invest in the firm as stockholders.

Questions:
1. Why do you think Toyco was willing to lend Terry $15,000?
2. Why did Terry reinvest all his profit in his firm?
3. What are the relative advantages and disadvantages of the sole proprietorship, partnership, and corporation in Terry's case?
4. Would you advise Terry to take Uncle Joe in as a general partner, to form a corporation, or to make the loan from Toyco? Explain the reasons for your recommendation.

4

Organization

4 ORGANIZATION

OBJECTIVES: After reading this chapter, you should be able to:

1. Tell the difference between personal and organizational objectives and explain how they are integrated.
2. Draw a figure that illustrates the hierarchy of organizational objectives.
3. List and give an example of the different bases for departmentation.
4. Relate the span-of-management concept to the levels of management.
5. Compare the "top-down" and "acceptance" views of authority.
6. Draw a figure that illustrates the top-down flow of authority.
7. List and discuss the three actions involved in delegation.
8. Relate the delegation process to the degree to which a firm is centralized or decentralized.
9. Tell the difference between line authority and staff authority and give an example of each.
10. Tell the difference between line function and staff function and give an example of each.
11. Identify several sources of line-staff conflict.
12. Draw an organization chart.
13. Compare formal and informal organizations.
14. Draw a figure that illustrates the hierarchy of human needs.
15. Explain why informal groups arise and discuss their basic characteristics.

KEY CONCEPTS: In reading the chapter, look for and understand these terms:

ORGANIZATION
HIERARCHY OF
   ORGANIZATIONAL
   OBJECTIVES
DEPARTMENTATION
SPAN OF MANAGEMENT
AUTHORITY
RESPONSIBILITY
ACCOUNTABILITY
DELEGATION
CENTRALIZATION
DECENTRALIZATION

LINE AUTHORITY
LINE FUNCTIONS
STAFF FUNCTIONS
STAFF
FUNCTIONAL AUTHORITY
ORGANIZATION CHART
HIERARCHY OF
   HUMAN NEEDS
SOCIALIZATION
INFORMAL GROUPS
INFORMAL
   ORGANIZATION

An organization is a combination of human, financial, and physical resources put together by management so that certain goals can be accomplished. All firms have goals that can be reached only if their human, financial, and physical resources are tied together logically. This is why businesses are formally structured.

Individuals can accomplish their own personal objectives and help to achieve company objectives if both types of objectives are carefully integrated. Otherwise, a person will not get much satisfaction from the organization. Thus, the formal organization is the structure which helps a firm and its employees to achieve their goals.

People are the most important resource of any organization. An organization can also be viewed in terms of how these people behave. A firm, or any other type of organization, contains a collection of smaller, informal groups. These informal groups are not created by management but by the group members themselves to satisfy some of their needs. The entire set of these small groups is the informal organization of a firm. It is separate and distinct from the formal organization.

WHAT IS AN ORGANIZATION?

Imagine two cars travelling in opposite directions on the same street. They approach an intersection with four stop signs. Both stop. No other cars are present. Suppose the driver of car *A* signals for a turn that will put his car in the path of the other car. Both drivers must interact to avoid a wreck. Avoiding a wreck is an objective of both parties. Traffic signals and rules lend structure so that one driver will always let the other pass first.

In the above example, the elements of organization are present. They are (1) human interaction, (2) actions towards an objective, and (3) structure. The two drivers must interact, and their activities must be structured to avoid a wreck.

Now consider the case of a firm whose objective is to make a profit. Its employees' objective is to make a living. Workers interact with one another and with the tools of production. They perform those tasks necessary for the firm to make a profit and for the employees to earn a living. These elements—people, tasks, and physical resources—must be "meshed together" into a structure. The structure permits both the firm and the employees to achieve their goals.

Organization

We have seen two examples of organization. They are very different types, however. They vary in size, length of life, complexity, and formality. In the first example there are only two drivers, who interact briefly and informally. The second example involves a greater number of people who interact over a longer period of time on the basis of rules and procedures. There is more formal structuring of the relationships among people, activities, and physical resources in the second case. In a business, accomplishing objectives requires continuing patterns of interaction instead of only one interaction. Both cases are examples of organization, however. **When a group of people and things interact to reach objectives and their behaviour is structured, we call this an organization.**

Why Do People Join Organizations?

People "belong" to organizations because they believe they can better achieve their goals within them than they could outside. Different people, however, have different personal objectives. No one organization could possibly satisfy them all. This is why people form so many different organizations in our society.

In a firm, people's needs are similar enough to enable them to satisfy some of their needs through the firm. If it ceases to satisfy some important needs, members will leave it.

Personal and Organizational Objectives

People contribute to an organization if they think it helps satisfy their personal objectives. Integrating the various personal objectives into a unified statement of the firm's objectives is a complex chore. An employee may easily accept a company's objective "to make a profit". Doing so may provide money to be shared in the company's profit-sharing plan. But that same employee may not accept a company objective "to be a good community citizen." Doing so could involve "extra costs" and leave less for pay raises.

Ideally, an organization would meet all the personal objectives of all the people associated with it. However, personal goals are often in conflict. As a result, a firm's objectives are usually something other than the sum of the personal objectives of its different publics (employees, suppliers, creditors, customers, etc.).

Formal organizations sub-optimize objectives. This means they settle for less than the total achievement of all the objectives of their publics. Not all the personal goals of all of a firm's publics will be fully satisfied. Nor will company goals be completely achieved. A trade-off is involved. For example, an employee may want to grow on the job. But the firm may need his or her services in a very specialized job. The worker's desire for more varied tasks conflicts with the firm's objectives. This is why the firm must sub-optimize objectives. A trade-off is necessary. Neither the worker's objective nor the firm's objective will be achieved 100 per cent. A compromise is worked out whereby the firm and the worker each get something but not all that each might want. The employee's job might be broadened to include a greater variety of tasks. This increases his or her job satisfaction. The employee may become more productive. This, of course, benefits the firm. (See Figure 4-1.)

ORGANIZATION AS STRUCTURE

All firms are structured to help achieve company and personal goals. How complex this structure is depends on many things. Small firms, for example, are less formally structured than are large corporations.

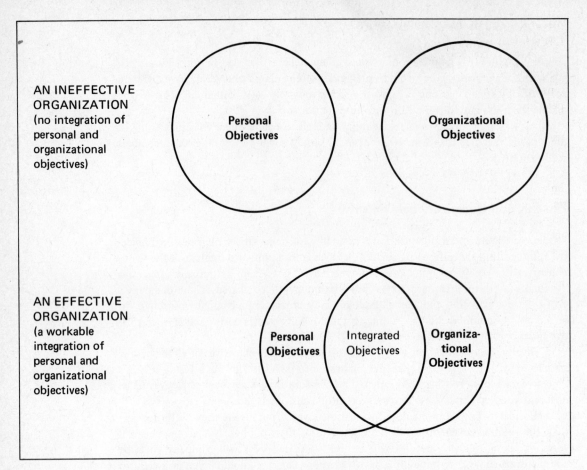

Figure 4-1. Personal objectives, organizational objectives, and organizational effectiveness

What Is Being Structured?

An organizational structure is created to help achieve goals. Within that structure, activities are performed which help achieve goals.

A primary goal of your school is to educate students. One of your goals is a good education. Relationships must be structured regarding the activities needed to reach these goals. Two essential activities here are teaching and learning. People are needed to perform these activities. Physical resources, such as textbooks and classrooms, are also involved.

Teachers must be assigned to courses. Texts must be related to courses. Classrooms must be assigned. Of course, students must also be assigned to teachers and courses. The three basic components—people, activities, and physical resources—must be related to each other through the process of organizing.

Objectives—The Underlying Reason for Organization

Activities are the "bridge" between company objectives and structure. Objectives (ends) are what we want to accomplish. Activities (means) must be performed by people working with resources if those objectives are to be achieved.

But what objective tells Harry the janitor to sweep the floor? The overall company objective "to supply customers with a quality product at a reasonable price" does not tell Harry what he should do to help the firm reach its goals.

Figure 4-2 shows a **hierarchy of organizational objectives. This concept involves breaking down broad company goals into specific goals for each person in the organization.** More specifically, broad company goals are first broken down into goals for company divisions. These, in turn, are broken down into departmental goals. Departmental goals are then broken down into goals for work groups within each department. Finally, work-group goals are broken down into goals for each individual worker. This lets each employee know what activities he or she must perform. The number of levels of objectives depends on the firm's size and complexity. There are fewer levels in very small firms than in very large ones.

As an example of this process, consider a manufacturing firm that has two divisions—industrial products and consumer products. The overall organizational objective might be "to profitably serve both the consumer and industrial market." The consumer division's objective would be stated more specifically and would deal strictly with the consumer market. The various departments within consumer products would

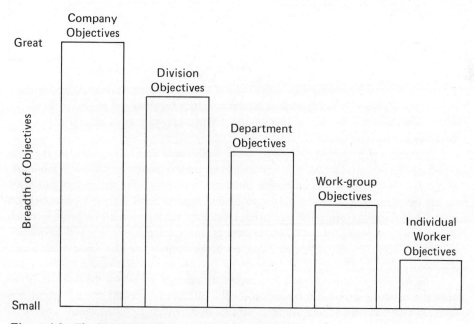

Figure 4-2. **The hierarchy of organizational objectives**

Hierarchy of organizational objectives

> **THINK ABOUT IT!**
>
> **Personal and Organizational Objectives and Values**
>
> The organizational structure in many firms often serves as a source of conflict between the personal objectives and values of its employees and the objectives and values of the organization itself.
>
> Many people today, for example, place a high value on personal freedom, control over their own lives, privacy, equal opportunity and treatment, self-expression, personal involvement, and personal growth. These personal values carry over to their world of work.
>
> But in many large organizations, employees face a bureaucratic structure whose objectives and values are very different. The management hierarchy is based on positional authority; jobs are highly specialized and often boring; policies and procedures spell out required behaviour on the job; and the activities and accomplishments of employees are reported in great detail to people higher up in the chain of command.

have even more specific objectives. At the lowest level of the organization, each individual worker would have very specific objectives which would be consistent with the overall organizational objectives.

GROUPING AND ASSIGNING ACTIVITIES

Departmentation

Most firms are broken down into departments through the process of departmentation. **Departmentation means grouping similar activities together and assigning them to specific departments.** The firm is divided into departments so that its objectives can be accomplished.

Activities are departmentalized to make work easier and to provide a means for controlling operations. A major factor leading to greater departmentation is modern technology. For example, the increasing importance of computer technology in business has led many firms to set up data processing departments. Your school probably has a data processing center. It performs specialized activities such as keeping employee records, printing students' grade reports, and storing many types of information. These activities are grouped together because they use the same computer equipment.

The number of subordinates a manager can manage is limited. Without departmentation, a firm's size would be limited to the number of persons the top manager could supervise directly. There are several bases upon which an organization could be departmentalized, as shown in Table 4-1. Regardless of which method is used, activities must be determined, grouped in some meaningful manner, and assigned to specific departments.

Activities must be identified first. There are two approaches. We could observe workers and classify activities like assembling parts, buying raw materials, and so on. The other approach is to start with company goals and determine what activities are needed to reach those goals. This method is the only one available to a brand-new firm.

Next, activities are grouped. Similar activities are grouped in the same department. For example, selling activities belong in the sales department, and producing activities belong in the production department. There are, however, exceptions. Typing is a common activity in most firms, but typists are not usually grouped into one department.

When problems arise that involve two or more departments, personnel from the departments involved are combined to solve them. After the problems are solved, these work groups are dissolved. The point is that departmentation is not intended to create "walls" around departments.

THE SPAN OF MANAGEMENT

As we have seen, one reason for departmentation is that there are limits to a manager's ability to supervise subordinates directly. **Span of management refers to the number of persons an individual manager supervises.** The wider the span of management, the

Span of management

Table 4-1. Bases for departmentalizing

Basis	Example
1. Functions	1. A manufacturing firm might be departmentalized into "producing" and "selling" departments; retailing firm into "buying" and "selling" departments; charitable organization into "collecting" and "disbursing" departments.
2. Geography	2. A manufacturing firm might organize its sales efforts in the Maritimes, Ontario, and the Prairie provinces.
3. Product	3. A company may have a manager in charge of sales for each major product which is produced.
4. Customer	4. A large meat packing firm may have one sales department serving retail accounts and another serving institutional accounts.
5. Process	5. A brewery with "cooking" and "aging" departments.
6. Type of equipment	6. Machines of one type are grouped into one department and those of another type are grouped into another department; e.g., grinding machines in one department and polishing machines in another.
7. Time	7. Some employees might work the day shift and others work the night shift.

THINK ABOUT IT!

How Many Subordinates Can a Manager Manage?

In our discussion about the span of management, we did not give any numbers for how many subordinates a manager can manage. The reason is that the number would vary among managers. Some important factors to consider in determining a manager's span of management are (1) how well-thought-out the manager's plans are; (2) how well-defined the subordinate's job is; (3) the quality of subordinates—their training, motivation, ability to work with others, and so on; (4) the pace of technological change in the firm; (5) manager and subordinate ability to communicate with each other; (6) the manager's ability to evaluate subordinates' job performance; (7) how much time a manager can devote to supervision; and (8) the manager's willingness and ability to delegate to subordinates. Can you add to this list? THINK ABOUT IT!

PRINCIPLES OF MANAGEMENT

During the twentieth century, many business managers and academics have spent a lot of time trying to develop some fundamental principles of managing people and things. Some of these "principles of management" are:

1. *Unity of command:* Subordinates should generally not have more than one boss so that they will not run the risk of receiving conflicting orders.
2. *Management by exception:* Managers should not continually check to see if subordinates are or are not performing their job. Only if something exceptional occurs should the manager become involved.
3. *Separation of doing and checking:* The people producing a product or a service should not have authority over the people who check to see if it has been done satisfactorily. For example, quality control inspectors should not be subordinate to production line foremen.
4. *Delegation:* Wherever possible, authority should be delegated to the lowest level possible in the organization, consistent with effective performance.
5. *Span of management (control):* The number of subordinates managers can handle should be kept within reasonable limits, given the job that is being done.
6. *Scalar chain:* There should be a clear line of authority from the top of the organization to the bottom. All people in the organization will, therefore, understand who they have authority over and to whom they report.

more subordinates are directly managed by a manager. Wider spans are more practical at lower levels in the firm.

Assembly-line foremen supervise workers who perform very similar and routine tasks. Modern production technology focuses on reducing complex tasks to more easily learned routine tasks. Thus, assembly-line foremen do not spend much time supervising any one subordinate.

Top-level managers, however, manage other managers. Top-level managers' work is not nearly so routine. The manager spends a lot of time managing each subordinate. As a result, only a few of them can be directly managed by one manager. This is why large firms usually have several echelons (layers or levels) of management—top, middle, and lower.

New technology is having an effect on the span of management. Computers in some firms now perform tasks which used to be performed by managers. For example, large retail chain stores use computers to record sales of various products in each outlet. This simplifies decisions about what to buy for different stores and makes it possible for a district manager to supervise a greater number of store managers.

The Right to Manage

A key ingredient in "the right to manage" is the concept of authority. **Authority is the right to take the action necessary to accomplish an assigned task.**   *Authority*

One view of authority holds that the right to manage springs from the right of private property. Authority flows from a firm's owners down to its managers. It's a "top-down" flow. (See Figure 4-3.) Another view, the "acceptance" view, holds that a manager has authority over a subordinate only because the subordinate accepts the manager's authority.

WHAT DO YOU THINK?

How Do You Manage in the Permissive Society?

Our society, over the years, has become more permissive with respect to authority. Women's liberation, children's liberation, "gay" liberation, and so on, are examples of "movements" to liberalize traditional ideas of authority.

The trend to greater permissiveness also affects the business world. In the past, subordinates routinely followed orders passed down the chain of command. This is not as true today. The right to manage or the validity of the top-down view of authority is being questioned in many cases. What might this mean to management in the future? WHAT DO YOU THINK?

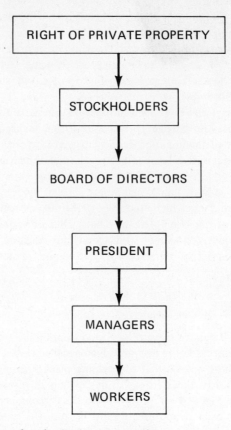

Figure 4-3. One view of authority in a corporation

Authority often is enjoyed by people because of personal qualities or expertise in a given area. A worker with "charisma" enjoys authority over fellow-workers. Likewise, the "old hand" who advises the new supervisor enjoys some degree of authority.

Responsibility   The counterpart of authority is responsibility. **Responsibility is the obligation of a subordinate to whom a task has been assigned to do it.**

Accountability comes from authority. Subordinates who are delegated adequate authority to fulfil their responsibility are accountable to their superiors for results.

Accountability   **Thus, accountability is the requirement of subordinates to report results to their superiors.**

Delegation

Delegation   Without delegation, a firm could not be departmentalized. **Delegation means a superior entrusts part of his or her job to a subordinate. The delegator (1) assigns duties, (2) grants authority for accomplishing them, and (3) creates a "moral compulsion" in the subordinate to perform them.**

Roy manages a car tire shop. Sam removes old tires, Tom switches them on the rims, and Bill balances and remounts them. These tasks were delegated to each of them by Roy. He also granted authority to Sam, Tom, and Bill to use the necessary tools to accomplish their tasks. He also created a moral compulsion on their parts to do their jobs. They are accountable or answerable to Roy. Notice, however, that responsibility has not been delegated. Delegation does not relieve Roy of the responsibility for seeing to it that his subordinates do their jobs. If a customer's tires wear out due to faulty balancing, it is Roy's responsibility.

The fact that delegation does not relieve a manager of responsibility makes some managers afraid to delegate. To delegate effectively, managers must have faith in their subordinates' abilities, be willing to let them learn by their mistakes, and be willing to follow up on how well they are doing their jobs.

Span of management and delegation are closely related. Managers who are afraid to delegate do everything themselves. Their span of management is likely to be very narrow. A sole proprietor who refuses to delegate limits the size of his or her firm. It cannot expand beyond what the proprietor is capable of doing alone.

The amount of delegation in a firm determines how much the power to decide is

---

YOU BE THE JUDGE!

Who Is Right Here?

Anna Scavo is a regional vice-president of sales for a large paint manufacturing firm. She is having some trouble with one of her district sales managers, Jim Clark.

This problem came to a head recently when several of Jim's sales representatives went "over his head" and complained to Anna. As they put it, "Jim has one rule of management—if it's an important decision, he'll make it. If it's a minor decision, he'll delegate it."

Anna talked this over with Jim. Jim's final statement on the matter was, "Delegation is a joke. You tell me to grant more authority to my subordinates. If I do it and they mess up, I get the blame. If I'm going to get the blame when things go wrong, I'm going to make the decision. At least that way I won't be taking the rap for somebody else's mistakes."

Despite her best efforts, Anna could not sell Jim on the delegation process. She even pointed out what she delegates to Jim. Jim agreed but said, "Sure, but you know I'll do what's right."

Last week, Anna hit on a new approach for dealing with Jim. She decided that the best way to "teach" Jim to delegate was to make his workload so heavy that he would have to delegate some decisions to his subordinates. Is Anna's approach sound? YOU BE THE JUDGE!

Centralization concentrated. **Centralization of authority means that decision-making authority is concentrated in the hands of a few people at the top level of a firm.** Such a firm is said to be highly centralized.

Decentralization **Decentralization of authority means that decision-making authority is spread throughout the firm.** Such a firm is said to be highly decentralized.

Centralization and decentralization of authority have nothing to do with geography. A firm with plants in many cities is not necessarily decentralized if decision-making power is concentrated at headquarters.

TYPES OF ORGANIZATION STRUCTURES

There are three basic types of organization structures:

- the line organization
- the line and staff organization
- the committee organization

The Line Organization

The line organization is the oldest and simplest type of organization structure. It has been used by military organizations and the Roman Catholic Church, and by business firms. In the military, the general gives orders to the colonel; the colonel gives orders to the major; the major gives orders to the captain; the captain gives orders to the lieutenant; the lieutenant gives orders to the sergeant; and the sergeant gives orders to the private. The Pope is the head of the Roman Catholic Church, and the chain of command extends downwards through cardinals, archbishops, bishops, and priests.

From one point of view, line authority is the authority relationship which exists between superiors and subordinates. **Line authority is the right to direct subordinates' work.** In the line organization, as we have seen, the chain of command extends from the top to the bottom of the organization. At any given position in the chain, a person takes orders from people higher in the chain and gives orders to people lower in the chain.

Line authority

Each superior has direct line authority over his or her subordinates, and each person in the organization reports directly to one boss. Each superior also has total authority over his or her assigned tasks. The major advantages of the line organization are:

- the organizational structure is easy to understand
- each person has only one direct supervisor
- decisions may be made faster because each supervisor is accountable to only one immediate supervisor
- authority, responsibility, and accountability are defined clearly and exactly, which makes it hard to "pass the buck" to someone else

Despite its advantages, the line organization suffers from some disadvantages which restrict its use in modern business to very small firms. The major disadvantages are:

- each supervisor must be an expert in all aspects of his or her subordinates' work because there are no "specialists" or advisors to turn to
- the paperwork required in directly supervising each subordinate is a burden on each supervisor's time
- the potential for the organization to become too inflexible and too bureaucratic

The Line and Staff Organization

Another view of authority involves the distinction between line functions and staff functions. **Line functions contribute directly to reaching primary firm goals.** For example, consider the case of a manufacturing firm whose primary goal is "to make a quality product and sell it at a fair price." The line functions are "production" and "marketing". For a retailer, "purchasing" and "selling" are the line functions. They are most directly concerned with achieving company goals. In a personal finance company, "lending" and "collecting" are the line functions.

Line functions

Staff functions help the line to achieve primary firm goals. In the manufacturing firm above, "quality control" and "market research" are staff functions. Quality control helps the production manager to produce a quality product. Market research helps the marketing manager to sell it.

Staff functions

The use of staff is one way to divide up the work of line managers. **Staff are people who advise and assist line managers in their work of achieving company objectives.**

Staff

The personal staff performs duties at the request of his or her line boss. The duties can range from opening the line manager's mail to representing the line manager at company meetings. Personal staff usually have the title of "assistant to".

Specialized staff serve the entire firm—not just one line manager. They have a high degree of expertise in their area of specialization. The marketing manager (a line executive) seeks advice from the director of marketing research (a staff executive) concerning whether or not to introduce a new product. The director of marketing research might also supply information to the production manager (a line executive) concerning sales forecasts so that production scheduling might be more efficient.

Staffpeople serve and advise line managers. In recent years, there has been a tendency in many firms to add new staff positions. Computer specialists, tax-law experts, and other advisors and analysts are examples. These staff people cannot issue orders to line managers. They can only give line managers advice and assistance. But the head of a staff department, such as the director of marketing research, does have the necessary line authority to run the marketing research department.

As you might suspect from the above discussion, there is a lot of potential for line-staff conflict in a business firm. Figure 4-4 discusses the nature of this conflict.

How staff sees line:

1. They get all the credit when things go right—we get the blame when things go wrong.
2. They have the final word, although we're trained experts in our fields.
3. They don't even want to try to see things from our point of view.
4. We may spend months developing recommendations, but they refuse even to listen to our advice, much less accept it.
5. They're older people who have been with the company so long that they're afraid to try new methods of doing things.
6. They resist change no matter what.

How line sees staff:

1. After all, I am the one who has the final word—they only give out advice.
2. There's only one right way—their way.
3. They can't even talk in ordinary language—it's always technical talk.
4. What he's doing now used to be part of my job.
5. They're know-it-all young college graduates who are always looking to change the way we do things around here.
6. They want to jump the gun without thinking through their ideas and recommendations.

Figure 4-4. Some potential sources of line-staff conflict

There are several ways to reduce this conflict. Some firms require line managers to consult staffpersons before making decisions on matters in the staff's area of expertise. In some cases, the line manager need only discuss the matter and is not required to follow the staff's advice. In other cases, the line manager must get the approval of the staff before making certain types of decisions. Thus, a production manager has to clear new recruits with the personnel department before hiring them.

In still other cases, the staff has authority to issue orders directly to line personnel. This is functional authority. It is granted only in the staff's area of expertise and only if it will benefit the firm. Thus, a plant manager's authority over safety matters may be removed by superiors and given to a safety inspector who has authority to shut down the plant if dangerous working conditions exist.

Automation and computers enable some firms to eliminate some lower-level management jobs. Suppose a computer can do the same tasks as a lower-level manager at a lower cost. Chances are that the manager will be replaced. But since top managers often lack the technical know-how for using computers, they hire specialists (staff) to advise them. Thus, instead of starting at the bottom of the line organization, many young technicians start in a staff position directly advising top management.

Staff status does not mean inferior status. Many staff specialists enjoy more pay and prestige than line managers. And, as we said earlier, staff executives do have line authority in their own departments.

The line and staff organization's major advantages are:

- staff specialists are available to advise, support, and serve line executives
- line executives need not get bogged down in technical matters and can devote more time and energy to their line functions

The line and staff organization is the most widely used organization form in contemporary business, especially among middle-size and larger firms. But it does have several potential disadvantages:

- the potential for line-staff conflict
- the potential for going overboard in creating and filling unnecessary staff positions

Functional authority

The Committee Organization

In the committee organization structure, several persons share authority and responsibility for accomplishing an objective. Instead of one manager, subordinates may report to several. This form of organization, where it does exist, usually exists within the overall line and staff organization.

Some committees, such as company policy committees, are standing, or permanent, committees. Others are *ad hoc* committees which are formed to accomplish a particular objective and are disbanded after accomplishing it. An example is a committee formed to select the site for a new plant.

Many manufacturing firms have new product committees. Such a committee is

> **PARKINSON'S LAW***
>
> As we have seen, one of the disadvantages of the line and staff organization is the potential for creating staff positions with little or no consideration to the need for those positions or the costs of staffing them.
>
> Some managers tend to measure their own importance in the organization in terms of the number of staff specialists reporting to them, the amount of paperwork they can generate and receive, and the amount of increase in their budgets they can get approval for.
>
> Paraphrasing two of Professor C. Northcote Parkinson's "laws" on bureaucracy:
>
> - Work expands to fill the time allowed for completing it.
> - As budgets increase, work expands to fit them.
>
> *C. Northcote Parkinson, *Parkinson's Law* (Boston: Houghton Mifflin, 1957; London: John Murray, 1958).

made up of line and staff executives from various departments within the firm—accounting, finance, production, marketing, engineering, and so on. The committee's job is to come up with new product ideas and develop them into marketable new products.

The major advantages of the committee form of organization are:

- decisions are based on the combined expertise and judgment of committee members
- participation by persons in different departments may increase their commitment to committee decisions
- there is less chance of one person's biases affecting committee decisions

The major disadvantages of the committee form of organization are:

- greater potential for buck-passing
- a tendency to take a long time in making a decision
- decisions often represent a compromise among the members rather than what might be "best" for the firm

THE ORGANIZATION CHART

Organization chart

A firm's structure can be quite complex. An organization chart helps us to understand a firm's structure. **An organization chart graphically depicts a firm's formal structure at a given point in time.** It indicates (1) the tasks which must be performed if the firm is to achieve its goals; (2) the lines of authority (chain of command); (3) how the firm is departmentalized; (4) how the departments relate to each other; (5) the various positions in the firm; and (6) the titles of those positions.

How complex these charts are depends on what management wants them to show. Some charts are simple, whereas others are rather complex. Generally, the larger the firm, the more complex the chart will be.

Figure 4-5 is an organization chart of a manufacturer. The top box shows the person who has final accountability for the firm's management. In this case it is the president. But it could have been the chairperson of the board of directors. The vice-presidents of production and marketing report directly to the president. Plant superintendents A and B report directly to the vice-president of production. The other reporting relationships (the chain of command or line authority) are indicated by the solid line. Notice the different echelons (levels) of management and wider spans of management as we move down from the top.

The staff organization is described by means of a broken line. The assistant to the president is personal staff. That assistant serves only the president. The personnel manager and the comptroller are specialized staff. They serve the entire firm. There is no functional authority in this firm. If the quality control manager did have functional authority over the plant superintendents, it would be shown on the chart by a dotted line.

From the chart we can infer something about the top-down delegation of authority. The president delegates to the vice-presidents of production and marketing who, in turn, delegate to the plant superintendent and regional sales managers.

There are many things an organization chart does not show. It does not tell the degree to which authority is delegated, what organizational objectives are, or the importance of various jobs in the organization. It also does not indicate the informal relationships which develop between people as they do their jobs. This is a real disadvantage of organization charts, because many important activities are done through the informal organization. This will be discussed shortly.

Organization Is Dynamic

We should point out that organizations are dynamic, ever-changing structures. We hinted at this when we said that an organization chart depicts an organization's formal structure at a given point in time.

Some critics of the traditional line and staff organization say that it often is too inflexible. It is built on hierarchies of power in which orders are passed down the chain of command. People at the lower levels have to wait for "orders" from people at the higher levels. This may reduce their initiative on the job. Meanwhile, rapid changes in technology, market conditions, government regulations, and so on, are taking up an increasing amount of top management's time. This slows down their passing of "orders" down the chain of command.

One of the newer approaches to organization that still is evolving is the matrix structure, or matrix management. Firms like Shell Oil and General Electric are using it to regain some of the flexibility of smaller firms. They are pushing decision-making

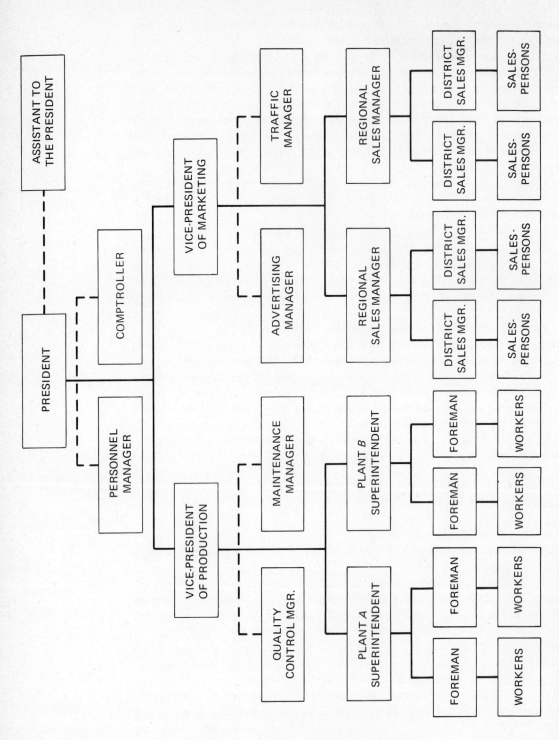

Figure 4-5. The line and staff organization

power down in the organization by structuring their organizations around objectives rather than around authority relationships and by placing emphasis on teamwork. Their structures tend to be flatter than the pyramidal shapes of many traditional line and staff types of organizations. But the matrix structure can cause conflict and uncertainty because there may be dual and even triple lines of authority. In other words, a subordinate may be accountable to two or three superiors.

As we have seen, the traditional line and staff organization chart depicts the chain of command. Rensis Likert, a management theorist, has suggested the concept of "linking pins" in structuring an organization. (See Figure 4-6.) Every manager, in

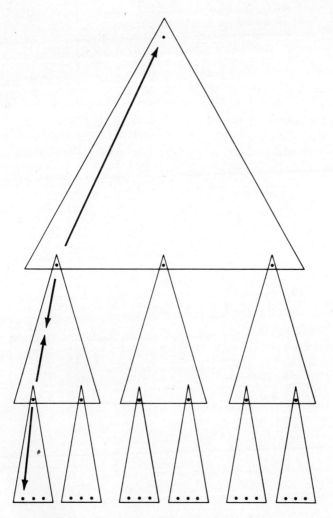

Figure 4-6. The linking-pin concept. *The dots indicate positions in the organization and the arrows indicate how they are linked together.*

reality, is a member of two groups. In one group, he or she interacts with superiors. In the other group, he or she interacts with subordinates. The manager can serve as a linking pin between these groups. The goal is to reduce the impersonality of hierarchical relationships among people in an organization and to integrate the work of people on different levels.

Some of the newer ideas on organization have led to some modifications of the traditional line and staff organization. Typical examples of these modifications are:

- wider spans of management
- greater decentralization
- less rigid chains of command
- authority based on knowledge instead of position

INDIVIDUALS AND GROUPS

Hierarchy of human needs

Psychologists tell us that a person is a unique being with a unique personality. No two people are exactly alike in terms of needs, wants, beliefs, values, and attitudes.

Although people are different, they all have certain basic needs. **These can be arranged into a hierarchy of human needs: physiological, safety, belonging, esteem, and self-actualization.** The hierarchy is shown in Figure 4-7.

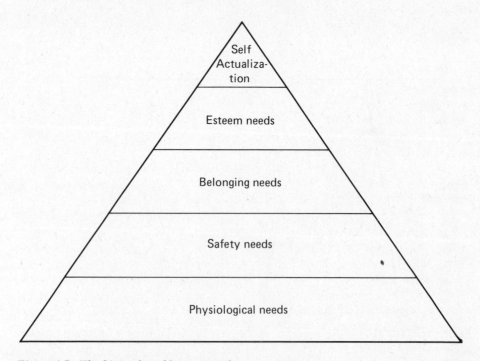

Figure 4-7. The hierarchy of human needs

The *physiological needs* (food, water, clothing, shelter) are the most prepotent of all needs. This means they must be satisfied before any of the higher-level needs can emerge and serve as *motivators* of behavior. If you are totally lacking in satisfaction of all needs in the hierarchy, your behaviour will be motivated by the physiological needs. The other higher-level needs, for all practical purposes, do not exist for you. But when the physiological needs are satisfied, they cease to be motivators of behaviour and higher-level needs emerge. Only unsatisfied needs can motivate behaviour. Need satisfaction, however, need not be 100 per cent. As lower-level needs become relatively satisfied, we are motivated mainly by the next higher level of unsatisfied need. The degree of need satisfaction required depends on the individual.

The *safety needs* are the next to emerge. These include the need to feel that you will survive and that your physiological needs will continue to be met. A cave dweller made a spear for protection against wild animals and other enemies. Our political and economic system helps to make most of us feel relatively safe from dangers such as starvation, tyrannical government, and severe economic collapse. The safety needs, for most adult North Americans, are not active motivators of behaviour. Of course, there are some people whose safety needs are relatively unsatisfied, and their lives are dominated by those needs. This may be true in some of our crime-ridden urban areas. But there are other examples of the security needs. Job security, for example, is a major issue in many contracts negotiated by labour unions and employers. Nevertheless, we probably can consider physiological and safety needs as basic survival needs.

The *belonging needs* are more social in nature. They include social belonging, love, affection, affiliation, and membership needs. The family provides much of this, but a business firm also can provide it by making the worker feel "needed". Co-workers also can provide it.

The *esteem needs* are of two basic types: *social-esteem* and *self-esteem*. Social-esteem (esteem from others) needs include prestige, status, and appreciation. They are your needs for relating to other people effectively. Self-esteem needs include competence and self-respect. These needs are very important, because success in life leads us to undertake new challenges, whereas repeated failures tend to reduce our willingness to undertake new challenges. Self-esteem and self-actualization needs are our needs for personal growth.

Self-actualization includes the drive for achievement, creativity, and developing your attributes and capabilities. It is your need to achieve your potential in life—to become what you are capable of becoming. People who are seeking to achieve self-actualization are very concerned about using their time in an effective manner to accomplish tasks which they believe are worthwhile and challenging.

As we mature, we tend to move up the hierarchy of needs. The behaviour of infants, for example, is dominated by physiological and safety needs. Older children and adolescents are more aware of others, and their behaviour usually is dominated by social needs. Their primary concern is "fitting in" and "belonging"—being like others with whom they associate instead of "being different". Normally, we tend to be motivated more by ego needs as we mature more fully. At some point in our teen-age years, therefore, we become more motivated by self-esteem and self-actualization needs. We want to assert our independence.

The challenge for business today is to provide employees jobs which offer rewards beyond those which satisfy only physiological and safety needs. Money, by itself, loses some of its power to motivate as we move up the need hierarchy. This, however, does not mean that money is not a motivator. Money can help in satisfying higher-level needs just as it can help in satisfying physiological and safety needs. For example, many people judge themselves in terms of how much money they earn. This means that, for them, money is related to the need for self-esteem. Likewise, money can give social esteem to the person who likes to brag about being a $100,000-a-year executive.

Money, however, is basically an extrinsic reward for work. This means that the reward does not come from the work itself. But there are other rewards which are part of the work itself. These are called intrinsic rewards. Social esteem can come from job titles and status symbols such as having a carpet on your office floor and a key to the executive washroom. Self-esteem comes from competent performance of your job.

In our society the need to achieve your potential in life (self-actualization) very often involves your work. Top-level executives and owner-managers, for example, often say that their work is their life. Work can satisfy a wide range of needs for these people. But this view of work is not shared by all people. Some people consider their work to be a sort of necessary evil which must be performed in order to satisfy their basic physiological and safety needs. They tend to look away from the job to satisfy their higher-level needs. As we'll see in our next chapter, managers use different approaches to leading their subordinates because managers know that work does not have the same meaning for all people.

Why Do Groups Exist?

People in our economy today have become less self-sufficient and more group dependent. Groups either enable us to do things we could not do alone or help us to do them more efficiently.

Whatever the reason—instinct, economics, and such—human beings are social animals. Our behaviour is influenced by our awareness of others. We tend to behave towards others as we think they expect us to behave.

Socialization

Socialization is the process by which human beings become social beings. It enables a person to become accepted by, and integrated into, groups. A worker becomes accepted as a member of a group by learning how to interact with its members.

Groups in the Modern Environment

In a large factory, a boss knows very little about the total personalities of subordinates, and vice-versa. As long as the job gets done, a boss ordinarily is not concerned about the other aspects of a subordinate's life. Life outside our most intimate, face-to-face groups has become more impersonal.

A person's role in some groups is very formal and governed by rules. This is true of the employer-employee relationship. In a sole proprietorship, the worker and the boss work out pay arrangements, hours of work, and other aspects of the job on a face-to-face basis. In a large corporation, most employees do not know the president. To the average factory worker, the employer is his or her immediate supervisor. Often, the employer is represented by hired managers, and the workers are represented by union negotiators.

All of us are subject to group influence and pressure. We are pressured to conform to group standards of conduct. This keeps the group together and working towards its goals.

THE INFORMAL ORGANIZATION

People on the job work and live together. They develop habits, customs, and patterns of behaviour and beliefs. This leads to the formation of informal groups. **Informal groups are small, face-to-face groups which spring up naturally as a result of human interaction on the job. They are created by their members to satisfy wants which are not being satisfied on the job by management.**

Informal groups

Workers are influenced in their attitudes and behaviour by the social groups they create to meet their personal needs. These interpersonal relationships cannot always be controlled by management. Employees interact in ways which are not always prescribed in the employee handbook. A business firm, therefore, is a social as well as an economic organization.

Within every formal organization there is an informal organization. **The informal organization is the entire complex of informal groups which exists within the framework of the formal organization.** It exists within the formal organization but is separate from it. Human interaction in the formal organization is structured by management. It arises naturally in the informal organization as a result of our social nature.

Informal organization

CAN YOU SETTLE THE ISSUE HERE?

The Top-level Woman Manager

Women are being promoted to top-level management jobs—jobs which used to be exclusively for men. But sometimes problems arise. Some women complain that their promotions show up on organization charts only, not in increased prestige and voice in company decision making.

Some female executives complain about being treated differently from the male executives on the same level in the organization. "The men quit telling off-color jokes, they rise from their chairs when we enter a room, and they open doors for us. We want them to forget that we are women. We want to be top-level managers—not top-level *women* managers." CAN YOU SETTLE THE ISSUE HERE?

How and with whom a worker chooses to interact can affect that worker's job performance. Some interactions may harm a person's job performance. Others may improve it. An employee may learn good work habits by associating with more experienced employees. Another employee may learn bad habits by associating with careless workers.

The Nature of Informal Groups

Harry Thompson is a new management trainee at Brigdon Corporation. This is his first job after graduating from a community college. As a new employee, Harry was introduced to his immediate boss and several co-workers. He was given information about the company, its objectives, and how he fits into the picture. After touring the plant, meeting a lot of people, and completing numerous personnel forms, Harry returned to his new office. His formal introduction to the job was over.

Suppose that Harry met two old friends who were former students at his college. They graduated one year before him and have been in the management trainee program at Brigdon since then. Harry will probably ask them questions. Now that he knows all the rules and what he's supposed to do, he asks his friends about "how things really are". If Harry had no old friends at Brigdon, he would try to talk to others who have been hired only recently. They, too, want to "talk it over" with people "on the same level".

Because of his need for information, Harry begins to interact with other people in ways which are not dictated by management. Although he was required to attend the formal orientation sessions, no one told him to seek out the advice of fellow-employees. No one told him to let those other employees influence his behaviour.

Informal Groups in the Firm

Workers learn customs and develop habits through their job experiences. Doing things "according to the book" is often abandoned in favour of shortcuts they learn on the job. The many informal, interpersonal relationships which exist in a firm lead to the formation of social groups. They are relatively small, and their members stick together because they share common values and goals. You become a group member when the group decides that you are a member. A young college graduate assigned to the production department may find it harder to be accepted than another person closer to the educational level of the group.

Work-group leaders are selected from their membership. Thus, even though Shannon Arnold may be the foreman appointed by management, Joe Smith is the person to whom the employees under Shannon go to "air their gripes". Joe has the "inside track" with Shannon's superiors. Joe has been around for a long time and knows "what's happening". He understands the workers' problems and knows how to "take care of them".

Groups develop norms or standards of behaviour, and they pressure their members to conform to them. Members who go against those standards are punished. Thus, a worker who produces "too much" may be rejected by the group.

> ## LINE OR STAFF?
>
> Let's assume you have resisted the urge to start your own business. Let's assume further that you choose to work for a medium- to large-size firm. This chapter discusses what staff and line positions are. What are *you,* a lineperson or a staffperson?
>
>   The big difference is that linepersons make the decisions and control the "mainstream" activities of the firm—producing and selling goods and services. Starting with the first-line supervisor and working up the organization, being "in the line" takes self-confidence and decisiveness—not unlike what it takes to run your own business. It will take a large measure of skills in human relations, too. The higher you get, the more breadth of vision it takes—being able to see the whole picture.
>
>   Staff positions are not so easy to describe because they are so varied. A personal staff-assistant position often requires helping to deal with important people and important problems, but only with those problems your boss (line) assigns you and without making major decisions of your own. If you are an efficient follower and don't mind "avoiding the limelight", a staff-assistant position may be for you.
>
>   If you have attained a special skill—as a lawyer, an accountant, a statistician, or creative artist—you may fill a staff-specialist position. This means you do special tasks a line officer hasn't the expertise or the time to do. You might have to write technical reports to help a line manager decide whether to sign an important contract or not. As you progress as a staff specialist, you might head a department of your own, such as the market research department of a firm. This will require more than a specialized skill in that area. It will take some of the same skills in human relations the lineperson needs.
>
>   What are you likely to be best at—line or staff work?

Since informal groups are based on interpersonal relationships, any changes which threaten to disrupt them are resisted. When changes in work assignments break up the coffee group, resistance will develop. It may take the form of workers asking why the change was made. It may even lead to a slow-down. (See Table 4-2.)

Member Benefits from Informal Groups

A worker usually belongs to several informal groups. Each satisfies different needs. Tom belongs to one group whose members work together in the receiving department. Their physical closeness on the job provides a basis for grouping.

Table 4-2. Characteristics of informal groups

1. small in size
2. created by their members
3. arise naturally and spontaneously
4. have leaders
5. resist change
6. cohesive
7. close contact among members
8. develop norms of behaviour
9. enjoy emotional commitment of members
10. rapid communication of facts and rumours through the communications grapevine

On Friday afternoons, Tom goes to the pub where he and several workers from other departments talk over their professional football and hockey clubs. They also talk over the "goings on" in their different departments.

Suppose Tom's boss recently hired someone with new ideas on running the department. If Tom and the older employees see this as a threat to their usual ways of doing things, they might form an informal group to "keep the new boss in line".

Informal groups help their members relate to the formal organization. Sometimes overspecialization of labour places workers in boring jobs. Membership in informal groups provides relief from such boredom. It enables members to feel human on the job even though they perform mechanical tasks. (See Figure 4-8.) Beth Welles works on an automobile assembly line in southern Ontario. She may not feel that she is a real part of the company. Her contacts and experiences in informal groups, however, help her to feel like somebody.

WHAT WOULD YOU DO?

Boredom on the Job

Boredom on the job is a major problem in many large factories. Many workers claim, for example, that the assembly line is inhuman in its requirements of workers on the line.

Assembly-line foremen are under pressure from their superiors to increase production. These foremen are also under pressure from their subordinates who want to improve the work environment.

Suppose you were a foreman on an assembly line and you wanted to try to satisfy the demands of both your superiors and your subordinates. WHAT WOULD YOU DO?

Characteristic	Informal	Formal
1. Nature of the business organization	A social organization	An economic organization
2. Central concern	Human interaction	Profit
3. Members seek to satisfy	Social needs	Economic needs
4. Structure	Informal, determined by voluntary patterns of interaction	Formal, determined by management
5. Communication	Face-to-face	Chain of command
6. Authority	Bottom-up	Top-down
7. Leaders	Chosen by members	Appointed by management
8. Commitment of members	Emotional	Limited personal involvement
9. Composed of	Small, social groups	Departments
10. Contact among members	Close, personal, face-to-face	Impersonal, often indirect contact through the chain of command
11. Human interaction	Spontaneous and natural	Determined by management

Figure 4-8. Informal and formal dimensions of the business organization

For some workers, the only needs satisfied by the company are physiological and safety needs. The pay cheque buys groceries. The union contract gives some job security. But the higher-level needs are often overlooked. The need to belong is an example.

Beth Welles may know that she is an important part of the lunch group. She is included in their plans. On the other hand, she may feel that the company would hardly miss her, even if she were to die. Nor does the job itself give her much self-esteem. She is "Beth" only to her immediate boss and close friends. The payroll clerk, the timekeeper, the plant superintendent, and others know her only as badge number 121. She feels unimportant to them. This is not true when she bowls with several co-workers. Thus, Beth needs to feel important to herself, and she wants others to look upon her as "somebody".

As individuals mature, they become increasingly conscious of their "selves". They want more independence. A worker who has to punch in, punch out, eat, rest, and wash up at a certain time is, in many respects, being treated like a child. That worker seeks an outlet for developing some degree of independence on the job.

Workers also want to feel that they are doing something "worthwhile" and that they are important for what they do. Workers who perform similar tasks recognize their particular contributions. An informal group may develop to help maintain a feeling of achievement. All the workers know and respect Jerry as the best person in the plant with an acetylene torch even though Jerry may receive no formal recognition from management. The realities of the informal organization are a vital part of the way a firm works—or doesn't work.

SUMMARY AND LOOK AHEAD

Business and their employees seek to accomplish objectives. Company and personal goals must be integrated so that employees will strive to accomplish company goals as well as personal goals. Because perfect integration is impossible, trade-offs are necessary.

Activities are the connecting link between the objectives and the structure of a company. Activities are determined by breaking down broad company goals into specific ones for each worker. Departmentation means these activities are grouped and assigned to various departments.

How many subordinates a manager can manage depends on many factors. Upper-level managers, however, manage fewer subordinates (a narrow span of management) than do lower-level managers.

According to one view, authority flows from the owner down through the various levels of management. Another view is that a manager's authority stems from subordinate acceptance.

Responsibility is an obligation of a subordinate to do a task which has been assigned to him or her. Accountability derives from authority. The subordinate to whom authority has been delegated is accountable to his or her superior for results.

Delegation involves (1) assigning tasks, (2) granting authority, and (3) creating a moral compulsion on the subordinate's part to do the task. In highly centralized organizations there is little delegation, since the decision-making power is concentrated at the top.

Line authority is the right to direct subordinate's work. Line functions contribute directly to achieving primary company goals. Staff functions advise and assist the line.

A firm has a social dimension which is not shown on its organization chart. Workers have needs which must be satisfied on the job. Many of these needs are social needs. If not met by the formal organization itself, these needs lead to the formation of informal groups. The groups are small, face-to-face groups. Their members stick together to achieve their goals.

The collection of small, informal groups in a firm is the informal organization. It exists within the formal organization but is separate from it. Its goals may conflict with those of the formal organization, or they may help the formal organization to accomplish its objectives.

In our next chapter, we look at the organization in terms of its management. We will study the management functions of planning, organizing, staffing, directing, and controlling.

KEY CONCEPTS

Accountability Derives from authority. Subordinates to whom adequate authority has been delegated to fulfil their responsibilities are accountable to their superiors for results. The requirement for subordinates to report results to their superiors.

Authority The right to take the action necessary to accomplish an assigned task.

Centralization Refers to concentration of decision-making power in an organization.

Decentralization Refers to dispersion of decision-making power in an organization.

Delegation Process of a superior entrusting part of his or her job to a subordinate. Involves assigning duties, granting authority for accomplishing them, and creating a "moral compulsion" in the subordinate to perform them.

Departmentation Grouping similar activities together and assigning them to specific departments.

Functional authority Staff issues orders to line personnel in the area of the staffperson's expertise.

Hierarchy of human needs A ranking of human needs into several layers. Those lower in the hierarchy must be satisfied first before higher-level needs can emerge.

Hierarchy of organizational objectives Broad overall organizational objectives are broken down into divisional, departmental, work-group, and individual worker objectives. The individual worker helps the firm to realize its objectives by helping his or her division, department, and work group to achieve their objectives.

Informal groups Face-to-face groups which spring up naturally as a result of human interaction on the job. Created by their members to satisfy wants not being satisfied on the job by management.

Informal organization The entire complex of informal groups which exists within the framework of the formal organization.

Line authority The authority relationship which exists between superiors and subordinates. The right to direct subordinates' work.

Line functions Functions which contribute directly to reaching primary firm goals. In a manufacturing firm, production and marketing are line functions.

Organization Something which is structured so that human activity can be coordinated to accomplish objectives. The three basic components of an organization are people, activities, and physical resources. They are related to each other through the process of organizing.

Organization chart A diagram showing an organization's formal structure at a given point in time.

Responsibility The obligation of a subordinate to do a task which has been assigned to him or her.

Socialization The process by which a person becomes a social being. Enables a person to become accepted by, and integrated into, groups.

Span of management The number of subordinates a manager supervises. It is limited.

Staff People who advise and assist line managers in their work.

Staff functions Functions which help the line to achieve primary firm goals. In a manufacturing firm, quality control and market research are staff functions.

QUESTIONS FOR DISCUSSION

1. Discuss the similarities and differences between the formal and informal organizations. Why is formal structure necessary in a business organization?
2. What is meant by the "hierarchy" of organizational objectives?
3. Are there any organizational objectives which are common to organizations in general (such as charitable, educational, religious, business, or fraternal organizations)? Are there any objectives which are common to all business organizations? Discuss.

INCIDENT:

Babson Homes

Babson Homes was founded 20 years ago in Calgary, Alberta, by Charles Babson. He foresaw tremendous future population growth in the province because of its vast natural resources.

Babson Homes enjoys a fine reputation as a builder of quality homes in the $75,000 to $125,000 price range. Although company headquarters are still in Calgary, the company builds homes within a 400-mile radius. Most of these homes are built for real estate developers.

From the beginning Mr. Babson has held a tight reign on his company. He makes on-site inspections of potential developments, negotiates contracts directly with the developers and sub-contractors, and personally supervises construction.

Mr. Babson usually works seven days a week. He takes great pride in the fact that he runs the company from the ground up, and he attributes his company's excellent reputation to the fact that he is personally involved in every aspect of its operations.

Babson's employees are non-unionized. They are skilled workers who take great personal pride in their work. They all believe in Babson's philosophy: "If we can't do it right, let's not do it at all." Babson is very proud of their loyalty to the company and their great skill.

Mr. Babson recently negotiated contracts with several large, nationally known real estate developers. Babson now finds himself constantly on the go and never having enough time to do "what needs to be done".

Questions:
1. Why is Mr. Babson in the position of not having enough time to do "what needs to be done"?
2. Would you recommend that Mr. Babson change his way of doing things? Discuss.
3. Do you think "the company" today is comparable to what it was 20 years ago? Explain.

4. Why are business organizations departmentalized? How far should departmentation be carried?
5. For years many provincial legislatures have talked about the need to "consolidate" many of the activities which are currently spread out over various departments. Why is this? Does this have any implications for business organizations? Explain.
6. How are the concepts of delegation, span of management, departmentation, and decentralization related?
7. What is meant by "line and staff"?
8. "Staff positions are generally dead-end jobs in most business organizations." Do you agree? Why or why not?
9. What is an organization chart?
10. Do informal groups exist only in large corporations? Discuss.
11. Discuss the characteristics of informal groups which spring up on the job.
12. What is the informal organization?
13. What "values" does an employee learn from his or her membership in informal groups?
14. How does a person become accepted as a member of an informal group?
15. Why is it typical for informal groups to resist change?
16. How might the informal group put pressure on its members to get them to conform to group standards?

INCIDENT:

Garvin Wicks

Garvin Wicks is a foreman for a large electrical contracting firm. Wicks has never been a popular foreman. His subordinates appear to tolerate him only because they can go to Van Booth, who is "in good" with the general superintendent.

Booth actually is under Wicks but has his "in" because he has been with the firm longer than Wicks or any of the other employees under Wicks. The workers turn to Booth for help when they need it. They routinely ignore Wicks.

Last week Wicks hired three new employees because of a heavy workload. He decided he would "set these new people straight" by having "a long talk" with them before they went to work for the company. He wanted them to know that he is the boss.

Questions:
1. Why do you think Booth enjoys so much status with the workers?
2. Who is the "real" boss—Wicks or Booth? Why?
3. What kind of problems might arise between Booth and Wicks?
4. What do you think will be the outcome of the "talk" Wicks gives to the new employees?

5 Management

5 MANAGEMENT

OBJECTIVES: After reading this chapter, you should be able to:

1. Explain why management is necessary in any organization which seeks to accomplish objectives.
2. Draw the management pyramid and show the different echelons of management.
3. Relate a manager's level in the management pyramid to the types of managerial skills he or she needs most at that level.
4. Discuss what personal characteristics are necessary to be a good manager.
5. List and define the functions of management.
6. Compare strategic and operational planning and relate them to the management pyramid.
7. Interrelate the systems concept to the practice of management.
8. Compare the democratic leader and the autocratic leader with respect to leadership styles.
9. List and define the various means of directing.
10. Illustrate the control process by means of a chart.
11. Identify and discuss the stages in the decision-making process.

KEY CONCEPTS: In reading the chapter, look for and understand these terms:

MANAGER
MANAGEMENT
ECHELONS OF
   MANAGEMENT
MANAGERIAL SKILLS
FUNCTIONS OF
   MANAGEMENT
PLANNING
STRATEGIC PLANNING
OPERATIONAL
   PLANNING
ORGANIZING
SYSTEMS CONCEPT

STAFFING
DIRECTING
LEADERSHIP
MOTIVATION
COMMUNICATION
PARTICIPATIVE
   MANAGEMENT
CONTROLLING
DECISION-MAKING
   PROCESS
ROUTINE DECISION
NON-ROUTINE DECISION

In the previous chapter we discussed the process of structuring the formal organization. This structure is the framework within which certain activities are performed. Any firm, however, is much more than a structure. It is a "living thing" in which effort is exerted to accomplish results.

Managers perform functions and non-managers perform tasks so that the firm can achieve its objectives. These functions of management are (1) planning, (2) organizing, (3) staffing, (4) directing, and (5) controlling. Let's begin our study of these functions by examining the nature of management.

THE NATURE OF MANAGEMENT

The terms "manager" and "management" are used in many ways. In any case, they both have to do with people.

Manager

A manager is a person who works through other people who perform operative tasks. A manager "brings together" their efforts to accomplish goals. Of course, non-human resources, such as money and materials, are also involved. Managers are not directly productive. They do not produce a finished product. Nor do they ordinarily perform operative tasks.

Management

Management can mean the process of managing, a collection of managers, a profession, or an area of study. **Our primary definition of management is the process of achieving goals through the efforts of others.** Management is necessary in any organization which seeks to accomplish objectives. Without it, an organization becomes a collection of individuals, each going in his or her own direction. There is no unifying guidance towards organizational goals.

Management is partly a science, because managers use organized knowledge in carrying out their functions. Production and marketing managers use formulas developed by statisticians to manage inventories, to schedule production, and to plan and control the distribution of their products. They also use knowledge in the fields of sociology and psychology in managing people. They borrow knowledge from other disciplines to improve their managerial skills.

But management also is an art. Through experience, managers develop judgment, insight, intuition, and a general "feel" for the management job. These are subjective skills which are learned through training and experience.

In general, management tends to be more science at the lower echelons and more art at the upper echelons. Staffpersons, for example, who advise top managers supply them with the "facts" they need to make decisions. In making their decisions, however, top managers often have to temper the objective facts with subjective judgment.

Echelons of Management

In a small firm with an owner-manager and a few workers, the owner-manager is the boss. In a large corporation the board of directors, the president, vice-presidents, department heads, and department supervisors are all managers. They work through

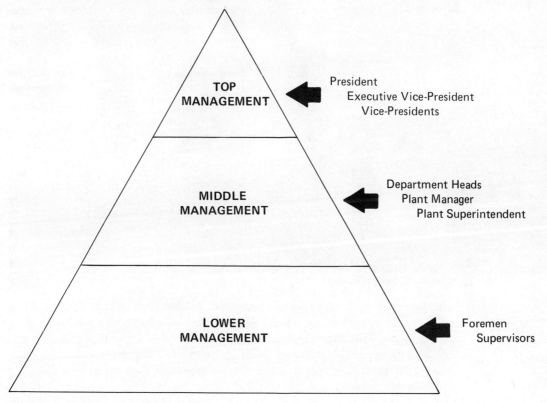

Figure 5-1. The management pyramid

various numbers of other people. The president works through a greater number of people than does the shop foreman.

There are, therefore, different echelons of management. **Echelons of management are the levels or layers of managers in a firm.** Upper-level managers work through a greater number of subordinates than do lower-level managers. (See Figure 5-1.)

Echelons of management

Managerial Skills

All managers must have certain managerial skills. These skills are:

Managerial skills

1. **Conceptual skills.** Overall ability to see the whole picture.
2. **Human relations skills.** Ability to get along and work with people.
3. **Technical skills.** Ability to understand how "things" operate; knowledge of the "nitty-gritty" of the job.

All managers need all three skills, but the level of management at which a manager is located affects the relative importance of each skill to that manager. Fore-

(Courtesy of Image Finders)

men, for example, need to know more about machinery (technical skills) than do top-level managers. Foremen directly manage workers who operate machinery. Top-level managers, however, are very far removed from most of the work which goes on at the "operations" level. Conceptual skills are much more important to top managers. Human relations skills, as you might expect, are important at all levels. This is why it is often said that "management is people!"

THE MANAGER AS A PERSON

The quality of a firm's managers is the most important ingredient in its ability to reach its goals. To look at it from another angle, poor management is the basic cause of business failures.

In any firm where the owners do not perform all the firm's work, workers are necessary. To keep them striving to accomplish company goals, management is necessary. Of course, the larger the firm, the greater the number of echelons of management.

The Manager's Basic Job

A manager's basic job is to accomplish goals through the efforts of others. To do this, a manager must create and preserve a work environment in which subordinates will put forth their best efforts to reach the firm's goals. Managers must make things happen. To do this, they must possess the managerial skills we discussed earlier. They also must seek to improve those skills constantly.

Managers must help their subordinates realize their potential. They must view their subordinates as their most important resource. Developing this resource is one of the most basic management challenges.

Managers also must satisfy the varying demands made by the firm's various publics—creditors, suppliers, owners, governments, superiors, subordinates, and others. Little wonder, then, that the manager's job is so demanding. What does it take to succeed in management?

What It Takes to Be a Manager

To be a manager, you must understand people and know how to work with and through them. You must be willing and able to learn constantly. You function in a dynamic environment. You must be able to be adaptive and innovative in that environment. You must know when to be adaptive and when to be innovative. You are adaptive when you are able to adjust yourself to the conditions in your environment. You are innovative when you introduce something new and when you introduce new ways of doing things.

The higher a manager is in the management hierarchy, the more important it is to have knowledge in many areas. He or she must be a generalist—one who has a broad general knowledge. One of the toughest tasks for some younger managers is to broaden their outlook—to stop looking at their job from the viewpoint of their former job. Fortunately, however, each of the managerial skills can be developed if a person is willing to work hard to learn them. They do not have to be "inborn".

Effective managers are willing to learn. They keep up with developments in their field and in related fields and seek to apply useful knowledge to the management job.

Effective managers are goal-oriented. They can set goals and put forth the effort needed to accomplish them. They also have a high achievement need and recognize the value of time. Time is one of the manager's scarcest resources, and he or she knows how to manage it productively. Managers can cram more into a twenty-four-hour day than people who have a lower achievement need. Hourly workers, such as carpenters and plumbers, receive overtime pay if they work overtime.

Managers must be able to cope with stress. Actually, non-managerial and managerial work both involve stress, because individuals must subordinate, to some degree, their individuality and their personal goals to the organization if the organization is to be effective. Since managers must work through the efforts of subordinates, they must help them to manage their stress. Line-staff conflict, personality conflicts, the potential for conflict between different departments, and the conflict between labour and management are only a few of the sources of organizational conflict.

Conflict can disrupt the smooth functioning of a firm and contribute to stress among workers and managers. Conflict, however, also can benefit the firm. It often leads to new ways of doing things; it brings deep-seated problems to the surface where they can be dealt with; and tension often stimulates the flow of creative new ideas. Management, therefore, should try to manage conflict, not eliminate it.

A person who perceives an assigned task to be important to accomplish but too demanding to accomplish comes under stress. It is stressful because the person expects that his or her costs and benefits for meeting the challenge are different from those of failing to meet it. It is a person's perception of the task's difficulty which influences the amount of stress which will emerge. People who have a high degree of self-

> ## THINK ABOUT IT!
>
> ### The Worker's View of Management
>
> Since a manager must work through others to accomplish goals, how those "others" view management is important to a manager's effectiveness. The following are some of the important things a worker expects of management:
>
> *Job and Working Conditions*
> - A job which is safe
> - A job which is not monotonous and boring
> - A job which enables a worker to use his or her acquired skills
> - A job with a future
> - A healthy job environment
> - Reasonable hours of work
> - Adequate physical facilities
> - Stable employment
>
> *Concern with the Worker as an Individual*
> - To be treated with dignity
> - To feel important and needed
> - To be managed by supervisors who can work with people
> - The right to be heard
> - The right to participate in decision making
>
> *Fair Treatment and Compensation*
> - To know what is expected in terms of performance
> - Objective basis for evaluating performance
> - No favouritism
> - Equal opportunity
> - Fair compensation system
> - Fringe benefits
> - Pay which reflects his or her contribution to the firm
>
> *Opportunity for Advancement*
> - Opportunity to learn new skills
> - Equal opportunity for promotion
> - Training and development programs
> - Recognition for past accomplishments
> - Opportunity to improve his or her standard of living

confidence usually will perceive less difficulty in accomplishing a task than people who lack self-confidence.

Stress on the job can come from within a person or from his or her work environment. Internal sources include low self-confidence, poor health, low tolerance for frustration, and a tendency to set unattainable goals for oneself. Examples of external sources are boring and monotonous work, too much responsibility, too little time to do the assigned work, and poor supervision.

Organizations are limited in their ability to control stress, however, because of its highly personal nature. A top manager who is under stress because of family problems, for example, does not want to "hang the laundry out" for all to see. There is, however, growing concern in business to help workers and managers cope with stress. "Workaholics" can be required to take time off for holidays. Many firms encourage employees to use coffee breaks as "exercise breaks" to help relieve tension on the job.

By and large, however, how a person copes with stress is an individual decision. Some people try to escape it by reducing their achievement drive, drinking excessively, overeating, or taking tranquillizers. Others face it in a more constructive way. Managers who feel under stress because they neglect their children may reorient their value systems and place limits on how much time and effort they are willing to give the company.

THE FUNCTIONS OF MANAGEMENT

To make the explanation simpler, we will break down managerial work into several functions. **The functions of management are planning, organizing, staffing, directing, and controlling.** Managerial work, however, cannot really be divided into component parts. A manager does not perform these functions one at a time. They are all performed at the same time. (See Figure 5-2.)

Functions of management

Planning

If managers are goal-oriented, they must have goals, or objectives. Objectives are the "whys" of managerial effort.

Organizational Objectives
A firm usually has more than one objective, and the objectives are likely to change over time as its environment changes. Since a firm is an economic and social organization, its objectives are both economic and social.

An economic objective of most firms is to produce and sell goods which people demand, at a profit to the firm. Accomplishing this brings profit to the firm and satisfies its customers. Another economic objective is survival and growth. During general economic slow-downs, more attention is usually given to survival. This might lead a firm to drop less profitable products. During "boom" periods, more attention is given to growth, possibly by adding new products.

A firm's economic objectives stem from its profit orientation. Examples are: to maximize profits; to achieve a 15 per cent rate of return on investment; or to increase market share by 10 per cent.

Figure 5-2. The functions of management

Greater awareness of the social dimension of business has led to greater attention to social goals. Large corporations especially have recognized that co-operation in attaining social objectives such as ethnic harmony and slum clearance is in their long-run interest. They have shown this by providing employment opportunities for the handicapped and job training for the disadvantaged unemployed.

The Nature of Planning

As a living thing with an instinct for survival, a firm is concerned with the future. This requires planning. **Planning is preparing a firm to cope with the future. Planning involves setting a firm's objectives over different time periods and deciding on the methods of achieving them.** In planning, managers rely on knowledge of past and present conditions in their environments. They use this to forecast probable future developments. They plan a course of action in accordance with this forecast.

No one knows for sure what the future holds. Managers, therefore, operate under conditions of uncertainty. The future, however, is not completely uncertain. Some conditions can be more or less taken for granted. We know that our government will not take over private property except under very unusual conditions. We project this knowledge into the future, and this reduces the number of planning "unknowns".

Certain internal facts are also under the firm's control. Planners plan in the face of something more than complete uncertainty and something less than complete certainty. They plan under conditions of risk—they have knowledge (or a good guess) about the likelihood of occurrence of some factors, but not all.

Planning and decision making are bound up in a future filled with risk. This is why some managers avoid planning. They argue that it takes them away from "doing" and accomplishing results—they prefer to live only in the present. They do not try to foresee problems—they "cross those bridges when they come to them".

Managers, however, cannot be effective in organizing and controlling if they do not know what they are trying to organize or control. Planning is really needed for effective performance of the other management functions.

This manager never did outgrow the need for "busywork". He buries himself in paperwork because he refuses to delegate to subordinates. How much planning do you think such a manager is likely to do?

Long-range planning is necessary for a firm to survive and grow. Adapting to its environment is easier if it predicts the future environment to some extent. But management also must be innovative so as not to be a victim of change.

Strategic planning

Innovation and adaptation both require continuous planning. **Strategic planning is concerned with a firm's long-range future and its overall strategy of growth. Strategic planning stresses innovation.** A top-level manager who thinks about expanding into new lines of business is engaging in strategic planning.

Operational planning

Operational (or tactical) planning is planning for day-to-day survival of the firm. Operational planning stresses adaptation. A sales manager who assigns salespersons to their territories is engaging in operational planning. Both types of planning are necessary for a firm to be effective. (See Figure 5-3.)

In summary, planning is managerial effort devoted to recognizing and exploiting opportunities. It enables management to anticipate and avoid problems before they occur.

Organizing

Organizing

A firm becomes a structure through the process of organizing. **The function of organizing involves (1) determining the tasks and duties the organization must perform, (2) establishing relationships between jobs, (3) writing job descriptions, and (4) setting up job methods.** All these activities should be carried out with the organization's objectives in mind.

As we already know, a firm has both formal and informal dimensions. Managers use both as instruments of management policy. A manager who considers the opinions of subordinates is more likely to gain acceptance of management policies. Such a manager works through informal leaders. The manager knows "what's happening" by listening to the employee grapevine. In this way, a manager lessens the chance that the formal and informal organizations will conflict.

The Systems Concept

A firm is something more than the sum of its parts. To understand this, the concept of system is helpful. A system is a complex of interacting parts. The parts are interrelated so that the unified whole is something more than the sum of its parts. The solar system, for example, is a complex of planets which revolve around the sun. From the systems approach, the earth is a part of that system, rather than independent. This way of looking at the earth gives us a different perspective of our planet.

Need for a Systems Orientation

Systems concept

The systems concept is also relevant to businesses. **According to the systems concept, a firm is not the accounting department or the marketing department. It consists of a network of interrelationships among the various departments and their environments.**

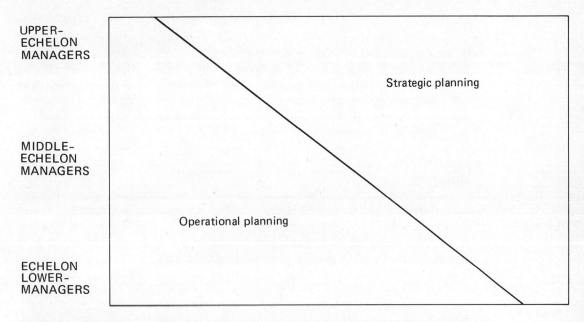

Figure 5-3. The relative importance of strategic and operational planning at various levels of the management hierarchy

The marketing research department is a sub-system of the firm. However, the firm is a sub-system of its industry, the industry is a sub-system of the total economic system, and so on. Management must integrate the various sub-systems so that overall system performance can be improved.

A manager's effectiveness is determined not by how well one management function is performed but by how well all of them are performed. A manager must recognize that a firm is really a complex of interacting parts.

The systems view underscores the importance of setting clearly defined company goals and communicating them to managers and workers. It is human nature for credit managers to judge their effectiveness in terms of reducing bad debts. Sales managers may think in terms of annual dollar sales increases. Credit managers and sales managers often view the company from different perspectives. But they should be striving to accomplish common goals.

The more that company personnel view the firm as a system, the less their actions will conflict and the more efficient the firm will become. The credit manager recognizes that some bad debts are acceptable in order to increase sales. The sales manager recognizes the need to deny credit to customers with poor credit ratings in order to keep bad debt losses down. This is the essence of the systems view. Another indicator of systems thinking is the responsiveness of a firm to social problems. When a firm accepts social responsibility, it is viewing itself as a sub-system of the larger socioeconomic system.

Staffing

A firm's organization is meaningless without people. Getting people to staff the organization is a topic in itself. This is the management function of staffing. **Staffing includes the recruitment, selection, training, and promotion of personnel to fill both managerial and operating positions in a company.** Because it is so important, we will study it in detail in Chapter 12. Remember, the quality of its managers and workers is probably a firm's single most important asset.

Directing

Assume that we have developed plans, created an organization, and staffed it. It now must be stimulated to action through the management function of directing. **Directing means encouraging subordinates to work towards achieving company objectives. It is sometimes called leading, guiding, motivating, or actuating.**

How Much Directing Is Necessary?

A manager's opinion of subordinates affects how they will be directed. If a manager thinks they are lazy, irresponsible, and immature, that manager will rely on rewards and punishments and will use formal authority to get things done.

On the other hand, a manager who thinks that the subordinates are responsible and are striving to achieve goals will likely "let them work". Thus, how much directing and what kind is needed depend largely on the manager's view of his or her subordinates.

Leadership as a Means of Directing

Leadership is a manager's ability to get subordinates to develop their capabilities by inspiring them to achieve. It is a means of motivating them.

Informal group leaders are selected by group members. These leaders influence group members' behaviour because the members believe it is in their best interests to follow them. In the role of leader, a manager is given authority over subordinates. A manager is usually more effective in motivating subordinates by using influence than by relying only on managerial authority to issue orders.

A manager should be a good leader and something more. Besides being able to inspire followers to want to achieve, a manager must also perform the other management functions.

Leadership is practised in different degrees by the people in a firm. The president is ultimately responsible for directing the entire company. He or she sets the style of leadership in a firm. If the president is a dictator, other managers are likely to be dictators, too.

Some managers make decisions on their own. Others allow subordinates to participate in decision making which will affect them. The "best" leadership style is one which considers the three elements in the leadership environment—the leader, the follower, and the situation. Some leaders do not have the personality needed to encourage followers to participate. Likewise, some followers do not want to participate—they want to "be told what, when, and how to do it." Situations which require speedy decisions limit the time available for true participation by followers. (See Figure 5-4.)

Motivating as a Means of Directing
What leaders do to inspire their followers to perform depends partly on how well-motivated the followers are. **Motivation comes from within the individual. Motivation is a person's drive to accomplish something.** A manager taps this by providing relevant

Motivation

Autocratic leader ←——————————→		*Democratic leader*
No	Willing to delegate?	Yes
No	Believes in two-way communication?	Yes
No	Believes in participative management?	Yes
No	Trusts subordinates?	Yes
No	Willing to let followers learn by their mistakes?	Yes
No	Believes subordinates' ideas are important?	Yes
No	Believes subordinates seek to do a good job?	Yes
No	Interested in subordinates' higher-level needs?	Yes
Yes	Believes subordinates are naturally lazy?	No
Yes	Believes subordinates must be "watched" at all times?	No
Yes	Believes subordinates are uninterested in their work?	No
Yes	Believes money is all that is necessary to motivate workers?	No
Yes	Believes that subordinates have to be treated like children?	No

Figure 5-4. A continuum of leadership styles

incentives. People, however, differ in their motivation to achieve. This is because of personality differences, past accomplishments, and many other factors.

A manager cannot always assume, therefore, that workers want to get ahead. A manager who relies only on a worker's built-in desire to get ahead may find that the desire is absent. Workers who believe that they were unfairly passed over for promotion in the past are likely to believe that getting ahead is impossible. Their outlook is different from other workers who have not experienced this. Incentives, therefore, must fit the individual.

For years, most managers thought that the key to better employee performance was to pay them more. Money is important in satisfying physiological and safety needs. While "fat" pay cheques may satisfy some people's need for self-esteem, the motivating power of money loses some of its "punch" as employees move beyond the basic needs.

THINK ABOUT IT!

What Does It Take to Be an Effective Leader?

A manager's ability to develop an effective leadership style is a major factor in determining his or her ability to perform the management functions. An effective leader:

1. is well-motivated
2. has imagination
3. is emotionally stable
4. has a practical orientation
5. uses good judgment
6. is willing to assume risk
7. can see the important variables in a complex situation (analytical skills)
8. is not fearful of decision making
9. has a strong desire to succeed
10. is a good communicator
11. has faith in subordinates
12. is flexible
13. is able to delegate
14. understands people
15. understands the need for limits on what he or she can do
16. has a high tolerance for frustration and ambiguity

Of course, the list probably could be expanded. What does it take to be an effective leader? THINK ABOUT IT!

An interesting view of motivation proposed in recent years has given managers added insight into how to motivate employees.[1] Pay and working conditions are called "hygiene factors". Poor pay and poor working conditions are dissatisfiers. But improving them will not provide positive motivation. It will only set the stage for true motivators to operate. The true motivators are recognition, the chance for advancement, and the chance for self-actualization.

An analogy may be in order. A lack of medical hygiene may cause disease. But supplying hygienic conditions will not cure disease. Poor pay may lead to dissatisfaction, but giving pay raises will not, by itself, motivate an employee to good performance.

The typical employee works at about 30 per cent of his or her capacity. This suggests the major importance of providing relevant incentives to employees. Imagine how much employee productivity would increase if management could get employees to work at 70 per cent or more of their capacity. Accomplishing this is a challenge for every manager.

All managers are responsible for motivating their subordinates. Each manager, however, is limited in what he or she can do. A foreman is limited by company policies on wage rates, fringe benefits, and so on. The president is limited by policies adopted by the board of directors.

Communication as a Means of Directing

A firm's effectiveness is closely related to how well orders are communicated down the chain of command. This is necessary if plans are to be implemented. This is one-way communication—passing orders down the line.

Modern managers, however, realize that subordinates need to communicate up the line also. This makes for two-way communication. In this way, communication can be used to motivate employees. **Communication, therefore, is a transfer of information between people which results in a common understanding between them.** When workers believe that they are not only "talked to" but can also "talk up", effective two-way communication exists.

Communication

Information which helps workers to see their relevance to company goals makes them feel needed. By being able to talk up the line, they feel important. Just as students want more input in decision making in colleges, workers want more input in decision making in firms.

Participation as a Means of Directing

One fault of big businesses is that employees often feel that they are not really part of the company. Encouraging them to participate in the decision making which will affect them helps overcome this fault. Participation leads to greater job satisfaction and morale. The company benefits from increased employee productivity.

Inspiring followers to achieve should be easier when they have their "say" than when they don't. But how much participation by subordinates is enough? This brings

---

[1] See Frederick Herzberg, *Work and the Nature of Man* (Cleveland: World Publishing Company, 1966).

Participative management

up a hotly debated issue—participative management. **Participative management means that the manager encourages and allows his or her subordinates to involve themselves actively in the decision making which will affect them.**

At one extreme, some managers argue that participative management amounts to a manager's delegating away decision-making power. At the other extreme are managers who view it as the best way to get subordinates to accept managers' decisions.

Many workers want to participate because it satisfies their esteem needs. Some, however, prefer not to participate.

Managers do or do not use participation, depending on their personal knowledge and assumptions about their subordinates. One management expert divides managers into two broad classes.[2] "Theory X" managers assume that their subordinates are lazy and have to be forced to work. "Theory Y" managers assume that their subordinates want to accomplish. Obviously, the "Theory X" manager would not allow subordinates much participation.

Leadership, motivation, communication, and participation all relate to the directing function—encouraging workers to work towards company goals. Let's turn now to the controlling function.

Controlling

Controlling

Managers must always check on operations (evaluate performance) to see if the firm is achieving its goals. This is the function of controlling. **Controlling involves measuring actual performance, comparing it with expected performance (a standard), and then taking corrective action, if necessary.**

Planning and controlling are closely related. When husband and wife prepare a budget, they are planning and setting up a control device. Suppose the budget plan is to save $1,000 by the end of the year. If their savings account shows a balance of $200 on July 1, the standard is not being met. Awareness of this should lead to corrective action.

Students want their exams graded and returned promptly. This lets them know "where they stand" and what changes in study habits might be needed. The greater the time between taking an exam and having it returned, the less effective it is as a control device. Faulty study habits continue longer and become harder to change. The same principle holds in business.

Like the other management functions, controlling is an ongoing process. Managers monitor subordinates' activities to make sure that their activities help the firm to reach its goals. The hierarchy of objectives breaks down overall company goals into specific goals for each worker. This makes controlling easier, since it sets goals for each worker.

Examples of controlling are an office manager's efforts to keep expenditures on typing paper in line with the budget, a plant manager's efforts to keep the number of

---

[2]See Douglas McGregor, *The Human Side of Enterprise* (New York: McGraw-Hill Book Company, 1960).

rejects down to an acceptable minimum, or a marketing manager's efforts to move a company into production and sale of a new product according to a time schedule.

Elements of Controlling

The elements of control are present in each of the above examples. First, there must be a definite idea of what we want to accomplish (a standard). The office manager, for example, cannot exceed the budget for paper. In practice, setting standards is not always simple. For example, how should management evaluate the production department's performance? On the basis of the number of rejects? On the basis of the average time required to produce an average unit of output? On the basis of the average cost of producing an average unit of output? Actually, all are important. A 3 per cent reduction in rejects along with tripled production costs is probably not desirable. But whether it is or is not depends on the relative importance of avoiding rejects and avoiding cost increases.

Second, a manager measures actual performance and compares it with the established standard. This is not so simple either. There are many problems in measuring employee performance. In some jobs only quantitative results (number of units produced) are important. In other jobs qualitative results (quality of the units produced) are the crucial basis for comparison. Evaluating the performance of higher-level managers is complex. We discuss this in Chapter 12.

The final element of control is taking corrective action. It is desirable to detect deviations from standards quickly. The longer corrective action takes, the more it will cost. (See Figure 5-5.)

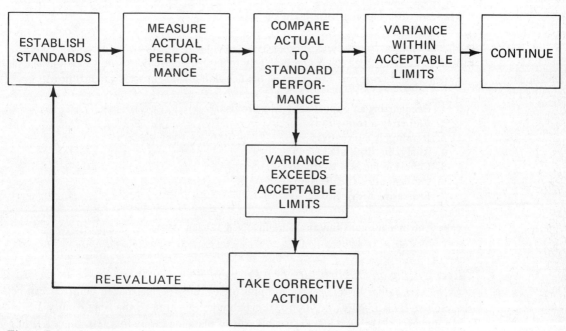

Figure 5-5. **The control process**

Practising Control

Every manager practises controlling. The foreman must "oversee" subordinates just as the president must "oversee" subordinates. In some cases, staff specialists aid in the control of line operations. The quality control inspector, for example, is the person who first spots deviations from standard in product quality.

Effective control requires a well-developed communication system. Control information should be sent immediately to the person responsible for the activity being controlled. A quality control inspector who does not have authority to halt the production line should know who does have it so that person can take corrective action immediately.

Great strides have been made in reducing the time-gap between deviation and correction. Much of this is due to the development of a systems approach to organization, electronic computers, and the development of quantitative tools to aid in decision making.

THE DECISION-MAKING PROCESS

One way to look at the management task is to think of it as decision making. Performance of the management functions requires that managers make decisions. In fact, everything a firm does is the result of decisions made by managers. Examples are deciding on company goals, what products to make, what equipment to buy, what advertising appeals to use, where to get funds, and where to sell its products.

Stages in the Decision-making Process

Decision-making process

Like the management functions, decision making is a process. It is a very complex process. In fact, there is no one best way to explain it. We'll look at it in a logical and straightforward manner. **The stages in the decision-making process are as follows:**

1. Recognizing an opportunity or a problem.
2. Gathering information.
3. Developing alternatives.
4. Analyzing alternatives.
5. Choosing the best alternative.
6. Implementing the decision.
7. Evaluating the decision.

The management functions are involved at each stage.

Recognizing an Opportunity or a Problem

As we saw earlier in our book, an opportunity must be recognized before it can be exploited. This requires decision making about the nature of the opportunity a firm can undertake. Decision making is a big part of planning in any business. Since a firm's

> ## TWO POINTS OF VIEW
>
> ### What Is Management?
>
> *A lower-level manager*
> "I'm a supervisor on a production line. To me, management means attending to detail—making sure that my subordinates do exactly as they are told. The really important work that can't be left up to somebody else to do is what it's all about. I spend a lot of time working right along with my subordinates. I'm not afraid to get my hands dirty!
>
> "As far as management functions go, and I'll admit I never thought of management in those exact terms, most of my time is spent on directing. Controlling would come in second. My job is to get production out of my workers and that means getting them to do a good job and making sure that they do it."
>
> *A top-level manager*
> "I'm president of a large firm that manufactures sporting goods. To me, management means keeping yourself free of detail work so that you can concentrate on thinking about your company's future—where your company will be 10 or 15 years from now. Of course, thinking by itself won't accomplish the job. That's why I've spent several years bringing together the best possible group of executives to carry out my plans for the future.
>
> "Undoubtedly, my most important function is planning."

resources are limited, management must decide what types of opportunity can be exploited. The actual exploitation of the opportunity also requires decision making.

Decision making is also needed when management recognizes a problem. Often, this is little more than a feeling that "something is not right". Decisions must be made regarding a clear definition of the problem and whether or not anything will be done about it. Actually, we could think of problems as opportunities. For example, recognizing that an advertising campaign is failing to produce sales presents the firm with an opportunity to correct it.

Gathering Information

Assume that the decision maker recognizes the opportunity or problem. The next step is gathering information about it. This involves talks with company personnel and persons outside the firm who might provide greater insight. The decision maker might also investigate company records and consult secondary sources of information such as libraries.

Because of the tremendous capacity of computers to store data, many firms now have management information systems. A firm's management information system (MIS) is made up of people and machines. People feed in the data needed for decision-making purposes. These data are processed, summarized, and reported to decision makers who need it. We will have more to say on this in Chapter 14.

Developing Alternatives

After developing a good "feel" for the opportunity or problem and its setting, the decision maker begins to formulate alternative courses of action. The support of others might be sought in brainstorming sessions in which the participants offer ways to deal with the opportunity or problem. Freewheeling creativity is important here, so no evaluation is made of the alternatives offered at this stage. The goal is to stimulate new ways of looking at the opportunity or problem and to develop alternative ways of dealing with it. Creativity is extremely important in this stage. Remember, a decision involves a choice. There should always be more than one alternative. If not, there is no decision.

Analyzing Alternatives

After making a list of alternatives, the decision maker begins to analyze the alternatives critically. It's a process of elimination. Alternatives which are unlikely to pay off are eliminated. Also eliminated are alternatives which involve high risk in comparison to expected pay-off. The remaining alternatives are often ranked in terms of their expected pay-off. The pay-off could be stated in terms such as least cost, maximum profit, or maximum customer service.

This might involve analyzing the projected consequences of each of the remaining alternatives. This is always tough because it involves forecasting the future. Never-

---

THINK ABOUT IT!

A Sure Way to Fail in Management

Decision making is needed in the performance of every management function, but decision making takes courage. In our dynamic business world, a manager seldom, if ever, has all the information he or she would like to make decisions. Nevertheless, some managers postpone making critical decisions on the grounds that they don't have enough information.

In many cases, getting all the information is practically impossible. In other cases it would cost more to get than it would be worth. Regardless, the manager who continually puts off decisions will fail as a manager. THINK ABOUT IT!

theless, the thought process required here helps to ensure that the decision maker considers the future consequences of a decision faced now.

Choosing the Best Alternative

In choosing the best alternative, a decision maker establishes a decision criterion and a decision rule. If the goal is to improve delivery service to customers, the decision criterion might be "fastest delivery". The decision rule would be to choose the transportation method which provides the fastest delivery to customers. The decision criterion, however, could have been "lowest cost". Choosing decision criteria is not easy. It takes good judgment concerning the firm's goals and an understanding of the risks involved.

Implementing the Decision

Once the decision has been made (the best alternative has been selected), the decision maker must move towards implementing it. This, of course, also requires decision making and performing the management functions. Planning, organizing, staffing, directing, and controlling are necessary to implement the decision.

Evaluating the Decision

After implementing the decision, the decision maker must evaluate it. Operations must be monitored, or checked, to see whether the decision is being properly implemented. This also gives the decision maker feed-back which helps in assessing whether the "right decision" was made and if corrective actions are needed.

Types of Decisions

One type of decision is the routine decision. **A routine decision is a recurring decision. It is a decision which must be faced over and over.** You probably do not consciously decide the route you will take to class each day. For most of us, this is a routine decision. Because we face that decision so many times, we develop a routine decision for it.

    Managers also face routine decisions. These are often set up as policies and standard operating procedures. An office manager who sets up a policy of "no smoking in the office" does away with the need to make a decision actively each time an office worker asks if it's alright to smoke in the office.

    Another important type of decision is the non-routine decision. **A non-routine decision is a non-recurring decision.** There are two types of non-routine decisions, strategic and tactical. Strategic decisions are made by top-level management. Examples are decision making regarding the types of opportunity a firm will attempt to exploit (what business the firm will be in) and whether or not the company should buy out another company which is currently a competitor. Strategic decisions have an important and long-run effect.

Margin notes: Routine decision; Non-routine decision

MANAGEMENT AS A CAREER

To become a manager is to accept responsibility, to work through others, and to make an organization work. The skills needed include human relations skills, the ability to plan, and the ability to make and stand by your decisions. Managers must also have the self-discipline needed to take orders.

Many of these managerial skills are transferable across a wide range of businesses and in many non-business enterprises. Specific areas include heavy industry, transportation, petroleum exploration, agribusiness, insurance, electronics, the hospitality industry, the health-care industry, local and federal government, and many others. In every case the manager organizes, plans, staffs, directs, and controls.

Of course, all industries are not expected to grow at the same pace. Chapter 20 reviews the growth prospects of all industries for you.

In nearly every field there are several levels of managers' jobs. It is normal to be promoted through the ranks. However, except in those rare cases where promotion depends mostly on seniority, the promotions get much tougher as you near the top of the organization. It is near the top that the exceptional management skills begin to show. This is where toughness of character, vision, and capacity to absorb and interpret complex information are required.

These days there is increasing emphasis on formal training for management. This usually means taking university or community college courses which will help you to better understand what management is all about. In many companies, you will find it easier to get a job if you have a degree or certificate in a business-related program from a university or community college.

Tactical decisions have less long-run effect. An example is a decision about where to locate a new warehouse. The dividing line between strategic and tactical decisions, however, is often an arbitrary one.

SUMMARY AND LOOK AHEAD

Businesses are living things which seek to accomplish objectives. Through performance of the management functions, a firm moves towards the realization of its objectives.

The functions of management—planning, organizing, staffing, directing, and controlling—are performed by managers in the process of achieving goals by bringing together people and other resources.

Management is necessary whenever results depend on group effort. There are different echelons (levels) of management in a firm. Its success or failure is traceable to the effectiveness of its managers.

In reality, management cannot be broken down into a series of separate functions. It is a process. This becomes clear when we think of the firm as a system.

We can also view the management task in terms of decision making. Managers make decisions in performing their functions.

There are seven stages in the decision-making process: (1) recognizing an opportunity or a problem; (2) gathering information; (3) developing alternatives; (4) analyzing alternatives; (5) choosing the best alternative; (6) implementing the decision; and (7) evaluating the decision.

Decisions can be routine or non-routine. Non-routine decisions can be either strategic or tactical. Truly strategic decisions are made by top management.

In the next section, we will study in greater detail the various areas of decision making in the firm. These are called the functional areas—production, marketing, and finance. Do not confuse the functions of management (planning, organizing, etc.) with the functional areas of business. The functions of management are performed in all the functional areas we will study.

KEY CONCEPTS

Communication A transfer of information between persons which results in a common understanding between them.

Controlling The management function of measuring actual performance, comparing it with expected performance, and taking corrective action, if necessary.

Decision-making process Involves recognizing an opportunity or a problem, gathering information, developing and analyzing alternatives, choosing the best one, implementing the decision, and evaluating the decision.

Directing The management function of encouraging subordinates to work towards achieving organizational objectives.

Echelons of management The different levels of management, i.e., upper, middle, and lower.

Functions of management Planning, organizing, staffing, directing, and controlling.

Leadership The ability of one person to inspire others to follow him or her.

Management Process of achieving goals by bringing together and co-ordinating the human, financial, and physical resources of an organization.

Manager A person who works through other people, who perform operative tasks, to accomplish objectives.

Managerial skills Conceptual skills, human relations skills, and technical skills. All managers need all three skills.

Motivation Comes from within a person. A person's drive to accomplish something.

Non-routine decision A non-recurring decision. Two types are strategic and tactical.

Operational planning Planning for the day-to-day survival of an organization.

Organizing The management function of relating people, tasks (or activities), and resources to each other so that an organization can accomplish its objectives.

Participative management The manager encourages and allows his or her subordinates to involve themselves actively in decision making which will affect them.

Planning The management function of preparing an organization to cope with the future by relying on knowledge of present and past conditions and forecasting probable future developments. Setting a firm's objectives over different time periods and deciding on the methods of achieving those objectives results in a plan which is to be followed in order to reach desired goals.

Routine decision A recurring decision. Often set up as a policy.

Staffing The management function of recruiting, selecting, training, and promoting personnel.

Strategic planning Planning for the long-range future. A broad overall strategy of growth.

Systems concept A way of looking at a company. A system is a complex of interacting parts. A socioeconomic system is made up of many smaller sub-systems. A given industry is one such sub-system. Firms in that industry are still other sub-systems. Each firm has several sub-systems, such as a production department and a marketing department. Management effort focuses on integrating the various sub-systems of the firm so that overall system performance can be improved.

QUESTIONS FOR DISCUSSION

1. Since it is typical that managers do not directly produce or sell anything, is management productive? Explain.
2. As a living organism, the business organization seeks to survive and grow. Explain how these objectives are accomplished.
3. Why did we describe management as a "process"?
4. List and describe the five functions of management.
5. When is management necessary?
6. Discuss the similarities and differences between operational and strategic planning. Which is more important? Why?
7. Picking apart the management process and breaking it down into its various functions is a convenient way to build a basic understanding of the nature of management. The danger in this approach is that we might come to "see the trees but not the forest". How might we avoid this danger?
8. It has been said that the president of a company need not be a good manager. All the president need do is possess an ability to select well-qualified subordinates. Do you agree? Why or why not?

9. In general, people in upper-management positions need to possess relatively few technical skills, whereas operatives must possess such skills. Does this mean that people in top management positions can transfer more easily from one type of industry to another than operatives can? Explain.
10. Do you think that a company president should direct subordinates in the same way that the boss of a work crew would direct subordinates? Discuss.
11. How are the management functions of planning and controlling related?
12. List and describe the seven stages in the decision-making process.
13. "If there is only one alternative, there is no decision." Do you agree? Why or why not?
14. Distinguish between a routine decision and a non-routine decision. Identify two types of non-routine decisions.
15. Explain how the functions of management are related to the decision-making process.

INCIDENT:

Marsha Thompson

Several years ago, Marsha Thompson quit her job with a large manufacturing firm and decided to go into business for herself. Although she was making a good salary as director of the firm's research department, she thought that she was required to do too much unnecessary "paperwork", which prevented her from applying her talents fully to her "real" job.

Marsha has been on her own for five years. Her firm has grown very rapidly as a result of several big contracts in addition to several patented processes which she developed in the field of pollution control.

Although Marsha's firm is very successful, she is not as happy as she once was. She and two other scientists formed the nucleus of her company, which has grown to include about 100 employees. Originally Marsha employed only a handful of employees and could devote most of her time to research. She now finds herself having to devote too much time to the affairs of her company.

Questions:
1. Why did Marsha quit her job as director of the research department in the manufacturing firm?
2. Is Marsha a good manager? Why or why not?
3. Why do you think Marsha is growing dissatisfied with her present situation?

INCIDENT:

Phil and Pete

Philip Herman is the director of environmental studies of a large oil company, and Peter Lucido is the vice-president of production. Phil and Pete have both been with the company for about ten years. Although they are very close friends, they have their differences concerning what management is all about. In fact, they have had many "friendly" debates.

Phil is a rather relaxed character. He is good-natured, draws people to him, has a cheerful outlook on life, and has a "live-for-today" philosophy of life. His subordinates respect him both as a manager and as a person. Phil believes that, given the chance, employees will want to advance themselves.

Pete, in many respects, is the opposite. Pete is best described as cautious and rather nervous. He attributes his ulcer to the fact that he doesn't know how to "take it easy". He is a constant "worrier". He is not nearly so sure as Phil that the average worker would pull his or her own weight if not coerced into doing so by a boss.

Despite their different personalities and outlooks on life, Phil and Pete are generally regarded as good managers by their superiors and their subordinates.

Questions:
1. Do you think that two persons who are so different can both be effective managers? Why or why not?
2. What similarities and differences would you expect between Phil's and Pete's performance of the functions of management? Discuss each function separately.

ROCKETOY COMPANY III

After carefully considering his options as to the legal form of ownership, Terry decided to form a small corporation. Rocketoy's common stockholders are Terry, Joe Phillips, the three Toronto investors (Richard Talley, Julia Rabinovitz, and Mike Schultz), and Terry's sister Pam. Terry owns 55 per cent of the common stock, Pam owns 5 per cent, and the others each own 10 per cent.

Terry was elected chairman of the board of directors. In a meeting of the owners, they agreed to fill the top management jobs themselves. Each department would be headed by that owner who had the most interest and experience in that particular type of work.

This informal arrangement reflected Terry's basic ideas about business. For example, Rocketoy's major organizational objective at the time of incorporation was "to make the highest-quality toys possible and to sell them at prices reasonable enough to earn a good return on the owners' investment in the firm." Terry suggested, and the other owners agreed, that this was enough to guide them in running the firm. As Terry said, "We are in business to serve customer wants so that we can make a profit. We don't have to act like a big business. We know each other's abilities and interests, and we are all dedicated to making Rocketoy a success. Let's not get bogged down in things like the chain of command and all the other things which take time away from doing. We can informally work out the details of operations as we gain more experience in working with each other. Let's not try to invent problems."

The owners decided on the following positions and titles:

President: Terry Phillips
Vice-president of finance: Pam Phillips
Vice-president of production: Joe Phillips
Vice-president of marketing: Julia Rabinovitz.

Richard Talley and Mike Schultz were heavily involved in several other business ventures, so they decided not to get actively involved in Rocketoy's management. The other owners agreed that this would be best. But because of their

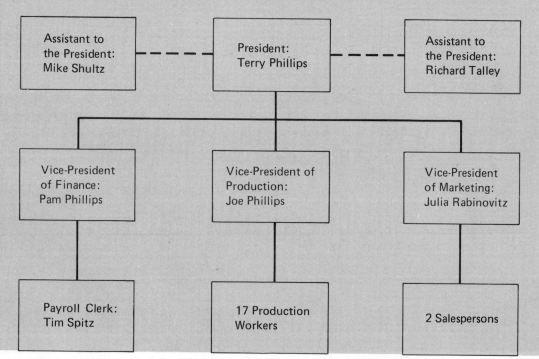

valuable business experience, Terry convinced them to serve as his advisors. Whenever Terry wanted advice, he would be able to turn to Richard and Mike.

Terry, Pam, Joe, and Julia each drew a salary from the business. Because they wanted to keep as much money invested in the business as they could for future growth, they drew modest salaries—Terry $11,000 a year; Pam, Joe, and Julia $9,500 per year. Richard and Mike got $2,000 each per year for their jobs as advisors to Terry.

Rocketoy's work force expanded to a total of 20 workers in addition to the managers. This is shown in the organization chart Terry prepared for Rocketoy. (See chart on previous page.)

Questions:
1. Why do you think Terry chose to incorporate his business?
2. Do you agree with Terry that Rocketoy's broad organizational objective is enough to guide its managers? Why or why not?
3. Is Terry a good chief executive? Why or why not?
4. Suppose you could step into Mike Shultz or Richard Talley's shoes. What advice would you give Terry at this point?

Section Three

Business Decisions

By now we know what a business firm is. We also know that it must be managed. In this section, we discuss the various activities required to achieve company objectives and the decisions managers must make to implement these activities.

Goods and services must be produced and marketed. Thus, production and marketing are two basic decision areas of business. Chapter 6 discusses production management. Chapter 7 looks at the firm's marketing task. Chapter 8 discusses marketing decisions faced while performing that task.

Accounting is a necessary tool of management. It helps managers to make better production and marketing decisions. Accounting is examined in Chapter 9.

The third major decision area is finance. A firm's production and marketing activities must be financed. Chapter 10 looks at financial institutions such as banks and securities exchanges, and Chapter 11 discusses financial decisions of the firm.

Because a firm's most important resource is its human resource, we devote two chapters to it. Chapter 12 explores the area of personnel administration. Chapter 13 looks at labour relations.

After you finish this section, you'll have a good idea of the types of work done by business managers. This may help you to develop a better idea of what types of business careers appeal to you.

6
Producing Goods and Services

6 PRODUCING GOODS AND SERVICES

OBJECTIVES: After reading this chapter, you should be able to:

1. Explain the nature of production.
2. Draw a chart illustrating the inputs, processes, and outputs of production.
3. Illustrate the several ways of classifying production processes.
4. Review the management functions and give an example of each as applied to production.
5. Develop a checklist for plant location decisions.
6. Evaluate the dangers of loss of human motivation in a big factory.
7. Explain the process of control as it applies to product quality.
8. Distinguish between value analysis and vendor analysis.

KEY CONCEPTS: In reading the chapter, look for and understand these terms:

PRODUCTION
COMBINATION
BREAKING DOWN
TREATMENT
INTERMITTENT
   PRODUCTION
CONTINUOUS PRODUCTION
LABOUR-INTENSIVE
CAPITAL-INTENSIVE
MAKE-OR-BUY DECISION
PLANT CAPACITY
PLANT LAYOUT

CONTROL CHART
PERT
QUALITY CONTROL
PREVENTIVE
   MAINTENANCE
OBSOLESCENCE
RECIPROCITY
VALUE ANALYSIS
VENDOR ANALYSIS
OPERATIONS
   MANAGEMENT

We begin our study of the functional areas of management with a look at production. The traditional management functions of planning, organizing, staffing, directing, and controlling are used as a framework for studying production management. Special emphasis is given to modern planning and control devices.

Production has a special place in Canadian business history. By doing good production work, and combining new technology with modern organizational techniques, Canadian firms have helped deliver a very high standard of living. Although the earliest application of production techniques was in factories, their use was soon extended to retailing, services, and other forms of business.

There has been considerable growth in production employment (See Table 6-1.) Using the year 1961 as typical (the index value for that year equals 100), we can compare the growth of various sectors of the economy from 1966 to 1977. Notice that some industries have grown faster than others.

WHAT IS PRODUCTION

Production

Production activity results in the creation of goods and services. Someone or something is "better off" because of production. Production can be viewed as a sequence. It starts with the input of resources. These are fed into one of several kinds of production processes. Finally there results an output of goods or services for use or sale. Figure 6-1 illustrates this view and makes it easier for us to explain the elements of production management.

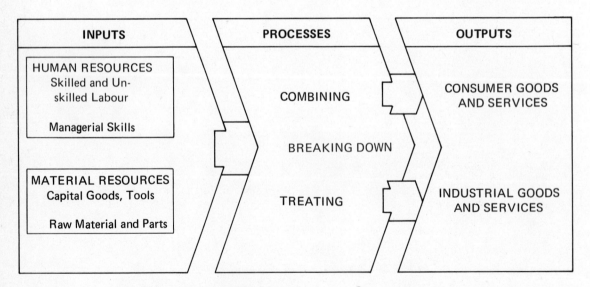

Figure 6-1. The production process

Table 6-1. Employment indexes in leading industries, 1966-77

	1977 (est.)	1976	1975	1974	1973	1972	1971	1970	1969	1968	1967	1966
Total	144.2	144.9	141.1	142.8	135.9	129.9	127.8	127.1	126.9	122.7	122.6	120.7
Forestry	78.8	76.2	76.0	87.4	86.4	76.3	79.4	84.3	88.7	91.1	102.7	106.2
Mining	123.6	118.3	114.1	115.5	111.3	110.4	114.8	115.3	107.9	109.8	109.1	106.9
Manufacturing	125.7	128.1	126.3	133.9	129.9	123.7	121.6	122.8	125.2	122.1	123.1	123.5
Durables	136.9	140.3	139.8	149.4	144.1	134.9	131.4	132.7	136.7	131.7	133.9	134.9
Non-durables	116.6	118.2	115.5	121.0	118.3	114.7	113.7	115.5	115.9	114.4	114.5	114.3
Construction	113.3	113.8	117.1	117.1	109.9	109.7	115.5	113.7	119.1	119.4	122.6	128.8
Transportation & Commerce	129.9	128.7	125.8	124.6	118.1	116.0	114.6	112.6	111.9	109.5	110.9	108.2
Wholesale & Retail Trade	170.0	172.2	168.5	165.7	155.3	146.2	140.3	139.3	136.6	129.4	125.8	122.1
Financial, Insurance, & Real Estate	193.1	183.9	175.0	167.3	157.0	148.7	145.9	143.6	138.8	131.4	126.0	120.5
Service	249.6	242.8	231.9	224.0	206.1	193.5	186.4	178.5	171.8	157.8	153.0	139.1

Source: *Financial Post.*

The Inputs to Production

To produce something, the following set of resources is usually needed as input: materials, labour, capital goods, and supervisory and managerial skills. Materials include raw materials such as raw cotton, corn, and crude oil; semi-manufactured products such as sheet steel and unfinished lumber; and manufactured parts such as spark plugs, bolts, and tires. Electric power and other energy sources are usually included. Capital goods include the plant investment, which can range from a huge refinery to a barber shop, and equipment such as a lathe or a typewriter.

The human input includes unskilled and skilled labour and supervisory and managerial skills. The main distinction between skilled and unskilled labour is ability to perform a special task which cannot be performed by all employees. Examples of skilled workers include carpenters, pipefitters, and technicians. Supervisors and managers are, of course, responsible for directing the activities of others.

Processes

There are several ways of classifying production processes. The first depends on the way that material inputs are dealt with. Thus, a process may combine material inputs, break down an input or inputs, or treat an input.

Combination

Combination means putting parts together. It is the most common process. Automobiles, pumps, and pencils must go through the process of combination.

Breaking down

Breaking down means removing or at least separating some of the original input, usually a raw material. When a log is cut into two-by-fours or when juice is taken from oranges, the process, which is also very common, is one of breaking down.

Treatment

Treatment is doing something to an input without adding or subtracting from it. It may involve hardening or softening or cooling or reshaping an input. Smoking a ham or moulding plastic toothbrush handles is a treatment process. Most production involves two of these processes, and many products require all three.

We can also classify production processes according to a time dimension. They can be intermittent or continuous. They are also either for stock or to order.

Intermittent production

An intermittent production process starts and stops and starts again, maybe several times. A specialty tool maker, for example, may nickel-plate certain batches of output and not others. Some products may require special heat treatments. The nickel-plating and heat-treating departments of the firm are used intermittently.

Continuous production

A continuous production process, as the name implies, goes on and on. A cola bottling plant repeats the same process countless thousands of times without interruption. The demand for this product is large and easy to predict. The process, as in the case of most continuous processes, is highly automated.

Most continuous processes produce goods for stock. This means that output is kept in inventory in anticipation of demand. This could not be done without losing money if the firm did not have good reason to expect a fairly steady demand. The reason is that keeping inventories costs money in several ways. They take up valuable space; they require financing; and they may be damaged or stolen or become obsolete in some cases.

The alternative to producing for stock is producing to order. This means that the firm waits until there is a specific order in hand before starting to produce. Such production is called jobbing. This kind of production usually uses general-purpose machines and tools—those which can do a variety of jobs.

Processes also vary in the amount of human input they need. **Labour-intensive processes depend more on people than on machines.** Some parts of the apparel trade are like this. Labour-intensive processes are most likely to occur when labour is cheap or when there is an artistic element in the work. There are some kinds of jobs, too, in which it is really hard to apply machines because the process varies a lot. Some kinds of farming are still highly labour-intensive as is high-quality jewelry making. *Labour-intensive*

The opposite situation exists when machines can do the job better than people. **This calls for capital-intensive processes where people may have little to do with production. Instead, investment in machinery is great.** The huge petroleum refinery is a classic example of this. *Capital-intensive*

An extreme example of intermittent production to order is a customized stereo-system producer. Such a firm waits for orders, produces a variety of outputs, and tends to be labour-intensive. Some parts of the shop may be idle for extended periods. The size of its labour force may vary. A similar situation exists in producing heavy machinery, ships, and high-fashion dresses.

A manufacturer of soap powder is at the other extreme. This firm has a heavy investment in special-purpose equipment and operates on a continuous "assembly-line" basis. It keeps a large inventory of finished goods. Other examples are producers of ball-point pens, paper towels, and gasoline.

The Outputs of Production

An almost unlimited variety of outputs can result from the production process. Some are very complex products which require both breaking down and combining the inputs. Examples are computers, cargo ships, and scientific equipment. Others are simple products, such as distilled water, which require a simple process. Services, such as preparation of tax returns, are a kind of product. Services represent an ever-increasing part of total Canadian production and employment. Their production is discussed more fully in a later section of this chapter.

WHAT IS PRODUCTION MANAGEMENT?

Production management is the application of managerial functions to production. The inputs, processes, and outputs we have just described require planning, organizing, staffing, and controlling. We will examine each in turn.

Production Planning

Planning is concerned with the future. It is a mapping out of how things are going to be done. It has short-run and long-run dimensions and requires a forecast of demand.

Planning for production is no exception. It includes planning the product (outputs) and planning for capital, labour, and material needs (inputs). We will emphasize strategic, long-run planning with an emphasis on capital goods planning and planning for the product itself.

Product Planning

The logical time to start planning for production is when planning the product. This planning really overlaps the production and marketing functions of the firm and is discussed more fully in our chapters on marketing. At this point we will describe just a few features of the product-planning process.

Product planning amounts to answering such questions as "what kind of products can be sold at a profit?" and "how much can be sold?" Other questions include what styles and sizes should be produced and what special features the product should include. The answers to these questions require study in the laboratory to determine the best inputs of raw materials and component parts. In many firms this kind of

---

THINK ABOUT IT!

New Products at Procter and Gamble*

In 1978 and 1979, the huge and successful firm of Procter and Gamble ($8.1 billion in sales in fiscal year 1978) had plans to introduce to large-scale distribution an unprecedented number of new products, according to Edward G. Harness, its chairman. The array of new products suggests just how broad Procter and Gamble's product mix is. New products include: Puritan cooking oil, a new disposable diaper, a new cake mix, Wondra hand and body lotion, Rely tampons, a line of disposable drapes and gowns for surgical patients, and new flavours of Pringles' potato chips.

Launching so many new products at the same time is costly, but Procter and Gamble already has good evidence of the new products' acceptability from market testing and other market research. (See Chapter 7 for a discussion of market research.) Heavy costs include capital spending to increase production and research and development spending to keep the new products coming.

Not many firms have product-planning activity on the scale of Procter and Gamble, but the nature of the activity and its importance remain the same.

*Used by permission of The Procter & Gamble Company.

thinking is done on a continuous basis by a product development department. The basic questions are always "will it sell?" and "can it be produced at a profit?"

The product mix of a firm, that is, the combination of products it produces, has both market and cost effects. Producing a line of related products has certain advantages from the selling standpoint. Producing several products, whether they are related or not, may affect the unit cost. The Jiffy Bottling Company could begin to produce root beer as well as cola. This would probably increase its total production volume without adding to plant or equipment. Thus, "fixed" cost would be spread over more units and the cost per bottle would be reduced.

One question which often arises in planning the product is the decision whether to make a product or a component part or to buy it—the "make-or-buy" decision. If we are talking about a firm that makes only one product, buying it rather than making it would take the firm out of the production business entirely. This could be the right decision if the firm could make more money by buying a product and reselling it than by making it and then selling it.

Make-or-buy decision

Other factors are involved in the make-or-buy question. If it is to be a question at all, there must be a reliable source from which to buy the product or part. There must be a supplier who is willing to meet the buyer's needs concerning quality, quantity, and delivery schedules. A decision to buy rather than to make often is made for the component parts of an assembled product. The major auto makers, for example, buy many of the component parts of the cars they make. In the construction industry, a contractor often uses one or more sub-contractors to produce various parts of the project.

Once a firm reaches a certain size, it might stop buying one or more component parts and begin to make them. This may reduce total costs of production and make the producer more secure about sources of inputs. It may also make the firm less flexible.

Planning the Plant

There are several important questions which relate to planning for the plant itself. These include the decisions of where to locate the plant (plant location), how large a plant to build (plant capacity), and how to arrange the plant once it is built (plant layout).

Plant location Plant location can affect overall cost, employee morale, and many other elements of a firm's operation. A manager should weigh carefully any location decision. A checklist such as the one in Table 6-2 could be a great help. The first question is, "in what area do we wish to locate?" This refers to a city or, perhaps, metropolitan area—but an even broader geographic location decision may be needed first. The factors to consider in selecting an area include input transportation, output transportation, and city and provincial inducements (such as tax exemptions) and deterrents (such as high real property taxes). These should be estimated well in advance of making a decision.

Table 6-2. Plant location checklist

Area Selection

1. Cost of materials and parts transportation
2. Cost of transportation of finished products to customers
3. Location inducements and deterrents by city and provincial governments
4. Quality and quantity of appropriate labour supply
5. Adequacy of power and water supply
6. Attractiveness as a place to live—climate, schools, safety, etc.

Specific Site Selection

1. Sufficient size
2. Accessibility to highways, railways, or water transport
3. Restrictions on land use, waste disposal, etc.
4. Land costs
5. Availability of leased facilities such as public warehouses

Input transportation cost depends on the distance of raw material and parts suppliers from the proposed area and the kind of transportation facilities connecting them. Output transportation cost depends on the expected location and density of customers as related to the proposed area. Sometimes these cost factors make it hard to meet competitors' prices.

A less tangible factor is the attractiveness of the area as a place to live. Important considerations here are climate, schools, housing, public parks, police protection, and taxes. A firm trying to decide between two possible locations might make estimates of their comparative profitability, taking into account the expected effects on sales and on the various cost components—especially taxes, transportation, and labour.

Once an area is selected, the next problem is to choose a specific site. The firm makes a survey of available parcels of land which are suitable in terms of size; zoning restrictions; drainage; and access by highway, railway, and water. If several sites are satisfactory, the decision might depend on cost. There is the added possibility of finding a suitable existing plant available for sale or lease. This could mean earlier availability for operations.

Specific site selection calls for compromise among the items on its checklist. A specific firm's choice may be determined by placing emphasis on access to a river or on distance from population centers. It may be selected because it is close to a large university or to a major industrial customer. Plant location decisions are becoming more difficult as land in and near major population centres becomes scarce.

Plant capacity

Planning for plant capacity and plant layout Once a site is chosen, the next step in planning is to design the building itself. With the help of production experts and architects, the firm must plan plant capacity. **Plant capacity is the production output limit of the facility.** A pocket calculator assembly plant, for example, may require 10,000 square feet of space to produce an expected maximum output of 2,000 units a day.

THINK ABOUT IT!

How Profit Expectation Helps Choose a Location

The Dogwood Company is about to choose a location for its new widget factory. With help from the accounting, production, and marketing experts, Ned Rink, head of the expansion committee, has come up with cost comparisons. They have narrowed the choice down to the cities of Saskatoon and Winnipeg because they are nearest to the market not presently being served. The sales and cost estimates below lead to the conclusion that Saskatoon is the better choice, assuming that land and plant construction costs are equivalent.

	Saskatoon*	Winnipeg*
Estimated annual sales in units	80,000	120,000
Selling price	$5.00	$5.00
Cost of materials	1.10	1.20
Labour	1.20	1.40
Average transportation to market	1.30	1.10
Taxes	.60	.80
Other costs	.20	.20
Total costs per unit	$4.40	$4.70
Net profit per unit	.60	.30
Expected total annual profit	$48,000	$36,000

*Cost figures and selling prices are hypothetical.

Although volume of sales from Winnipeg is greater, unit profit is twice as high in Saskatoon. It is dollar profit that counts. THINK ABOUT IT!

Plant layout describes the relative location of the different parts of the production process in the building or buildings. The planner must answer questions such as those in Table 6-3. The correct answers depend on the kind of production process involved—whether it is labour-intensive or capital-intensive—and on the size and perishability of the units of output. The calculator assembly plant is a continuous, for-stock, labour-intensive process. It might well be laid out on one floor in a straight line. The parts could be stocked at one end; the several-step assembly process could occur in the centre; and the inspection, testing, packaging, and shipping could be done at the other end.

Plant layout

Table 6-3. Questions for planning a new plant

What are output expectations?
Can we operate around the clock with a smaller plant?
Will we operate year-round?
Is inventory storage easy?
What are our financing limitations?
Do we arrange machines by product or by process?
How many storeys are practical?
How much space must be allowed for expansion?

Organizing for Production

Some of the departmentation process in manufacturing depends on the type of layout. A plant may be arranged according to process or according to product, depending on the number of products being produced and the volume of production. A process-organized plant might have a grinding department, buffing department, stamping department, and so on. A product-organized plant might have a ball-point pen department and a mechanical pencil department, each of which is arranged on an assembly-line basis. How machines are related to each other often determines the organizational structure, including the number of persons in each department.

The Human Dimension in Production

A classic problem in organizing a factory is that of combining the human resource with huge capital investments without losing human enthusiasm for work, as the following situation shows. John Bivona's job is to press the green button every time the red light lights up in the boiler room. It is hard for John to see his role in the productive process. This kind of job presents a real challenge to the production manager to make full use of John's ability. Machines can reduce the "humanity" of an organization when they seem to "tell a person what to do".

A related problem is conflict in the assignment of activities among work groups. Introduction of new technology in the plant may cause work to be distributed in a new way. What the machines "dictate" is not necessarily accepted by workers or by their unions. In organizing a factory, traditional labour union definitions of job responsibility—what kind of work a worker can or cannot do in the production process—cannot be overlooked.

Production Organization Structure

The organization as it relates to production may be on product or on process lines. Sometimes it is a combination of the two. The organization chart for the Dunn-Products Corporation in Figure 6-2 shows such a case in which there are two products, each run by a superintendent. Each product group has two or more processes, each of

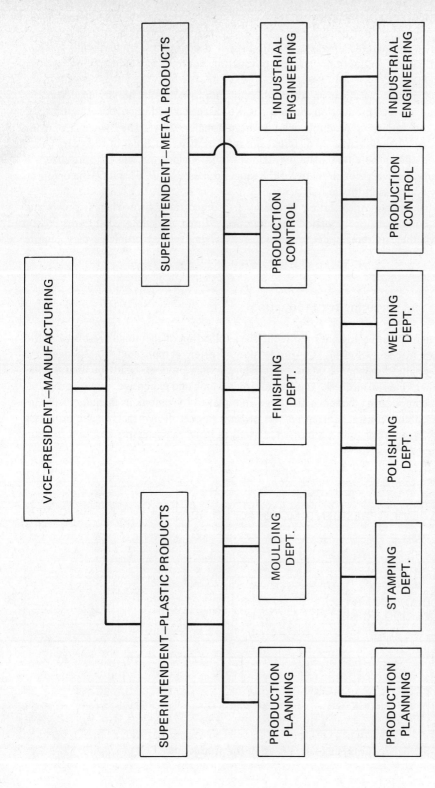

Figure 6-2. Dunn-Products Corporation—organization for production

which is in a separate department. This firm is decentralized in that each product organization has production-planning, production control, and industrial engineering departments.

The specific organization for production in a firm may be heavily influenced by the nature of the production process, as we discussed earlier in this chapter. The number and variety of products will also have a major effect. The Switzer Company produces small amounts of 20 different products at different times of the year. This is the reason that production planning, control, and engineering will be centralized. If the 20 products go through many of the same processes, A, B, C, and D, the organization may appear as in Figure 6-3.

Other special organization problems are forced on a firm when it grows and acquires competing or unrelated manufacturing firms. Deciding what can be combined, whether to remain decentralized, and solving related problems may require experimentation.

Staffing and Directing for Production

Staffing and directing together represent the application of the human resource to the production process. Staffing for production is often complicated by the need for highly specialized personnel. More extensive searching processes must be used than in many other kinds of business. The alternative is a long and expensive training program. Another complicating factor is the large role played by unions in manufacturing industries. Labour contracts often restrict or limit personnel managers in the processes of hiring, transferring, and promotion, as well as in job assignments and in setting up working conditions.

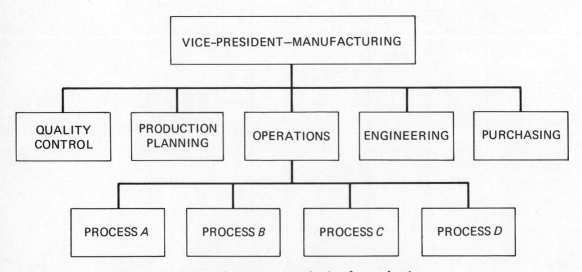

Figure 6-3. Switzer Company—organization for production

A production manager must cope with the fact that machines, their timing, their co-ordination, and their very high costs determine how the production process is carried out. This includes determining the relationship between jobs, the span of control of supervisors, and other factors which must influence the kind of direction a supervisor can give.

It is easier to motivate workers to outproduce co-workers than to motivate them to operate at the pace dictated by a machine. How does a manager set a wage rate which all workers think is fair when the principal contributor to productivity is a machine? Also, many production workers fear that automation will put them out of work. This makes it hard to motivate them, especially when their unions oppose automation.

In such situations, harmony is hard to achieve. It may be done by means of co-operation between labour and management. Both sides can agree that greater productivity is good for all. If automation can bring about greater productivity and management can assure the labour unions and their members that workers will share fairly in this increased productivity, a good working relationship can be built. It takes planning to prepare for the period of change to automation. This planning includes worker retraining and relocation and other guarantees of security for those workers most affected. Wage rates will rise because workers expect to share in the benefits of increased productivity.

Controlling Production

Controlling production involves setting production standards and developing systems for comparing production performance with those standards. There are several different types of production control, including order control, product quality control, controlling the plant's operating effectiveness, and inventory control.

Order Control

When a plant is engaged in continuous, high-volume production of standard products, close control of individual orders is not very important. But in those plants using intermittent production, it is vital to create systems to control the flow of orders. New orders which have never been processed before must be checked to see what operations must be performed. They must be checked for correct sequence and to see that the right tools are on hand. This may require the use of a control chart. **A control chart is a device which shows the standard set of steps to be taken in the performance of a procedure.** It ensures that things will be done as planned.

Control chart

Figure 6-4 shows a control chart. The left column lists thirteen documents of different kinds. The labels at the top show the points (departments, individuals) at which the documents originate, are acted upon, and are filed (temporarily or permanently). The top line shows that purchase orders originate with the customer, are acted upon by the sales and production departments, and are filed in the sales department. A copy is also filed by the customer. A failure in any of these illustrated paperwork

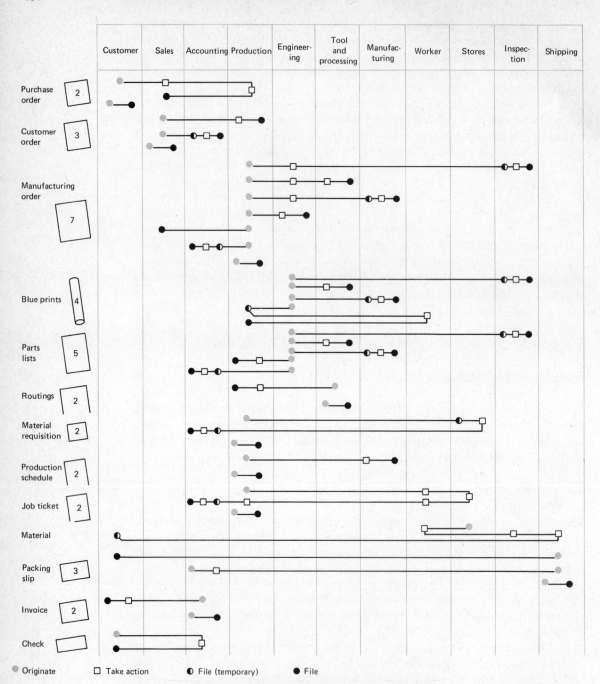

Figure 6-4. Flow of paperwork in a production control procedure. Courtesy of Professor John G. Carlson, University of Southern California. Reproduced by permission from Elwood S. Buffa, *Modern Production Management* (New York: John Wiley & Sons, 1973).

6 PRODUCING GOODS AND SERVICES

flows could mean production delays and unnecessary costs. Similar charts were first conceived by Henry Gantt in the early 1900s, and today a wide variety of commercial variations of Gantt Charts is available.

Plants which produce a variety of products on an intermittent basis establish a system of decision rules to guide the movement of orders through various manufacturing processes. Assume there is a question of which of two orders should be processed first by a given machine. It is not always logical to say that the order with the earliest due date should be processed first. Some firms use a simple "first-come, first-served" rule; others use a "first-come, first-served within priority class" rule for making these decisions. Which rule is best depends on the kind of processing going on in the plant. Regardless, some form of decision rule is necessary in intermittent process production to guarantee a smooth flow. In recent years computer simulation techniques have been used to speed up production. These are discussed in Chapter 14.

Important tools used in order to assure a smooth flow of operations include PERT (Program Evaluation and Review Technique) and CPM (Critical Path Method). **PERT is a planning and control tool focusing on the timing of the occurrence of many operations included in a project. It helps identify and remove bottlenecks.** This tool was first applied by the U.S. government to speed up completion of vital weapons systems.

PERT

A CPM diagram is illustrated in Figure 6-5. It is much like PERT except that specific estimates rather than variable estimates of time elapsed in operations are used. Wetherall Utility Buildings plans construction of a standard 20,000-square-foot warehouse building by listing the major stages of construction, the time needed for each

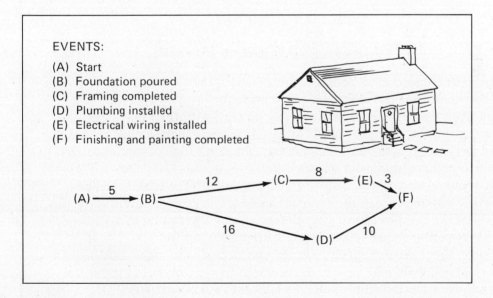

Figure 6-5. A CPM diagram

stage, and the sequence in which they must occur. The circled letters are events. The sequence of events is indicated by arrows. Branching occurs whenever two events can be worked on at the same time. The numbers next to arrow segments show how many days are needed. A manager can follow a "path" to completion and can tell which chain of events is the longest or "critical" path. It is this chain of events which the manager must focus on to cut production schedules or to avoid delays.

Order control systems must include rapid communication systems, such as special mail service or teletypes or even direct feed-back to a central computer. They help management to take action quickly enough to correct flaws in the operation. This is the essence of control.

Product Quality Control

Quality control

Clearly, there is a need to control the movements of orders through a production process. There is also a need for quality control. **A quality control system sets up a standard for an input or output and makes comparisons against this standard to prevent non-standard items from going into or coming out of the production process.** This, however, does not necessarily mean that quality must be kept high. It means that the level of quality must be known and checked so that it can be kept within a certain range of acceptable quality standards. Obviously, a manufacturer does not always seek to avoid high quality. Lower quality, however, may be acceptable when higher quality control would be too costly. A firm may seek a quality level far below what could be achieved.

WHAT DO YOU THINK?

How Much Quality Should Be Built into a Product?

You have probably heard some people complain about the "low quality" of the products they have bought. Perhaps you have made the same complaint. These complaints mean that customers are not really satisfied with their purchases.

Take a car muffler, for example. If Midas Muffler can make a muffler that is guaranteed to last as long as you own your car, why don't the car manufacturers install them as original equipment on new cars?

What about an electric percolator for making coffee? The top-of-the-line models usually last longer than the cheaper versions. The manufacturers know this, but they still make cheap versions. They defend this by saying that a lot of consumers want the cheap model—"and they get what they pay for".

Decisions about product quality are not easy decisions to make. But they are very important decisions, and they are becoming even more crucial in the age of consumerism. How much quality should be built into a product? WHAT DO YOU THINK?

The Good Time Corporation makes children's outdoor play products. It produces a backyard slide selling for $19.95. The production manager has specified an inexpensive rust-resistant galvanized nut-and-bolt for fastening the slide's major parts. These nuts-and-bolts sell for one cent apiece in large quantities. The manager could have specified chrome-plated nuts-and-bolts costing eight cents apiece and have been able to guarantee against rust. But customers paying $19.95 for a backyard slide just don't expect chrome-plated parts, so the added cost is unjustified. This does not mean, however, that the production manager would accept even cheaper, lower-quality bolts than the ones selected just to reduce costs by a fraction of a cent. They might cause the slide to fall apart, and the firm would lose its reputation as a maker of good, low-priced play equipment.

CAREERS IN PRODUCTION

If you like the idea of working with a manufacturing company, there are a wide variety of jobs available. They require a wide range of skills and education. We will describe a few.

Purchasing agents need to understand the materials and equipment requirements of the firm they work for. They must have some technical skill and a good "business sense". They must have the ability to communicate effectively with operating department heads and the skill to appraise alternate sources of supply. They must have some training and skill in accounting. Often a college degree in business is required. Information is available from the Purchasing Management Association of Canada, 80 Richmond Street, W., Toronto, Ontario M5H 2A4.

Inspectors are needed in every manufacturing operation. They can enter with little technical training and learn on the job. Persons who work well with numbers and measurement, who value accuracy, and who are not afraid of performing repetitive tasks may want such a career. Those who perform well may be promoted to jobs as skilled inspectors or quality control technicians. Opportunities exist in this type of job despite increasing use of automated inspection systems. For further information you might contact the Society for Quality Control, 161 West Wisconsin Avenue, Milwaukee, Wisconsin 53203.

Traffic managers are also needed in manufacturing and wholesaling. They are responsible for physical movement of goods to their destination. They are experts on rates, schedules, and availability of alternative modes of transportation. College graduates make up an increasing proportion of traffic managers. Skills needed are similar to those of purchasing agents. For additional information contact the Canadian Institute of Traffic and Transportation, 44 Victoria Street, Suite 2120, Toronto, Ontario M5C 1Y2.

Production quality control applies to raw materials, parts purchased, the finished product, and to various stages in the production process. Thus, certain testing standards and procedures may be used in the purchasing department for all raw materials and parts, while other testing standards and procedures may be used in the production process to guarantee the quality of the finished product.

The Brakewell Bike Company buys tires, tubes, and handle grips from the Mutual Rubber Supply Company in large lots. These components are inspected upon arrival to see if they meet specifications (standards of quality). During the process of manufacture, welded joints of the bicycle frame are inspected and the braking assembly is tested. After an entire bicycle is assembled, it is lubricated and "test-driven" for a quarter of a mile. These tests cost money, but they help build Brakewell's reputation for quality.

It is practically impossible to guarantee 100 per cent quality of all inputs and finished goods. Ordinarily, this would be too expensive. Most manufacturers, therefore, use sampling in quality control inspection. When a shipment of steel tubing arrives for processing, the purchasing department at Brakewell may require that a sample of 50 twenty-foot sections be examined. Instructions might require that a shipment be accepted if no more than one defective section is found in the sample. This is called acceptance sampling. In certain types of production, of course, a 100 per cent inspection is necessary because the occurrence of even one imperfect unit could have very serious results.

The basis for classifying a unit as defective depends on what characteristics of the unit are "critical". For steel tubing, perhaps hardness and strength are important. Applying these standards may require mechanical assistance, or perhaps it can be done visually. Welds, for example, may be examined visually or by using x rays. A wide variety of measuring and testing machines exists in industry today. Often, however, a simple visual inspection is sufficient. A balance between the cost of inspection and the cost of failure to meet standards must be achieved.

Maintenance and the Control of Plant and Equipment

In a manufacturing plant—particularly a continuous-process, assembly-line plant—one critical machine breakdown means high costs. During this "downtime" most of the other machinery is idled, and all the workers on the line are being paid even though their machines are idle.

Preventive maintenance

This points up the need for maintenance. **Many firms practise preventive maintenance. This means that they inspect and/or replace certain critical machines and parts on a regular basis to avoid downtime.** Not all firms practise preventive maintenance, however. It may cost more than a firm is willing to pay. The units may be very expensive to replace. Also, the cost of interrupting the production process for maintenance work may be too high. Some production managers prefer to "leave well enough alone". They install the highest-quality equipment to start with. This decision is a matter of balancing costs.

Control of Inventory Levels

A fourth area of production control is the control of inventory levels. This includes inventories of raw materials, parts, and finished and partly finished goods.

There are some good reasons for keeping high levels of inventory. A firm is less likely to "run out" of parts and partly finished goods where assembly-line production is employed. Running out of inventory can mean expensive downtime. This also applies to finished goods. Big orders (and possibly big customers) can be lost if deliveries cannot be made as promised. Large stocks protect against this.

There are some equally good reasons for keeping inventories low, however. First, inventories require an investment of funds. A factory which operates with a lower inventory is operating more efficiently—that is, it is producing profits with a smaller investment than a factory with large inventories. Second, inventories take up scarce space. Third, goods in inventory may decrease in value because of deterioration, theft, or damage. **Inventory goods also may be subject to obsolescence. This is what happens when something is out-of-date or not as efficient as newer products.** New inputs or new finished goods may be invented or found which would make inventories obsolete.   *Obsolescence*

Some firms use mathematical formulas to determine the best inventory levels to maintain. These formulas include such things as order loss risks, storage costs, interest, expected delivery time, and other factors. They also aid in determining the best quantity to order and the time interval between orders.

Other Production Controls

As we will see in Chapter 9 on accounting, there are other important control devices relating to production costs. These include cost systems and various budgeting devices.

PRODUCTION AND ECOLOGY

Many production processes cause problems of ecological balance in the areas where a product is produced and consumed. Extracting coal by strip mining has sometimes caused long-lasting environmental damage. The use of aluminum and glass containers for certain consumer products has created a solid waste disposal problem. This has brought unsightly litter to our cities and countryside. The manufacturing process itself has poured smoke into our air and poisonous wastes into our waters. A certain type of aerosol may pose a threat to the ozone belt, which protects us from deadly rays. Our most modern nuclear power plants are viewed by significant numbers of people as dangerous.

For years, our factories have made only limited use of waste materials and by-products. They have used these materials only where there was obvious profit in it. There are many recyclable inputs which have little dollar value but which should be recycled for the sake of our ecological balance. The government is setting new standards, and it is up to production management to do its work within the new limits at a cost which will permit the sale of products at competitive prices.

The effect of business on the environment is discussed in detail in Chapter 17.

MATERIALS MANAGEMENT

All of the variety of materials purchased by a firm—including raw materials, partly finished products, finished products, supplies, and capital goods—require a special managerial effort. These products must be purchased (or made), physically handled, and stored. Capital goods also need to be maintained in working order.

The most highly developed managerial activity in dealing with materials—one which often requires setting up a special department—is purchasing.

THE PURCHASING TASK

The purchasing task can have a critical effect on profitability for manufacturers, retailers, and wholesalers. In manufacturing, this is especially true when materials and parts are a major part of total manufacturing cost. Many firms establish a separate purchasing department with its own manager. Many large purchasing departments have divisions which specialize in specific types of purchases and divisions for records management and follow-up.

A typical large manufacturing firm must buy fuels, cleaning supplies, lumber, sheet steel, dyes, handtools, nails, electronic calculators, stationery, electrical equipment, paint, food for the company canteen trucks, and many other items. They must stay ahead of these needs by locating adequate sources of supply and actually buying the products.

Sources must be adequate in terms of volume, delivery schedules, and quality levels. The purchasing department must be in constant contact with their suppliers' sales offices to keep up with their pricing policies, new product features, and delivery schedules.

Many firms, especially large ones, practise centralized purchasing rather than allow individual departments and divisions to make their own purchases. This can result in cost savings from large-volume buying, greater co-ordination of purchasing and receiving functions, and a more uniform application of standard purchase specifications.

Depending on the nature of the good or service, the purchasing department may provide different types of advice and perform different types of services. When specialized and expensive machinery is needed, the purchasing department must work closely with the production manager whose department will use it.

Often, finance, purchasing, production, and engineering people work together to develop a list of specifications for the equipment and to plan for its financing and procurement. This team approach is helpful in make-or-buy decisions, and it ensures that the right machine will be bought. The production manager alone might specify higher quality than is actually needed. If the finance department were to make the decision on its own, perhaps quality would be compromised in order to save money. The team objective is to get the level of quality needed by the using department at the best price. This requires a systems approach to decision making in which the "net" welfare of the firm is the guiding principle.

For less expensive items the purchasing department assumes a larger role. For items such as paper clips, the purchasing department makes its own price and quality decisions. Most routine purchases are handled in this way.

Purchasing Policies

Over time, firms usually develop standard purchasing policies. For example, some follow a policy of building up inventories when prices are right. The purchasing agent is a "professional purchaser" who generally has a very good idea of when prices are right.

Frank McGee, purchasing agent for the Claxton Art Supplies Manufacturing Company, is in constant contact with several suppliers of quality natural bristle for artists' paintbrushes. He buys only when prices are down, because storage costs and production deterioration are not a problem. In March he received notice and samples of a huge shipment available immediately from South Korea at a price 20 per cent below the typical price. Since he had dealt with this importer before and had been favourably impressed, he ordered a two-year supply of the product.

Some purchasing agents follow a policy of concentrating all their purchases for a specific good or service with one supplier. This is often because that supplier's past performance has been excellent. Other purchasing agents avoid this for fear of "being taken for granted" or "putting all their eggs in one basket". A strike, for example, at a supplier's plant may place the buyer firm in a bad position.

Other policies deal with such matters as taking discounts offered by suppliers. If a supplier offers a cumulative quantity discount, all purchases made during a certain period are subject to a discount based on the total volume purchased during that period. This builds buyer loyalty but probably reduces the average size of orders. A non-cumulative discount probably leads to larger orders but may not do much to develop customer loyalty.

CAN YOU SETTLE THE ISSUE HERE?

The Purchasing Agent, the Salesperson, and Business Ethics

The purchasing agent is the firm's professional buyer. He or she is a very important person as far as salespeople are concerned. That is why purchasing agents are frequently offered "special favours" from salespersons who want to sell to them. Christmas gifts, free "samples" for the purchasing agent's personal use, and other "favours" are not uncommon.

Is it unethical for purchasing agents to accept "favours" and for salespersons to offer them? CAN YOU SETTLE THE ISSUE HERE?

Reciprocity

Some firms follow a policy of leasing equipment rather than purchasing it whenever possible. Leasing often offers a tax advantage, since lease payments are deductible business expenses. It also shifts part of the risk of equipment obsolescence to the leasing firm and ties up less of the lessee's capital. Just as firms face make-or-buy decisions, they also face buy-or-lease decisions.

An important purchasing policy involves reciprocity—"you buy from me and I'll buy from you." This is widely practised by industrial marketers and buyers. It makes buyer and seller interdependent—it "guarantees" the seller a customer but limits the sources of supply. It may also cause buyers to become too lazy in their search for "the best quality at the lowest price".

Value Analysis and Vendor Analysis

The purchasing function is being handled increasingly by professionals. Two tools which are receiving growing attention are value analysis and vendor analysis.

Value analysis

Value analysis starts by reviewing existing product specifications as set by user departments. Attention then focuses on identifying and eliminating inessential cost factors. It may involve a committee including engineers, cost accountants, production

---

YOU BE THE JUDGE!

Sources of Supply

Paul Block, production manager at Acme Small Motors Ltd., has an important decision to make. For years Acme has been producing electric motors for pumps, fans, and large industrial door-opening devices. Ever since it started production Acme has purchased the steel casing of its motors from Specialty Steel Products. This supplier is located within two miles of Acme's main assembly plant in Hamilton. The relationship has been excellent until this year.

Acme's volume is growing rapidly. On the other hand, Specialty Steel is near bankruptcy and has notified Acme that it is shutting down the Hamilton plant. They want to continue supplying the casings through another plant in Michigan but will have to increase prices 10 per cent and limit deliveries to once every two weeks.

Paul Block sees his alternatives as follows: first, to accept the new offer by Specialty Steel; second, to invite bids from the only two nearby suppliers (with whom Acme has never done business); and third, to invest $200,000 in additional equipment and produce the parts in the Acme plant. Acme has never produced such parts before, but the technology is not very complex. Which of these three alternatives makes the most sense to you? YOU BE THE JUDGE!

representatives, and others. They review the specifications set by the user department. Wherever a specification is thought to add unnecessary cost, the function of that "spec" is examined to see if it can be eliminated or if a cheaper way of doing it can be found. Such a review requires close contact with potential vendors to verify cost.

Vendor analysis is an approach to evaluation of potential suppliers. It considers the technical, financial, and managerial abilities of vendors and rates them in terms of their past performance in these areas. It is a method of substituting facts for feelings in the selection of suppliers.

Vendor analysis

When a purchasing department has made its analysis of possible suppliers, it is in a position to sign a contract. It makes a decision and sends a purchase order to the supplier. Some purchasing departments, however, invite sellers to submit bids. In some cases, the buyer elects to award the purchase contract to the lowest bidder. This competitive bidding requires that the buyer specify in detail what it is which he or she wants to purchase. In other cases, specifications are not so exact, and bids received are subject to further buyer-seller negotiation over price and quality. Bids, however, are not used in all cases. The buyer might contact and deal with only one supplier.

A purchasing department is also accountable for following up purchases already made. Elaborate file systems are used to ensure that deliveries are made on time. This includes the follow-up on transportation details and expected delivery dates. A final responsibility is for the physical receipt of goods. This involves checking contents against invoices before giving approval to the accounting department to make payment.

OPERATIONS MANAGEMENT—A BROADER VIEW

Up to now we have emphasized the use of management techniques for the production of goods—production as opposed to marketing or finance, and physical goods as opposed to services. The bias was justified on two grounds. First, scientific management was developed in factories, not in retail stores; and second, the next few chapters will be devoted to marketing and finance.

Students of modern business should recognize that most of the techniques we have seen so far in this chapter can be made to work quite well outside the factory. These methods work in service firms like beauty parlours or repair shops; they also work in distribution firms like wholesalers or truckers.

Operations management is a new, expanded version of the idea of production management. It represents a systems approach to all business functions with an emphasis on current operations and control rather than on long-range planning.

Operations management

This broader view includes such business problems as controlling the billing process for a large dental practice, designing the physical layout of a restaurant, or locating a new supermarket for a large chain. Let's look briefly at how we might apply operations management techniques to each of these.

Dr. Jenny Wilson opened her office in a western city soon after completing her internship. The practice grew very quickly. It was a great success after a few years, but she wasn't doing too well when it came to collecting her fees. A management consultant could have pointed out several specific problems. First, she was not regularly

GUIDELINES

An Operations Management View of Key Decisions in the Life of a Productive System*

BIRTH of the System	What are the goals of the firm? What products or service will be offered?
PRODUCT DESIGN and *PROCESS SELECTION*	What are the form and appearance of the product? Technologically, how should the product be made?
DESIGN of the System	Where should the facility be located? What physical arrangement is best to use? How do you maintain desired quality? How do you determine demand for the product or service?
MANNING the System	What job is each worker to perform? How will the job be performed, measured; how will the workers be compensated?
STARTUP of the System	How do you get the system into operation? How long will it take to reach desired rate of output?
The System in *STEADY STATE*	How do you run the system? How can you improve the system? How do you deal with day-to-day problems?
REVISION of the System	How do you revise the system in light of external changes?
TERMINATION of the System	How does a system die? What can be done to salvage resources?

*Adapted from Richard Chase and Nicholas Aquilano, *Production and Operations Management*, rev. ed. (Homewood, Ill.: Richard D. Irwin, 1977), p. 14. © 1977 by Richard D. Irwin, Inc.

recording and filing fees for the oral hygienist. Second, her office assistant was often absent, and his replacements made a lot of billing errors. Finally, there was no routine for following up on slow or non-paying accounts.

The practice of good operations management methods would have resulted in the design of a routine for recording oral hygiene charges; the hiring of a regular replacement for the office assistant; and the establishment of a past-due bill follow-up procedure and use of a collection agent for extreme cases.

A restaurant needed to bring its 50-year-old building up to date. An operations management approach to the problem might have started by asking the question, "what has changed between 1930 and 1980 concerning the needs of our customers?" Solutions might have included providing a warmer atmosphere, improved parking, and drive-in facilities.

A systems approach often requires that human inputs be modified along with the plant itself. Good operations management would have recommended such things only after interviewing customers and employees and checking the timing and flow path of patrons inside the building as well as their modes of transportation to the restaurant.

Similar analysis and application of operations management techniques could work for the Safeway supermarket chain in choosing a new retail location. Management might have started by checking census data and other sources to find those areas of population growth in the city. They might also check the location of competition, the cost of land, traffic flows, zoning information, and the availability of the right kind of land and/or buildings for lease. The final decision would be made in much the same way as the plant location example earlier in this chapter.

Management of Safeway would consider factors likely to draw customers and factors affecting costs. The motivation, as in the case of all business management, is mostly profit. The methods of estimating profit found in operations management are quite similar to those a large factory might use.

SUMMARY AND LOOK AHEAD

Production creates goods and services by a variety of processes out of human and material inputs. The management of such a process begins with the planning of the product, the plant, and its location. Production includes special organizational and staffing problems related to the impact of technology. It requires the application of a variety of control devices to assure uniformity of quality of output and efficient production scheduling.

Purchasing, we have seen, has evolved into a science in itself. Centralized purchasing departments develop purchasing policies and procedures to help assure that the firm gets goods and services of required quality at a minimum price.

The mountain of products generated by a giant production system demands an ingenious marketing effort to move them into Canadian markets. Financing a manufacturing plant and its related facilities is also a complex undertaking which requires up-to-date accounting methods. Computers are needed to support the production, finance, accounting, and marketing. All of these will be examined in the next few chapters. Marketing is the first to be examined. It is such a large subject that we will devote two chapters to it. The first chapter takes an overview of the marketing task.

KEY CONCEPTS

Breaking down A production process which involves separation of raw materials into parts.

Capital-intensive Productive processes depending more heavily on machinery, plant, and equipment than on labour.

Combination A production process which involves bringing things together into a new arrangement.

Continuous production Production carried out on a day-to-day basis without interruption.

Control chart Any of a variety of charts or illustrations guiding the scheduling of production.

Intermittent production Production undertaken when inventory stocks fall below a critical value; also, may include production for stock when machines and workers would otherwise be idle.

Labour-intensive Productive processes involving relatively large amounts of human input.

Make-or-buy decisions Decisions whether it is more economical to make a product or a component part or to buy it.

Obsolescence Growing old, becoming outdated; applied to technological or fashion goods.

Operations management A general term applied to production, financial, and marketing management with an emphasis on the treatment of all three as parts of a system which is subject to improvement by scientific management methods.

PERT (Program Evaluation and Review Technique) A planning tool used by managers to estimate how much time it takes to complete various parts of a project so that bottlenecks can be avoided and the project can be done on time.

Plant capacity Output limits of a production facility.

Plant layout The internal design of a factory, including the arrangement of the machines used in the manufacturing process.

Preventive maintenance The inspection and/or replacement of certain critical machines and parts on a regular basis to avoid downtime.

Production An activity which results in the creation of goods and services.

Quality control Means by which the level of quality is known and monitored so that it can be kept within a certain range of acceptable quality standards.

Reciprocity A "you buy from me and I'll buy from you" purchasing policy.

Treatment A production process which doesn't add parts to, or remove them from, something, such as cooking meat.

Value analysis The reviewing of existing product specifications as set by user departments; the identification of costs and the attempt to cut them where possible.

Vendor analysis An approach to evaluation of potential suppliers which considers the capabilities of vendors and may include a specific vendor-rating system. Vendors are rated in terms of their past performance.

QUESTIONS FOR DISCUSSION

1. Who was Frederick W. Taylor? Check an encyclopaedia.
2. Distinguish between production management and operations management.
3. Does continuous production or intermittent production justify larger capital expenditure? Why?
4. To locate a plant near suppliers but far away from customers would imply what about the nature of inputs and outputs and their transportation costs?
5. Give one example of how a Gantt Chart may be used.
6. Distinguish between production organization along product lines and along process lines.
7. Does production management have human relations problems not found in financial or marketing management? Explain.
8. Consult Fraser's Canadian Trade Directory and give a brief description of its contents.
9. What is reciprocity? Is it a sound basis for purchase decisions? Why or why not?
10. What are the pros and cons of centralized purchasing?
11. What is the essential idea of value analysis?
12. Give an example of the use of operations management in the control of the quality of hamburgers at a McDonald's restaurant.

INCIDENT:

Woodrow Shimmick

Woodrow Shimmick, the purchasing agent for Samuels Farm Equipment Ltd., had dealt with several different suppliers of high-grade grommets for years. The King Grommet Company, one of these suppliers, has been losing Samuels' business lately because casual inspection has proved the quality of King's grommets to be uneven. Shimmick has recently written King that unless the next order meets specifications, no more orders from the Samuels firm will be given them.

Questions:
1. What alternatives are available to King?
2. Could sampling be used to help assure quality control? How?
3. How could a program of value analysis affect the Samuels firm's choice of grommets to be purchased?

INCIDENT:

Sam Gold

Sam Gold has founded Mahogany Enterprises Ltd., to manufacture small wooden art objects for the tourist trade in Central America. He purchases mahogany logs from local lumber dealers in Guatemala and hires native artisans to carve the finished products from pieces of various sizes pre-cut in his mill. After having been carved, the objects are lacquered and sold in the local markets.

Questions:
1. What type or types of production processes has Mr. Gold employed?
2. Explain why the location of Mr. Gold's mill is advantageous. What are major factors in the selection of plant location? Discuss.

ROCKETOY COMPANY IV

Terrence Phillips, by now a reasonably experienced corporate executive, began to realize that corporate growth brings with it serious problems. The production facilities of Rocketoy were becoming clearly inadequate. The number and the variety of orders were so large that the firm was two weeks behind schedule on deliveries. Equipment was rapidly "falling apart" despite good maintenance and supervision by Mark Flynn, the assistant production manager. A new plant was essential and Terrence called in Uchello and Jaynes, consulting architects, to design it.

Looking ahead at what Rocketoy expected to be the future production level of various products, the seasonal factors, and chances of obsolescence, T. Phillips, J. Phillips, Flynn, and the architects agreed on a one-level, 60,000-square-foot plant with nine "shops" or processing areas as well as office, inspection, packaging, shipping, and receiving areas. The nine processing areas included painting, wood assembly, plastic moulding, metal stamping, and five other areas. Each was capable of performing one of the major production processes for one or more of the three types of toys produced—wood, metal, and plastic.

A site was selected near Windsor with access to major highways. The site was also close to suppliers of many of Rocketoy's parts and raw materials.

Some differences of opinion arose among Rocketoy's managers. Ms. Rabinovitz and Ms. Carter (formerly Pam Phillips) argued that moving the plant to Windsor from Toronto would create great morale problems among their loyal production workers who disliked moving. Finding replacements in Windsor would, at least in the short run, cause lots of problems, according to Ms. Rabinovitz and Ms. Carter.

Questions:
1. Classify some of the production processes at Rocketoy according to the process classifications discussed in the text.
2. Do you agree with the kind of plant arrangement worked out by the architects? Why or why not?
3. Can you think of any alternatives to building a new plant? Explain.
4. What are the criteria for site selection which were not mentioned in the case?
5. How should Rocketoy deal with the problem of assuring quality labour in the new location? What morale problems could arise in addition to those mentioned?

7
Marketing

7 MARKETING

OBJECTIVES: After reading this chapter, you should be able to:

1. Illustrate the fact that marketing is a matching process.
2. Explain the role of the consumer in the marketing process.
3. Distinguish among four kinds of utility in a product.
4. Give an example of a marketing mix for a specific product or a service.
5. Show what justifies the existence of middlemen.
6. Describe the principal characteristics of the industrial goods market.
7. Draw up a list of consumer goods and tell which are convenience, shopping, and specialty goods.
8. Compare the strategies of differentiation and market segmentation.
9. Name and explain two different approaches to market research.
10. Describe a case of conspicuous consumption.
11. Find an example in the newspaper of consumerist activity and criticize it from the consumer's and the firm's points of view.

KEY CONCEPTS: In reading the chapter, look for and understand these terms:

MARKETING
DISCRETIONARY INCOME
FORM UTILITY
PLACE UTILITY
TIME UTILITY
OWNERSHIP UTILITY
MANAGERIAL APPROACH
   TO MARKETING
MARKETING CONCEPT
MARKETING MIX
PRODUCT
PRICE
PROMOTION
DISTRIBUTION
TARGET MARKET

INDUSTRIAL GOODS
CONSUMER GOODS
MIDDLEMEN
CONVENIENCE GOODS
SHOPPING GOODS
SPECIALTY GOODS
PRODUCT DIFFERENTIATION
MARKET SEGMENTATION
MARKETING RESEARCH
SECONDARY RESEARCH
PRIMARY RESEARCH
CONSPICUOUS
   CONSUMPTION
CONSUMERISM

This chapter introduces the second important area of decision making for firms—marketing. It builds on concepts developed in earlier chapters on economics and management. It also prepares us for the detailed discussion of marketing activities which will be presented in the next chapter. We'll see why a "marketing" orientation has been adopted by many firms and how this is reflected in their approach to the customer. We'll also see the difference between utility creation in production and in marketing and learn about the special interdependence of production and marketing in a high-level economy. At the outset, it will prove helpful to define what is meant by "marketing".

WHAT IS MARKETING?

Marketing

Marketing is a whole set of activities needed to find, build, and serve markets for products and services. It often consists of finding out what products and services people already need or want and then proceeding to design, promote, and distribute them. Sometimes it involves stimulating new wants, and other times it tries to satisfy established wants. Marketing activities include such things as marketing research, retailing, sales force management, advertising, and transportation. These activities will be examined in the pages to follow.

The greater the production ability of a nation or a firm, the more important it becomes to improve its marketing ability. The benefits of production technology are wasted in a free economy if the wrong things are produced or if people don't know about the product. It is vital that every firm design a good marketing program to sell what it makes.

In a rich nation like Canada, there are many choices available to consumers. The more money a family or individual has, the smaller is the proportion of income required for absolute necessities. Affluent buyers can shift their spending patterns around. **We call what a buyer has to spend but which is not earmarked for necessities discretionary income.** Figure 7-1 shows that a family, the Collinses, has 67 per cent more income ($20,000 vs. $12,000) than another family, the Browns. However, it also demonstrates that the discretionary part of the Collins' income is 500 per cent greater than that of the Browns ($6,000 vs. $1,000). Rising income, then, means businesses find it harder to predict what will be bought. This complicates the marketing task and underlines the importance of watching the consumer closely. The consumer's tastes and preferences can change quickly.

Discretionary income

The marketing process must also cope with risks related to technology. As we will see in greater detail in Chapter 18, a firm never really knows when a competitor will devise a new product which outdates its present product or which makes its present product unnecessary. Frozen orange concentrate has greatly reduced the market for orange-squeezing appliances in the kitchen.

A full explanation of the function of marketing is helped by expanding on a concept introduced in Chapter 1—the concept of utility. The next section explores the relationship between marketing and utility.

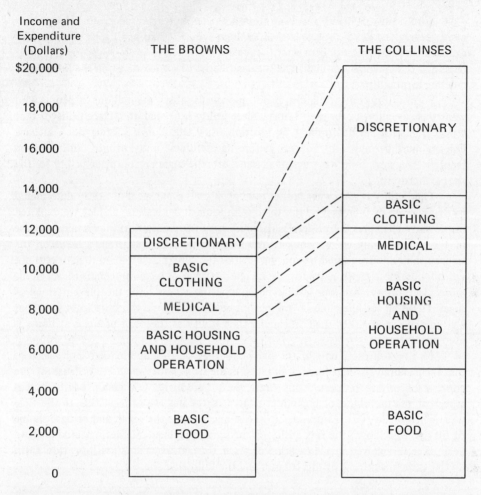

Figure 7-1. Discretionary income of two families. *The Browns and the Collinses have five members in each family—two adults and three children between the ages of 5 and 15. The Collins' income after taxes is $20,000—$8,000 more than that of the Browns. The Collinses spend $3,000 more than the Browns for necessities—basic food, clothing, housing, and medical expenses. This leaves them $6,000 to spend as they prefer—discretionary income. The Browns have only $1,000 in discretionary income. As income increases, discretionary income increases at a faster rate.*

MARKETING AND UTILITY

The utility of goods and services is at the heart of the marketing problem. We can distinguish the following kinds of utility: form utility, place utility, time utility, and ownership utility. We will show how these four aspects of utility relate to the household purchase of sugar.

Form utility   **Form utility is utility resulting from a change in form.** It is produced by treatment or breaking-down processes such as those we described in Chapter 6. Sugar, for example, becomes more useful after the juice is extracted from the sugar cane or sugar beets and is cooked and refined. The sugar refineries of Louisiana are in the business of creating form utility.

Form utility, unfortunately, is not enough to satisfy the millions of people who want this sugar for use in their homes. Place utility is needed, too. **Place utility is that aspect of usefulness determined by location.** Raw sugar on a loading dock in New Orleans must be moved to Toronto before its usefulness to a Toronto family can be realized. The ship, train, or truck that transports the sugar creates place utility for the people in Toronto.

Place utility

Time utility   **Time utility is somewhat harder to explain. It is that aspect of utility determined by the passage of time as it relates to consumption.** It depends on an idea we explored in the early chapters, the principle of diminishing marginal utility. Let's take the case of the Jones family in Vancouver. They have a pound of sugar in a bowl on the kitchen table. A five-pound bag on the shelf of the supermarket down the street is not yet fully useful to the Joneses for several reasons. One, as we have seen, is that it needs more place utility. Another is that more time must pass until the present stock of sugar is used up and they "need" more sugar—enough to go to the store to buy it. That bag on the supermarket shelf is gaining time utility as the bowl of sugar at home is used up.

Finally, full usefulness of the five-pound bag on the supermarket shelf, as far as the Jones family is concerned, can't be reached until they own it. Because of the concept of private property, this sugar must be bought. **Ownership utility is that aspect of the usefulness of a product related to the passage of legal title to the final user.** When Mr. or Mrs. Jones goes to the store, pays for the sugar, and brings it home to fill the sugar bowl, the full utility of the sugar is realized. Marketing activities have been directly involved in the creation of place, time, and ownership utility. How firms do this is the main topic of this chapter and the following one.

Ownership utility

THE MARKETING CONCEPT AND THE MANAGERIAL APPROACH

The broad field of marketing has been analyzed in many ways. Some people study the major institutions (wholesalers, retailers, etc.) involved in the marketing process. Others study how different commodities vary in the way they are marketed (marketing of grain, meat, hardware, etc.). Still another method is to divide the total marketing task into its separate functions (buying, selling, risking, etc.).

Managerial approach to marketing   Today the most common way of studying the subject is called the managerial approach to marketing. This approach takes the point of view of a firm which must make a variety of marketing decisions, each of which may affect the profit of that firm. We'll use this approach because it is the most popular one today and because it fits the scheme of this book.

THINK ABOUT IT!

The Marketing Functions

The process of marketing can be divided into many separate "marketing functions". These are identifiable tasks which must be performed before the complete marketing task—bringing goods or services from the producer to the consumer—can be done.

The marketing functions have often been divided into those relating to buying and selling and those relating to physical movement and storage. In the first group are such functions as creating demand, determining needs, negotiating prices, transferring title, advising buyers, and finding buyers and sellers. In the second group are transporting, storing, classifying, packaging, assembling, and dividing.

Think of a particular product with which you are familiar. How does it get from the producer to the consumer? Who performs the function of creating demand? the function of transportation? of negotiating prices? THINK ABOUT IT!

The managerial approach makes use of the marketing concept. This is the idea that a firm must co-ordinate all its activities to satisfy customers and to make a profit in the process. The customer is at the center of the firm's planning. This kind of thinking gives a firm the goal it needs to operate as a system. If there is a disagreement within a firm, it should be resolved in terms of the marketing concept. You do what best serves the customer.

Marketing concept

The Marketing Mix

To apply the marketing concept effectively, a firm must understand the major factors or tools it may use to meet (and to influence) consumers' wants. These factors are called the marketing mix. The marketing mix includes four elements: product, price, promotion, and distribution. (See Figure 7-2.)

Marketing mix

The product element is concerned with the firm's product offering. **The product, however, includes much more than a physical product. It includes the guarantee, service, brand, package, installation, alteration, etc., which go with it. The product can best be thought of as a bundle of utilities.** A person who buys a car is not only buying steel, glass, and nuts and bolts. That person is buying satisfaction such as transportation or prestige.

Product

The price element of the marketing mix is also important. **Price is the amount of money for which the product sells.** How a firm sets this price is discussed in our next chapter. It involves setting a basic price and then adjusting it through discounts and markups to arrive at a price target customers are willing and able to pay.

Price

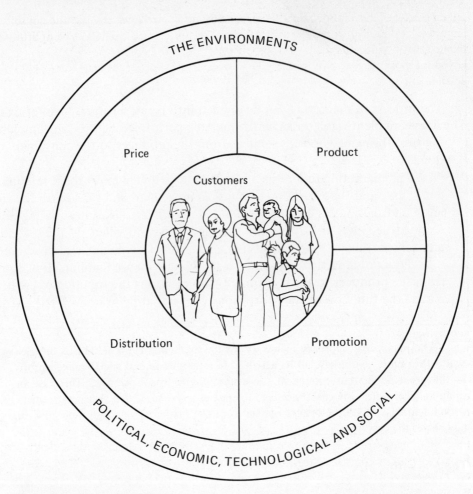

Figure 7-2. **The marketing mix**

Promotion

The third element in the mix is promotion. **Promotion is concerned with persuading people to buy the firm's products—turning people into customers.** It includes advertising, personal selling, publicity, public relations, and sales promotion. As we will see, promotion is basically concerned with a firm's efforts to communicate with its customers.

Distribution

The final element in the mix is distribution. **Distribution (place) means getting things from where they are made to where they are used.** Distribution requires transportation, warehousing, and, in many cases, middlemen.

All four elements of the marketing mix are focused on the consumer and are influenced by factors in the environment such as the society, technology, the economy, and the law.

There are many possible combinations of those four major elements. Price, for instance, might play a big role in the mix for selling fresh meat but a very small role in selling perfume. Distribution might be very important in marketing gasoline and not so important for lumber. Promotion could be vital in toy marketing and yet be of little importance in marketing nails. The product is important in every case but probably less so for toothpaste than for cars. We'll discuss the four parts of the marketing mix in detail in our next chapter.

The Target Market

If a firm is to adopt the marketing concept and design a good marketing mix, it must define the characteristics of its customers. This set of customers is called the target market. Aiming at this target guides the firm in designing its marketing mix. Often there are certain factors, such as existing investment in production facilities or experience of personnel, which restrict the kind of target at which the firm "aims". In other words, there may be a compromise between the resources already available for designing a "mix" and the setting of target markets. — Target market

When choosing a target, one thing is usually easy—deciding whether what we produce is used for its own sake or whether it is used to make something else or to provide a service. This is an important distinction.

Industrial goods are goods which will be used by a firm or an institution to make another good or to provide a service. Manufacturers, hospitals, and lawyers buy products and services for reasons which are different from those which motivate ordinary consumers to buy. — Industrial goods

Consumer goods are goods which people buy for their own use—to wear or to eat or to look at or to live in. We usually call these buyers ultimate consumers. Their motives and buying behaviour are quite different from those of industrial firms. — Consumer goods

Besides industrial buyers and ultimate consumers, firms must also consider a third kind of customer. **These are called middlemen because they usually hold goods for a while in the process of bringing them from their producer to their user. Retailers and wholesalers are examples of middlemen.** We'll study middlemen more in Chapter 8 when we see how a producer uses them in marketing a product. — Middlemen

Meanwhile, let's clarify the importance of defining the target market by comparing the industrial goods market with the consumer goods market and with other distinctive markets.

The Industrial Market

What are the features of industrial goods markets? First of all, the target market is generally smaller than it is for consumer goods. A maker of shoemaking machines has fewer customers than a candymaker or a tuna canner or a TV manufacturer.

Industrial customers are often more concentrated geographically than are household or individual customers. Many industries which are the sole users of certain products are centered in one or a few areas. The manufacture of agricultural implements, for example, is concentrated in the Hamilton-Brantford-Toronto area.

Industrial buyers are also different from household or individual buyers because they have more formal systems for buying. This is illustrated in Figure 7-3 in the purchase of an air-conditioning system by a large manufacturer.

A firm also has more clearly defined and profit-oriented purchase motives. Industrial markets can also be especially "risky" because of the dynamic nature of technology. One change in technology can cause the sudden death of many industrial products (parts, supplies, etc.) which go into the production of one newly obsolete major product. Conversion from the older type of electron tube to solid-state parts in TVs and radios hurt many small producers of the older tubes.

Industrial goods include installations, accessory equipment, supplies, raw materials, and component parts and materials. Services, of course, are also sold to industrial firms. Examples are security services, banking services, and janitorial services. These goods and services have either narrow or broad target markets. The target depends on how widely the goods or services are used in industry. Many types of supplies (stationery and fuel), accessory equipment (typewriters), and services (legal assistance) are used by nearly all firms. On the other hand, most types of major equipment or installations, raw materials, and parts have a much narrower market.

The breadth of a market for an industrial product is limited by the product's nature. Who the customers of a given firm are also depends on its location, its experience and good name, its financial strength, and the size and the strength of its distribution system.

Governmental and Institutional Markets

Goods sold to non-profit institutions such as federal, provincial, and municipal governments, hospitals, and schools are also industrial goods. Many of the same products sold to businesses are also sold to non-profit institutions. These are often like industrial firms because they use a formal, usually professional purchasing system. They draw product specifications and request bids from several suppliers.

Marketing to the federal government is a special case because of the complex purchasing system it uses. Some firms sell exclusively to the government. Large defence purchases may involve years of parliamentary debate and lobbying. Marketing for such products is in a class by itself.

The Consumer Market

Manufacturers which produce goods for ultimate consumers often face a huge, tricky target market. The "household" buyer is not as professional or formal as the industrial firm. However, the high level of income among Canadian consumers leads to a fantastic number of different products and services being bought. This affluence makes possible frequent changes in taste. These changes, together with the amazing rate of technological progress, cause a rapid "turnover" of consumer products. New products and new brands of products appear daily on retail shelves. Nearly as many soon

1. In early May workers began to complain that it was too hot in the main assembly plant.
2. In mid-May the production manager noted a decrease in productivity and further complaints about the inadequacy of the old air-conditioning system.
3. At a conference between the executive vice-president, the production manager, the comptroller, and the plant engineer on June 1, it was decided to replace the system immediately.
4. The plant engineer prepared a description of the system Sweeny Ltd. needs and a time schedule for installation. These specifications were sent to five local industrial air-conditioning contractors on June 14, with a request for installation bids and proposals.
5. Four proposals are received by July 4. Each is checked by the plant engineer, the purchasing agent, and the production manager. These three confer with the comptroller and the executive vice-president and award the contract to the Acme Company. Their bid was somewhat higher than one of the others, but their reputation for quality is very high and their service and warranty are at least equal to those of all other bidders.
6. On July 10 contracts are prepared by the purchasing agent and signed both by him and by an Acme Company representative. Work is begun the same day.

Figure 7-3. Sweeny Ltd. buys an air-conditioning system

disappear. This tougher competition for the consumer dollar represents the main challenge in marketing consumer goods. A number of different marketing strategies are used to meet the challenge.

Classes of Consumer Goods

Although many specific goods are hard to classify, it is useful to distinguish three kinds of consumer goods: convenience goods, shopping goods, and specialty goods. This classification depends on frequency of purchase, the product's significance to the buyer, and the buyer's pre-selection of a specific product brand.

Convenience goods are items bought frequently, demanded on short notice, and often purchased by habit. Cigarettes and many foods and drugs are convenience goods. These are usually low-priced goods which people don't think much about when buying. They don't make very careful comparisons. *— Convenience goods*

Shopping goods are products which are taken seriously enough to require comparison and study. Most clothing, appliances, and cars fall into this category. Gifts are almost always shopping goods. Stores which sell shopping goods are often grouped together to help customers make comparisons. *— Shopping goods*

Specialty goods are those for which strong conviction as to brand, style, or type already exists in the buyer's mind. The buyer will make a great effort to locate and purchase the specific brand. Usually such goods are high in value and aren't purchased frequently. For some customers, however, a can of soup could be a specialty good. *— Specialty goods*

> **CAN YOU EXPLAIN THIS?**
>
> **Why Are There So Many Brands of Bath Soap?**
>
> On your next trip to the supermarket, pass by the shelf which displays bath soaps. You'll see many different brands. Some are packaged in fancy foil wrappers. Notice how many different sizes there are. Some soaps are deodorant soaps, some are mild enough for babies, and some have ingredients that promise to give you youthful-looking skin.
>
> But don't we all buy bath soap for the same reason? Isn't the reason we buy soap to wash ourselves? Why are there so many brands of bath soap? CAN YOU EXPLAIN THIS?

The class depends on the individual consumer's attitudes. A certain item (a shirt, for example) could be classed in three different ways by three different people. However, there is enough agreement among consumers for most goods to make this goods classification scheme useful to firms in decision making.

How a particular firm classifies a product greatly influences the way it is sold. A manufacturer who considers a product a convenience good will want it to be sold in as many places as possible. If, however, the product is viewed as a shopping good, it is likely to be placed in stores which are near stores selling similar shopping goods. A specialty good manufacturer doesn't have to worry as much about the retail location. Since such buyers will go out of their way, the firm's distribution channel problem is simplified. These are only a few examples of how the way that consumers classify goods may affect the way they are marketed. Figure 7-4 summarizes the features of the three classes of goods.

	Convenience	*Shopping*	*Specialty*
How far will a buyer travel?	short distance	reasonable distance	long distance
How much does it cost?	usually low price	usually middle to high price	usually high price
How often purchased?	frequently	occasionally	infrequently
Emphasis on comparison?	no	yes	no
Purchased habitually?	often	never	not usually
Which advertising media?	television, newspapers, and general magazines	television, newspapers, and general magazines	special-interest magazines and catalogues

Figure 7-4. **Classes of consumer goods**

In summary, the consumer goods producer must put himself or herself in the shoes of the buyer to figure out the probable class in which most buyers will place a given product. This is an application of the marketing concept and makes it more likely that the marketing mix will be truly matched to what consumers want.

PRODUCT DIFFERENTIATION AND MARKET SEGMENTATION

Consumer knowledge is also involved in the choice of two important types of marketing strategies—product differentiation and market segmentation. **Product differentiation is a simple strategy of getting customers to think of our firm's product as different from (and better than) the competition.** It can be done by stressing distinctive product features or the product's guarantee, its service, or its availability. It can also be done by an advertising campaign which emphasizes product features or creates the impression of special product advantages. (The government, of course, regulates advertising to prevent false claims.)

Product differentiation

A strategy of product differentiation may treat customers as one general target group to be aimed at with one common marketing mix. **A strategy of market segmentation, however, calls for making a special marketing mix for a special segment of the market or several different mixes for several different segments.** The idea is that there is really more than one set of needs to be satisfied within the general market for a product. If a firm believes it will improve its market position, it might make a different version of the product to satisfy the special needs of each group or market segment.

Market segmentation

YOU MAKE THE DECISION!

To Segment or Not to Segment

The Willow Candy Company has made one candy bar under the brand name DELECTO for 25 years. Sales growth was great for 15 years, but since then there has been only little growth. Dan Glass, the marketing vice-president, thinks that market segmentation is the only way to start growing again. He proposes two kinds of candy—one bar for kids and one bar for weight-conscious working people. They would be similar except for reducing the sugar content of the second bar. The wrappers would be different colours, but they would use the same basic brand name.

Assuming that unit costs would not increase much and that retail prices would remain the same, would you agree to this market segmentation? YOU MAKE THE DECISION!

Consider the case of a producer of wristwatches. Such a firm might design one marketing mix to satisfy the "jewelry" watch market and another to satisfy the "time-telling" watch market. There might be important differences in product design—perhaps gold cases with gems for one and waterproof steel cases for the other. Advertising themes for one would emphasize romance or prestige, while the other might emphasize accuracy, durability, and price. They might even use two different sets of retail stores. The expensive ones might be sold in jewelry stores and the cheaper ones in drug and variety stores. Deciding which of these means of segmentation to use requires a study of the market and the cost of segmenting.

Other bases of segmenting markets for certain products might be age, sex, income, educational level, or personality traits. At the broadest level, segmentation can be done on a cultural or language level. In Canada, for example, there are significant differences in consumer attitudes and behaviour between English- and French-speaking Canadians, and prudent marketers must take these differences into account in order to be effective. Segmentation by culture or language is also practised by many businesses with respect to certain regions of the country and to geographic concentrations of various ethnic groups.

At the other extreme, segmentation would provide a "custom" design for each buyer. In this case, each individual customer is a "segment". Profitable segmentation requires good consumer knowledge on the part of the seller.

KNOWING THE CONSUMER

Marketing research

Intelligent decisions about marketing strategies—whether they relate to product, price, promotion, or channel of distribution—require a clear understanding of those who are or might become customers. This kind of decision calls for marketing research. **Marketing research means applying methods of science to marketing problems. These techniques are largely directed at the people who make up the market.**

Two general approaches to the study of the market are the "demographic" approach and the "behavioural" approach. The former collects facts about people, families, or firms who are thought likely to be customers. It concentrates on counting and tabulating. The "behavioural" approach uses ideas from psychology about human behaviour and attitudes. It seeks to understand why buyers feel and behave the way they do.

The demographic features of a market of ultimate customers are such things as the age, sex, race, and income of its members. Often, a firm makes assumptions about the relationship between these characteristics and the likelihood of a person buying a product. These assumptions may be based on past knowledge about customers. For example, we might assume that only persons over age 60 are interested in a certain health tonic. We might also assume that only persons with incomes of $40,000 or more will be interested in a trip around the world.

Secondary research

Assume that a firm decided which characteristics are related to consumption of its product. Now it seeks information about the number of such prospects in the area to be served. **It may find this information in Canadian census publications or other government or private sources. This is called secondary research.**

> ## THINK ABOUT IT!
> ### The Quebec Market
>
> During the last fifteen years a number of studies have compared the behaviour and attitudes of French and English Canadians. Some of the interesting findings are as follows:
>
> 1. Quebec has the highest per capita sales of sweets in Canada—soft drinks, syrup, etc.
> 2. French Canadians drink very little lager, whereas English Canadians drink both ale and lager.
> 3. The French-Canadian woman is more involved with home and family than is her English-Canadian counterpart.
> 4. Quebec leads all provinces in per capita expenditures on clothing and cosmetics.
> 5. French Canadians drink three times more fresh orange juice than the rest of Canada, and eat less frozen food.
> 6. The French-Canadian woman is less oriented towards convenience food than her English counterpart.
>
> What implications do these differences have for marketing? THINK ABOUT IT!

The firm might also do some primary research. **Primary research is getting new facts for a specific purpose.** For example, the Swiss Clock Company may include in each package of its product a "registration" card. They request the buyer to fill in facts about herself or himself and mail it in to the company. These cards tell the firm about the people who buy their clocks.

Primary research

Instead of finding out who the customers are, what they buy, when they buy, and so on, the behavioural approach asks the question, "why do they buy?" This approach assumes that what people buy often depends on complex motives which can be understood only by psychological probing. Experts in human motivation test a sample of people to find out the basis for their product choice. The researcher might try to find, for example, what a particular brand name "means" to certain people. These researchers often use techniques borrowed from psychologists to discover motives and/or attitudes which customers might ordinarily try to hide.

Regardless of the approach, market research must be undertaken on a continuous basis. You need only look at the changes in products for sale in the last five years to see how dynamic the market is. Consumer tastes and values change so fast that a firm must keep its eyes open to the future. This requires market research.

PATTERNS OF CONSUMPTION

In the last thirty years in Canada we have witnessed wide swings in attitudes towards discretionary spending (spending beyond basic needs). **For some people there has been a lot of conspicuous consumption. This means spending in a visible way so that your neighbours will be aware of your wealth and "good taste".** This kind of "show-off"

Conspicuous consumption

buying often has been widespread among people when they first have extra buying power. They're trying to tell others that they have "made it".

A large number of our younger adults in the late 1960s reacted strongly to "conspicuous consumption". Many rejected the "middle-class" consumption pattern. Of course, this rejection was not limited to the kinds of products purchased. It included the entire life style, the politics, the religion, and most of the values of their parents. This rejection caused many people of all ages to question the values of the majority, including the value of high-level consumption. These doubts are still alive.

Even stronger doubts about high-level consumption and the accompanying waste were caused by the oil crisis. It began to make people wonder about the good sense of "gas-guzzling" cars and other wasteful practises we have accepted because of wealth. If oil is running out, what about our other scarce resources? Does our "throw-away" mentality make sense? Will Canadians really have to reduce their scale of living? Some believe that businesses may have to engage in demarketing—marketing activities aimed at making adjustments to shorter supplies of certain materials.

CONSUMERISM

Consumerism

Consumerism is a movement to strengthen the power of product users in relation to the power of product makers and sellers. This movement is still strong and centers around the idea of certain basic consumer rights—the right to choose, the right to be informed, the right to be heard, and the right to safety. (See Figure 7-5.)

Throughout the 1960s and into the 1970s, the consumer movement grew and began to include an ever-larger list of objectives. Leadership in the U.S. movement has been provided primarily by Ralph Nader, who first became well-known when he wrote a book critical of General Motors and the Corvair in 1966. By pointing out the defects in the Corvair, Nader succeeded in gathering support for the consumerism movement throughout the United States and Canada. The news media have been an important aid in making the consumerism movement well-known.

In Canada, the consumerism movement has resulted in many laws being passed to protect consumers. The Consumers Association of Canada (CAC) carries on many of the same activities as the Consumers Union in the U.S. (see second box on page 182).

With the appearance of the energy crisis and economic difficulties of the mid-1970s, there has been some weakening, or at least a shift in emphasis, in the consumerism movement. Certainly auto emission control has taken a back seat to better mileage. More emphasis has also been given to protection of consumers from prices which are higher than they should be—especially in the area of health care.

In any case, the consumerism movement is still alive and well. It still has some very important meaning to marketers of products and services. Reaction to the movement by firms has varied a lot. Some view it as a threat to free enterprise. Some see it as a passing annoyance which can be solved by an extra effort in public relations and lobbying. Many, however, see it as a true reflection of customer anger over abuses of market power by firms. Such firms have taken steps to correct the problem and to bring greater credibility to the marketing concept.

THE RIGHT TO CHOOSE from an adequate number of products and brands

THE RIGHT TO BE INFORMED of all the important facts about the product or service (price, durability, health, and safety hazards, etc.)

THE RIGHT TO BE HEARD by producers and government when treated unfairly or when a question or complaint arises

THE RIGHT TO SAFETY in the use of all products and services

Figure 7-5. The basic rights of consumers

TWO POINTS OF VIEW

Consumer Protection

The Firm:
This consumer movement is going to ruin us! We just get our new model on the market and bang! The self-appointed guardians of the people are starting a campaign against us. They're writing to Consumer and Corporate Affairs, and they're even giving TV interviews saying that the ZINGER is a death trap and that it has a faulty braking system.

We market-researched that car and know we've built in the features which customers said they wanted—super styling, bucket seats, stereo system—the works. It's got all the latest accessories, too. I call that real consumer responsiveness! We have been pioneers in applying the marketing concept.

The Consumer Advocate:
The marketplace does not provide consumers with guarantees to protect them from unscrupulous manufacturers like the ZINGER Company. The average consumer doesn't know enough about technical products like cars to know if he or she is getting gypped. All it takes is a high-pressure ad campaign and a lot of superficial gadgets to convince him or her to buy.

Consumers need protection. They need product safety codes with heavy penalties for those firms who don't follow them. At present, only the consumer movement is around to prevent consumers from making dumb mistakes. Without us they are almost helpless against the marketing skills of the giant corporations. Corporations don't care who gets hurt as long as they make their profit.

CAN YOU EXPLAIN THIS?

Private Organizations and Consumerism

The Consumers Association of Canada (CAC) is a private consumer organization. Its bi-monthly magazine, *Canadian Consumer,* gives information on product tests which are conducted by the CAC. The organization originally received a $415,000 federal grant in 1966 to finance its operations. However, it has experienced a chronic shortage of funds, and is currently appealing for funds to the general public, on whose behalf it has been working for many years. Why has this happened? CAN YOU EXPLAIN THIS?

The Gillette Company, for example, has appointed a vice-president for product integrity. This officer is directly responsible for the safety and quality of Gillette's more than 800 products. He has power to remove them from the market any time they fail to meet standards he sets. He can also modify advertising claims or prevent new product introduction. In 1973 this executive caused the recall of more than $1 million of anti-perspirants because there was a question about the safety of one of the ingredients.

Many firms have taken similar steps which do much more than "pay lip service" to consumerism. This, they believe, does not hurt the long-run profit of their firms. Rather, they wonder about the survival of firms which do not react positively to public complaints and thereby bring on oppressive government controls.

MARKETING OF SERVICES

Employment in services has been growing at a rapid rate in recent years. (See Figure 7-6.) A service is like a product in that it provides a benefit, but it is unlike a product in that it is not a concrete, physical object. Services represent a big part of what consumers and business firms buy. A service may be quite personal, such as the service provided by a physician or beautician. It can be impersonal as well. Banks, insurance agents, and people who do repair work provide impersonal services in the sense that the human tie between the producer and consumer is not close. In any case, it is usually hard to separate the service from the person who provides it. It almost always

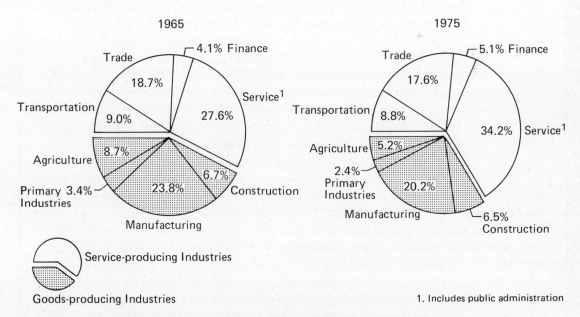

Figure 7-6. Employment by industrial group, 1965 and 1975. (Source: *Canada Year Book*, p. 375. Reproduced by permission of the Minister of Supply and Services Canada.)

B CAREERS IN RETAILING AND WHOLESALING

Like most career areas, retailing offers opportunity at many levels and in a variety of jobs. Most people enter retailing as a floor salesperson or as a worker in the stockroom. Usually this does not require college work or even a high school diploma. As a matter of fact, distributive education programs in many cities allow high school students to work in retailing while they study related courses. Similar programs also exist in community colleges where students train more specifically for higher positions. A person from a community college may become a suburban sales manager, an assistant buyer, a buyer, or a credit manager. Many four-year university graduates begin their retail careers in selling. They often spend short periods in several different sales departments and store branches during their time as trainees.

Retail selling requires skills in customer relations, simple mathematics, an eye for merchandise and display, and a good memory for inventory information. It also requires physical endurance to stand for long periods and to cope with hectic "sales".

Branch suburban managers require supervisory skills and a willingness to work long hours. They usually do not have buying responsibilities but must be skilled in customer relations. Assistant buyers and buyers must understand merchandising and be able to spot changes in customer tastes. They must be skilful negotiators.

Higher education is not essential for promotion to buyer or executive positions, but competition for such positions is great and the more education you have, the better your chances are. A university graduate with a marketing or retailing specialty is especially likely to be promoted. For information, write the Retail Council of Canada, Ste. 525, 74 Victoria Street, Toronto, Ontario M5C 2A5.

Wholesaling also offers opportunities in selling. A high school diploma is usually required, although a community college or university background is extremely helpful for success and promotion, especially where the product is technical in nature. Without some college training a person will not begin in wholesaling as a salesperson, but rather as a clerk, a stockperson, or a driver.

Wholesale selling involves learning the product line, the competitors' lines, and the needs of potential customers in your territory. You will spend a lot of time in the field, working alone. It can be physically and emotionally demanding. Often the wholesale salesperson spends many hours driving, trying to make contacts, entertaining clients, and making out reports.

A wholesale salesperson may be "promoted" by being assigned to territories with less travel requirement or higher sales potential. Successful salespersons may become local, regional, or general sales managers or may move into buying, credit work, or public relations for their firms. Information is available from the National Association of Wholesale Distributors, 1725 K Street, N.W., Washington, D.C. 20006.

requires that the producer be a specialist. Guy Cafleur and Dave Cutler provide the service of entertainment in their respective sports. They are highly paid because they are specialists. The same is true of a heart surgeon or an actress. Insurance agents are specialists, too, as are plumbers and electricians.

Since we have defined the product part of a marketing mix in terms of the "bundle" of services it provides, it follows that marketing of services is not very different from marketing a product. There are a few differences, though. Perhaps the biggest difference is the simple fact that most service producers have not paid as much attention to marketing as they should. Only recently have bankers, for example, realized the importance of the marketing concept. They have begun to think in terms of attracting and pleasing customers and of differentiating their services.

Legal services are limited in the use of promotion by their ethical codes. However, there is often a need for developing the marketing mix by improving the human dimension of the service (product) and by improving access to lawyer offices and the courts (distribution). Pricing, in the competitive sense, has been discouraged by ethical codes in medicine, law, and other professions. However, for some specialists, high fees are viewed as a positive marketing factor. The whole question of lack of price competition in the professions is hotly debated these days and may be changing.

For most services all four parts of the marketing mix—product, distribution, promotion, and price—should be developed. Success in marketing services, even more than in marketing tangible products, depends upon knowing the buyer well and serving his or her needs. This is especially true for personal services where success means treating each client as an individual market segment.

For many services the distribution or "place" part of marketing is a little different. Usually the producer and distributor are the same. In most cases there is no "channel" of distribution as we normally think of it. Yet, the location of the bank branch or the watch repair shop or even the doctor's office must be considered carefully. A successful TV repair shop often provides home pickup and delivery. A marketer of rented apartments in a distant suburb might offer private commuter bus service to tenants. This is just another way of "distributing a product".

The marketing of services has a long way to go, however. The sooner service producers realize that the marketing concept applies to their "product", the better off the consumer will be.

SUMMARY AND LOOK AHEAD

Marketing is a very large part of the economic activity in Canada. Its role is becoming central to business planning as firms adopt "the marketing concept". The reason for this is that business profit and the national rate of economic growth depend on the growth of consumption. Businesses must learn more about their customers and concentrate on satisfying them if they are to succeed.

It is convenient to think of a set of customers as a "target market". This target might consist of an industrial market or of a market of ultimate consumers. Consumer goods can be classed as convenience goods, shopping goods, and specialty goods.

The kind of good and the consumer help to define the "marketing mix" or combination of product, distribution, promotion, and price a firm will use. A successful marketing mix depends on consumer knowledge gained from market research.

Successful marketing also requires that firms pay attention to spending behaviour such as "conspicuous consumption". This includes a positive response to the spirit of consumerism. If firms reject consumerism, it could possibly trigger widespread attacks on our marketing system.

In the next chapter, we'll show how business firms can approach the problem of designing a marketing mix and the policies which accompany such a mix.

KEY CONCEPTS

Conspicuous consumption Spending in a visible way so that neighbours will be aware of the wealth and "good taste" of the spender.

Consumer goods Goods or services provided for the consumption of ultimate consumers, as opposed to firms.

Consumerism Any action taken or legislation enacted showing concern for the welfare of consumers.

Convenience goods Items bought frequently, demanded on short notice, and often habitually purchased, like cigarettes or gasoline.

Discretionary income The amount of income a household has left after paying taxes and making expenditures for necessary goods and services.

Distribution An element in the marketing mix concerned with the movement of goods through a channel from producer to consumer.

Form utility Goods become more useful because form has been changed or separate parts are combined. Value has been added.

Industrial goods Goods sold to business firms, such as installations, accessory equipment, component parts and materials, supplies, and raw materials. Industrial services are sometimes included.

Managerial approach to marketing An approach to marketing which views marketing decision making as the central problem.

Marketing A set of activities to find, build, and serve markets for products and services.

Marketing concept Guiding philosophy of a firm which says a firm must be consumer-oriented and the entire firm must recognize this fact in order to make a profit.

Marketing mix A certain combination of the elements of price, product, promotion, and place (or channel of distribution) manipulated by a firm in order to achieve its marketing objectives.

Marketing research The systematic gathering, recording, and analyzing of data concerning the marketing of goods and services.

Market segmentation Designing a special marketing mix for a special segment of the market or several mixes for several different segments.

Middlemen Firms which participate in buying and selling goods as part of a channel of distribution. Middlemen who take title are merchant middlemen. Those who do not are agent middlemen.

Ownership utility The actual transfer of title to a product creates the utility of ownership.

Place utility Increase in the value of a good by changing its location.

Price An element in the marketing mix which relates to monetary cost and the terms of sale of a product or service.

Primary research Obtaining original information for a specific research objective.

Product An element in the marketing mix consisting of the entire "bundle of services" made available for sale. It includes the package, brand, and physical characteristics.

Product differentiation Making products or services appear different from those of the competitors, thereby attracting buyers or commanding higher prices.

Promotion All communication a firm has with its customers or its potential customers for the purpose of expanding sales, directly or indirectly.

Secondary research Research derived from sources other than those provided personally by the researcher. Library research is an example.

Shopping goods Consumer goods whose purchase is taken seriously enough to require comparison and deliberation.

Specialty goods Goods for which strong conviction as to brand, style, or type already exists in the buyer's mind and for which he or she will make a great effort to locate and purchase a specific brand.

Target market The set of customers to which a firm directs its attention when it designs its marketing mix and the policies associated with it.

Time utility Increase in value resulting from the passage of time. To a family of two, a dozen eggs is worth more over a period of time than all at once.

QUESTIONS FOR DISCUSSION

1. Why is the marketing task more complex in an economy where there are high levels of discretionary income than in one where the people live at the subsistence level?
2. Do you think that high mass consumption is desirable? Why or why not?
3. How is the marketing concept related to the economic problem we discussed in Section 1?
4. If a firm adopts and implements the marketing concept, all of its actions are oriented to the satisfaction of its target market. Is that desirable? Discuss.
5. If a product possesses only form utility, is it useful to a customer? Why or why not?
6. Why is there a demand for industrial goods?
7. What is product differentiation? Give an example.
8. Why is market research necessary?

9. Does marketing activity induce or persuade people to want things they do not really need? Explain.
10. Two divergent views of the proper relationship between buyer and seller are "let the buyer beware" and "let the seller beware". Which is the "proper" view? Why?
11. Why is a consumerism movement under way in Canada when consumer sovereignty is a basic characteristic of capitalism?

INCIDENT:

Bentham

The town of Bentham, Saskatchewan, was selected for the location of a new, large factory which employs 500 people in high-paying jobs. Before the plant was built Bentham had only 2,200 citizens, 500 of whom were employed in low-paying, farm-related jobs and retailing. The new prosperity brought many new retail stores and a whole new urban middle-class subdivision to the previously sleepy farm town.

Questions:
1. What do you think happened to the pattern of consumption in Bentham?
2. Explain how the idea of "conspicuous consumption" might fit into the changes in the way of life in Bentham.
3. Could the idea of "consumer sovereignty" explain the new kinds of retail stores which spring up from time to time? Why?

INCIDENT:

Weatherford Drug Store

Right after Roger Weatherford got his degree in pharmacy, he opened the Weatherford Drug Store in Halifax. He had taken a course in marketing while he was in university, so he was especially concerned about the kind of marketing mix his store should have.

Questions:
1. How can he define his target market?
2. Which elements in the marketing mix do you think are the most important for Weatherford Drug Store? Why?
3. Would your answer above be different for a gasoline service station? Why?

8
Marketing Decisions

SECTION THREE BUSINESS DECISIONS

OBJECTIVES: After reading this chapter, you should be able to:

1. Identify the bundle of services offered by a product to its user.
2. Draw a chart illustrating the life cycle of a product.
3. Present arguments for and against a broad product mix.
4. Explain the functions which a package performs.
5. Draw a chart which illustrates how a middleman may bring about economies in distribution.
6. Distinguish between an integrated and a non-integrated channel of distribution.
7. Provide an illustration of the total cost concept.
8. Describe ideal distribution task conditions for each of the major modes of transport.
9. Describe how an advertising agency serves a large seller of consumer goods.
10. Compare the advantages of the various media.
11. Act out with another student an example of personal selling as it should be done.
12. Show the difference between the cost and the demand approaches to pricing in a small dress shop.
13. Describe two possible strategies for introducing a new product.

KEY CONCEPTS: In reading the chapter, look for and understand these terms:

PRODUCT LIFE CYCLE
PLANNED OBSOLESCENCE
PRODUCT MIX
BRAND
PATENT
TRADEMARK
CHANNEL OF
   DISTRIBUTION
MANUFACTURER'S
   AGENT
FRANCHISED RETAILER
PHYSICAL DISTRIBUTION
TOTAL COST CONCEPT
COMMON CARRIER
CONTRACT CARRIER
PRIVATE CARRIER
CONTAINERIZATION
ADVERTISING AGENCY
ADVERTISING MEDIA
AIDA PROCESS

PERSONAL SELLING
PUBLICITY
PUBLIC RELATIONS
SALES PROMOTION
OLIGOPOLY
PRICE LEADERSHIP
MONOPOLISTIC
   COMPETITION
MARKUP
PRICING MODEL
TRADE POSITION
   DISCOUNT
FUNCTIONAL DISCOUNT
MARKET PENETRATION
   PRICING
SKIMMING PRICING
INVENTORY TURNOVER
   RATE
PRICE LINING

In this chapter we will expand our discussion of the four elements in the marketing mix. In its broadest sense, the product is the basis of any marketing program. The flow of products to customers requires a distribution system which is often complex, and relationships between participants in the channel of distribution are of critical importance to that system. Promotion stimulates this flow, and price often interacts with promotion, distribution, and product in determining marketing success.

PRODUCT

Decisions about what and how much to produce represent the first step in the product-planning process. Good product planning requires coordination with other marketing mix elements and strong customer orientation. The relationship between a firm and its customers focuses mainly on the product (or service). A product is a "bundle of services", which might include a variety of things. For example, when a mother buys a sweater for her baby at a department store, she buys warmth and comfort. She is also buying assurance that the product will last and the right to return it if it does not fit. When a man buys a cartridge of razor blades, he is buying comfortable shaves and convenient blade replacement. Buyers of ice cream think of the good taste and nutrition they are buying. This is the product. It is very important for a firm to know what "bundle of services" customers expect from its product offering.

A retailer's "bundle of services" includes those services which might be provided by each of the products sold in the store. It also includes the convenience of the location, the parking facilities, credit, assortment, skill of the salespeople, returned merchandise policy, and all those things which may attract customers. The enjoyment of shopping itself is for many customers a very important part of the "bundle of services".

The Product Life Cycle

The "life history" of a product is called the product life cycle. It is especially easy to see in the case of appliances and fashion goods.

Product life cycle

Figure 8-1 indicates a typical life cycle for an appliance. Notice that the sales volume is very low during the introductory phase of the cycle. There is a steep increase in sales volume during the growth phase, but it begins to level off in the maturity phase, the longest part of the cycle. It is hard to tell how long the maturity phase will be, because we can't predict technological change. The maturity phase ends when a better product appears and the old product enters the period of decline. The console radio was at one time the chief source of entertainment for millions of Canadians. Today it has been replaced by the television and by other home entertainment devices.

Although product classes have life cycles, individual brands of products have life cycles, too. The brand's lifetime is shorter than that of the product class. People will

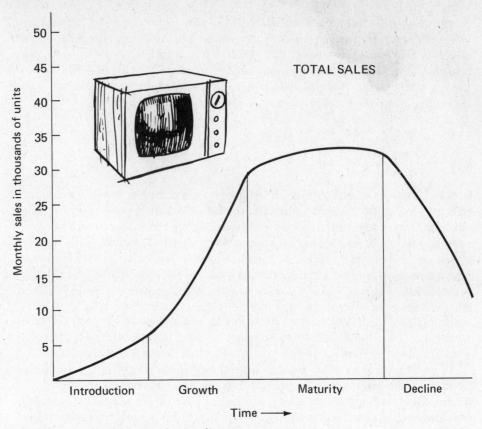

Figure 8-1. The life cycle of an appliance

continue to use dish detergents, for example, much longer than they will use "Sudso". The makers of "Sudso" will try to extend the maturity phase of its cycle as long as it is a profitable item. They can do this by making small changes in "Sudso", such as adding purple bleach crystals. If sales continue to decline, then maybe it is time to "phase out" "Sudso" from the product line and concentrate on a new brand with a different sales feature. A good example of a life-cycle extension strategy is baking soda. Despite the decreased use of the product for baking purposes, sales have continued to rise as new uses and new users for the product have been found. New uses include use as a deodorizer for refrigerators and "kitty litter" boxes.

When a product is no longer purchased, it is obsolete. Obsolescence is usually the result of a change in consumer tastes or in technology. A firm spends money on research and development (R&D) to improve its product so that it won't become obsolete. Sometimes R&D leads to an entirely new product, which takes the place of the old product, making it obsolete. **When a firm intends to replace its products, it is called planned obsolescence.** When it occurs because of new technical features, it is technological obsolescence. (See Figure 8-2.) Where the appearance or style is changed,

Planned obsolescence

Figure 8-2. *The change from mechanical technology to electronics has revolutionized the business machine market. These units, spanning sixty years of progress, suggest the importance of research, engineering and market analysis in avoiding product obsolescence.* (Courtesy of Burroughs Corporation.)

it is fashion obsolescence. Promotional activities help to create this type of obsolescence by making customers dissatisfied with "old" products.

One of the most critical and delicate marketing tasks occurs when a firm introduces a new product. Only one out of perhaps twenty new products is successful. Most firms try to avoid such expensive failures by using marketing research. This reduces, but doesn't by any means prevent, the chance of loss.

The success of a new product depends on its ability to perform a service customers will buy at a price which more than covers cost. It also depends on good introductory promotion and strong distribution. Chevrolet's huge dealership organization was extremely helpful in introducing the subcompact Chevette. The competitive advantages of Chevrolet did not stop them from doing a lot of consumer research before introducing the Chevette. They wanted to ensure a long life cycle by getting the product mix just right.

Services also require innovation to succeed. Recently, Japan Air Lines introduced "Sky Sleeper Service" on many transpacific flights. For an additional charge, passengers from New York to Tokyo may have the use of one of the five beds installed in the airline's Boeing 747's. This is a significant improvement in long-distance air travel for many customers.

The Product Mix

Product mix

A manufacturer's or a retailer's product mix is the combination of products it produces or sells. General Foods produces hundreds, while Coca-Cola produces a much smaller number. The size of the product mix affects marketing policy.

First of all, there is safety in numbers. A firm with a broad product mix has a kind of insurance against the dangers of obsolescence. Also, economies of scale often make the difference between success and failure. A firm with many products can spread its "overhead" cost over the entire product mix. This means savings in production costs if the products are manufactured in the same factory. It also cuts unit distribution cost. A firm can save on distribution costs by using the same salespeople or transportation system for all products.

At the retail level and, to some extent, at the wholesale level, firms with many products have an advantage in the form of product exposure. When a shopper goes into a store to buy shaving cream, he or she will see razors and hair conditioners. The shopper may become a regular customer for hair conditioner as well as for shaving cream. If a store sells only luggage, this sort of thing cannot happen.

There are advantages, sometimes, to having a single product, or a "narrow" product mix. A firm with a narrow product mix can be more easily promoted as a specialist than a firm that produces many kinds of products. On the other hand, it is usually the multi-product firm which has the competitive advantages. The users of each of its products may hear about, and become interested in, the other products made by the same firm. This is most likely if the products all bear the same brand.

Packaging

All the elements which constitute the broad concept of "product" must be considered in developing the product mix. Among these is the package. In recent years, packaging has become a more important part of product policy. Packaging does several things. It protects the product. It also divides the product into convenient units, such as a six-pack of cola. The package is often a part of the product itself. Razor blade dispensers and deodorant aerosol spray cans are good examples. Finally, it helps in promoting. The distinctive, identifiable package plays a major part in advertising and self-service retailing. A good package is a silent kind of salesperson.

Brands and Labels

Brand

A brand is "a name, term, symbol, or design, or a combination of them, which is intended to identify the goods or services of one seller or group of sellers and to differentiate them from those of competitors."[1] Brands usually include both a name

---

[1] Committee on Definitions, Ralph S. Alexander, Chairperson, *Marketing Definitions: A Glossary of Marketing Terms* (Chicago: American Marketing Association, 1960), p. 8

and a symbol. The key to successful branding is making a lasting impression in customers' minds. A good brand name such as Bic or Seven-Up is distinctive and easy to remember. A manufacturer's advertising program achieves utmost success when the brand becomes a "household word" like Scotch tape.

Brands also play a role in the marketing strategy of wholesalers and retailers. Brands developed by such middlemen are called distributor's brands or private brands. Sears' Kenmore appliances and Craftsman tools are examples. They are produced by other firms for Sears. Large grocery chains do the same thing. They generally make a larger profit per unit on private brands.

Many major retailers of grocery and related products have begun to sell unbranded products, sometimes called "generics". These products, such as paper towels, dog food, green beans, and laundry detergents, are priced even below private brands and may have some features which make them less attractive than manufacturer's or private brand items. String beans may be uneven or peas may not be uniform in generic packages, although the nutrition is equal to that of branded items.

YOU MAKE THE DECISION!

Convenience vs. Ecology

Fairchild Bottling Company is a leading soft drink bottler in a Western province. You are the president and a major stockholder. Until 1965 you used exclusively returnable bottles in three sizes. At that time there were loud complaints by large supermarket chains and slight indication of consumer concern over the inconvenience of returning bottles. Along with many bottlers, Fairchild switched to "one-way" bottles and increased average prices by 10 per cent. Sales were unaffected except that one large grocery chain, which had dropped the product earlier, picked it up again.

The ecology movement in the province has made many people, especially the young, very conscious of litter in the streets and the parks and beaches. Environmental groups have made many public demonstrations against "one-way" bottles. They have succeeded to the extent that Fairchild still receives about 100 letters a week complaining about the problem Also, for the last five years a private member's bill to ban non-returnable bottles has been introduced in the legislature, and there is concern that the party in power may support the bill this year.

Assuming that Fairchild's cost of converting back to returnables would be about the same as the long-run savings in bottle cost, what would you do? Is the possible loss of some retail outlets outweighed by the expected gain of public support and the threat of government-enforced conversion later on? As president, you make the decision.

Patent

Legal protection is available for products and brands. This is especially important for small firms which cannot spend a lot on advertising. **A patent protects an invention, a chemical formula, or a new way of doing something from imitation.** It makes it very hard for a competitor to copy this new product or new idea for a period of seventeen years. The commissioner of patents (Department of the Secretary of State) administers this activity. It accepts applications for patents, and if an idea is "patentable" it is registered and protected. **The patent office also protects a name or symbol which, when registered, is called a trademark.** Brut cologne and Dream Whip are examples of trademarks. They are also, of course, brands.

Trademark

One of consumerism's goals is the informative labelling of products. This would help a buyer to make a more informed choice among products—particularly when self-service is involved. The label could describe in simple terms the content, nutrition, durability, precautions, and other special features of a product. It might also indicate the grade or standard of quality of the item as established by industry-wide agreement or by law.

Informative labelling can lead to better purchase decisions only if the labels are read. For example, how many consumers bother to read the list of ingredients found on boxes of breakfast food? For those who do read them, how do the items listed affect their purchasing behaviour? Some consumers want a massive consumer education program to teach buyers the benefits of using label information.

Much of a firm's product policy is governed by its decision to segment or not to segment the market. As explained earlier, it is often wise to design special marketing mixes for each segment. The product is usually at the centre of a segmentation strategy.

DISTRIBUTION

Both time and space separate a manufacturer from its customers. These "gaps" occur as a result of differences in production rates and consumption rates and because consumers are more widely scattered than producers. A firm compensates for these gaps by means of a channel of distribution.

What Is a Channel of Distribution?

Channel of distribution

A channel of distribution is the firm or firms directly involved in getting the product from the manufacturer to the consumer. Channels also make up for the difference between one manufacturer's product mix and the product mix a consumer wants. Since these two mixes rarely match, other firms are often needed to complete the marketing process. These firms, except for the producer, are usually called middlemen. The type and number of such middlemen describe the typical distribution channel. (See Figure 8-3.)

If a manufacturer of a consumer good were completely customer-oriented and cost were no object, it would probably send out salespeople to call directly on every

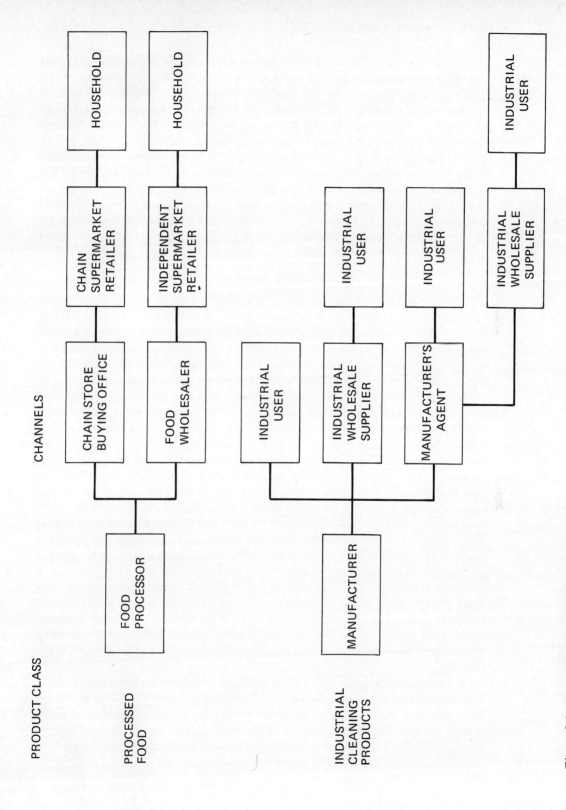

Figure 8-3. Typical channels of distribution

household. The cost of such careful attention to customers is almost always too great. The final decision about how to get products to customers requires a compromise between cost and providing the best service and convenience.

The final development of a channel depends on (1) the number of customers; (2) the functions which the channel is expected to perform; (3) the costs of alternative channels; and (4) the importance of controlling the marketing process. Thus, a maker of lathes selling to a handful of industrial firms may sell directly to customers. A maker of toothbrushes selling to millions of customers, by contrast, will need a long (several middlemen) channel. A firm making several different appliances with high unit profit margins, like Singer, can operate its own retail stores because the overhead cost is spread over the several products. A manufacturer of blankets could never afford to do so. A gasoline manufacturer which wants to treat consumers with great care is likely to operate its own retail stores.

The Principal Middlemen

Middlemen can be classified as merchant and agent middlemen. Merchant middlemen actually take title to the goods they sell, while agent middlemen do not. A third class of middlemen is the facilitating middleman who simply participates in the transportation and storage of the product without actually buying and selling it.

Examples of merchant middlemen include wholesalers and retailers. These firms regularly buy stocks of goods and resell them. Retailers are involved in selling a wide variety of consumer goods. Food, for example, is sold mostly through food and grocery retailers known as supermarkets, some of which are parts of large chains (Loblaws, Safeway, Dominion), and some of which are independent. Independent supermarkets usually buy from independent wholesalers, while chain stores receive goods from their own firm's central distribution points, which perform the wholesaling function. The top part of Figure 8-3 shows a typical channel of distribution used by a processor of a food product. This firm sells to a central buying office of a large food chain which, in turn, distributes to its retail stores. These, of course, sell to households. Other important retailer types include department stores, drugstores, variety stores, discount houses, and vending machine operators.

Manufacturer's agent

Agent middlemen include manufacturer's agents, brokers, and selling agents. These firms are involved in selling a wide variety of products, including consumer goods and industrial goods. **A manufacturer's agent is paid a commission to represent manufacturers of several non-competitive lines in a limited geographic territory. Without taking possession of goods, such an agent aggressively seeks to establish these products in this territory.**

A manufacturer's agent may help a producer of industrial cleaning products to introduce the line in British Columbia. Since the agent gets paid a commission only for the actual sales made, this can be a more efficient way of distributing in a new territory, rather than dealing directly or through wholesale industrial suppliers. The bottom part of Figure 8-3 shows the "set" of channels which a manufacturer of industrial cleaners might use. In some cases, this firm sells directly to large industrial users, using its own sales force. In other cases, the producer sells to industrial suppliers (wholesalers) who sell to industrial users (usually the smaller customers). Another method is to use the manufacturer's agent to reach either users or wholesale industrial suppliers. It is common for a national manufacturer to use different channels of distribution in different parts of the country.

The more direct a channel is, the more control a manufacturer has over its distribution. Control is improved by means of franchising. **A franchised retailer is tied closely by contract to a manufacturer, and its operations are strictly supervised. Examples of franchised outlets include McDonald's and other fast-food firms, as well as auto dealers and many gasoline stations.** Franchising is discussed in detail in Chapter 15.

Franchised retailer

Facilitating middlemen are not pictured in Figure 8-3, although they may be involved in each case. These are railroads, warehouse companies, insurers, and other firms which facilitate or help in distribution but are not directly engaged in buying or selling.

Are Middlemen Necessary?

At some time you might have heard a friend say that she got a great bargain from a store because she bought it "direct from the factory" or "eliminated the middleman". Your friend assumed that the price was reduced by the profit which would have gone to the middleman. Such a claim would lead you to believe that middleman are unnecessary and expensive and should be abolished. Let's examine this claim.

The Glammer Company is a hardware wholesaler. Glammer buys saws, nails, electrical fixtures, adhesives, and camping equipment from five different manufacturers. Glammer buys a rail carload of each product each month. It sells most of these five products to 50 retail hardware stores in southern Ontario. The retailers each make small purchases of some of the five products every week or two from Glammer. The hardware retailers also buy from three to six other wholesalers in order to keep a complete inventory.

Why don't the retailers buy all the items they carry direct from the manufacturers? The answer is clear—they save money by buying through the wholesaler. The Glammer Company simplifies the number of purchases and sales and reduces unit shopping costs and record keeping. It provides quicker delivery to the retailers than they could get by dealing directly with the manufacturers. Figure 8-4 illustrates this principle. Each connecting line represents a transaction. Clearly, shipping in carload lots is much cheaper than the many small shipments which would be needed for direct sales from manufacturers to retailers.

There are great economic advantages in a system of distribution which uses a wholesaler. This middleman (1) buys in large quantities and sells in small quantities; (2) makes it possible for the retailer to simplify its buying process by carrying a broad line of hardware items; (3) often takes credit risks which manufacturers might not accept; and (4) guarantees delivery on short notice so that retailers need not keep large

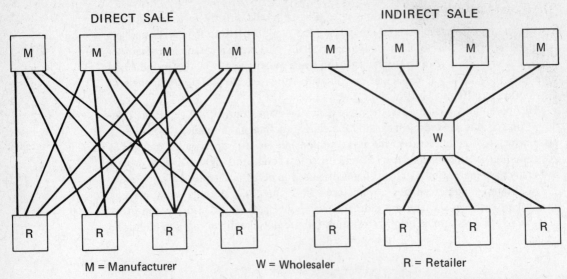

Figure 8-4. *In the absence of the wholesaler, each retailer would find it necessary to deal with all four manufacturers. Each manufacturer would also have to deal with all four retailers. With the introduction of the wholesaler, each retailer and each manufacturer need deal with only one intermediary—the wholesaler. Thus, the wholesaler reduces the number of transactions (represented by the connecting lines) and makes it possible to increase the average size of transactions. This increase brings about economies of scale and thereby justifies the existence and the profit margin of the middleman.*

stocks. In most cases, wholesalers and their margins (the difference between the costs of goods to them and what they sell them for) are justified.

If your friend got a real bargain from the retailer who "eliminated the middleman", she probably made the purchase under very special circumstances. Maybe the retailer is practising "leader" pricing with this one product. Practising leader pricing means setting a price low on a widely bought item to attract buyers to a store. The retailer also could be part of a chain operation with a very high volume of sales. In any case, the functions usually performed by the wholesaler have to be performed by someone. This costs money.

Physical Distribution

Physical distribution

The growth in volume and variety of goods sold, together with new transport technology, has turned the attention of manufacturers towards the problem of physical distribution or logistics. **Physical distribution is concerned with the physical movement of raw materials into the plant, the in-plant movement and storage of raw materials and semi-manufactured goods, and the movement of finished goods out of the plant to the ultimate consumer or industrial user.**

The Total Cost Concept

At one time physical distribution management was mainly concerned with minimizing the cost of transportation. This is a narrow view, since the transportation cost may be less than half of the total cost of physical distribution. **Modern firms apply the total cost concept, which considers all costs related to a particular means of physical distribution.** In addition to transportation there are storage costs and "out-of-stock" costs. Concentrating only on transportation rates is shortsighted.

Some firms have developed distribution systems which depend on electronic computers to schedule the flow of goods from manufacturer to consumer or industrial user. They select the best location of intermediate storage points and the best means of transportation. These computer-based systems take into account the costs of transporting and storing as well as the cost of "running out" of merchandise. Such accurate cost systems are common where the channel is under the control of a retailer or manufacturer. The objective of modern physical distribution management is to achieve a balance between costs and service.

Total cost concept

The Modes of Transportation

Firms have a choice among railways, motor trucks, air freight, and, in some cases, ships and barges or pipelines to move their products. Decisions like this are in the hands of the traffic manager. This important decision maker keeps track of the in-and-out flows of materials, delivery dates, and storage space. Recent history, as can be seen in Figure 8-5, has seen some dramatic shifts in the relative importance of the various modes of transportation.

The legal relationship between a firm which owns goods which need to be shipped and the carrier (railway, truck, barge, etc.) may vary. **A common carrier offers its services to the general public at uniform, published rates. These are supervised by public agencies.**

Common carrier

When a firm needs to move goods which can't be moved by common carrier, it may call on a contract carrier. **A contract carrier is a firm, such as a trucking company, that negotiates long- or short-term contracts with shippers to handle their freight.** It is a private contract. The shipper may want customized service and may want a guarantee of availability without investing in its own truck or barge fleet.

Contract carrier

If a manufacturer or middleman owns and operates its own transportation, it is called a private carrier. This kind of operation is justified when a large, predictable volume exists and common carriers are not as economical or can't do exactly what the shipper needs.

Private carrier

Each of the major modes of transportation has its good and bad points. The railway's major advantages are its cost advantages for "long hauls" and for bulky items where water transportation is not available. The railway also provides reliable service, since varying weather conditions do not often affect it. Economies of scale are evident in railroading, since one diesel engine can pull one or one hundred loaded cars. This enables the railroad to spread the cost of the motive power over a large number of shipments.

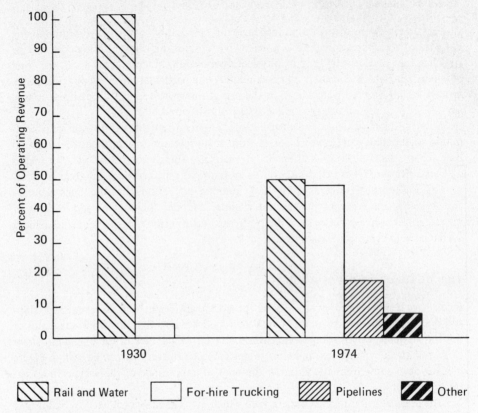

Figure 8-5. Per cent distribution of operating revenues by mode of transportation: 1930 and 1974. (Source: *Canada Year Book,* 1976–77, p. 731. Reproduced by permission of the Minister of Supply and Services Canada.)

The major limitations, however, also relate to economies of scale. Less-than-carload lots LCL are not well-suited to rail movement. Modern railways use mechanized loading and unloading equipment which is designed to handle single, large units. Small shipments require very costly manual handling. Furthermore, the more cars which can be moved by one engine, the lower the cost to move each car. Therefore, the small shipper may find its shipment waiting on the siding while a large train is being made up.

The major advantage of motor truck transport is flexibility. Trucks can go anywhere there is a road. Thus, by truck the shipper can reach many more potential customers than by other modes.

As in the case of the rails, there are many different types of trucks, some of which are highly specialized. Since the required investment is rather small, many shippers own and operate their own trucks. Door-to-door service is possible with trucks, and service is speedy.

Water transport is important in both domestic and foreign commerce. The major advantage is low cost. Low-value, bulky goods move at very low rates by barge. As bulk goes up and value goes down, the advantage of water transport increases. On the other hand, as delivery time becomes more important, water transport becomes less attractive. Accessibility to waterways and ports is, of course, necessary.

Pipelines are the most "invisible" of the modes, although they move many millions of tons of goods over many miles. Thousands of miles of pipelines move crude and refined petroleum, chemicals, and natural gas. Pipelines are almost completely unaffected by weather and, once they are installed, the cost of operation is very low. Very little labour is involved in operating them.

WHAT DO YOU THINK?

Are the Railways Doomed?

During the nineteenth and early twentieth centuries, railway lines moved most of the goods in Canada. This included high-value, low-bulk goods (manufactured goods) and low-value, high-bulk goods (raw materials). But the truck came along and took a lot of the traffic in manufactured goods. The rails are now mainly movers of low-value, high-bulk goods. They fought barges and pipelines for this business, sometimes winning, sometimes losing.

Today you read and hear a lot about deteriorating passenger service on the railroads. This may overshadow the progress railways are making in other areas. Fast-freight service is offered to shippers who must move perishable goods. It amounts to express service over long distances. Containerized terminals permit piggyback (rail-truck) and fishyback (rail-ship) service. The in-transit privilege enables shippers to interrupt the movement of their goods for further processing for only a small fee. The shipper is still charged the lower "through" rate for the movement. Thus, a cattle shipper can have the cattle off-loaded for feeding while they are moving towards the buyer. Diversion in transit lets a shipper change the destination of the shipment. A produce farmer who shipped goods from British Columbia to Montreal can divert the shipment to Windsor if the market price for the produce is higher there.

Railways have turned to computerized and automated operations to speed up their service, to reduce damage claims, and to provide overall higher-quality service. Rolling stock is being up-dated with new types of cars such as the tri-level car which moves new automobiles.

The unit train is an entire train which moves one bulk commodity from one point to another without stopping. Rent-a-train service permits a grain shipper to rent an entire train for moving grain to ports for shipment overseas. Are the railways doomed? WHAT DO YOU THINK?

In the not-too-distant past, airplanes were considered basically "people carriers". At best, they could move only very high-value, low-bulk cargo. The arrival of the jet age and "jumbo jets", the increased number of airports, and containerization have changed many of these ideas. Air transport is speedy, safe, and can help the shipper in reducing other elements of total distribution costs. Many airlines appeal to the shipper on the basis that if it is willing to spend a little more money on transportation, its other distribution costs can be reduced. One major problem is the great rise in fuel costs.

Containerization

Our discussion of physical distribution would be incomplete without further explanation of containerization. **Containerization is the practise of using standard large containers, preloaded by the seller, to move goods.**

Modern containers move many types of goods. They are loaded at the shipper's plant, sealed, and moved to the receiver's plant. Instead of many individual items being individually handled, the entire container is mechanically handled. They move with great efficiency from truck to train to ship or barge. The savings in distribution cost can be great. They also reduce theft and waste in transit.

PROMOTION

Promotion is probably the most dynamic, aggressive, and persuasive element of the marketing mix. It includes all communication by a firm with its customers or potential customers for the purpose of expanding sales, directly or indirectly. Promotion is communication which gains attention, informs, teaches, reminds, persuades, or reassures others about a product, a service, or an idea.

Because of the dynamic nature of competition, promotion must be viewed as a process. It is the process of sending messages through a variety of media. A cartoon commercial message, for example, is sent to families through the medium of television. The choice of a medium can influence the way the message is received. The principal methods of promotion are advertising and personal selling. Of somewhat less importance are the methods of sales promotion, publicity, and public relations. Let's discuss these methods and how they relate to one another.

Advertising

Modern Canadians are familiar with the process of advertising. They are being subjected to it during a large part of every day of their lives—much of it through TV. The special feature of advertising is its ability to reach large numbers of people at the same time at a moderate cost per contact. It is done through the principal advertising media of television, consumer magazines, newspapers, business publications, radio, direct mail, and billboards. It is carried on by such institutions as advertising departments of firms, advertising agencies, and the media themselves. The volume of advertising in each of the major media is shown in Figure 8-6.

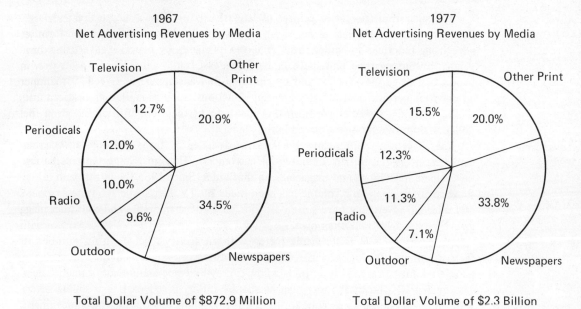

Figure 8-6. Share of net advertising revenue by media, 1967 and 1977. (Source: 1967–Statistics Canada, reproduced by permission of the Minister of Supply and Services Canada; 1977–Report on Advertising Revenue in Canada prepared by McLean-Hunter Research Bureau.)

Individual firms spend most of their advertising budgets on brand or selective advertising. This means that the purpose is to promote a particular brand of product sold by that firm. Sometimes firms also engage in institutional advertising, which promotes the good name of the firm as a whole. When a firm or group of firms advertise a general class of product without mentioning brands, this is primary demand advertising. You have probably seen ads urging you to drink more milk or eat more eggs.

Advertising Institutions

The principal institutions involved in making advertising decisions are the advertising departments of firms, advertising agencies, the media—newspapers, television stations, magazines, and so on—and research organizations employed to evaluate the effectiveness of advertising. Let's examine the functions of each.

Most large- and medium-size firms have a separate department to oversee advertising activities. If the firm in question has adopted the marketing concept, the advertising department is under the authority of the top marketing executive. This provides for co-ordination of advertising with other promotional activities and with the rest of the marketing mix. Many advertising departments serve as communicators between the firm and the advertising agency.

Advertising agency

The operational, creative centres of advertising for most medium-size and large firms are their advertising agencies. **An advertising agency specializes in performing advertising functions for other firms.** It serves its clients by planning advertising campaigns, by buying time and space in the broadcast (radio and TV) and print (newspapers and magazines) media, and by checking that ads appear as agreed. Sometimes ad agencies perform additional marketing functions, such as marketing research and public relations. Agencies are normally paid a 15 per cent commission based on the dollar amount of advertising placed in the media.

Advertising media

The advertising media carry the message designed by firms and their agencies to many receivers (customers or potential customers). The most important media are newspapers, television, and magazines, in that order. Since 1955 the largest gain in the share of total advertising volume has been made by TV. Newspapers, business papers, and outdoor advertising have suffered major losses of share; radio and consumer magazines have roughly held their own.

An advertiser selects its media (often with an agency's help) with a number of factors in mind. The marketing executive must first ask "which medium will reach the people I want to reach?" If a firm is selling turkey-breeding equipment, it might select *Turkey World,* a business paper read mostly by turkey breeders. If it is selling silverware, it might choose a magazine for brides. If it is selling toothpaste, the choice might be a general audience television program or a general audience magazine like *Reader's Digest.*

Another important factor in media choice is the medium's ability to "deliver" a particular message effectively. Some messages need visual communication, and some need the added dimension of colour. Foods are an example of products that benefit from colour in communication. Selling an electric organ requires a sound-oriented medium. Some messages need colour, sound, and motion. This is available only in colour television—the medium with the greatest set of "communicating tools".

There are other special considerations in media selection. For example, print media are more permanent. They can communicate several times or can be taken to the store as a shopping aid. Some media (radio, television, and daily newspaper) provide frequent communication and relatively short lag-time before the ad will appear. Finally, there is the cost per contact with a customer or prospect. This will vary greatly, but the one-dimensional media, such as radio, are usually inexpensive per contact.

How an Advertisement Works

AIDA process

Much of what advertising can do can be shown by careful examination of a particular advertisement. **A promotional process can be thought of in terms of how it works on a particular receiver or prospective customer, leading him or her through the stages of attention, interest, desire, and action—the AIDA process.**

The advertisements on pages 218-220 illustrate how a printed ad may apply to this process. Headlines are usually attention-getters and interest-builders. Sometimes they go a long way towards building desire, too. Copy, in many ads, is the desire-builder or convincer because it gives facts and anticipates objections by the reader. The signature is usually the familiar company trademark or brand name. It says who is sending the message. Sometimes it is accompanied by a coupon or an action-inducing offer.

The ads shown at the end of the chapter display a variety of appeals or themes used to "reach" customers. The advertiser uses the appeal which fits the product and the customer.

The arrangement of parts in a print ad—illustration, headline, and copy—is called the layout. Ideally, the layout makes it easy to "carry the reader through" the phases of attention, interest, desire, and action. A similar AIDA "game plan" can also be used in personal selling.

Advertising Complaints and Regulation

Because advertising is such an obvious part of the lives of Canadians, it is not surprising that it is subject to much criticism and government regulation. Complaints come from

---

CAREERS IN ADVERTISING

If you are interested in advertising, you should consider positions like account executive, copy writer, and media buyer.

Advertising account executives work for advertising agencies and co-ordinate the services the agencies provide for their clients. They are planners, budget makers, salespersons, and "idea people" all rolled into one. This requires a rare combination of creativity and good "business sense". It usually takes a university degree and at least a few years working in related advertising jobs. As a career it can have drastic ups and downs, depending on the gain and loss of large accounts by the agencies. Levels of pay vary widely, depending on the size of the agency and the success of ad campaigns created under the account executive's control.

Among those persons who work with account executives in advertising agencies are copy writers and media buyers. Copy writers are creative people who have a way with words. They are often, but not always, college graduates with backgrounds in English or journalism. Pay depends on the degree of success of their creations. Copy writers also often work for the advertisers or the medium (newspaper or TV stations) instead of the agency, but they do the same type of work.

Media buyers work for ad agencies or for advertisers themselves. They are experts in finding space and time in the media for their employers. They analyze rates and availability of media to maximize the advertising dollar. A B.Comm. helps but is not essential. Information is available from the Canadian Advertising and Sales Association, Suite 369, Hotel Queen Elizabeth, Montreal, Quebec H3B 1X8.

consumer groups, conservationists, sociologists, and economists about the effects and some of the methods of advertising.

Some complaints relate to truth in advertising. Some exaggeration about the quality of products has always been permitted, but there are limits to such exaggeration. The Combines Investigation Act has, in recent years, been amended so that false and misleading advertising is more closely controlled. Its ground rules provide that all statements of fact be supported by evidence.

Other questions about advertising that are worth discussing are (1) Are we being brainwashed? (2) Is too much advertising wasteful? (3) Does advertising lead to monopoly and high prices? and (4) What effect is advertising having on the values of our people? While some of these questions have merit, it is hard to imagine how the great productivity and wealth of our economy could have happened without the stimulating effect of modern advertising.

Personal Selling and Sales Management

In some situations there is no substitute for one-to-one human persuasion. All of us experience it nearly every day. There is a lot one person can do to convince another of a point of view. This could mean one's willingness to try a new brand of beer or change an attitude towards a politician. However, persuasive talent cannot serve a business effectively unless it is properly managed.

Personal selling

Personal selling includes any direct human communication for the purpose of increasing, directly or indirectly, a firm's sales. The special quality of personal selling is its individuality—the one-to-one relationship between the seller and the buyer—and the fact that the seller may give very special attention to the buyer's needs. The "tone" of the personal selling can vary widely. It can be like a sideshow barker or like a skilled computer salesperson. The style is different, but the goal is quite similar—to sell. Both hope to guide the receiver through the AIDA process. (See Figure 8-7 for a typical example of how a salesperson might follow this process.) Advertising and personal selling are complementary. This means that they work well together.

There are special problems in managing a sales force which are not usually as serious in the management of other personnel. To be successful, salespeople must be confident in themselves and in their product. Their morale must be kept high. Sales managers are often handicapped in trying to maintain high morale because they usually lack continuous personal contact with sales personnel.

The sales manager's responsibilities include:

- building an effective sales force
- directing the sales force
- monitoring the sales effort

To build a sales force, the sales manager must develop recruiting sources and techniques, devise methods of selecting from the recruits, and maintain an effective sales training program for those selected. Directing the salespeople includes developing workable pay plans, which might include salaries, commissions, and/or bonuses, and

Salesperson (S)	Prospect (P)
Enters situation with thorough knowledge of product, incomplete knowledge of prospect's needs. Is confident.	Enters situation with a poorly formed idea of need, very little information about salesperson or product. Some distrust and hostility towards salesperson.
1. Attracts attention of P by setting up appointment and, perhaps, by indicating some awareness of the needs of P.	2. Greets S, attitude improved by the pleasant, interested manner of S.
3. Begins to show how product can solve a problem P has.	4. Becomes more deeply interested but brings up certain objections regarding price and quality.
5. Answers P's objections by describing credit plan and explaining how the service department of S's firm can overcome the problems in (4) by P. Asks for the order.	6. Finds another reason to object to signing order, but the objection is mild.
7. Answers last objection and closes with, "If you'll just give me your okay on this, we can make delivery on the first of the month."	8. Agrees to buy on a trial basis.
9. Thanks P, checks over details of order. Later, checks on P's satisfaction.	10. May experience some post-sale doubt, but reassures self of wise decision.

Figure 8-7. A typical personal selling sequence

programs for appealing to higher-order motivations such as those discussed in Chapter 4. The third task, that of monitoring sales effort, is a special form of control. The sales manager must, therefore, set up standards—such as sales quotas and sales expense budgets for sales territories, products, and individual salespersons—and make regular comparisons of actual sales results to such standards and make the necessary corrections. Corrective action may range from redefining sales territories to additional sales training to dismissal of ineffective sales employees.

Other Promotional Methods

Publicity is communication through the news media as a legitimate part of the news. It is an inexpensive means of promotion, because its only cost is preparing the news story or "press release". But only items considered "newsworthy" by the press are used, and very often carefully prepared items are never printed or broadcast. Often, news stories (like the announcement of the new car models), may be cut down by newspapers for lack of space. Thus, publicity is a promotional method over which the firm has little control. Also, it is limited to reporting facts.

Publicity

Public relations

Public relations includes any personal communication with the public or with government (lobbying) which seeks to create goodwill for the firm. Its effect on sales is usually indirect and long run in nature.

Sales promotion

Sales promotion includes special events directed at increasing sales. Special sales, coupon offers, contests, games, entertainment features, and trading stamps are examples. Some would include specialty advertising devices such as matches, calendars, and ball-point pens. Others might add "PMs" or "spiffs", which are cash payments to retailers and their employees to promote a firm's products. The use of sales promotion methods varies greatly by industry.

PRICE

In a competitive world dominated by promotion and product differentiation, price still plays a part. Most firms still concentrate much of their effort on the problem of setting price. How this process fits into the firm's planning is a different story for almost every firm. The role of price in the mix is not as great as it once was. However, the importance of careful pricing cannot be ignored.

Market Conditions and Pricing

The key concepts of price, market, supply, and demand were introduced in Chapter 2. These explanations were limited to a rather narrow range of market conditions (a large number of buyers and sellers, a homogeneous product, easy entry into the market, and market information in the hands of buyers and sellers). A homogeneous product means that the buyers do not make any distinction between the products for sale. Easy market entry occurs when no large capital investment is needed for new sellers to emerge. The necessary market information includes facts about the numbers of competitors and supply and demand conditions.

Oligopoly

As consumers, we know that all these conditions rarely exist. The more common kinds of market conditions are oligopoly and monopolistic competition. **An oligopoly means there are only a few sellers of the same or slightly different products.** The market for automatic washing machines, for example, might be classified as oligopolistic. There are few sellers and prices are generally stable.

Price leadership

In oligopoly, one of the stronger competitors may sometimes raise price, and it is likely that others will follow suit. This is called price leadership.

Monopolistic competition

Monopolistic competition is a market in which many somewhat differentiated products or services are sold. This is most common in retail and service firms. Location and quality of service are the differentiating factors. New competitors can easily enter the market. This prevents the typical competitor from making large profits.

There is such a great array of markets and products in the developed Western nations that labels such as "oligopoly" and "monopolistic competition" are not enough. They help us understand some of the price behaviour, but they are not enough to guide price decisions. Such theoretical market concepts don't account for the use of the marketing mix elements other than price. Price is only one of the competitive tools.

The issue of pricing is further complicated by government involvement in pricing in both the private and the public sector. For example, provincial utilities must get approval for rate increases from a provincial board. Government also affects price by subsidizing activities like public transportation. Marketing boards establish prices for agricultural products, and this further influences prices consumers have to pay.

Setting Basic Price

Under the guidance of pricing objectives, firms must set basic prices. There are two different approaches to setting basic price which must be understood. These are the cost approach and the demand approach. (See Figure 8-8.)

The cost approach of setting basic price involves "building" unit selling prices on the basis of cost. This approach is simple when the cost of one unit is easy to identify. **A markup is an addition to cost to reach a selling price.** It is usually expressed as a percentage. Mr. Schultz, who operates a men's clothing store, uses a percentage "markup" applied to his unit costs for an item or group of items. Thus, he might buy 100 suits at $40 apiece and apply a 50 per cent "markup on cost", resulting in prices of $60 per item for his customers (150 per cent of $40). Schultz might use this same percentage markup on all items in his store. If so, his basic price policy is a very simple one with a cost basis. He probably has allowed demand factors to influence his markups only in an indirect way.

Markup

Mr. Schultz probably knows from past experience about how much his customers are willing to pay. But he does not look at customer price attitudes very closely because cost-plus pricing is so easy for him. He also knows that, over the years, the policy has allowed him enough "gross margin" (the difference between total dollar sales and costs of goods sold) to pay his rent, his clerk's salary, and other costs of operation, and to leave him a fair profit.

Manufacturers also often use a cost approach to pricing. However, it is not always easy for them to identify costs attached to a given unit of output. Cost accounting systems have been developed that aid in this task.

Manufacturers who sell to governments often use cost-based pricing because government buyers frequently specify such pricing, sometimes called "cost-plus". In some cases the federal government is the only customer of a firm. Pricing in such a case may be required by the government to be on a cost-plus basis. This allows the contractor to cover cost and to make a certain profit.

In its most extreme form, the demand approach to pricing neglects the role of cost. It is more in tune with the marketing concept than the extreme cost approach because it considers possible customer reaction. More precisely, it estimates the amounts which are likely to be sold at different prices. The most extreme form of demand-approach pricing would set prices very low to sell the greatest number of units.

The Sayles Company, for example, is owned by a former salesperson whose past experience (remote from production cost considerations) leads him to accept the demand approach, giving only some thought to costs. Mr. Sayles realizes that costs, in the long run, must be covered if he is to stay in business. He is, however, more likely to accept lower prices than a cost-oriented firm would, because he is accustomed to

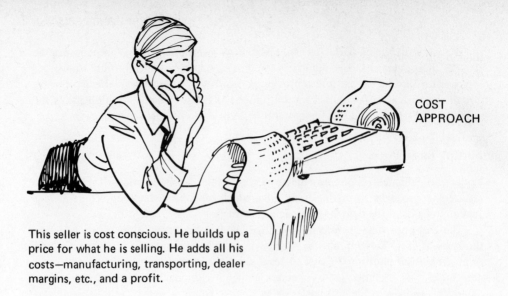

COST APPROACH

This seller is cost conscious. He builds up a price for what he is selling. He adds all his costs—manufacturing, transporting, dealer margins, etc., and a profit.

DEMAND APPROACH

This seller wants to sell! She examines the market and estimates the quantity demanded for her product at various prices—with an eye on her competitors.

Figure 8-8. Two approaches to price

thinking in terms of sales growth. He knows that customers and competition must be considered in order to build sales.

Many firms set prices by making both cost and demand estimates. These are translated into profit estimates at various unit price and sales levels. This often involves marginal analysis, which estimates the "best" price and level of production from a profit standpoint.

In large firms which have the services of economists and large computers, pricing models often are created. **A pricing model is an equation or set of equations which represents all the important things in a pricing situation to help decide on the "best" price.** Past experience in pricing and knowledge of the market conditions and cost factors help to determine the equation which best predicts pricing results. This process is known as model-building. No matter how carefully it is done, it still requires judgment about what other human beings will do and what their tastes and needs will be.

Pricing model

Discounts

Specific prices charged by a manufacturing firm in actual sales often vary from the basic price. Such variations generally result from an established discount policy. Discounts from "list" prices are granted for a number of reasons. For instance, it is common for a firm to offer small discounts for prompt payment of bills. These are called cash discounts. Another typical discount is the trade position discount. A wholesaler who normally sells to retailers, for example, may make a special sale to another wholesaler at a discount from the regular price to retailers. **Any discount granted because of a difference of position in the distribution channel is called a trade position discount.**

Trade position discount

Sometimes a "functional" discount is granted to a customer in return for services rendered. A retail grocer may, for example, receive a discount or allowance from a detergent manufacturer if the grocer features the manufacturer's brand in local newspaper ads. Some would argue that this practise is not really a discount, but rather a simple purchase of a service.

Functional discount

Still another frequent discount is the quantity discount described in the previous chapter. A firm's discount policy makes its pricing more flexible in special competitive situations. However, it is often also the cause of serious legal problems stemming from price discrimination laws.

Pricing New Products

Some special considerations arise when a new product is introduced. Two typical alternative approaches may be used. **To feature low price when introducing a new product is called market penetration pricing.** The firm's goal is to build a large initial market share and to build brand loyalty before competitors can enter the market. The initial low price discourages some competitors who foresee smaller profit at such a low price.

Market penetration pricing

> # WHAT DO YOU THINK?
>
> ### Consumer Credit—Curse or Blessing?
>
> If the typical Canadian household were denied credit, its standard of living likely would be lowered. Ours is a credit economy. Consumers can enjoy goods during the time they are paying for them. This runs the gamut from houses to cars to clothing—to practically anything. But some people believe that some of us go overboard with credit. In some cases, this leads to personal bankruptcy.
>
> To reduce this possibility, each province has enacted legislation to protect consumers. Two types of laws are most prominent—those which require lenders and sellers to state the annual interest rate which is being charged, and those which forbid credit reporting agencies to include certain kinds of information about consumers on their credit reports.
>
> Is consumer credit a curse or a blessing to the consumer? What about sellers? What about our economy? WHAT DO YOU THINK?

Skimming pricing

Some firms, however, prefer to charge a high initial price and put greater stress on the other marketing mix elements. The goal is to get the greatest early revenue from sales to recover product development costs before competitors enter the market. **This is called skimming pricing.** It is often used by small firms, by firms with large development costs, and by firms which are not well-protected by patents and good reputations. This policy amounts to "getting it while the getting's good". A manufacturer of small plastic toys who introduces a new and inexpensive toy is likely to practice "skimming". It may charge $1.95 at retail on an easily imitated item which costs 23 cents to make. This enables it to maximize immediate return on its investment. A drug manufacturer is likely to do the same with a new antibiotic.

Retail Pricing

Most prices charged in retail stores are determined by a markup mechanism. Ms. Jill Gladney runs a jewelry store. She knows that, on the average, her costs of doing business—including salaries, rent, and desired profit—have amounted to about 50 per cent of sales revenue. She also knows that the cost of goods she buys for resale accounts for the other 50 per cent. She might plan prices so that, considering special "sales" to sell slow-moving items, she would realize a 50 per cent gross margin (difference between gross sales and cost of goods). Thus, her average initial prices might need to be more than double the cost of goods. A shipment of rings costing her $100 apiece might be marked up to $250 and finally sold at $200—just enough to provide a gross margin of 50 per cent of sales.

Jill Gladney's initial "markup" could be expressed two ways. It represents $150/$250 or 60 per cent of the originally established sales price and $150/$100 or

150 per cent of the original cost price. It should always be made clear whether a percentage markup is expressed in terms of cost or in terms of expected selling price. Of course, Ms. Gladney may assign different markups to different classes of items in her inventory, depending on competitors' practice and her experience with the "turnover" rate of various items.

The inventory turnover rate in a particular period of time is determined by dividing cost of goods sold by the average inventory value. It can be expressed in units or in dollars. Thus, if inventory is worth $20 on January 1 and $30 on December 31 and if the cost of goods sold in that year amounted to $250, the average inventory is $25 and the turnover rate is 10.

Inventory turnover rate

If Ms. Gladney uses different markups on different items, she will try to average them out to provide her desired "gross margin". Sometimes retailers use very low markups on certain items to attract customers. This is known as a "leader" item. The goal is to increase sales of items carrying higher margins. Such practices may be illegal if it is shown that the item is sold below cost. Some firms use illegal "bait-and-switch" schemes. This means advertising one inexpensive item which is really not available and then convincing people to buy a more expensive substitute when they arrive.

Many retailers use price lining. An apparel retailer may make ten purchases of men's sportcoats in a season. Marking up each of these by the same percentage might result in ten different prices for sportcoats. **Partly to simplify choices for customers and partly to simplify the job of the salespeople, the retailer may use price lining—grouping the costs at three or four sales price levels.** Thus, Bill's Men's Wear might make ten purchases of coats in a cost range of from $11.50 to $38.50 per unit and present them to customers in the $29.95, $39.95, and $69.95 price lines. This makes it easier for the salesperson to get the customer to "trade up". He comes into the store expecting to buy the $29.95 sportcoat he saw in the newspaper ad and finally buys the $39.95 or $69.95 coat, after comparing quality and listening to the salesperson's advice.

Price lining

The prices selected by Bill's Men's Wear are "odd" amounts, and they were close to the next "$10 break". Partly out of tradition and partly because of a slight psychological effect, retailers tend to set prices at $39.95 rather than $40. A price starting with thirty "sounds" like more of a bargain than one starting with forty.

Special promotions such as "one-cent sales" and "two-for-one sales" also are a part of the art of pricing. They require careful estimation and experience in order to result in overall profits for the retailer.

SUMMARY AND LOOK AHEAD

The full meaning of the idea of a marketing mix is now apparent. A firm adds to its product offering and distribution system a program of promotion and decides upon a basic pricing system. These four mix components together describe a firm's marketing policies and strategy. They cannot be developed independently of each other, however.

Product policy is the focal point of the marketing mix of many firms. They watch the sales trends of products in their product mix to determine how far along each is in its "life cycle". This permits them to take corrective action to prolong product life if possible or to replace them with new products. Research and development make it possible to postpone failure of a product or to replace a dying one.

Another key marketing mix element is the selection of distribution channels. These channels are made up of middlemen who assist in bringing goods to customers. Where possible, manufacturers select channels which bring the necessary elements of service and convenience to customers without raising costs so high as to bring prices above competitive levels.

Promotion is primarily composed of advertising and personal selling activities which are complements of one another. Advertising reaches large numbers of people, while personal selling is more aggressive and personally tailored to the customer. Both seek to convince the potential customer that a product or service is worth buying. They use a wide variety of appeals to do so.

Price, we have seen, is capable of administration. Firms emphasize a cost approach, a demand approach, or a combination approach in setting basic price. They may offer a variety of discounts and try to control prices at the retail level. There are special pricing problems related to new products and to retail stores.

In the next few chapters we will turn to the subjects of finance, accounting, and personnel and labour relations. We will describe financial institutions which use accounting and financial techniques as well as the fantastic capabilities of computers. These topics will give us yet another perspective of the dynamic world of business management.

KEY CONCEPTS

Advertising agency A firm which produces and places advertisements in media and may arrange total programs of advertising for other firms.

Advertising media Any means by which advertising can be carried: radio, television, print, etc.

AIDA Short for attention, interest, desire, and action—the process of selling sought by advertising and personal selling.

Brand A name, term, symbol, or design, or a combination of them to identify the goods or services of one seller or a group of sellers and to differentiate them from those of competitors.

Channel of distribution The path goods take from producer to final user, e.g., manufacturer to wholesaler to retailer to consumer.

Common carrier The carrier which offers its services to the general public at uniform rates to all, e.g., railways.

Containerization Method of shipping products in standard container units. This simplifies handling at points of transfer, loading, and unloading.

Contract carrier The carrier which makes temporary contracts with shippers to move their goods.

Franchised retailer A retail firm operating under a licensing agreement with a franchise grantor. Examples include car dealerships and fast-food restaurants.

Functional discount Discount given for performance of a specific marketing function. Example is an advertising allowance.

Inventory turnover rate The cost of goods sold divided by the average inventory. A high turnover rate indicates efficiency in the use of resources.

Manufacturer's agent An agent middleman representing a manufacturer in a specific region for a given commission.

Market penetration pricing Setting a price low to secure a market share for the product.

Markup An addition to cost by a middleman. A middleman adds a markup to the cost of an item in order to compute selling price. It can be expressed in money terms or as a percentage.

Monopolistic competition Many sellers compete for customers by offering differentiated products and services.

Oligopoly A few firms selling highly similar products and dominating a market.

Patent Legal protection of a process, invention, or formula by the federal government. Protects against imitation for a period of 17 years.

Personal selling Direct contact of a prospective buyer with a salesperson and a face-to-face sales effort.

Physical distribution The physical movement and storage of goods. This has provided the main opportunity to apply technology to reduce costs of marketing.

Planned obsolescence Intentionally scheduled replacement of a previous product or model with a new one by one firm.

Price leadership A situation common in oligopoly markets in which one firm sets a price and others follow suit in the interest of price stability and avoidance of price competition.

Price lining Setting retail prices at standard preset levels such as $6.95, $8.00, etc., for items bought at a variety of wholesale prices to simplify consumer decisions.

Pricing model A mathematical formula to help firms set the best basic price.

Private carrier A transportation system owned by a shipping firm for the exclusive use in moving its own products.

Product life cycle The evolution of a product through periods of introduction, growth, maturity, and decline-obsolescence.

Product mix All products offered for sale by a firm.

Publicity Information about a company or its product which is considered "news" by the media and is reported by them with no charge to the company.

Public relations Communication with the public or with the government oriented towards the creation of goodwill for the firm as a whole.

HOW ADS WORK

Examples of different advertisements appear on pages 218–220. Consider the format of each advertisement. What different appeals does each use to attract its particular audience? Are they effective?

Catch our Fall.

This is the perfect time to take a holiday in British Columbia.

The weather's good, the summer crowds are gone *and*, best of all, you can take advantage of off-season rates.

Many hotels and resort areas have special fall packages. So give your favourite holiday spot a call.

Or see your travel agent. His services won't cost you anything and he can be a big help. He'll tell you about everything from a lovely 3-day city stay to a grand 16-day Circle Tour.

You can play ride'em cowboy on a guest ranch.

Get in hot water at one of our hot springs.

Or go sailing, sailing up the spectacular Inside Passage.

Tennis anyone? Or how about a round of golf?

If you'd rather be fishing, the big ones are waiting.

Dig gold? You'll love exploring the gold rush country.

If the city is your beat, sample the first class restaurants, interesting shops and diverse cultural activities of our two cities by the sea — Vancouver and Victoria.

British Columbia is one of the best vacation values in the world.

And it's all yours.

So this fall, enjoy it.

There's a fall holiday package that's perfect for you.

Super, Natural British Columbia, Canada.

TOURISM BRITISH COLUMBIA
HON. DON PHILLIPS, MINISTER.

This ad is part of the ongoing campaign to convince residents and non-residents of British Columbia to vacation there.

The Gobbledegook.

We're doing our best to make it extinct.

If you still think that life insurance policies are full of gobbledegook—the "fine print" and the "lawyers' language"—chances are you haven't bought any life insurance lately.

Many companies have made great strides in simplifying the wording of their policies so you can understand them, and most companies are working on this task.

And, if you still see some "lawyers' language" in a policy, remember: it's in there for your benefit too. Because your policy is a legal contract—so it must spell out your rights and the company's obligations to you.

Simplified policies. It's an improvement you wanted.

You also said you wanted easier access to information about life insurance from time to time, and advice and assistance when you have a problem or complaint.

The Life Insurance Information Centre has been established, as a public service, by The Canadian Life Insurance Association—to provide consumer information and to respond to consumer complaints and problems. The people there won't sell you anything; they won't give your name to any company or person, unless you specifically request them to do so.

You can write to the Centre at this address: The Life Insurance Information Centre, Box 11331, Edmonton, Alta., T5J 2K6. Or call it, toll-free, at the number listed below— any time between 7:00 a.m. and 3:00 p.m., Monday to Friday.

Call or write about anything that's on your mind concerning life insurance in general, or your life insurance in particular.

THE INFORMATION CENTRE: 1-800-268-8663 (Toll-free)

A public service of The Canadian Life Insurance Association

This ad was part of a test advertising campaign. The advertising objectives of the campaign were: (1) to promote the life insurance industry as one which is concerned with today's needs and tomorrow's security; and (2) to demonstrate that the life insurance industry is accessible, communicative and responsive to the consumer.

"The Bank of Commerce probably has a better concept of my operation than I do."

W. Clarke Campbell, President, Consolidated Canadian Faraday Limited, Bancroft area, Ontario.

"When you set out to find an orebody you know that you're going to spend millions even before you get anything out of the ground.

"Now no bank I know is going to get into the gamble of exploration. Not the Commerce. Nor any other bank.

"But once a discovery has been made, feasibility studies have to be prepared by independent experts. They have to be assessed. And your bank then can make a judgment of the risk factor and move accordingly.

"There's still an element of risk on the bank's part, but naturally they minimize it as much as they can.

"To do this, a bank making a mining loan must develop and rely on their own mining expertise. And that's where the Commerce comes in.

"As far as I know, the Commerce handles more mining ventures than any other chartered bank in Canada. It's recognized as Canada's Mining Bank.

"Why, I bet you could mention just about any new development in Canadian mining and the men at the Commerce would be aware of it. They keep up with the technology. And they have talent. Let me show you why I deal with the Commerce.

"Some years ago, I was acting on a situation where another bank was negotiating an interim loan of several million dollars against royalties from an iron ore situation. The aggregate part investment in the project was several hundred million dollars.

"A big thing.

"I was dealing with this banker used to mortgage loans. And he has no concept of mining. I try to explain the situation to him, but he can't appreciate the security he's getting. So finally, I talked to some people from the company and suggested that we show him the project.

"So we took him to see the whole operation. To speak to the manager. And at the end he asked, 'Who pays all these people?' When I told him who did, he couldn't believe it. You see? Even though he saw everything, he still had no appreciation of what it was all about.

"Well, that's the opposite of the Commerce."

CANADIAN IMPERIAL BANK OF COMMERCE

This is one of a series of ads; the company's Communication Objective is to illustrate the expertise the CIBC has in dealing with the many problems that arise in the banking requirements of oil, gas, and mining companies. This is done by using actual officers of companies who deal with the CIBC in testimonial format.

Sales promotion Any of a variety of devices to raise sales including trading stamps, contests, and special attractions.

Skimming pricing Setting a high initial price in order to get maximum quick return on product development costs; often anticipates entry of competition into the market within a short period.

Total cost concept Procedure for planning transportation and storage so as to minimize total costs.

Trademark A legally protected name, symbol, or combination of name and symbol.

Trade position discount A discount given to middlemen in proportion to their position in the channel of distribution. A wholesaler usually gets a larger one than a retailer.

QUESTIONS FOR DISCUSSION

1. When you purchase a new toaster, what "bundle of services" are you buying? How about a tube of toothpaste?
2. How long does a product "live"? Why does a fashion product "live" a shorter time than a hardware item like a wrench? Explain the concept of "life cycle" as applied to these two types of products.
3. Has packaging become more important or less important in selling consumer goods in recent years? Discuss.
4. Name six brand names which you think have good "memory value". Name six which you think are poor. What makes the difference?
5. Why does the distribution channel for an industrial good usually differ from a convenience consumer good channel? How does it differ?
6. How is it that "middlemen" exist even though they must cover their cost of operation and make a profit? Are their functions necessary? Why or why not?
7. Draw a chart illustrating four different channels of distribution which a manufacturer of toys might employ.
8. What are advertising media? Which ones are most important to a department store? Why?
9. Write a brief piece of copy for a television commercial announcing a new chain of sandwich shops called "Margy's" to be introduced in your city next month.
10. Write a paragraph giving advice to a salesperson for a hardware wholesaler going out on his or her route for the first time.
11. How can a sales manager tell if Tom Raincheck is a good salesperson before hiring him and six months after hiring him?
12. What is basic price? Give an example in which cost-plus basic pricing is a sensible alternative.
13. If you want a gross margin on sales of 40 per cent, would your average initial markup (as a percentage of selling price) have to be more or less than 40 per cent? Why? If the markup on selling price is 40 per cent, what is the equivalent markup expressed as a percentage of cost?
14. Is price as important a part of the marketing mix today as it was 50 years ago? Discuss.

INCIDENT:

Wick Wardlow

Wick Wardlow invented a game suitable for teen-agers and adults involving a game board, dice, etc., and pertaining to financial investment. He has no marketing experience, but his rich uncle says he will back him in the production and distribution of the game in eastern Canada.

Questions:
1. Does Wick need legal protection? Why? How can he get it?
2. What aspects of product policy are critical?
3. Would you employ market segmentation? Explain.

INCIDENT:

Gallman and Sons

Gallman and Sons operate a jewelry store in Brandon, Manitoba. They price at retail by adding 80 per cent to the cost price (invoice price) of every item purchased. Their inventory includes greeting cards, ball-point pens, a complete line of silverware, watches, and diamond rings. Nick Gallman, Jr., now admitted by his father into the partnership, questions the value of such a pricing policy.

Questions:
1. What kind of pricing policy is this? What are its advantages?
2. What are the problems created by such a pricing policy?
3. Would Gallman and Sons make less profit if they converted to a variable markup policy? Explain.

ROCKETOY COMPANY V

Before Rocketoy's new plant in Windsor was half completed, sales reports reaching Terrence's desk showed that sales of two of Rocketoy's most successful non-seasonal plastic products were down 20 per cent from the previous year. Terrence called in the vice-president of marketing, Julia Rabinovitz, for an explanation. Julia gave several possible causes of the sales decline.

First, she had been told by several of Rocketoy's salespersons that some of the customers were placing trial orders with a toy manufacturer in West Germany. This manufacturer produced two high-quality toys which were very similar to Rocketoy's. The major difference was price. The imported toys sold at retail in Canada for 75 per cent of the retail price of Rocketoy's two toys. Julia suggested that retail toy buyers were growing more price conscious.

Julia also suggested that recent cost increases for plastic had led to a reduction in the thickness of the plastic used to make several Rocketoy products, including the two "problem" products. She thought that this hurt Rocketoy's reputation for making quality toys. In fact, Rocketoy had received 150 letters from retail customers complaining about the "shoddy" toys. Three customers said they were reporting the problem to the Department of Consumer and Corporate Affairs. They claimed that their children had received cuts on their hands. Jagged pieces of plastic became exposed when the toys broke during normal play activities. Julia and Terry worried about this problem. They knew that three competitors had had to remove several of their toys from the market as a result of safety problems. Julia suggested, however, that this was a production problem beyond her control.

Julia's third reason for declining sales was the declining birth rate in Canada. The two "problem" toys appealed to children between the ages of 2 and 4 years. As Julia said, the number of potential users of these toys is declining. We can only expect sales of these toys to decline as well.

Julia also suggested that many parents were complaining about the tremendous volume of advertising aimed at children. Rocketoy had once concentrated its advertising on Saturday morning children's television shows. Many parents had complained that the constant bombardment of television commercials was "bad" for their children. Several parents had written to Rocketoy and had accused the company of "taking advantage" of children. Julia also knew that the government was investigating these types of complaints from parents. As a result, Rocketoy decided to reduce its advertising several months ago.

Finally, Julia said that more and more toys were being sold through discount stores. Rocketoy had always refused to sell through discount stores because many toy store owners said they would stop buying Rocketoy products if Rocketoy sold to "price-cutting" discount stores. Julia said that this had become a big problem for Rocketoy and something had to be done about it.

Questions:
1. Should Rocketoy reconsider its plans for the plant layout? Why or why not?
2. What action, if any, should Rocketoy take regarding the new competition from West Germany?
3. Who are Rocketoy's target market, children or their parents? Discuss. Are there any other targets?
4. With respect to the marketing mix, Rocketoy has problems with product, place, price, and promotion. Discuss the nature of these problems and how you would deal with each of them for each element in the marketing mix.

9
Accounting

9 CHAPTER

OBJECTIVES: After reading this chapter, you should be able to:

1. Distinguish between financial and managerial accounting processes.
2. Identify the three principal tasks of accounting.
3. Describe what a CA does.
4. Prepare a chart showing the major information flows of accounting.
5. Explain the relationship between transactions and accounts.
6. Complete both of the principal accounting equations.
7. Draw up a simple example of the two principal financial statements.
8. Explain the purpose of the two principal financial statements.
9. Demonstrate how an investor might use the statements of a firm he or she may wish to invest in.
10. Explain and use at least one of the "key" ratios.
11. Show how a budget may be used in internal control.

KEY CONCEPTS: In reading the chapter, look for and understand these terms:

ACCOUNTING	CURRENT ASSET
FINANCIAL ACCOUNTING	FIXED ASSET
MANAGERIAL ACCOUNTING	DEPRECIATION
ACCOUNTANTS	CURRENT LIABILITY
ACCOUNT	ACCRUED EXPENSE
ASSETS	INCOME STATEMENT
EQUITY	GROSS PROFIT
LIABILITY	KEY RATIO
OWNER'S EQUITY	TANGIBLE NET WORTH
REVENUE	CURRENT RATIO
EXPENSE	BUDGET
TRANSACTION	SALES FORECAST
BASIC ACCOUNTING EQUATIONS	RESPONSIBILITY ACCOUNTING
STATEMENT OF FINANCIAL POSITION	PRODUCT COST ACCOUNTING

If you were asked to describe or evaluate a firm, how would you do it? A lot would depend on why the request was made and who made it. If the request came from a new employee, you might want to describe the wage and promotion policies and the working conditions of the firm. If it came from a new customer, you might want to describe the quality of your product and your delivery, price, and credit policies. But what if you were the executive vice-president of a firm and the president asked you to evaluate the sales growth of the firm over the last five years? Or what if the president asked you to make a report on the firm's financial position for a bank that may lend the firm a million dollars?

The last two requests could not be met completely by the use of words. These two requests are common, however, and they both require the use of accounting.

WHAT IS ACCOUNTING?

Accounting

Accounting has been defined in many ways. We will define it as a process of recording, gathering, manipulating, reporting, and interpreting information which describes the status and operation of a firm and aids in decision making.

This process is guided by certain widely accepted principles and rules. These principles are especially important when a manager must report to people outside the firm. They play a smaller role when accounting is for internal use. We call the internal processes managerial accounting and the external processes financial accounting. Both financial and managerial accounting are useful to managers.

Financial accounting

Financial accounting helps the manager to "keep score" for the firm. It watches the flow of resources and lets those who have an interest in them know where they stand.

Managerial accounting

Managerial accounting calls attention to problems and the need for action. It also aids in planning and decision making. It is aimed more at control and less at valuation than financial accounting. It is also less traditional.

Like any tool, accounting must be designed to do its various jobs (scorekeeping, calling attention, and helping in decision making) quickly and at a fair cost. The accounting system must provide clear and efficient estimates of financial facts. What it produces must be, above all, relevant. Accounting is relevant when it is useful to managers, creditors, investors, or government agencies in doing their jobs. Accounting is a much broader term than simple bookkeeping. Bookkeeping is simply the mechanics of accounting—the recording of financial data.

What Is an Accountant?

An accountant is much more than a person who keeps the books. Accountants know basic procedures for recording transactions quickly, accurately, and with maximum security. They know how to summarize data so that all kinds of users can understand it. They know enough about the law to build a system of accounts which reflects those laws—especially tax laws. They know where to find specialized information about law and other tough questions—especially as they relate to the firm they serve. For example, an accountant for a forest products company would keep a library relating to land valuation and the use of natural resources. He or she must also be aware of the history and policies of the firm.

Accountants are individuals who have satisfied the knowledge and experience requirements of a professional group and have been admitted as members of the profession. Accountants may be employed by public auditing firms (companies which express an opinion as to the fairness of the financial statements prepared by the accountant of a firm). Auditors are just as expert as the accountants employed by industrial or commercial firms. Often their knowledge must be broader, because they deal with the accounting processes of many different firms.

Accountants are also employed by individual firms in both the private and the public sector. In these positions, accountants keep track of revenues and expenditures for the company as well as doing many of the functions which are the subject of this chapter. Some accountants serve government directly and develop a much more specialized skill in reporting and checking the spending of public funds. Careers in accounting, then, are available in public, private, or governmental accounting.

Accountants

Accounting to Whom and for What?

Accounting traces a sequence of information flows. Figure 9-1 indicates something about who is "accounted to". Employees at the operating level use accounting to "account to" managers (A) who must use accounting to "account to" owners (B). The firm's manager must also "account to" creditors and future creditors (C) and to government agencies (D).

The flow of information is quite varied—and some of the bits of information reaching managers do not go any further, but are retained for internal purposes (E). Firms have many uses of their own, as we will see when we discuss managerial accounting.

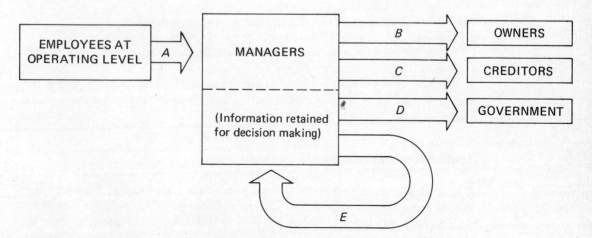

Figure 9-1. The flow of accounting information

> **THEN AND NOW**
>
> **The Accountant's Role Changes**
>
> In the past, when you mentioned the term "accountant", many people pictured a person seated at a desk making handwritten entries in journals and ledgers. That person's job was to "keep the books". He or she was confined to the "back office".
>
> Modern accountants are anything but confined to the "back office". They play a key role in management decision making. Their services are crucial in helping management to make more informed decisions. Public accounting firms, too, are branching out. They no longer confine themselves to "auditing the books". Many such firms now offer management services to their clients. They serve their clients as management consultants.

Much of the same information is contained in flows B, C, D, and E, although it takes different forms and emphasizes different kinds of facts, depending on who is to read it.

What are accounted for are the firm's resources, expressed in dollars and cents (usually cost) terms. In some cases, units other than money are accounted for, but it's usually money.

FINANCIAL ACCOUNTING

Financial accounting is an old and traditional practise designed to "keep tabs" on a firm's assets and to protect its owners' property rights. It also has many built-in safeguards for outsiders who want to know of the firm's financial condition. Financial accounting is a "scorekeeping" process.

Financial accounting is a general system because it includes the entire firm (or entity). Managerial accounting, on the other hand, usually focuses on one activity within the firm.

Transactions and Accounts

Account A basic idea in accounting is that of an account. This is a register of financial value. The set of accounts kept by a firm represents all changes in value which occur. There are four principal kinds of accounts: asset accounts, equity accounts, revenue accounts, and expense accounts.

Assets are things of value which a firm owns. Their value is registered in asset accounts. Usually a large number of asset accounts are kept. Examples are Land, Cash, and Accounts Receivable (money owed us by customers).

An equity account is a register of claims or rights of different groups to a firm's assets. These include the claims of outsiders and the claims of owners.

The claims of outsiders are called a firm's liabilities. An example of such an account is Notes Payable, which shows what we owe in the form of promises or orders to pay.

The claims of insiders, or owners, are kept in owner's equity or capital accounts. Examples of these are Retained Earnings and Common Stock Outstanding.

A revenue account is a register of gross earnings or inflows of value to a firm during a given time period. The most important revenue account is Sales. It includes the total selling price of all goods or services sold during a given time period.

Expense accounts are measures of the "using up" of resources in the normal course of business in a time period. A typical example is Wages Paid.

The term "transaction" is used to describe any change in an asset or an equity. If we buy raw material for cash, we must reduce the Cash account balance and increase the Raw Material account balance.

For a better idea about how the basic types of accounts are related, we will introduce some basic accounting equations which underlie the financial statements.

Some Accounting Equations

An equation represents the fact that two expressions are equal. One expression is placed on the left side of the equality sign and the other expression is placed on the right. For example, $3x + 4 = 16$. **Basic accounting equations are equations which explain the basic system of relationships in financial accounting.**

The first basic accounting equation represents the fact that the assets equal the sum of the equities (claims on assets) and that these equities include liabilities and capital (or owner's equity). This basic accounting equation, then, is

$$\text{Assets} = \text{Liabilities} + \text{Capital}$$

or

$$\text{Assets} = \text{Equities}$$

If one side of the equation (assets) is increased or reduced, then the other side must be increased or reduced by exactly the same amount, just as in algebra. (Some transactions affect only two assets or only two equity accounts. They do not affect the totals on either side of the equation, as in our earlier example concerning the cash purchase of raw materials.)

A second equation reflects current operations. It is as follows:

$$\text{Revenues} - \text{Expenses} = \text{Net profit}$$

YOU BE THE JUDGE!

Should Auditors Be Liable for Failing to Spot Fraud?

The main function of auditors is to analyze a firm's accounting records and to express an opinion on the fairness of those records. Suppose that a team of auditors from a large CA firm audits a client's books. They follow generally accepted accounting principles and standards in checking the books, and they find them to be in good order.

But one month after the audit, the client firm goes bankrupt due to top management's fraudulent activities. The stockholders want to sue the auditing firm. But the auditors claim that they uncovered nothing in their audit to tip them off about the fraud. Therefore, they believe that they were not negligent. Should the auditors be held liable for the investors' losses? YOU BE THE JUDGE!

Current revenue minus the expenses incurred in gaining that revenue equals net profit. Net profit measures the success of the firm's current operations during the period. When expenses exceed revenues, there is a net loss for the period. Net profit is also a measurement of the change in the owner's equity which occurred during the period.

WHAT DO YOU THINK?

Auditors' Responsibility for Uncovering Fraud

Auditors are responsible for checking to see if a client has followed accepted accounting principles in reporting its results of operations. On various occasions, discovery of fraud in the financial reporting of major firms has led to some serious questioning by the government and by the accounting profession itself about the adequacy of present auditing practises. The possibility of sueing auditing firms for not uncovering fraud has also been raised.

A large CA firm, Peat, Marwick and Mitchell and Company, has pointed out that traditional audit practises are not enough to detect management fraud in world-wide businesses tied together by computer. New tools, quite different from ordinary auditing, are needed. Some believe lawyers should accompany CAs; others recommend that auditors should automatically "blow the whistle" to government when suspicions arise instead of bringing discoveries to the firm's management.
WHAT DO YOU THINK AUDITORS SHOULD DO?

The capital account, Retained Earnings, is increased by profitable operations and reduced by payments to owners. Net profit is added to or net loss is deducted from retained earnings. The basic equations are summarized in Figure 9-2.

FINANCIAL STATEMENTS

You have probably seen financial statements in a newspaper or in a firm's annual report. Two of them, the statement of financial position and the income statement, have been widely used for more than a century. These statements are useful for managers, investors, and creditors. They are central to financial accounting. Let's examine them both.

The Statement of Financial Position

A business is a living, functioning entity. **A statement of financial position (also known as a balance sheet) presents a picture of a firm at one point in time.** Family albums filled with snapshots of children as they grow represent a record of that growth for

Statement of financial position

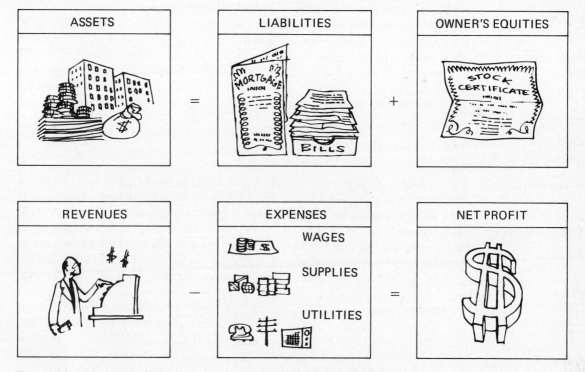

Figure 9-2. Two basic equations which sum up financial accounting

their parents. A file of x rays kept by a doctor records the progress a fractured bone makes in healing. Likewise, a set of statements of financial position, drawn at the first of each year, over a period of years, depicts the rate of growth and nature of the growth of that firm.

Figure 9-3 is a statement of financial position for Gloria's Dress Shop as of January 1, 19*9. This "balance sheet" is a more detailed way of expressing the first basic accounting equation Assets = Equities. The firm's assets are divided into three major classes: current assets, fixed assets, and other assets.

Current Assets

Current asset A current asset is one which the firm normally expects to hold no longer than a year. Examples are cash (currency and chequing account), accounts receivable, merchandise inventories, and prepaid expenses. Accounts receivable are amounts owed to the firm from its normal operations. In this case, they are amounts owed by customers for dresses purchased from the firm recently. As of January 1, 19*9, 43 different customers owed the firm the total sum of $18,000 "on account".

Merchandise inventory consists of all goods purchased for resale but not yet sold. Most retailers must take an inventory of stock at the end of the year to determine the value of their goods for balance sheet purposes. At the end of the year, Gloria's counted $10,000 worth of goods for sale.

Prepaid expenses might include prepaid insurance premiums which have not yet been used by the firm. Gloria's purchased a fire insurance policy on July 1, 19*8, and the premiums were paid one year in advance. Thus, the firm now owns something of value, that is, a prepaid insurance policy, only half of which has been "used".

Two current assets not shown in Figure 9-3 are marketable securities and notes receivable. The former are stocks and bonds of other firms and government bonds held by a firm as short-term investments. The latter represents short-term loans to customers or others.

Fixed Assets

Fixed asset A fixed asset is a tangible resource which is expected to remain useful for more than a year. Such an asset is valued at its cost to the firm. When a firm buys a building, it is listed among the firm's fixed asset accounts at a value equal to its purchase price.

Depreciation As an asset loses value, it suffers depreciation. This loss of value is charged off as an expense, and the stated value of the fixed asset is reduced on the statement of financial position (balance sheet). In Figure 9-3 we have deducted $4,000 from the value of our building for depreciation. This means that this fixed asset is "half used up".

There are many acceptable methods of figuring depreciation in traditional accounting. The simplest is called straight line depreciation. It provides for charging equal parts of the original cost of an asset in each year of its expected life. Thus, the life of the depreciable fixed assets of Gloria's is half over as of the end of the year. They have four years of "life" left.

Current assets:			Current liabilities:		
Cash	$ 6,000		Accounts payable	$10,000	
Accounts receivable	18,000		Accrued expenses payable	1,000	
Merchandise inventories	10,000		Estimated tax liability	7,000	
Prepaid expenses	2,000				
Total current assets		$36,000	Total current liabilities		$18,000
Fixed assets:			Other liabilities:		
Land		4,000	Bonds payable		10,000
Building	8,000		Stockholders equity:		
Less depreciation	4,000	4,000	Common stock	$25,000	
			Retained earnings	1,000	26,000
Other assets:					
Goodwill		10,000			
Total assets		$54,000	Total equities		$54,000

Figure 9-3. Gloria's Dress Shop Ltd. Statement of Financial Position (January 1, 19*9)

One other general class of asset is included in Figure 9-3. It is an intangible asset known as goodwill, which results from years of good business reputation. According to accepted accounting principles, goodwill is assigned a dollar value only when it is bought by the firm. In other words, when the corporation bought the assets of the previous sole proprietorship, it was estimated that the corporation paid $10,000 more than the tangible net worth of the proprietorship (the difference between the value of tangible assets and the liabilities).

Current Liabilities

Under the liabilities section of Figure 9-3, the current liabilities total $18,000. This figure includes the $10,000 due to suppliers, $1,000 of accrued expenses, and $7,000 in estimated taxes owed. **These liabilities are "current" because they will be paid off within a year.**

Accrued expenses illustrate a major accounting principle—accrual. Expenses are charged against revenue in the period in which the firm benefits from them. **An accrued expense is used up but not paid for yet.** Gloria's accrued expenses are the result of some work performed by a salesclerk during the Christmas season who is not yet paid as of January 1, 19*9.

Current liability

Accrued expense

Long-term Liability and Owner's Equity

Gloria's owes bondholders $10,000. This is a long-term liability because it won't be paid off within a year. Stockholders' equity is listed at $26,000 including the original "stated value" of the stock when it was issued and $1,000 in retained earnings (earnings of previous years which have been put back into the firm). Together these add up to what the owner's claims on assets are—owner's equity. If the firm were still a

THEN AND NOW

Certification Programs in Accounting and Finance

Not too long ago, when you talked about a professional certificate in accounting, you were talking about the CA, or Chartered Accountant. To become a CA, a person must first earn a university degree, then complete an accounting-oriented educational program, and then pass a national exam. About half of all CAs work in public accounting firms (CA firms). These firms give external opinions on their client's financial statements. The other half work in business, government, and other non-profit organizations. The main emphasis in CA work is on financial accounting, auditing, and taxation accounting.

In recent years, accounting and financial skills have become increasingly specialized. In addition to the CA, the following certification programs are now available:

1. **Certified General Accountant (CGA)**—To become a CGA, a person must complete an educational program and pass a national exam. A university degree is not required for admission to the program. To be eligible for the program, a person must have an accounting job with a company. There are fewer CGAs than CAs, and in most provinces they are not allowed to give opinions on financial statements of publicly held companies. Almost all CGAs work in private companies, but there are a few CGA firms. Some CGAs work in CA firms.

2. **Registered Industrial Accountant (RIA)**—The goal of the RIAs is to train accountants for industry. To become an RIA, a person must have an accounting position with a company and must complete an educational program (a university degree is not required for admission). Unlike CAs, RIAs have management accounting as their focus, that is, they are concerned about internal uses of accounting data, not external uses as the CAs are.

3. **Certified Financial Analyst (CFA)**—This program is relevant for people in investment jobs. To earn the CFA designation a person must complete an educational program and pass a national exam dealing with securities regulations, investments, and related topics.

---

proprietorship, the owner's equity would simply be listed on the statement of financial position as Gloria Smith, Capital. The owner's equity, then, is the claim of the owner against the firm's resources.

In any case, the sum of the equities is always equal to the sum of the assets. The basic accounting equation always holds. The $54,000 in current, fixed, and other assets have claims upon them (equities) in the amounts of $18,000 (current liabilities), $10,000 (bondholders), and $26,000 (stockholders or owners).

The Income Statement

The balance sheet or statement of financial position shows a "cross-section" of a firm's resources and equities at one point in time. **The income statement, on the other hand, shows what actually happened over a period of time to explain some of the differences between successive balance sheets.**

The income statement summarizes the revenue and expense accounts, just as the balance sheet or statement of financial position summarizes the asset and equity accounts. Figure 9-4 illustrates Gloria's income statement for the period ending December 31, 19*9—one year after the statement in Figure 9-3.

Income statement

Revenues

Gloria's sold $267,000 worth of dresses this year. The selling price of the dresses is used in this valuation rather than the original cost. Sales are "net" because any discounts or returns and allowances granted to customers have been subtracted from the gross sales.

From net sales we deduct the actual cost of goods sold. This cost may be calculated by taking a physical inventory of goods in stock. The cost of this is subtracted from the cost of the previous inventory plus the purchases made during the period. Gloria's had $40,000 in inventory at the beginning of the year, bought $150,000 more during the year, and had $38,000 remaining when the closing inventory was taken ($40,000 + $150,000 − $38,000 = $152,000).

Expenses

The difference between net sales and cost of goods sold is gross profit. In Figure 9-4 we show a few expense accounts including wages, general and administrative expenses,

Gross profit

Net Sales		$267,000 (100%)
Less: Cost of goods sold		152,000 (56.9%)
Gross profit		$115,000 (43.1%)
Less expenses:		
Wages paid	$68,200 (25.5%)	
General and administrative expenses	38,000 (14.2%)	
Interest expenses	1,500 (0.6%)	107,700 (40.4%)
Net profit before taxes		$ 7,300 (2.7%)
Taxes (paid and accrued)		2,000 (0.7%)
Net profit after taxes		$ 5,300 (2.0%)
Less dividends		4,000 (1.5%)
Added to retained earnings		$ 1,300 (0.5%)

Figure 9-4. Gloria's Dress Shop Ltd. Income Statement (year ending 12/31/19*9)

and interest expenses. Wages include Gloria's salary, wages of a bookkeeper and a janitor, and wages and commissions paid to salespersons. General and Administrative Expense includes depreciation, office expenses, utilities, and insurance. Interest Expense includes interest paid to the bank for a loan made and repaid during the year plus interest paid to bondholders. Net Profit Before Taxes, then, is $7,300, from which taxes paid or accrued are deducted in the amount of $2,000. Notice the taxes which apply to this year's operations—whether paid or not—are rightfully deducted from this year's revenue. This is another example of the principal of accrual. Figure 9-5, the new statement of financial position, shows an estimated tax liability in the amount of $1,000. This means that, of the tax bill for the previous year, only half had been paid by the close of the year.

Returning to Figure 9-4, the final deduction ($4,000) is that made for dividends paid to stockholders, leaving $1,300 of the net profit for the year in the business as retained earnings. Thus we have traced the summarized revenue and expense transactions during the year.

The usefulness of the income statement—especially for internal purposes—is increased when it includes a "percentage of net sales" column as does Figure 9-4. This feature makes the income statement easier to compare with those of earlier years, to those of other firms, and to industry averages.

Financial Accounting—Users and Uses

Before we begin a discussion of key ratios, we will describe two specific cases tying in accounting processes and their use.

How Dr. Feldspar Used Accounting

William Feldspar is a retired doctor who has invested a large part of his savings in the common stock of the Marshall Corporation, a producer of steel tubing. He owns 2,000

Current assets:			Current liabilities:			
Cash	$17,000		Accounts payable	$10,700		
Accounts receivable	22,000		Accrued expenses payable	1,000		
Merchandise inventories	38,000		Estimated tax liability	1,000		
Total current assets		$77,000	Total current liabilities		$12,700	
Fixed assets:			Other liabilities:			
Land		4,000	Bonds payable		10,000	
Building	8,000					
Less depreciation	5,000	3,000				
Other assets:			Stockholders equity:			
Goodwill		10,000	Common stock	$69,000		
			Retained earnings	2,300	71,300	
Total assets		$94,000	Total equities		$94,000	

Figure 9-5. Gloria's Dress Shop Ltd. statement of financial position (January 1, 19*0)

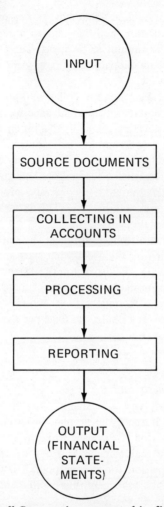

Figure 9-6. How the Marshall Corporation generated its financial statements

shares, which represent about 3 per cent of Marshall's outstanding stock. Dr. Feldspar is interested in getting a reasonable return on his investment in the form of common stock dividends.

This investor does not know any of the corporation's officers or managers personally, and he lives in a town in which none of the firm's plants is located. He needs information, so he must rely on the firm's financial statements to judge the quality of his investment.

Let's review how these statements came to be. Dr. Feldspar could not have made a wise decision about his investment if someone (probably Marshall's treasurer or comptroller) had not set up an information collecting, processing, and reporting system (see Figure 9-6) which did the following things. First, it scanned the firm's operations to identify financially significant events (those having a bearing on the firm's profit) and entered them on some kind of source document. For example, when the office manager bought an order of stationery, he or she signed a purchase order describing the items to be purchased and the amount to be spent. The purchase order, or perhaps a copy of the invoice (list of items shipped) made out by the stationery store, is a source document for the purchase event.

The second step involved recording the dollar amount and the nature of the event in some form of register (account) set up in advance by the comptroller. This is the classifying function. The basic facts found in the purchase order were entered on a punched card, together with a code number indicating the kind of expense. The card (along with many others) was fed into the firm's computer and stored for later use.

At the end of a quarter, Marshall's accounting department took all the stored bits of data, such as the stationery purchase record, and processed them. This included adding up all company expenditures by type. This processing resulted in computing Marshall's quarterly financial statements.

The reporting function has also been fulfilled. The statements, first printed by the firm's computer, were checked by the accounting staff and published for distribution to stockholders. This is how Dr. Feldspar got the financial statement he needed to evaluate his investment. He can calculate earnings per share (total profit divided by the number of outstanding shares of common stock) and other financial ratios from these statements. He can, of course, get similar earnings per share and dividend data about other firms from his stockbroker.

Checking a Prospective Customer

Suppose Hydraspace Company wishes to sign a long-term contract with the Marshall Company that will make Hydraspace the supplier of an important part for Marshall's major product. It's important for Hydraspace to know about Marshall's financial condition so that it can be sure that Marshall can pay on time. Hydraspace will use a number of sources of information for this purpose. They will depend a lot on Marshall's past financial statements—especially those statements which indicate Marshall's ability to pay its current bills.

Marshall's accounting system should be able to provide a summary of its past payment behaviour if the firm wishes to give such information to Hydraspace. In practice, this kind of information is accumulated by independent credit reporting services, such as Dun & Bradstreet Canada Ltd. and sold to their customers.

Once the contract is negotiated and the first shipment of parts has been made, Hydraspace becomes a trade creditor of Marshall. For further insight about how a creditor might interpret financial statements, we now turn to the subject of ratio analysis.

Important Financial Ratios

The numbers on the financial statements take on more meaning when they are related to each other. For instance, the net profit of a firm is more meaningful when it is mathematically related to that firm's sales or to the stockholder's equity. Such relationships are usually expressed as financial ratios or "key" ratios.

A key ratio is a value obtained by dividing one value on a financial statement by another value. A particular firm's financial condition can be judged by comparing several important key ratios of items from its financial statements with typical key ratios of similar types of firms.

Dun & Bradstreet publishes typical key ratios for a variety of types of firms. Such typical ratios are presented in Figure 9-7.

Let's look at several of these ratios and see how Gloria's Dress Shop compares with other women's clothing stores as reported by Dun & Bradstreet. First, let's look at a ratio which measures overall performance—net profit to tangible net worth.

First of all, tangible net worth is equal to stockholders' equity minus goodwill (goodwill is an intangible asset). From Figure 9-5 we see that Gloria's tangible net worth equals $61,300. From Figure 9-4 we see that net profit after taxes is $5,300. The ratio, then, is 5,300/61,300 = .0865 or 8.65 per cent. Now, turn to the "typical ratio" of net profit to tangible net worth in women's ready-to-wear stores as found in Figure 9-7. The sixth item in the table pertains to women's clothing stores. The circled figure in the fifth column represents the median or typical net profit/tangible net worth ratio for such stores in that year as reported by Dun & Bradstreet. Gloria's ratio is lower than average. This suggests that Gloria's is not performing as well as the average store of its type.

Gloria's will compute this ratio and others each year to measure its financial strength. Banks or investors will compute such ratios to see whether they should lend money to Gloria's when the firm requests it. Let's look at some other ratios which measure specific things about a firm.

A short-term key credit ratio which is widely used is the current ratio. It is computed by dividing current assets by current liabilities. On January 1, 19*9 (Figure 9-5), Gloria's current ratio was 77,000/12,700 = 6.06. This is excellent and it means that Gloria's is quite solvent—it can easily pay off current debt, since it has $6.06 of current assets for every $1.00 of current liabilities. Compare this ratio with the typical one in the third column (circled) of Figure 9-7. The average women's ready-to-wear store had a current ratio of only 1.90, that is, only $1.90 of current assets for every dollar of current liabilities.

The sales-to-inventory ratio (also known as stock turnover) is computed from Figures 9-3, 9-4, and 9-5:

$$\frac{\$267,000 \text{ (sales)}}{\frac{1}{2}(\$10,000 + \$38,000) \text{ (average inventory)}} = \frac{267}{24} = 11.1$$

This is better than the average ratio of 5.8 shown in column 8 of Figure 9-7.

Dun & Bradstreet has been a pioneer in the development and analysis of key ratios. For many years, it has published "industry average" ratios for many kinds of firms. This provides benchmarks by which the financial status of similar firms may be evaluated.

An overview of modern accounting must also include discussion of some of the internal management tools. Although financial accounting helps, a good manager needs some managerial accounting tools, too.

Key ratio

Tangible net worth

Current ratio

KEY BUSINESS RATIOS CANADA-CORPORATIONS

LINE OF BUSINESS (and number of concerns reporting)	Cost of Goods Sold Per Cent	Gross Margin Per Cent	Current Assets to Current Debt Times	Profits on Sales Per Cent	Profits on Tangible Net Worth Per Cent	Sales to Tangible Net Worth Times	Collection Period Days	Sales to Inventory Times	Fixed Assets to Tangible Net Worth Per Cent	Current Debt to Tangible Net Worth Per Cent	Total Debt to Tangible Net Worth Per Cent
ALL COMPANIES (221,243)	69.4	30.6	1.09	5.64	12.82	2.27	60	6.6	105.5	170.9	280.1
RETAIL TRADE (39,944)	73.9	26.1	1.46	3.94	26.84	6.80	16	7.0	47.1	107.6	135.0
Auto Acc. & Parts (1,566)	72.0	28.0	1.32	1.90	14.90	7.84	21	5.1	55.3	179.1	248.6
Book & Stat. Stores (474)	57.0	43.0	2.07	2.58	10.44	4.05	86	4.5	22.3	104.4	162.0
Clothing, Men's (1,358)	67.4	32.6	1.65	3.36	11.86	3.53	26	4.0	23.2	84.0	106.2
Clothing, Women's (1,812)	62.7	37.3	1.90	3.25	16.32	5.02	20	5.8	36.5	97.2	151.9
Dept. Stores (141)	66.4	33.6	1.51	1.60	7.34	4.59	14	6.1	34.7	78.7	111.1
Drug Stores (2,339)	69.1	30.9	1.57	2.49	14.35	5.76	11	4.4	35.7	116.2	159.2
Dry Goods (1,514)	67.6	32.4	1.67	2.56	10.99	4.30	23	4.3	30.2	90.3	122.1
Elec. Appliances (225)	68.7	31.3	1.48	2.17	16.28	7.51	44	5.2	51.2	174.4	253.5
Florists (508)	48.7	51.3	1.12	3.54	18.52	5.24	34	17.2	83.0	101.5	159.3
Food Stores (4,070)	82.3	17.7	1.24	1.01	9.10	9.02	4	15.6	58.3	69.7	123.8
Fuel Dealers (635)	69.2	30.8	1.28	3.28	8.18	2.50	62	17.7	44.8	62.1	109.9
Furniture (3,633)	69.0	31.0	1.58	2.57	13.92	5.42	43	5.0	31.2	133.2	175.9
Gas Serv. Stns. (3,554)	74.8	25.2	1.08	2.09	18.42	8.80	24	10.9	95.0	198.3	294.6
General Mdse. (1,712)	80.7	19.3	1.69	2.83	15.04	5.28	22	6.7	55.7	80.2	134.3
Hardware (1,452)	66.9	33.1	1.94	3.01	13.81	4.58	31	4.3	28.6	87.6	133.4
Jewelry Stores (800)	53.0	47.0	1.99	5.62	15.32	2.73	45	3.0	20.2	73.5	97.9
Motor Veh. Dealers (4,113)	83.2	16.8	1.19	1.13	16.82	14.91	13	7.4	75.5	243.4	308.6
Motor Veh. Repairs (2,081)	64.2	35.8	1.25	3.28	19.26	5.87	30	11.6	64.9	112.1	160.6
Shoe Stores (832)	61.4	38.6	1.72	2.06	8.08	3.92	14	4.0	31.5	85.9	119.9
Tobacconists (118)	83.4	16.6	1.48	1.42	13.04	9.16	3	9.3	60.9	123.9	330.4
Variety Stores (518)	64.9	35.1	2.20	3.34	12.90	3.86	18	4.8	49.4	51.6	82.8

Figure 9-7a. Dun & Bradstreet key ratios. (Reprinted by permission of Dun & Bradstreet Ltd.)

HOW THE RATIOS ARE FIGURED—WHAT THEY MEAN

These ratios are based on analysis of a composite sample of audited financial statements published by Business Finance Division of Statistics Canada. The statements were filed by Corporations with Dept. of National Revenue for income tax purposes for the taxation year 1973. These ratios are averages and include both profitable and unprofitable concerns.

COST OF GOODS SOLD

This includes the cost of inventory which has been sold or used, freight or transportation, customs duties, direct labor and factory overhead. Discounts on purchases are deducted. The ratio is a percentage of sales.

GROSS MARGIN

This ratio is derived by deducting the cost of goods sold from the sales figure. It answers the question "Is the markup on cost to selling price sufficient to show a profit?"

CURRENT ASSETS TO CURRENT DEBT

Current Assets are divided by total Current Debt. Current Assets are the sum of cash, accounts receivable, inventories including supplies, and Government securities. Current Debt is the total of bank loans, accounts payable, tax liabilities and amounts due to shareholders. This ratio is one test of solvency.

CURRENT YEAR PROFITS ON SALES

Obtained by dividing the profit declared by the companies, by total sales. This important yardstick in measuring profitability should be related to the ratio which follows. Profits are shown after taxes.

CURRENT YEAR PROFITS ON TANGIBLE NET WORTH

Tangible Net Worth is the equity of stockholders in the business, as obtained by adding preferred and common stock plus surplus (less deficits) and then deducting intangibles. The ratio is obtained by dividing Profits by Tangible Net Worth. The tendency is to look increasingly to this ratio as a final criterion of profitability. Generally, a relationship of at least 10% is regarded as a desirable objective for providing dividends plus funds for future growth.

SALES TO TANGIBLE NET WORTH

Sales are divided by Tangible Net Worth. This gives a measure of the relative turnover of invested capital.

COLLECTION PERIOD

Annual sales are divided by 365 days to obtain average daily credit sales and then the average daily credit sales are divided into accounts receivable. This ratio is helpful in analyzing the collectability of receivables. Many feel the collection period should not exceed the net maturity indicated by selling terms by more than 10 to 15 days. When comparing the collection period of one concern with that of another, allowances should be made for possible variations in selling terms.

SALES TO INVENTORY

Dividing annual Sales by Inventories. This quotient does not yield an actual physical turnover. It provides a yardstick for comparing stock-to-sales ratios of one concern with another or with those for the industry.

FIXED ASSETS TO TANGIBLE NET WORTH

Fixed Assets are divided by Tangible Net Worth. Fixed Assets represent depreciated book values of buildings, leasehold improvements, machinery, furniture, fixtures, tools, and other physical equipment, plus land. Ordinarily, this relationship should not exceed 100% for a manufacturer, and 75% for a wholesaler or retailer.

CURRENT DEBT TO TANGIBLE NET WORTH

Derived by dividing Current Debt by Tangible Net Worth. Ordinarily, a business begins to pile up trouble when this relationship exceeds 80%.

TOTAL DEBT TO TANGIBLE NET WORTH

Obtained by dividing total current debt plus mortgage and other funded debt by Tangible Net Worth. When this relationship exceeds 100%, the equity of creditors in the assets of the corporation exceeds that of owners.

Figure 9-7b. Dun & Bradstreet key ratios. (Reprinted by permission of Dun & Bradstreet Ltd.)

MANAGERIAL ACCOUNTING

Managerial accounting provides information for a manager's own use. It helps management to plan, to measure and control performance, to set prices, and to analyze situations. The biggest difference between managerial and financial accounting is the lack of traditional rules and principles in managerial accounting. Management is free to make up its own systems.

Because managerial accounting practises are less rigid, we will find different systems in every firm. The idea is to keep any kind of record or summary of costs and revenues managers need for planning or control purposes. They might want to evaluate other managers or to judge the success of new products or a new piece of equipment.

Such special accounting is needed because regular financial accounts aren't enough to measure the performance of departments or products or managers within the firm as a whole. They focus, rather, on overall firm profit.

If the Norton Sales Company wishes to evaluate the performance of its sales force, it must maintain adequate records on its activities. Suppose those records show that the average travel expenditure per person has been 11.2 cents per month per square mile of sales territory. After analysis, the sales manager decides to adopt this average amount as a standard. Any salesperson whose expenses exceed the average would be checked out. This shows how accounting can be used to control costs of selling.

Managerial accounting can be used to set minimum order sizes, to decide whether to shut down a production line, or to decide whether a certain retail store staff is over-worked. Other systems could help a manager allocate funds for growth among territories or to set standards for the entertainment of customers. Such managerial accounting activities fall under one of two headings: budgeting or cost accounting.

Budgeting and Cost Accounting

By tradition, financial accounting is not expected to provide predictions of a firm's condition. Predicting is a risky business, but it must be done. Managers use special managerial accounting tools to help them make predictions. One such device is the budget.

Budget   **A budget is a formal dollar-and-cents statement of expected performance.** It is a means of (1) requiring managers to plan carefully for the future; (2) causing managers to examine present and past performance critically; and (3) helping to co-ordinate the plans made by different parts of the firm. A budget may be very specialized, or it may be general. It may be a short-term (one year or less) or a long-term budget.

Sales forecast   **The sales forecast is the starting point for a general (master) budget. It predicts what sales will be over a certain period of time. This forecast depends on what effect the marketing manager thinks the planned changes in the marketing mix will have on sales. Sometimes the sales forecast is tied to a projection of Gross National Product or to industry sales forecasts.** Larger firms, employing a staff of economists and computer facilities, often construct models to predict sales. Whether it be by such a method—or by a simple assumption of a 10 per cent increase over the current year—the sales forecast is a keystone for planning.

Cost accounting includes responsibility accounting and product cost accounting. **Responsibility accounting involves setting up responsibility centres in a firm. These are used to classify cost information so as to evaluate the performance of various parts of the firm and their managers.** The costs of operating the shipping department, for example, may be collected in a shipping department responsibility centre. Figure 9-8 shows how responsibility accounting works.

Responsibility accounting

Product cost accounting systems also use cost centres to allocate all costs to the various products made by a firm. This gives a firm a better idea about which products are profitable and which are not. Some firms use standard product cost accounting systems. Standard costs assigned in such a system are those which should have been

Product cost accounting

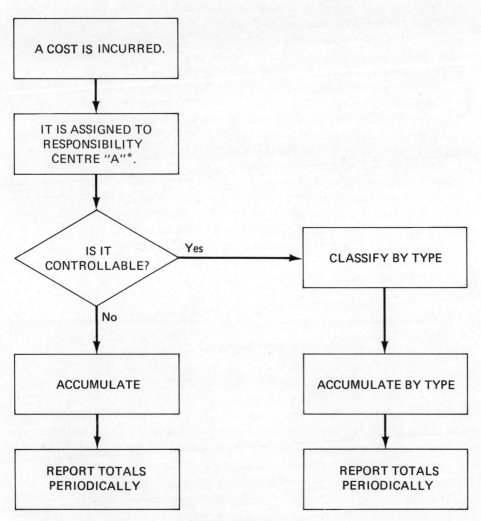

*This same process occurs for *each* responsibility center.

Figure 9-8. The responsibility accounting process

incurred, not those actually incurred. Differences between actual and standard costs are called variances and are charged to variance accounts.

Manual vs. Computer-based Accounting Systems

For many years accounting activities were restricted in what they could do because they were manual. They started with a handwritten record of a transaction which was later copied by hand into a summary book of some kind for monthly or weekly tallying purposes.

Such simplified manual systems still exist in small businesses, but they are gradually being replaced by machine-based systems. Some of these systems depend on modern cash register equipment and some depend on punched cards and computers. Although every accounting system is a "data processing" system, we now restrict use of the term "data processing" to machine-based systems.

A data processing system must do the following: (1) select relevant data describing the business transaction and prepare a document containing such a description; (2) classify and store these input items in appropriate places and summarize them for future use; (3) convert such information into proper form for use by decision makers; and (4) prepare reports, such as financial statements.

A computer-based accounting system requires still another step to convert the source document to a form which can be handled by the system. In the typical case the original document is converted to a punched card and stored either in card form, in the memory of a computer, or on tape. In Chapter 14 we will see how such computer storage makes it easier to classify, summarize, and report.

Mechanical and computer processing are quicker and much more flexible and accurate than manual processing. For very small firms, however, it is doubtful whether these advantages outweigh the cost.

THINK ABOUT IT!

How Should Oil Companies Evaluate Reserves?

One of the important questions in accounting circles today is how to evaluate oil reserves held by firms which explore for and produce petroleum. The two methods used by the industry traditionally have been the "successful" efforts method and the "full-cost" method. The first method takes the cost of the unsuccessful efforts ("dry holes") and charges off these immediately. The second method lumps in all developmental costs (regardless of success of effort) and assigns this value to oil reserves, to be written off over the life of the reserves. WHICH SYSTEM IS MOST REASONABLE? THINK ABOUT IT!

CAREERS IN ACCOUNTING

B

Accounting and related careers may be divided into two categories according to the amount of formal training required. Some, like bookkeeping jobs, cashiers, credit officials, and bank clerks, do not require college preparation. To do these jobs, you must like to work with numbers and be good at detail work and simple mathematics. In most of these jobs personal dependability ranks above human relations skills. Bookkeepers record financial information, make out invoices, check the accuracy of financial records, and prepare tax reports. Cashiers and bank clerks do a lot of counting, checking, and tracing of financial data. Credit officials interview credit applicants; they also check and evaluate credit records. All of these positions may lead to promotion to higher-paying jobs or supervisory-level positions, such as bank teller, head bookkeeper, or credit manager.

The second category does require a university or community college degree. Private accounting jobs such as internal auditor, tax accountant, budget analyst, or cost accountant require considerable technical skill and knowledge of law and business commercial practises. They require some understanding of computer information systems as well as some appreciation of facts peculiar to the specific type of firm they work for. A cost accountant for a furniture manufacturer, for example, would need to understand its manufacturing processes. An internal auditor for a bank would have to know the clerical and control procedures of that bank. Such positions may lead to financial executive posts or to top management jobs in corporations.

Rapidly expanding opportunities exist in private accounting. There is also a great demand for governmental accountants and for public accountants. Government accountants do a variety of jobs similar to private accounting, except that they deal with expenditure and control of public funds. Budget preparation and auditing are major areas of government accounting. Government accountants are also involved in the regulation of private industry and in taxation. Revenue Canada, for example, employs many accountants.

Public accounting firms usually hire graduates of community colleges (CGAs) and universities (CAs and CGAs). Public accountants do auditing work and offer tax and management services. Promotion within these firms often requires that the employee first pass a national exam. A young CA or CGA may progress from junior accountant to auditor to senior accountant and possibly up to manager or partner. Partners are the top positions in firms. In smaller firms the number of steps in promotion is not so great, and travel requirements may not be as great as with large national firms. Earning potential is high for skilled CAs and CGAs. Information is available from the Canadian Institute of Chartered Accountants, 250 Bloor Street East, Toronto, Ontario, Canada M4W 1G5, and from the Certified General Accountants Association of Canada, 535 Thurlow Street, Room 800, Vancouver, British Columbia V6E 3L2.

SUMMARY AND LOOK AHEAD

Accounting's task is three-fold: (1) to "keep score" of the use of financial resources; (2) to draw attention to problems; and (3) to assist in decision making. Financial accounting does the first task; managerial accounting, the second and third.

Accounting principles guide managers in making and interpreting financial statements. Two principal financial statements represent the focus of the financial accounting processes: (1) the statement of financial position (balance sheet) and (2) the income statement. Key financial ratios are applied to values reported on the financial statement to help make inter-firm comparisons and comparisons of financial conditions over time.

Managerial accounting is internally oriented. Planning for financial (and other) resources can be greatly assisted by a variety of budgeting techniques. Cost accounting is another broad component of managerial accounting. It usually takes the form of responsibility accounting or product cost accounting. In either case, costs must be identified and assigned.

A basis has now been established for the subject to be analyzed in our next two chapters—"Financial Institutions" and "Financial Decisions and Insurance". With a

---

YOU MAKE THE DECISION!

Install a New Computer?

The growth of the Dantron Manufacturing Company has been fantastic over the past five years. Sales have jumped from $20,000 per year to more than $2 million. The sales force has grown from two to thirty. The payroll is up, too, and just processing invoices and keeping accounts receivable up-to-date has become nearly impossible under the present manual accounting system. The head bookkeeper recently resigned and it is possible the firm will be fined by Revenue Canada for not keeping up with reports and payments related to tax withholding for its employees.

Dantron's vice-president for finance has been conferring with a representative of a large manufacturer of business computers for the last six months about installation of an electronic data processing system which will do the job.

There are two big problems, however. One is company morale. The clerical and accounting staff people are afraid they will lose their jobs if a computer system is installed. Also, the rental cost of the hardware is more than $60,000 per year to start—more than half of last year's profits. This is a big problem because Dantron is counting on profits to put back into further plant expansion.

How would you make this critical decision? Can you think of any compromise solutions?

basic knowledge of the financial institutions and the basic language of financial statements, it will be possible for us to examine the kinds of financial decisions made by financial managers and how they make them. Let's look at the financial institutions first.

KEY CONCEPTS

Account A register of financial value. There are four basic kinds of accounts: asset accounts, equity accounts, revenue accounts, and expense accounts. Account balances are changed by transactions.

Accountants Individuals who have satisfied the knowledge and experience requirements of a professional group and have been admitted as members of the profession.

Accounting Recording, classifying, summarizing, and interpreting of financial data. The two main types of accounting are financial and managerial.

Accrued expense A liability created by incurring an expense in one period but not paying for it until the next.

Assets A firm's resources such as land, cash, and accounts receivable.

Basic accounting equations Two equations which explain the basic system of financial accounting: (1) Assets = Equities and (2) Revenues − Expenses = Profit.

Budget A financial forecast showing expected income and expenditures for a given period of time.

Current assets Cash or property which can be quickly converted to cash, that is, accounts receivable, inventories, short-term notes receivable.

Current liability Debts which must be paid in less than one year.

Current ratio Measure of liquidity. Current assets divided by current liabilities.

Depreciation A deduction in the balance sheet to indicate the proportion of the cost of a fixed asset which has been treated as an expense over time.

Equity Claims on resources or assets. The two major types of equity are owner's equity and liabilities.

Expense Using up of resources. Expenses are deducted from the revenues in the pursuit of which they were incurred. The result is profit or loss.

Financial accounting The accounting process directed towards communication with people outside the firm.

Fixed asset An asset which a firm expects to have a life of more than one year when it is originally acquired.

Gross profit Sales minus cost of goods sold. Profit before operating expenses are deducted to compute net operating profit.

Income statement A financial statement showing revenues, expenses, and profits of a firm during a given period of time.

Key ratio A financial ratio computed from items on the financial statements of a firm and used to evaluate the credit risk or financial strength of a firm. Typical ratios are published by Dun & Bradstreet.

Liability The claim of a non-owner against a firm.

> # INCIDENT:
>
> ## Alma Wintergreen
>
> Alma Wintergreen's husband recently died and left her $40,000. She wishes to invest it in a good small corporation. The Alpha Corporation has assets of $5 million and liabilities of $4.2 million. For the last five years Alpha Corporation earnings have averaged $10,000. Beta Corporation has assets of $10 million and liabilities of $7 million. Average earnings for the Beta Corporation for the last five years have been $20,000.
>
> Questions:
> 1. Would you advise Mrs. Wintergreen to invest in either of these corporations? Explain your answer.
> 2. Which of the critical ratios are you able to compute for Alpha and Beta?
> 3. What other information could you suggest that she get before making a decision?

Managerial accounting The accounting process when internally directed to facilitate the firm's management.

Owner's equity Claim of owners against resources of the firm. Often referred to as "capital".

Product cost accounting Systems for allocating costs to products produced by a firm.

Responsibility accounting A method of classifying cost information and thereby evaluating the performance of the components of the firm (responsibility centres) and their managers.

Revenue Inward flow of value to a firm—mostly from sales.

Sales forecast An estimate of the sales which will be made in a future period of time. Makes budget construction easier.

Statement of financial position A statement or list of the assets and equities of a firm at one point in time. Also called a balance sheet.

Tangible net worth A conservative measure of owner's equity which excludes goodwill from assets.

Transaction Any financially significant event. A transaction causes a change in the balance of a firm's accounts (usually two at a time).

QUESTIONS FOR DISCUSSION

1. What kinds of information are communicated by means of accounting? Give two examples.
2. To whom is the accounting process directed? What are its three principal tasks?

3. Must all accountants be CAs? Discuss.
4. What do we mean when we say that financial accounting is relevant?
5. Is it possible for a transaction to occur without affecting the balance of any account? Discuss.
6. What are the basic accounting equations? Show how a change in one side must result in a change in the other.
7. What is the functional relationship between a statement of financial position (balance sheet) and an income statement?
8. Define working capital. In what sense is it "working"?
9. What kind of business is Dun & Bradstreet in? Ask any manager what his or her relationship with this firm is, if any.
10. What is meant by the principle of accrual? Explain by giving an example involving rental expense.
11. Give an example of a simple budget. Your weekly expenses will do.
12. In what way could an accounting system help to control salespeople's entertainment expenses? Explain.

INCIDENT:

Mary Plunkett

When Mary Plunkett attended her first stockholders' meeting of the Dingley Corporation, she came in waving a copy of the financial statements for the previous year at the board's chairperson. "How do you expect a retired schoolteacher to make heads or tails out of these things?" she shouted. There was general laughter and nothing was done about her complaint. However, many of the other stockholders, after leaving the meeting, realized that Mrs. Plunkett was right. More than fifty letters of agreement with her position reached the president's office the next week.

Questions:
1. Examine a large corporation's financial statements (they are inside the Annual Report) and look them over carefully. Do you get the same feeling as Mrs. Plunkett? Explain.
2. What suggestions can you make to improve the understandability of these statements?
3. Why is it that such standard technical language is used?

10

Financial Institutions

10 FINANCIAL INSTITUTIONS

OBJECTIVES: After reading this chapter you should be able to:

1. Compare the services offered to business and non-business customers by chartered banks, life insurance companies, trust companies, credit unions or caisses populaires, factoring companies, sales finance companies, consumer finance companies, government financial institutions, venture capital funds, and pension funds.
2. Explain why the chartered banks are important to business firms.
3. List the different types of money in Canada.
4. Develop an example showing how the banking system creates money.
5. Demonstrate how the Bank of Canada controls the money supply.
6. List and identify the major types of credit instruments used by business firms.
7. Tell the difference between common stock and preferred stock.
8. Identify three concepts of value with respect to common stock.
9. Give examples of value with respect to common stock.
10. Give examples of several types of preferred stock.
11. Compare the underwriting and brokerage function of investment dealers.
12. Tell the difference between listed and unlisted securities.
13. Compare the workings of the over-the-counter market with the workings of the securities exchange.
14. Translate a quote for a stock or bond as reported on the financial pages of newspapers into dollar-and-cents terms.
15. Compare speculating with investing.
16. Compare the workings of commodity exchanges with the workings of securities exchanges.

KEY CONCEPTS: In reading the chapter look for and understand these terms:

MONEY
CHARTERED BANKS
DEMAND DEPOSITS
TIME DEPOSITS
PRIME RATE OF INTEREST
BANK RATE
RESERVE REQUIREMENT
TRUST COMPANY
CREDIT UNION
LIFE INSURANCE COMPANY
FACTORING COMPANY
SALES FINANCE COMPANY
CONSUMER FINANCE COMPANY
VENTURE CAPITAL FIRM
PENSION FUNDS
INDUSTRIAL DEVELOPMENT BANK
PROMISSORY NOTE
DRAFT
TRADE ACCEPTANCE
CERTIFIED CHEQUE
PUBLIC MARKET
COMMON STOCK
PREFERRED STOCK
BOND
UNDERWRITING
BROKERAGE FUNCTION
SECURITIES EXCHANGES
OVER-THE-COUNTER MARKET
SPECULATIVE TRADING
MARGIN TRADING
SHORT SELLING
MUTUAL FUND
PROSPECTUS
COMMODITY EXCHANGES
CASH TRADING
FUTURES MARKETS

All firms need money to begin a business and to remain in it. Money is the life-blood of a firm—it enables a firm to buy buildings, equipment, inventories, and pay its employees. Money flows into a firm as sales revenues and flows out as the expense of doing business and as a return to the owners.

But where does a firm get money? Some of it comes from the owner's (or owners') original investment in the firm. Some of it comes from profits which are reinvested in the firm. Often, however, that is not enough. Most firms need the services of financial institutions to provide them with the money they need.

We examine these financial institutions in this chapter, starting with financial intermediaries. There are many different types of financial intermediaries, each of which serves business firms and individuals in different ways. Two of the most important financial intermediaries are banks and life insurance companies.

We also examine the workings of investment dealers and securities exchanges to see how they help business firms get the money they need. These financial institutions facilitate the growth of corporations. In fact, most firms would stay very small if they could not get money from people who are willing to lend it as bondholders and/or from people who are willing to invest their money to become owners or stockholders.

THE FINANCIAL SYSTEM

The financial system is very important to business firms in Canada. A financial system is composed of organizations and individuals who are sources of funds, users of funds, and/or who facilitate the flow of funds from sources to users. (See Figure 10-1.) People or organizations with surplus funds will want to earn a return on them. An individual may put his or her money into a savings deposit, buy a bond or a life insurance policy, or invest in the stock market. A business firm, depending upon how long the money will be available, may deposit its money in a bank on a short-term basis, buy treasury bills, or buy long-term securities. Charitable organizations and governments may also have surplus funds they wish to invest for certain periods of time.

Without money, business as we know it could not exist. Money enables us to progress beyond barter (exchanging goods for goods) because it permits specialization and exchange.

Money is anything people will generally accept in payment of debts. Paper currency and coins issued by a recognized government are legal tender in that nation. If you owe a debt which is stated in money terms, your creditor must accept your payment in legal tender.

But that is not all there is to money. Most Canadian workers, for example, are paid with cheques. Many workers deposit their cheques in their chequing accounts and pay their bills with cheques. This is possible because of our modern banking system. **Money, therefore, means paper currency, coins, and chequing account balances.**

Money

All people and organizations are users of funds. Sometimes they are unable to raise all the funds they require from personal or internal sources. In this situation, money must be raised from other people or institutions who have surplus funds. Individuals

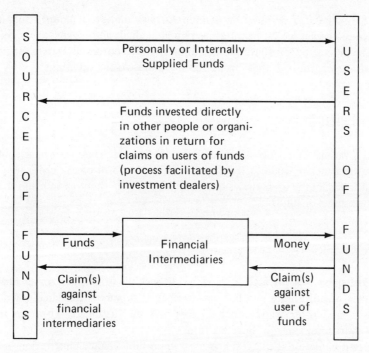

Figure 10-1. Sources of funds, users of funds, financial intermediaries, and investment dealers.

borrow money for certain purposes. Business firms raise money to finance projects. Governments also borrow money.

Financial institutions facilitate the flow of funds from sources to users. There are many different types of financial institutions. It is important to understand them if you are in an organization which either has surplus funds to invest or is in a deficit position and must raise funds. It is also important to understand the institutions if you are interested in a career in one of them.

There are two basic types of financial institutions. One type is the financial intermediary. Financial intermediaries issue claims against their assets to sources of funds and in return provide funds to users in exchange for claims against them. For example, a bank will accept funds (deposits) from people and organizations. The depositor has a claim against the bank for the value of the money he or she has deposited. The bank will then make loans to individuals and organizations in return for claims against these borrowers. Life insurance companies offer a different type of claim against their assets to the public, and they in turn also invest the money they have accumulated.

The other basic type of financial institution facilitates the direct transfer of funds from sources of funds to users. The source, or supplier of funds, receives a claim against the user. A common example is Canada Savings Bonds. The sources of funds (individuals who buy bonds) receive claims against the Government of Canada (Savings

Bonds). Similarly, if a corporation wanted to raise money, it could issue stocks or bonds (claims) which would be purchased by individuals and organizations (sources of funds). The process of transferring funds directly from sources to users is facilitated by investment dealers through their brokerage and underwriting functions.

FINANCIAL INTERMEDIARIES

There are a variety of financial intermediaries in Canada. They vary in size, in importance, in the types of sources they appeal to, in the form of the claim they give to sources of funds, in the users they supply credit to, and in the type of claim they make against these users of funds.

Chartered Banks

Chartered bank

A chartered bank is a privately owned, profit-seeking firm which serves individuals, non-business organizations, and businesses. Chartered banks offer chequing accounts, make loans, and offer many other services to their customers. They are the main source of short-term loans for business firms.

Chartered banks are the largest and most important financial institution in Canada. They offer a unique service. Their liability instruments, or claims against their assets, are generally accepted by the public and by business as money or as legal tender. Initially these liability instruments took the form of bank notes issued by individual banks. The Bank Act changes of 1944 removed the right to issue bank notes.

Canada has a branch banking system. Unlike the United States, where there are hundreds of banks, each having a few branches, in Canada there are only eleven banks, each with many branches. Of these, five account for about 90 per cent of total bank assets. (See Table 10-1.) Some of them also have branches in other countries. There are more than 7,000 branch bank offices in Canada, or about one for every 3,300 people.

Services Offered by Banks

Table 10-2 lists several types of services offered by chartered banks. Banks are chartered by the federal government and are closely regulated when they provide these services.

Chartered banks provide a financial intermediary service by accepting deposits and making loans with this money. Banks make various types of loans to businesses. When applying for a business loan, it is wise for the manager to remember that the banker is interested in making the loan because it will make money for the bank. The banker is also interested, however, in how the loan will be repaid and how it will be secured. When making an application, it is important to indicate what the money is required for and how it will be repaid. A brief written statement accompanied by a cash-flow analysis is a useful approach when applying for a loan.

Table 10-1. Characteristics of chartered banks in Canada

	Assets		Revenue 1976	Balance of Revenue (Deficit) 1976		Branches in Canada 1975	Canadian Employees 1975
	1967	Oct. 1976		Before Tax	After Tax		
	(Millions of dollars)						
The Royal Bank of Canada	7,810	28,832	2,437	267.4	157.4	1,432	28,655
Canadian Imperial Bank of Commerce	7,516	26,104	2,208	273.9	145.9	1,660	27,855
Bank of Montreal	6,345	20,492	1,799	174.8	95.9	1,243	25,770
Bank of Nova Scotia	4,303	18,181	1,508	213.5	116.9	924	18,430
The Toronto Dominion Bank	3,568	16,192	1,298	170.3	92.2	891	14,766
Banque Canadienne Nationale	1,281	5,675	498	45.1	24.5	479	7,304
La Banque Provinciale du Canada	627	3,624	316	31.8	17.0	324	4,573
The Mercantile Bank of Canada	198	1,708	150	20.6	13.9	12	483
Bank of British Columbia	**	844	77	5.9	3.1	30	888
Unity Bank of Canada	**	180	18	(3.5)	(3.5)	23	303
Canadian Commercial & Industrial Bank	**	16	—	—	—	—	—
Total	31,648	121,848	10,309	1,199.8	663.3	7,018	129,027

Source: Government of Canada, *Royal Commission on Corporate Concentration,* 1978. Reproduced by permission of the Minister of Supply and Services Canada.

Table 10-2. Some important services provided by banks to business firms

1. Offer chequing accounts.
2. Offer savings deposits.
3. Offer safe-deposit boxes.
4. Offer personal term deposits.
5. Store idle cash in certificates of deposit.
6. Make short-term loans (1–16 months).
7. Make long-term loans.
8. Make loans to a firm's customers.
9. Exchange Canadian dollars for foreign currencies.
10. Exchange foreign currencies for Canadian dollars.
11. Finance export operations.
12. Give advice to businesspeople on financial matters.
13. Handle details of registration of corporations' stocks.
14. Buy and sell securities.
15. Safeguard property entrusted to them.

Bank Deposits

Demand deposit

One type of deposit a customer can make in a bank is a demand deposit. **A demand deposit is a chequing account.** Customers who deposit coins, paper currency, or other cheques in their chequing accounts can write cheques against the balance in their accounts. Their banks must honor these cheques immediately. That is why chequing accounts are called demand deposits.

Time deposit

The other type of deposit a customer can make in a chartered bank is a time deposit. **A time deposit is one which is to remain with the bank for a period of time. Interest is paid to depositors for the use of their funds.**

There are two types of time deposits. The most popular is the regular passbook account. Although a bank can require notice before withdrawals can be made, this is seldom done. These accounts are intended primarily for small individual savers and non-profit organizations.

DID YOU KNOW?

Banks in Canada*

Since 1820, 159 banks have been chartered in Canada. Ninety-eight of these have actually become active banks, 45 failed or were closed down, and 41 merged with other banks. There are currently 11 active banks in Canada.

*Based upon statistics compiled in E. P. Neufeld, *The Financial System of Canada* (Toronto: Macmillan of Canada, 1972). Dr. Neufeld's book provides an excellent discussion of chartered banks and other financial institutions in Canada.

> ## THEN AND NOW
>
> ### Banking and Credit
>
> Because people used to believe more strongly in thrift and staying out of debt and because people were not generally as affluent, most purchases were made for cash. Most people, even in the early twentieth century, believed that you should "save up" enough money to "pay cash" for your purchases—even the purchase of a house.
>
> Today, many Canadian households use Master Charge and/or VISA cards to pay for purchases. Most households have chequing accounts. The "cashless" society is getting closer.
>
> The age of electronic banking is dawning. Already, some people use cards to deposit or withdraw cash, transfer funds among accounts, and borrow money. Computer terminals located in retail stores and shopping centres can serve as branch banks of financial institutions. The "chequeless" society is arriving. These and other developments will revolutionize banking in the near future.

Another type of time deposit is the savings certificate. This is a deposit which is made for a certain period of time. It can range from 28 days to several years. Savings certificates are available to all savers. The interest rate paid on a savings certificate is higher than that paid on a regular passbook account, but a saver must give up interest if a savings certificate is cashed in before its maturity date.

Bank Loans

Banks are the major source of short-term loans for business. Although banks make long-term loans to some firms, they prefer to specialize in providing short-term funds to finance inventories and accounts receivable. Many loans made to businesses are secured by inventory under section 88 of the Bank Act. Section 86 of the Bank Act allows banks to make loans against the security of bills of lading and warehouse receipts. Section 82 allows banks to take as security hydrocarbons in store or under the ground.

Borrowers pay interest on their loans. Large firms with excellent credit records pay the prime rate of interest. **The prime rate of interest is the lowest rate charged to borrowers.** This rate changes from time to time owing to changes in the demand for and supply of loanable funds and also to policies of the Bank of Canada.

Prime rate of interest

A secured loan is backed by collateral such as accounts receivable or a life insurance policy. If the borrower cannot repay the loan, the bank sells the collateral. An unsecured loan is backed only by the borrower's promise to repay it. Only the most creditworthy borrowers can get unsecured loans.

TWO POINTS OF VIEW

What a Bank Should Be

McDonald Stevenson:
McDonald Stevenson is the manager of a large bank branch in the Maritimes. He believes that a bank's main job is to protect the funds entrusted to it by its depositors. Loans should be granted only to firms with proven profit records and ability to pay. Mr. Stevenson is very skeptical of trying to market bank services in the same way that some of the consumer goods firms like Procter and Gamble do. "We're a different breed of business firm. Our depositors expect us to be cautious with their money. After all, it is their money. We can't take the risks that other firms can take. For us, slow growth is the best type of growth. Swinger-type bankers will sooner or later wind up in trouble—writing off hundreds of thousands of dollars in bad debts and damaging their chances for promotion within the bank."

Rosalie Hedgerow:
Rosalie Hedgerow is the manager of a bank branch in Ontario. She believes that a bank's main mission is to serve the financial needs of borrowers and savers. To her, a bank is a sort of middleman operation between people who have money to lend (depositors) and people who need loans (borrowers). The greater the variety of services a bank can offer, the better it is fulfilling its mission, according to Ms. Hedgerow. "We're not all that different from other firms. We need to offer new services that meet our customers' needs. The days of bars on teller cages are gone forever. We have a responsibility to serve the financial needs of new firms as well as those of older firms. The trouble with some bankers is that they are so conservative about lending they are willing to lend only to firms which really don't need loans. At my bank, we are heavily involved in consumer finance, lending to new firms with good ideas, and many other types of programs to better serve our business and non-business customers. Bankers who crow about the virtues of slow growth are really making excuses for their own failures as executives."

Deposit Expansion
Suppose you saved $100, took it to a bank, and opened a chequing account. Some portion of your $100 is likely to stay in your account. Your bank can earn interest by lending some of it to borrowers.
    Banks must keep some portion of their demand deposits in vault cash or as deposits with the Bank of Canada. These are legal reserves. Let's assume that the reserve

requirement is 10 per cent. Your bank, then, must keep $10 of your $100 deposits in legal reserves. It therefore has $90 to lend.

Now, suppose Tom Powers borrows that $90 from your bank. Tom has $90 added to his chequing account. Assume that Tom writes a cheque for $90 payable to the Acme Stores. Acme's bank ends up with a $90 deposit. But Acme's bank has to keep only 10 per cent of $90 ($9.00) in legal reserves. Acme's bank, therefore, can lend out $81.00.

This is the process of deposit expansion. It can continue as shown in Table 10-3. The banking system creates money in the form of demand deposits. Of course, the process of deposit expansion is much more complex in practice. General economic conditions, for example, influence the willingness of bankers to make loans and the willingness of borrowers to borrow.

As you can see from Table 10-3, your original deposit of $100 could result in an increase of $1,000 in new deposits for all banks in the commercial banking system. Remember, we are assuming a reserve requirement of 10 per cent. Thus, your original deposit of $100 could expand by 10 times (the reciprocal of the reserve requirement, 100/10), or to $1,000. Our example assumes that no borrower takes part of his or her loan in cash and that the banks want to lend as much as they legally can. Otherwise, the increase would be less than $1,000.

The Bank of Canada

The Bank of Canada was formed in 1935 and is Canada's central bank. It has an important role to play in management of the Canadian economy and in regulation of certain aspects of chartered bank operations.

Table 10-3. How the banking system creates money*

Bank	New deposit	New loan	Legal reserve
Your bank	$ 100.00	$ 90.00	$ 10.00
Bank 2	90.00	81.00	9.00
Bank 3	81.00	72.90	8.10
Bank 4	72.90	65.61	7.29
Bank 5	65.61	59.05	6.56
Bank 6	59.05	53.14	5.91
Bank 7	53.14	47.83	5.31
Bank 8	47.83	43.05	4.78
Bank 9	43.05	38.74	4.31
Total for first nine banks	$ 612.58	$551.32	$ 61.26
Total for entire banking system	$1,000.00	$900.00	$100.00

*Assuming a reserve requirement of 10 per cent.

The Bank of Canada is managed by a Board of Governors composed of a governor, a deputy governor, and twelve directors who are appointed from different regions of Canada. The directors, with Cabinet approval, appoint the governor and deputy governor. The deputy minister of finance is also a non-voting member of the board. Between meetings of the board, which are normally held eight times a year, an executive committee acts for the board. This committee is composed of the governor, the deputy governor, two directors, and the deputy minister of finance. The executive committee meets at least once a week.

Operation of the Bank of Canada

The Bank of Canada plays a very important role in management of the money supply in Canada (see Table 10-4).

If the Bank of Canada wants to increase the money supply, it purchases government securities. People will sell their bonds for money, which they deposit in their banks. This increases bank reserves and their ability to make loans.

If the Bank of Canada wants to decrease the money supply, it sells government securities. People spend money to buy bonds. This draws down bank reserves and reduces their ability to make loans.

Member Bank Borrowing from the Bank of Canada

Bank rate

The Bank of Canada is the lender of last resort for chartered banks. **The rate at which chartered banks can borrow from the Bank of Canada is called the rediscount, or bank rate.** It serves as the basis for establishing the chartered banks' prime interest rate. By raising the rediscount rate the Bank of Canada can depress the demand for money, and by lowering it the demand for money can be increased. In practice, chartered banks seldom have to borrow from the Bank of Canada, however, the bank rate is an important instrument of monetary policy as a determinant of interest rates.

Setting the Reserve Requirement

Reserve requirement

The Board of Governors sets the reserve requirement. **The reserve requirement is the percentage of its deposits member banks have to keep in vault cash or as deposits with the Bank of Canada.** Lowering the reserve requirement increases the money supply. Raising it decreases the money supply.

The reserve requirement differs for demand and time deposits. For demand deposits it is 4 per cent and for time deposits it is 12 per cent.

Trust Companies

Trust company

Another financial intermediary which serves individuals and businesses is the trust company. **A trust company safeguards property—funds and estates—entrusted to it.** It

Table 10-4. Bank of Canada actions and business activity

To stimulate the economy, the Bank of Canada:
1. Buys government securities.
2. Lowers the bank rate.
3. Lowers the reserve requirement.

To slow down the economy, the Bank of Canada:
1. Sells government securities.
2. Raises the bank rate.
3. Raises the reserve requirement.

also may serve as trustee, transfer agent, and registrar for corporations and provide other services.

A corporation selling bonds to many investors appoints a trustee, usually a trust company, to protect the bondholders' interests. A trust company can also serve as a transfer agent and registrar for corporations. A transfer agent records changes in ownership of a corporation's stock. A registrar certifies to the investing public that stock issues are correctly stated and in compliance with the corporate charter. Other services include preparing and issuing dividend cheques to stockholders and serving as trustee for employee profit-sharing funds. Trust companies also accept deposits and pay interest on them.

Caisses Populaires and Credit Unions

Credit unions are important to business because they lend money to consumers to buy durable goods like cars and furniture. They also lend money to businesses. **A caisse populaire, or a credit union, is a cooperative savings and lending association formed by a group with common interests.** Members (owners) can add to their savings accounts by authorizing deductions from their pay cheques or by making direct deposits. Members can also borrow short-term, long-term, or mortgage funds from the credit union. Credit unions also invest substantial amounts of money in corporate and government securities.

Credit union

Life Insurance Companies

An important source of funds for individuals, non-business organizations, and businesses is the life insurance company. **A life insurance company is a mutual or stock company which shares risk with its policyholders for payment of a premium.** Some of the money it collects as premiums is loaned to borrowers. Life insurance companies are substantial investors in real estate mortgages and in corporate and government bonds. Next to chartered banks, they are the largest financial intermediaries in Canada.

Life insurance company

Factoring Companies

Factoring company

An important source of short-term funds for many firms is the factoring company. **A factoring company (or factor) buys accounts receivable (amounts due from credit customers) from a firm.** It pays less than the face value of the accounts but collects the face value of the accounts. The difference, minus the cost of doing business, is the factor's profit.

A firm which sells its accounts receivable to a factor "without recourse" shifts the risk of credit loss to the factor. If an account turns out to be uncollectable, the factor suffers the loss. However, a factor is a specialist in credit and collection activities. Using a factor may enable a client to expand sales beyond what would be practical without the factor. The client trades accounts receivable for cash. The factor notifies the client's customers to make their payments to the factor.

Financial Corporations

There are two types of financial intermediaries referred to as financial corporations. These are sales finance companies and consumer loan companies.

Sales Finance Companies

Sales finance company

A major source of credit for many firms and their customers is the sales finance company. **A sales finance company specializes in financing instalment purchases made by individuals and firms.**

When you buy a durable good from a retailer on an instalment plan with a sales finance company, the loan is made directly to you. The item bought serves as security for the loan. Sales finance companies enable many firms to sell on credit, even though the firms could not afford to finance credit sales on their own.

General Motors Acceptance Corporation (GMAC) is a sales finance company. It is a "captive" company because it exists to finance instalment contracts resulting from sales made by General Motors. Industrial Acceptance Corporation is a large Canadian sales finance company.

Sales finance companies also finance instalment sales to business firms. Many banks also have instalment loan departments.

Consumer Finance Companies

Consumer finance company

An important source of credit for many consumers is the consumer finance company. **A consumer finance company makes personal loans to consumers.** Often these loans are made on a "signature basis," and the borrower pledges no security (collateral) for the loan. For larger loans, collateral may be required, such as a car or furniture.

These companies do not make loans to businesses. But they do provide the financing which turns many would-be customers into actual paying customers. Household Finance Corporation is an example of a consumer finance company.

Venture Capital or Development Firms

A venture capital, or development firm will provide funds for new or expanding firms which are thought to have significant potential. Venture capital firms obtain their funds from initial capital subscriptions, loans from other financial intermediaries, and from retained earnings.

Venture capital firms may provide either equity or debt funds to firms. Financing new, untested businesses is risky, so venture capital firms want to earn a higher-than-normal return on their investment. The ideal situation would be an equity investment in a company which became very successful and experienced substantial appreciation in its stock value.

Venture capital firm

Pension Funds

Pension plans accumulate money which will be paid out to plan subscribers at some time in the future. The money collected is then invested in corporate stocks and bonds, government bonds, or mortgages.

Pension funds

Government Financial Institutions and Granting Agencies

In Canada there are a number of government suppliers of funds which can be important to business. In general, they supply funds to new and/or growing companies. However, established firms can also use some of them.

The Industrial Development Bank (IDB), a subsidiary of the Bank of Canada, makes loans to business firms. The Federal Business Development Bank (FBDB) took over operation of the IDB in 1975. The IDB was set up to make term loans, primarily to smaller firms which are judged to have growth potential but are unable to secure funds at reasonable terms from traditional sources. It also expanded IDB services by providing proportionally more equity financing and more management counselling services.

Industrial Development Bank

There are also a variety of provincial industrial development corporations which provide funds to developing business firms in the hope that they will provide jobs in the province. These are discussed in Chapter 15.

The federal government's Export Development Corporation can finance and insure export sales of Canadian companies. The Canadian Mortgage and Housing Corporation (CMHC) is involved in providing mortgages and in guaranteeing them. The CMHC is, therefore, very important to the construction industry.

There are a number of federal and provincial programs specifically designed to provide loans to agricultural operators. Most of these, with the exception of farm improvement loans which guarantee bank loans to farmers, are long-term loans for land purchase.

In addition to these activities, governments are also involved in providing grants to business operations. The Department of Regional Economic Expansion (DREE) is one of the more notable examples. It provides grants for certain types of business expan-

International Sources of Funds

It should be noted that not all of the financing requirements of Canadian businesses and governments are met from within Canada. Foreign sources of funds are very important. The financial institutions of Canada—financial intermediaries and investment dealers—play a role in facilitating the flow of funds into the country.

The Canadian capital market is one part of the international capital market. Canadian provinces borrow extensively in foreign markets such as in New York. Canadian corporations likewise find it attractive to borrow in foreign markets.

Foreign sources of funds have been important to the economic development of Canada. We are now at a stage where we are concerned about foreign ownership of Canadian firms. However, projections of Canada's future capital requirements indicate that we will continue to need foreign sources of funds. Canadian financial institutions will continue to play a large role in making these funds available.

CREDIT INSTRUMENTS

A variety of credit instruments, or claims against users, are offered by users of funds. Some are used for short-term funds and some for long-term funds. Some of the more common forms of credit instruments are discussed below.

Promissory Notes

Promissory note

A common credit instrument is the promissory note. **A promissory note is a written promise to pay a certain sum of money (principal and interest) to a certain person at a certain time in the future.** The borrower is the "maker". The lender is the "payee". When a firm borrows from a bank, it gives the bank a promissory note. (See Figure 10-2.)

If a promissory note is negotiable, the payee (lender) can sell it to another party. Suppose Firm A sells goods to Buyer B and takes B's promissory note for $1,000. The note is to be paid in 180 days and carries an interest rate of 10 per cent per year. If Firm A wanted to get its money sooner, it could sell the note to its bank. Firm A gets the $1,000 (or perhaps less if the bank wants to earn more than 10 per cent per year on a loan). Buyer B pays $1,000 to the bank on the due date plus interest. If Firm A sells the note with recourse, the bank can collect from Firm A if Buyer B fails to make good on the note.

Usually, however, banks prefer to discount notes. Suppose Buyer B signed a note for the $1,000 loan with the bank. The bank discounts the note by deducting its interest in advance. Buyer B gets $950, not $1,000. Buyer B pays 10 per cent interest on a note of $1,000 but only gets $950 in actual money. The effective rate of interest earned by the bank is roughly 10.5 per cent.

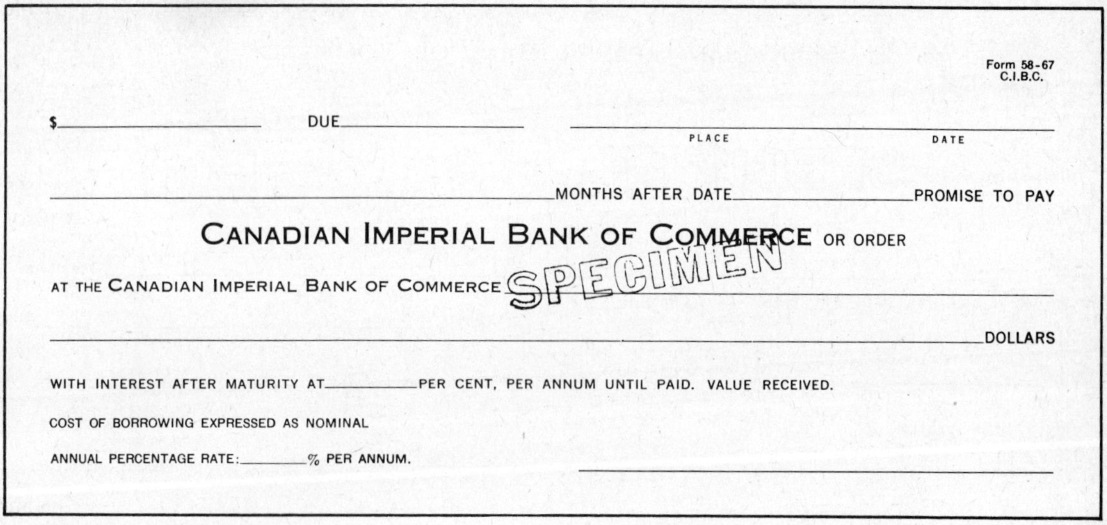

```
                                                              Form 58-67
                                                              C.I.B.C.
$_____        DUE_____        _____    _____
                                           PLACE               DATE
_____MONTHS AFTER DATE_____PROMISE TO PAY
              CANADIAN IMPERIAL BANK OF COMMERCE OR ORDER
AT THE CANADIAN IMPERIAL BANK OF COMMERCE  SPECIMEN  _____
_____DOLLARS
WITH INTEREST AFTER MATURITY AT_____PER CENT, PER ANNUM UNTIL PAID. VALUE RECEIVED.
COST OF BORROWING EXPRESSED AS NOMINAL
ANNUAL PERCENTAGE RATE:_____% PER ANNUM.
```

Figure 10-2. A Promissory Note

Drafts

Another credit instrument is the bill of exchange, or draft. **A draft is a written order made by one party ("the drawer"), addressed to a second party ("the drawee"), ordering the drawee to pay a third party ("the payee") a certain amount of money.** You use a draft when you write a cheque on your chequing account. You (the drawer) order your bank (the drawee) to pay a certain amount of money to a third party (the payee). (See Figure 10-3.)

Draft

A draft is an order to pay, not a promise to pay. A demand or sight draft is payable whenever the payee presents it to the drawee. A time draft orders payment at a specified date.

As you know, you can make a cheque payable to "cash". The cheque is not made out to a specified payee. It is called a bearer draft. Anyone who holds the cheque can cash it.

Drafts often are used by firms that sell to customers with unknown or risky credit records. This is better than selling to them on account. Drafts also are used to encourage customers to pay overdue accounts.

Acceptances

One type of time draft used in selling goods is a trade acceptance. If Seller Company is selling goods to Buyer Company, Seller can draw a draft and send it to Buyer. When Buyer signs it, it becomes a trade acceptance. It is binding on Buyer. It is also negotiable, which means that Seller can sell the draft. **A trade acceptance, therefore, is a time draft drawn by a drawer and sent to the drawee.** When the drawee (Buyer) accepts it by signing it, it becomes binding on Buyer.

Trade Acceptance

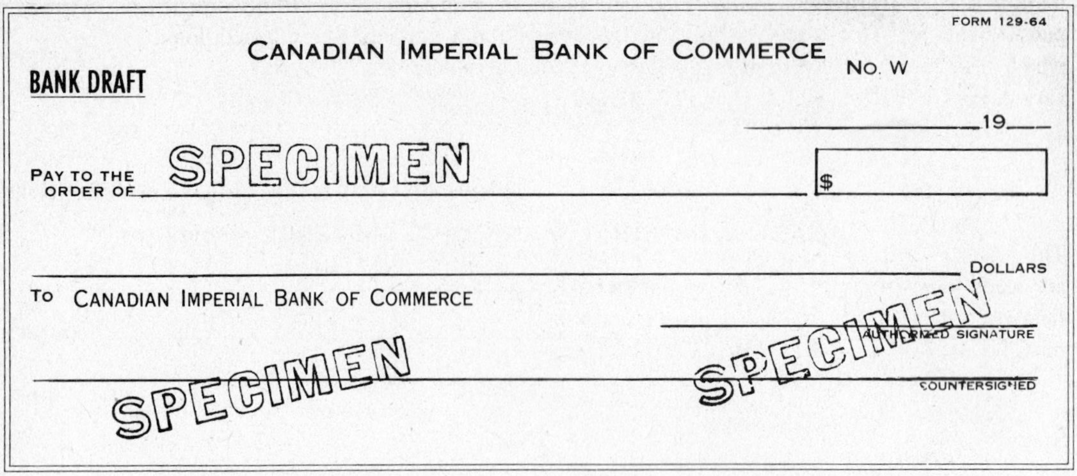

Figure 10-3. A Bank Draft

Another type of time draft is a bank acceptance. Before ordering goods from Seller, Buyer arranges with its bank to accept drafts drawn against the bank to pay for the goods. The bank issues a letter of credit saying that it will accept the seller's draft. Seller sends its draft along with the bill of lading (which shows ownership of goods in transit) to Buyer's bank. When the goods reach Buyer, the bank accepts the draft. Buyer signs a note in favour of the bank. The bank acceptance enables Buyer to borrow the exact amount needed for the purchase at the exact time the loan is needed. **A bank acceptance is a time draft drawn by a drawer and sent to the drawee.** When the drawee (a bank) accepts it, it becomes binding on the bank.

Certified Cheques

Certified cheque

If a seller is unsure that a buyer is creditworthy, the seller may refuse to accept the buyer's cheque. One way for the buyer to deal with this is to pay with a certified cheque. **A certified cheque is a cheque stamped by a bank as certified.** The bank immediately deducts the amount of the cheque from the drawer's account. This ensures that the cheque will not "bounce". The bank charges a small fee for this service.

THE PUBLIC MARKET

Banks are the major source of short-term funds for businesses. Insurance companies, pension funds, and other financial institutions discussed previously are important sources of long-term funds for some firms. In the discussion which follows, we look at stocks and bonds and the workings of securities exchanges.

With very few exceptions, most large- and medium-size corporations use the public market as a source of long-term funds. **The public market is made up of millions of people who buy stocks and bonds and the business and non-business organizations which also invest in corporate securities.** Also included in the public market are the various securities "middlemen" who bring buyers and sellers of securities together.

Public market

STOCKS AND BONDS

The two major financial instruments by which a firm gains access to the public market are stocks and bonds. As we saw in Chapter 3, only corporations issue stock. While a firm need not be a corporation to issue bonds, a form of long-term debt, it usually must be a well-financed and sound firm if it is to attract any buyers of its bonds.

Stocks

All corporations in the private sector issue shares of stock. Stock represents ownership of a corporation. There are two basic types—common and preferred.

Common Stock

Common stock is a certificate showing ownership in a corporation. All shares of common stock are equal in value, and all common stockholders enjoy the same rights. Common stock is voting stock. Common stockholders are the residual owners of a corporation. They own what is "left over" after all debts have been paid.

Common stock

The book value of a share of common stock is the difference between the dollar value of what a company owns (its assets) and what it owes (its debts or liabilities) divided by the number of shares of common stock. The market value is the price the stock is selling for on the market. This changes daily in response to supply and demand. If the corporation which originally issued the stock certificate printed a value on it, that is the par value.

A stock split gives a stockholder a greater number of shares which represent the same proportionate ownership in the corporation. Sue Adams owns 100 shares of IBM common which is selling at $400 per share. The market value of her shares is $40,000. If the directors vote for a four-for-one stock split, Sue will have 400 shares valued at $100 per share. The total market value of her shares after the split is still $40,000. The purpose of the split is to reduce the selling price per share. This may make the stock attractive to more buyers, increase the demand for it, and raise its selling price.

A cash dividend is a payment of cash to stockholders. It rewards them for their investment in the corporation.

Preferred Stock

Preferred stock is a certificate showing ownership in a corporation. Preferred stockholders, however, usually cannot vote their shares. But they do enjoy certain preferences.

Preferred stock

Preferred stockholders have a right to receive a stated dividend (indicated on the stock certificate) before common stockholders receive any dividends. This dividend, however, is not owed until declared by the board of directors. If a corporation goes out of business and sells its assets, the preferred stockholders share in these proceeds before common stockholders. Table 10-5 discusses several important types of preferred stock.

Bonds

Bond

Although all private sector corporations issue common stock, not all issue bonds. Stockholders provide equity (ownership) capital, while bondholders are lenders. Stock certificates indicate ownership, while bond certificates indicate indebtedness.

All levels of government issue bonds, as do many non-businesses. **A bond is a written promise to pay. It indicates that the borrower will pay the lender, at some stated future date, a sum of money (the principal) and a stated rate of interest.** Bondholders have a claim on a corporation's assets and earnings which comes before that of common and preferred stockholders.

Most bond issues are sold to individuals. The agreement under which they are issued (the indenture) names a trustee to represent the bondholders' interests. This trustee is usually a large bank or trust company. Table 10-6 describes several important types of bonds.

Table 10-5. Types and characteristics of preferred stock

Type	Characteristics
1. Cumulative preferred	1. Dividends not paid in one or more years cumulate and must be paid before common stockholders receive any dividends.
2. Non-cumulative preferred	2. Dividends not paid in one or more years need not be paid in future years but, in a given year, must be paid before common stockholders receive any dividends.
3. Fully participating preferred	3. Once the dividend stated on the stock certificate is paid and the common stockholders receive the same sum, preferred shareholders share in any remaining dividends.
4. Non-participating preferred	4. Shareholders are entitled only to the dividend stated on the stock certificate.
5. Convertible preferred	5. Preferred shareholders can convert their preferred stock to common stock at their option.
6. Redeemable preferred	6. Preferred stock issued with a call price, at which price the issuing corporation can legally require the holder to sell his or her shares back to the corporation.

INVESTMENT DEALERS

Two very important financial functions are performed by investment dealers. They are both crucial to the purchase and sale of stocks and bonds. One function is the primary distribution of new stock and bond issues (underwriting). The second function is facilitating secondary trading of stocks and bonds both on stock exchanges and on "over-the-counter" stock and bond markets (the brokerage function).

Table 10-6. Types and characteristics of bonds

Types	*Characteristics*
1. Secured bonds	1. Backed by security pledged by the issuing corporation. This can be sold by the trustee and the proceeds used to pay off the bondholders if the corporation fails to pay principal and/or interest.
(a) Real estate mortgage bonds	(a) Secured by real property.
(b) Chattel mortgage bonds	(b) Secured by movable property.
(c) Collateral trust bonds	(c) Secured by stocks and bonds in other corporations which are owned by the issuing corporation.
2. Unsecured bonds (debentures)	2. Not secured or backed by specific assets but by the general credit and strength of the issuing corporation.
3. Registered bonds	3. Owner's name is registered with the issuing corporation and is printed on the certificate. Interest is mailed to him or her by the corporation or its trustee.
4. Coupon bonds	4. Owner's name is not registered and does not appear on the certificate. Owner must clip coupons from the bond and present them to the corporation's bank.
5. Convertible bonds	5. Can be converted to common stock at the bondholder's option.
6. Serial bonds	6. The issuing corporation issues a large block of bonds which mature at different dates.
7. Sinking-fund bonds	7. The issuing corporation makes annual deposits with the trustee so that those deposits, along with earned interest, will be available to redeem the bonds upon maturity.
8. Redeemable or callable bonds	8. Can be called in or redeemed prior to maturity.

Underwriting

Underwriting

Underwriting is quite different from banking and the other financial institutions we discussed earlier. **Underwriting involves helping corporations and governments to sell new issues of stocks and bonds.**

Suppose the Jaron Corporation decides to expand its plant and wants to sell $10 million worth of bonds to finance the expansion. It might contact an investment dealer to help with the sale. If its study of Jaron's financial condition is favourable, the underwriter would offer, in effect, to buy Jaron's bonds. If Jaron accepts, the cash is made available to it. The investment dealer then sells the bonds. The underwriter earns a profit by charging a commission for its services or by selling the securities at a higher price than it paid for them.

If the risk of selling a large issue of stocks and bonds is too great for one underwriter, several may combine in a syndicate to underwrite the issue. Each agrees to take a portion of the securities offered for sale. In other instances, the underwriter may arrange to sell the entire issue to a financial intermediary such as an insurance company. This is called a "private placement". In this case, the firm is acting as an agent rather than an underwriter, because no financial risk is assumed.

The Brokerage Function

Brokerage function

The brokerage function involves buying and selling securities which previously have been issued by businesses and governments.

Large investment dealers perform many brokerage-related functions for corporations and investors in addition to underwriting. They perform a brokerage function when they buy and sell previously issued securities on behalf of their investor-clients. Secondly, they perform a credit function when they finance purchases made on credit by securities buyers (margin purchases). Thirdly, they perform a research function when they compile information about firms. Fourthly, they perform an advisory function when they use the information gathered through research to advise their corporate clients on issuing new securities and when they advise their investor-clients on buying and selling securities.

SECURITIES EXCHANGES

Securities exchanges

Many stocks are traded on securities exchanges. **Securities exchanges are places where buyers and sellers deal with each other through members of the exchanges.** The exchanges are set up by investment dealers to reduce the cost and increase the efficiency of the brokerage function. Members of an exchange own "seats" on that exchange. Only members can trade on an exchange.

Most large brokerage firms hold seats on all of the exchanges in Canada and some in the United States. Of course, there are securities exchanges in other countries also.

A corporation does not receive any money from the sale of its securities on stock exchanges. If Joe Smith buys 100 shares of Bell Telephone common on an exchange, the money goes to the party who sold the shares, not to Bell.

Listed Securities

Securities traded on organized stock exchanges such as the Toronto Stock Exchange (TSE) and the Montreal Stock Exchange (MSE) are called listed securities.

The TSE is made up of about 100 individual members who hold seats on the exchange. Owning a seat enables a brokerage firm to buy and sell on the TSE floor. The securities of most major corporations are listed here. A fee must be paid before a security can be listed on an exchange.

Buying a Listed Security

Suppose you want to buy a listed security. If you have never "dabbled" in the market, your first step is to go to a branch office of an investment dealer and open an account. A corporation has only a certain number of outstanding shares (issued by the corporation and owned by investors). If you want to buy some of those shares, you must deal with people who own them. The investment dealer brings you (the buyer) and someone else (the seller) together.

When you go to the investment dealer, you will be introduced to an account executive. This person is often called a stockbroker, because he or she is involved in the brokerage function. If you are serious about becoming an investor, take the time to become familiar with your account executive. Be truthful about your investment goals and your financial situation. Because all of your dealings with the investment dealers will be handled by and through your account executive, you must know and understand each other.

After talking with your account executive, Ms. Perkins, you decide to buy some MacMillan Bloedel common stock. You ask her what the selling price is. Ms. Perkins uses an electronic device on her desk to tell her the last price at which the stock sold. Now you must make a decision. If you tell her to buy "at market", she will buy the number of shares you want at the lowest price offered. If that price is $20 per share, you would pay $2,000 plus commission, if you buy 100 shares.

But suppose you want to pay no more than $19 per share. You can place a "limit order" with Ms. Perkins. Your order would not be filled unless she could find someone willing to sell for $19 or less per share.

If you placed an "at-market" order, Ms. Perkins contacts her firm's Toronto office. That office contacts its representative on the Toronto Stock Exchange floor who goes to the "post" where MacMillan Bloedel stock is traded. That floorperson buys the shares at the offering price. No delay is involved, since someone is always willing to sell if a buyer is willing to pay the seller's asking price.

Within minutes, Ms. Perkins will get an electronic message direct from the exchange floor telling her that the transaction is complete. Meanwhile, the seller's account executive sends his or her client's stock certificate to MacMillan Bloedel's transfer agent, who cancels it and issues a new certificate in your name. This may be held by your account executive for safekeeping or sent to you.

THE OVER-THE-COUNTER MARKET (OTC)

Over-the-counter market

Many securities are not listed on any of the organized securities exchanges. Making a market in these securities is the third most important function of investment dealers. They are traded in the over-the-counter (OTC) market. In reality this is an over-the-telephone market. **The over-the-counter market is a complex of dealers who are in constant touch with one another.** Stocks and bonds of some smaller corporations are traded on the OTC market. All fixed-income securities including bonds and debentures are traded in this manner.

Security dealers in the OTC market often buy securities in their own name. They must maintain an inventory of securities in order to make a market in them. They hope to sell them to their clients at a higher price. These dealers also buy shares at the request of their clients. Dealers receive a commission for this. Dealers selling to one another charge a wholesale price and sell to their customers at a retail price.

STOCK AND BOND PRICES

Stocks and bonds traded on the exchanges and the OTC market are listed and reported in the financial section of many daily newspapers. Major newspapers all give daily detailed coverage of stock price changes and trading volumes.

Stock Prices

Figure 10-4 indicates the type of information newspapers give about daily market transactions of individual stocks. The corporation's name is shown along with the number of shares sold (expressed in round lots of 100 shares). Prices are quoted in

High Low		Dividend	Volume	High Low	Close	Ch'ge
50 1/2 35	Ajax Co.	1.25	1500	45 1/8 43	45	4 1/2
The highest and lowest price paid for one share during the current year.	Company name.	$1.25 per share dividends annually	Total shares traded on this date.	The highest and lowest prices paid per share for this day.	Price per share at the closing of the market for this day.	The increase or decrease between the closing price per share today and the closing price per share on the previous day.

Figure 10-4. Reading the financial section of the newspaper

dollars and fractions of a dollar ranging from 1/8 to 7/8. A quote of 50 5/8 means that the price per share is $50.63. Figure 10-4 explains the meaning of the various columns.

Bond Prices

Bond prices also change from day to day. These changes provide information for firms about the cost of borrowing funds.

Prices of domestic corporation bonds, Canadian government bonds, and foreign bonds are reported separately. Bond prices are expressed in terms of 100 even though most have a face value of $1,000. Thus, a quote of 85 means that the bond's price is 85 per cent of par, or $850.

A corporation bond selling at 155 1/4 would cost a buyer $1,552.50 ($1,000 par value times 1.5525) plus commission. The interest rate on bonds is also quoted as a percentage of par. Thus, "6 1/2s" pay 6.5 per cent of par value per year.

The market value (selling price) of a bond at any given time depends on (1) its stated interest rate; (2) the "going rate" of interest in the market; and (3) its redemption or maturity date.

If a bond carries a higher stated interest rate than the "going rate" on similar quality bonds, it will probably sell at a premium above its face value—its selling price will be above its redemption price. If a bond carries a lower stated interest rate than the "going rate" on similar quality bonds, it will probably sell at a discount—its selling price will be below its redemption price. How much the premium or discount is depends largely on how far off in the future the maturity date is. The maturity date is indicated after the interest rate.

Bond Yield

Suppose you bought a $1,000 par value bond in 1977 for $650. Its stated interest rate is 6 per cent, and its maturity or redemption date is 1997.

You paid $650 for the bond and its interest rate is 6 per cent per year of par value. You get $60 per year in interest. Based on your actual investment of $650, your yield is 9.2 per cent. If you hold it to maturity, you get $1,000 for a bond that originally cost you only $650. This "extra" $350, of course, increases your true, or effective, yield.

Stock and Bond Averages

To give investors an overall idea of the behaviour of security prices, several types of stock and bond averages are reported. The TSE index gives an average for stocks. Common stocks on the TSE are averaged so that an investor can tell in dollars and cents how much an average share changed in price on a given day.

SPECULATING AND INVESTING

Speculating

Speculative trading

Some people think that buying stocks and bonds is a way to get rich quick. They buy on the basis of "hot tips". This is called speculative trading. **Speculative trading means buying or selling securities in the hope of profiting from near-term future changes in their selling prices.**

Sometimes, amateur speculators do "strike it rich", but the losers far outnumber the winners. Speculating is most popular during a bull market, when stock prices as a whole are rising and there is a great deal of optimism among speculators. Speculating is less popular in a bear market, when stock prices as a whole are falling and there is a great deal of pessimism among speculators. Some people are successful speculators. Successful speculation requires courage, persistence, judgment, and the analysis of all available facts.

Margin Trading

A speculator has to pay cash for securities bought only when the margin requirement is 100 per cent. Otherwise, the speculator buys partly on credit.

Margin trading

Margin trading enables speculators to buy more shares for a given amount of money because they are buying partly on credit. Brokers put up the shares they sell on margin as collateral for the loans they make to finance their clients' margin purchases. As long as the price of a stock bought on margin rises, there is no problem. The banker's collateral increases in value. But if its price falls, the banker wants more cash from the broker or wants to sell the shares.

In the 1920s, many speculators were buying on 10 per cent margin. When stock prices began falling, bankers started selling, in large volume, the stocks they held as collateral. This helped to bring on the eventual collapse of the stock market.

Short Selling

Speculators may also make a profit from selling stocks when prices are falling. Martha Todd, an established client of Broker *B,* believes that the selling price of ABC common stock will fall in the next few weeks. It is now selling at 65. Martha does not own any ABC stock but "borrows" several shares from her broker. Many investors do not take possession of the stock certificates they own. They let their brokers keep them for them. Thus, brokers can "lend" some of this stock to their other clients.

Short selling

Martha tells her broker to sell 500 of these borrowed shares at 65. If the price subsequently falls, Martha buys the shares to "cover" her earlier sale. She buys in, say, at 55. She thus makes a $10 profit on each of the 500 shares (less commission). But if the price went up instead of down, Martha would have incurred a loss. This practice is called short selling. **Short selling means selling a security which you do not own by borrowing it from your broker.** At some time in the future, you must buy the security to "cover" the short sale.

Investing

Unlike a speculator, an investor invests in securities for the longer haul. Before even considering investing, much less speculating, you should have a cushion of cash reserves and adequate insurance. You should be able to choose when you want to sell your shares and not be forced to sell them because you need cash for an emergency.

Your approach to buying and selling securities should be logical. Your investment goals should guide your buying and selling decisions. The kinds of goals may vary among investors, but each investor should have definite goals.

An important goal for investors is to protect their invested dollars. You could do this by putting your money in a safe-deposit box. But this earns nothing, because of inflation the buying power of those dollars declines. You would be wiser to put your money in an insured savings account. You might also buy government bonds. All those are highly liquid investments which can be quickly converted into cash. Furthermore, they are very safe investments. In fact, they involve almost no risk at all. But to increase your earning potential, you will have to make riskier investments.

How Much Risk?

Of course, different investment strategies involve different degrees of risk. Investing in preferred stocks of established and profitable corporations is less risky, for example, than investing in common stocks of new and risky ventures. But in terms of return, the new venture might prove to be the better investment. In other words, risk and return are directly related.

There is no one answer to the question of how much risk you should assume in your investment program. You have to consider your financial situation, age, investment goals, patience, self-discipline, and so on. To put it simply, if your goal is to get rich quick, you will have to take a lot more risk than someone else whose goal is to get rich more slowly.

Balancing Objectives

The typical investor wants a safe investment which will return regular earnings and has a lot of potential for future growth. But it's hard to satisfy all three objectives.

Investing in securities involves keeping up with developments in the economy and in the industries and firms in which you invest. **If you don't have the time or "know-how" to do this, you might invest in a mutual fund. The owners of a mutual fund pool their investment dollars and buy securities in other businesses. Buying one share in a mutual fund makes you part owner of all the securities owned by the fund.** You spread your risk over a broad range of securities. Mutual funds are professionally managed. Before they were created, only people with large sums to invest could afford to hire professional managers to oversee their portfolios (the stocks and bonds they own).

Figure 10-5 lists various types of mutual funds and explains how they work.

[margin: Mutual fund]

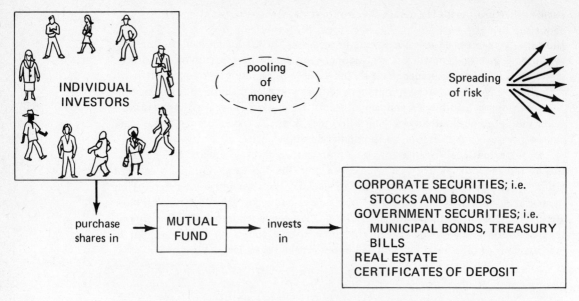

- Stock fund.   Invests mainly in common stocks.
- Bond fund.   Invests mainly in government and corporation bonds.
- Balanced fund.   Invests in common and preferred stocks and bonds.
- Specialty fund.   Invests in particular fields such as real estate, banking, natural resources, etc.
- Growth fund.   Primary objective is long-term capital appreciation.
- Income fund.   Primary objective is dividends.
- Load fund.   Purchase price of a share includes a sales commission.
- No-load fund.   Purchase price of a share is net asset value (no sales commission).
- Open-end fund.   Sale of shares is not limited. Shares are offered as long as buyers are willing to buy them. The fund will repurchase the shares (redeem them) if an investor wants to sell them.
- Closed-end fund.   Sale of shares is limited. Once they are sold, no more shares are offered. The fund does not redeem shares. The investor must sell his or her shares on the open market.

Figure 10-5. How a mutual fund works. *Investors in a mutual fund send their money to the trustee for the mutual fund, and the trustee is told which securities to buy or sell by the fund's investment advisor. A management firm performs the recordkeeping for the fund.*

SECURITIES REGULATION

Canada, unlike the United States with its Securities and Exchange Commission (SEC), does not have comprehensive federal securities legislation or a federal regulatory body. Government regulation is primarily provincial. There is also self-regulation through the various securities exchanges.

In 1912 the Manitoba government pioneered in Canada with "blue sky" laws applying mainly to the sale of new securities. Issuing corporations must back up securities with something more than just the "blue sky". Similar laws were passed in other provinces of Canada. Provincial laws also generally require the licensing of stockbrokers and the registration of securities before they can be sold. In each province, issuers of proposed new securities must file a prospectus with the provincial securities exchange. **A prospectus is a detailed registration statement which includes information about the firm, its operation, its management, the purpose of the proposed issue, and any other things which would be helpful to a potential buyer of these securities.** The prospectus must be made available to prospective investors.

Prospectus

CAREERS IN SECURITIES AND INSURANCE

Securities salespersons or brokers work for investment dealers. They study the investment potential of stocks and bonds and analyze the investment or speculative goals of their clients. They recommend purchases and sales of securities, place orders, and keep their clients informed of opportunities.

Their income depends primarily upon commissions earned from trading securities for their clients. Technical skills in interpreting financial statements and comparing financial performance of firms are required. Human relations skills are also needed.

Large investment dealers provide training programs for new salespersons. The demand for such people is expected to grow moderately. Income potential depends greatly on personal skill and market conditions. Information is available from the personnel department of any large investment dealer, or from the Investment Dealers Association of Canada, P.O. Box 217, Commerce Court South, Toronto, Ontario, Canada M5G 1E8.

Insurance-related careers include those of agent, broker, underwriter, claims adjuster, claims examiner, and actuary. Agents sell insurance for one firm or for several firms. Brokers serve their clients by selecting and buying insurance for them. Underwriters evaluate and select risks the insurance firm will take on. Claims adjusters check and settle claims against insurance companies. Claims examiners investigate large or suspicious claims. College degrees are helpful for agents and brokers and are highly desirable for underwriters, adjusters, and examiners.

Actuarial jobs involve the assembling and analysis of statistics on expected losses. This is done in order to fix premiums for policyholders so that costs and profit are covered. Tough exams are given for certification as an actuary.

All the insurance-related careers we have described require human relations skills, particularly the agent-broker careers.

All require computational skills, factual observation, and reporting skills. Of the jobs discussed here, actuaries earn the highest average incomes. The most successful of agents can earn even more. Perhaps the greatest financial rewards result from operation of an insurance agency. Agency owners, of course, assume the same risks as any business owner. Information is available from the Insurance Institute of Canada, 220 Bay Street, Toronto, Ontario, Canada M5J 1P3.

In recent times, Ontario is regarded as having the most progressive securities legislation in Canada. The Ontario Securities Act contains disclosure provisions for new and existing issues, prevention of fraud, regulation of the Toronto Stock Exchange, insider trading, takeover bids, and others.

The Toronto Stock Exchange provides an example of self-regulation by the industry. The TSE has regulations concerning listing and delisting of securities, disclosure requirements, and issuing of prospectuses for new securities, among others.

Commodity exchanges

Commodity exchanges provide a market for commodities much as securities exchanges provide a market for stocks and bonds. They are voluntary trade associations whose members must follow specified trading rules. The Winnipeg Commodity Exchange is the largest commodity exchange in Canada. The Chicago Board of Trade is the world's largest commodity exchange.

Cash trading

Commodity exchanges deal in cash trading. **Cash trading involves the actual buying and selling of commodities for delivery.** A sales contract may call for immediate delivery or delivery at a specified date in the future. The contract is fulfilled upon delivery.

Futures markets

The larger commodity exchanges also have futures markets. **In futures markets, traders buy and sell contracts to receive or deliver a certain quantity and grade of commodity at a specified future date.** Prices are set on the exchange floor by traders. Most futures trading does not result in the physical exchange of goods.

Suppose you expect the price of flax to go down. You sell a futures contract for flax you don't actually have. You sell at $5.00 per bushel. The price goes down to $4.90, and you buy to "cover" your sale. You make 10 cents per bushel in profit (less commission). But if the price had gone up instead of down, you would have incurred a loss.

Suppose instead that you expect the price to go up. You buy a futures contract in June at $5.00 per bushel for delivery in October. In July, October futures are selling at $5.25 per bushel. You sell in July at $5.25. Your purchase and sales contracts cancel each other. But you make 25 cents per bushel in profit (less commission). If, however, the price had gone down instead of up, you would have incurred a loss.

Although many people have made fortunes on the commodities exchanges, it is no place for amateurs.

SUMMARY AND LOOK AHEAD

The financial institutions we discussed are vital to modern businesses. Ours is a modern banking system. The chartered bank, which is at the heart of this system, is the most important source of short-term funds for business firms. Our chartered banking system creates money in the form of demand deposits. The Bank of Canada is the central bank of Canada. Its main job is to control the nation's money supply.

In addition to chartered banks, other financial intermediaries include trust companies, caisses populaires and credit unions, life insurance companies, sales and consumer finance companies, venture capital funds, government financial intermediaries, and pension funds. While the others are not as important to businesses as the chartered bank, they play an important role in providing services to their business and non-business customers.

Important credit instruments used in everyday business are promissory notes and drafts. The need for cash is declining in proportion to the number of transactions. Electronic banking may someday even make ours a "chequeless" society.

The public market is made up of the millions of people who invest in corporate securities and the "middlemen" who bring them together for buying and selling. A corporation which issues new stock may use the services of an investment dealer underwriting department to help sell the securities. Through their brokerage function, investment dealers buy and sell previously issued securities for their clients.

A corporation's common stockholders are its residual owners. Common and preferred stock indicate ownership in a firm. Bonds indicate indebtedness.

A speculator looks mainly for short-term profits from buying and selling securities—in some cases (short selling) by selling something he or she does not own. An investor takes a longer-run view. Speculating is very popular during bull markets and much less popular in bear markets.

Commodities such as grain and copper are traded on commodity exchanges such as the Winnipeg Commodities Exchanges. The cash market involves the actual buying and selling of commodities, whereas most futures trading does not result in the physical exchange of goods.

Just as banks and the commodity exchanges are regulated, so are the securities exchanges, brokerage houses, and stockbrokers.

In our next chapter, we will see how businesses use these financial institutions. Our topic there is financial management.

KEY CONCEPTS

Bank rate The rate at which chartered banks can borrow from the Bank of Canada.

Bond A certificate of indebtedness. A written promise to pay made by the borrower (government, business firm, other organization) to its bondholders.

Brokerage function Buying and selling securities which previously have been issued by businesses and governments.

Cash trading In the commodities market, cash trading means the actual buying and selling of commodities for delivery in the cash market. In the securities exchanges, it means purchases of securities for cash.

Certified cheque A cheque which is stamped "certified" by a bank, which immediately withdraws the amount of the cheque from the drawer's chequing account.

Chartered bank A privately owned, profit-seeking firm which serves individuals, businesses, and other organizations. Major services are accepting demand deposits and making short-term loans. It also provides many other services such as accepting time deposits, exchanging foreign currency for Canadian currency, and giving financial advice.

Commodity exchanges Voluntary trade associations whose members engage in trading commodities. Provides a market for commodities such as flax, rapeseed, cattle, gold, etc.

Common stock A certificate showing ownership in a corporation. All shares are equal in value, and all common stockholders enjoy the same rights. Common stockholders are the residual owners of a corporation.

Consumer finance company Makes personal loans to individuals, not to business firms.

Credit union A co-operative savings association which is owned by its depositors. Also makes loans to members.

Demand deposit Chequing accounts of individuals, businesses, and other organizations.

Draft A written order made by one party (the drawer) addressed to a second party (the drawee) ordering the drawee to pay a certain amount of money to a third party (the payee). Examples are ordinary cheques drawn on banks, trade acceptances, bank acceptances, certified cheques, and cashier's cheques.

Factoring company Also called a factor. Buys accounts receivable from a firm. It pays less than the face value of the accounts (it buys them at a discount) but collects the face value of the accounts from its client's customers. The difference, minus the factor's cost of doing business, is the factor's profit.

Futures markets On the larger commodities exchanges, traders buy and sell contracts to receive or deliver a certain quantity and grade of commodity at a specified future date. Most futures trading does not result in the physical exchange of goods. Buyers of futures contracts try to protect themselves against adverse future price changes or try to make speculative profits from those price changes.

Industrial Development Bank A Bank of Canada subsidiary set up to make loans to small firms.

Life insurance company A mutual or stock company which shares risk with its policyholders for payment of a premium. A source of long-term funds for some business firms.

Margin trading A person who buys securities partly on credit financed by his or her stockbroker engages in margin trading.

Money Anything people will generally accept in payment of debts. In Canada, coins, paper currency, and demand deposits are considered money.

Mutual fund An investment company. The owners of a mutual fund pool their investment dollars and buy securities in other businesses and government securities.

Over-the-counter market (OTC) A complex of securities dealers who buy and sell securities for investors without using a securities exchange.

Pension fund A fund that accumulates money for payment to people upon retirement. Money from the fund is invested in corporate stocks and bonds.

Preferred stock A certificate showing ownership in a corporation. Has preference over common stock in dividends and liquidation of the corporation's assets. Preferred stockholders may or may not have the right to vote for members of the board of directors.

Prime rate of interest The lowest rate of interest charged by commercial banks to their most creditworthy borrowers.

Promissory note A written promise to pay a certain sum of money (principal and interest) made by a borrower (maker) to a lender (payee) at a certain time in the future.

Prospectus A statement filed when a firm proposes to issue new securities. It must be made available to prospective buyers before they buy the issuing corporation's securities.

Public market Composed of individual and institutional investors who buy stocks and bonds. Also includes the various securities middlemen who bring together buyers and sellers of securities.

Reserve requirement The percentage of its deposits which chartered banks must keep in vault cash or as deposits with the Bank of Canada. It is set by the Board of Governors of the Bank of Canada.

Sales finance companies Organizations specializing in financing instalment purchases of individuals and firms.

Securities exchanges Places where buyers and sellers of securities deal with each other through members of the exchanges. The exchanges are set up by investment dealers to reduce the cost and increase the efficiency of financial investment.

Short selling Occurs when a person sells a security which he or she does not own by borrowing it from his or her broker. At some time in the future, the person must buy the security to "cover" the short sale.

Speculative trading Buying or selling securities in the hope of profiting from near-term future changes in their selling prices. Common especially in bull markets.

Time deposit A deposit which is made for a period of time. Savings accounts and certificates of deposit are examples. Interest is usually paid on a time deposit.

Trade acceptance A time draft drawn by the drawer and sent to the drawee. When the drawee (a business firm) accepts it by signing it, it becomes binding on the firm.

Trust company Safeguards funds and estates entrusted to it. Also may serve as trustee, transfer agent, and registrar for corporations.

Underwriting Helping corporations and governments to sell new issues of stocks and bonds.

Venture capital firm A firm which specializes in providing debt and equity funds for newer and rapidly growing firms.

QUESTIONS FOR DISCUSSION

1. Contrast the following: (a) a chartered bank; (b) an investment dealer; (c) a trust company; (d) a pension fund; and (e) a credit union.
2. Define: (a) sales finance company; and (b) consumer finance company.
3. Explain how the banking system creates money.
4. Why do some banks give their demand depositors "free" chequing accounts (no service charge) when they maintain minimum monthly balances of $300? Why do some give "free" chequing accounts regardless of the depositors' balances in their chequing accounts?
5. What is a "branch" banking system?
6. Give two examples of actions the Bank of Canada uses to control the nation's money supply.
7. What is the major difference between a promissory note and a draft?
8. Explain why business firms sometimes use cashier's cheques.
9. List and define three different concepts of "value" for common stock.

INCIDENT:

Russ Madison

Russ Madison is a 43-year-old, middle-level manager of a medium-size corporation. He currently earns $30,000 a year and expects that he will be promoted to higher management because of his good work record.

The Madisons have a daughter who is in grade ten and a son who is in his first year at a university. They have no other children.

The Madisons have $20,000 in savings certificates. They have a joint savings account with a balance of $4,500. They also have $5,000 in Government of Canada bonds, and the mortgage on their home will be paid off in ten years. Russ has a $100,000 life insurance policy on his life. Finally, ten years ago, Russ began buying shares in a no-load mutual fund which emphasizes capital appreciation. The current market value of those shares is $15,000.

The major financial obligation the Madisons face in the future is the expense of putting their son and daughter through college. Russ and his wife began planning for this at the birth of their children by opening savings accounts for each of them. The balances in those accounts are now adequate to pay their children's tuition and expenses. Obviously, the Madisons are in good financial shape—in fact, quite enviable shape!

Russ and his wife believe that they can assume a little more risk in their approach to investing.

Questions:
1. Why do the Madisons believe they can assume greater risk in their investment program now than when their children were born?
2. Why do you think that Russ and his wife chose a no-load mutual fund whose objective is capital appreciation?
3. What investment advice would you offer the Madisons at this time? Explain your reasoning.

10. What is the purpose of (a) stock split, (b) a stock dividend, and (c) a cash dividend?
11. What is the major difference between bonds and common stock?
12. What is a brokerage house?
13. What is the purpose of a stock exchange?
14. How do listed securities differ from unlisted securities?
15. What is a stockbroker?
16. Is speculation more likely to exist during a "bear market" or a "bull market"? Explain.

17. What is a mutual fund? Explain how it works.
18. Does the Securities and Exchange Commission guarantee the "value" of a corporation's stocks and bonds to investors? Explain.
19. What is the purpose of a commodity exchange?

INCIDENT:

Gorgeous Day Boutique

The Gorgeous Day Boutique is a small partnership owned by Doris Gammel and Gloria Hyde. The partners have been in business for nine months. They carry a medium-priced line of women's sportswear and casual wear. Sales are running about $30,000 a month.

Doris and Gloria invested all their personal savings in the firm when they started out. Business has been better than they expected, but they believe that it could be even better.

The partners started out selling on a "cash-only" basis. But they are now reconsidering the decision not to sell on credit. They estimate that one in three persons who comes into the store does not buy when the customer learns that Gorgeous Day does not offer credit.

Doris and Gloria are now trying to decide how they should go about offering credit to their customers. They both lack experience in dealing with credit, but each has her own ideas on the matter.

Questions:
1. Why do you think that the partners originally decided not to sell on credit?
2. Why do you think the lack of charge privileges caused some shoppers not to buy from Gorgeous Day?
3. What advice would you give to Doris and Gloria with respect to offering credit? Justify your advice.

11
Financial Decisions and Insurance

11 FINANCIAL DECISIONS AND INSURANCE

OBJECTIVES: After reading this chapter, you should be able to:

1. Identify the three principal duties of a financial manager.
2. Demonstrate a case in which a manager balances the twin objectives of liquidity and profit in the use of working capital.
3. Draw a chart showing the normal flows of working capital in a manufacturing firm.
4. Explain the relationship between technology and fixed capital needs.
5. Illustrate the use of two major sources of short-term credit—trade credit and bank loans.
6. Contrast the advantages of issuing bonds with the advantages of issuing preferred stock for getting long-term funds.
7. List four kinds of risks which can be reduced by good management techniques.
8. Distinguish between pure and speculative risk.
9. Show how the law of large numbers is used to figure insurance premiums.
10. Contrast the procedures of voluntary and involuntary bankruptcy.

KEY CONCEPTS: In reading the chapter, look for and understand these terms:

WORKING (SHORT-TERM) CAPITAL
LIQUIDITY
OPPORTUNITY COSTS
LONG-TERM CAPITAL
CAPITAL BUDGET
LEASE
DEBT FINANCING
EQUITY FINANCING
MATURITY
TRADE CREDIT
LINE OF CREDIT
REVOLVING CREDIT
SECURED LOANS
FLOOR PLANNING
SINKING-FUND
SELF-INSURANCE
PURE RISK
SPECULATIVE RISK
LAW OF LARGE NUMBERS
MERGER
AMALGAMATION
HOLDING COMPANY
RECAPITALIZATION
REORGANIZATION
BANKRUPTCY

In the two previous chapters we studied the communications devices of accounting and the financial institutions available to financial managers. In this chapter, we'll see how a manager uses the services of chartered banks, stockbrokers, and insurance companies in managing the firm's financial affairs. Accounting systems provide the basis for most of these financial decisions.

THE FINANCIAL MANAGER AND FINANCIAL PLANNING

Financial decisions are the task of the financial manager (who may be called the comptroller or the vice-president for finance). This executive projects the firm's long- and short-term financial needs and meets them with the help of banks and others. He or she is the chief guardian of the owners' equity. The financial manager's job is to get the best return on the owners' investment without taking unnecessary risks.

Company presidents may include non-economic or social objectives in their decision making, but financial managers must think of dollars and profit. Financial managers specialize in funds and their allocation. They help to provide the president of a firm with a purely economic, profit-maximizing point of view.

A financial manager must do three things. The first duty is to meet the firm's short-term (working) capital and long-term capital needs in the face of uncertainty. A second duty is to evaluate and select from several sources of funds. Finally, it is the financial manager's duty to protect the owners' resources while helping to maximize their return.

The financial manager's first task is financial planning. This means identifying the firm's basic financial needs for working capital and for long-term capital. Let's begin by checking the uses for these funds. Then we'll discuss the sources of funds.

Uses of Funds

As we have seen, the two basic financial needs are working capital and long-term capital needs. Each has its special characteristics.

Working Capital

Working (short-term) capital

Working capital is a term applied to a firm's investment in short-term assets—the current assets we discussed in Chapter 9. It includes those assets which flow regularly in the day-to-day operations of a firm—cash, accounts receivable, and inventories. (See Figure 11-1.) If we deduct the amount owed as current liabilities from the gross working capital (the sum of current assets), we are left with net working capital.

Liquidity

Working capital must be handled carefully by the financial manager so as not to interrupt or slow the regular operations of the business. **The firm needs to have enough cash coming in to meet bills, wages, and other current payments. This ability to make payments which are due is the test of a firm's liquidity.** If the Mangham Feed Store

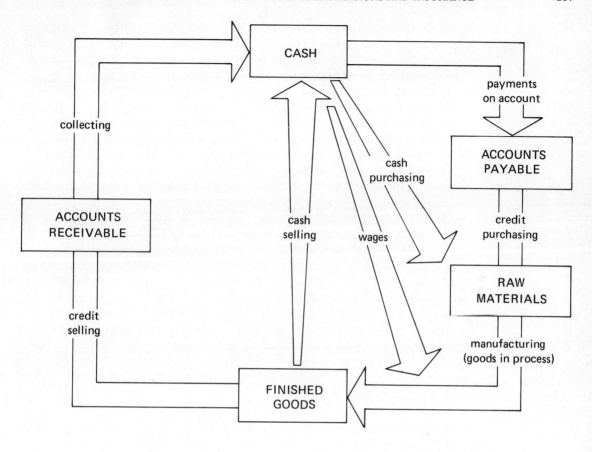

has a payroll of $800 due next Monday as well as a repair bill of $500 due on the same day, the manager must examine Mangham's liquidity. If the firm has only $200 in its chequing account and expects no significant cash inflow before Monday, some borrowing may be in order—maybe from the bank. If Mangham borrows $1,500 from the bank, the firm increases its gross working capital—but its net working capital stays the same because it has created a new current debt—a note payable to the bank. The firm could have found some temporary cash in other ways, as we will see.

The financial manager seeks to balance liquidity with profit. The goal is to minimize idle cash balances by keeping "near cash" on hand. This means money which would otherwise earn no return is invested and becomes an earning asset. A manager seeks stable short-term investments which are readily convertible into cash. Examples are certificates of deposit (CDs) in banks or short-term government securities such as treasury bills. Tying up cash in long-term investments such as bonds of another firm does not meet the goal of balancing liquidity with profit. Long-term investments of cash do not qualify as "near cash", because they might not be convertible into cash quickly or they might involve some loss due to changes in their market value.

Credit sales represent another use of short-term funds. A firm which sells "on credit" uses its funds to finance its customers' operations. The credit manager and the sales manager often disagree on credit policies. The sales manager sees this as a means of increasing sales. The credit manager sees it as leading to more bad debts.

The financial manager seeks to achieve a balance. The firm wants profits to increase if more working capital must be tied up in accounts receivable. If the increase in receivables results from purchases by proven paying customers, profits will increase. The Mangham Feed Store may be wise to avoid selling on credit to a young farmer whose farm is poorly managed. Even if the debt will be repaid, it might take a long time to collect—tying up valuable working capital. The financial manager uses Dun & Bradstreet Canada Ltd. or other credit reporting services to judge possible credit customers. An example of a credit report is presented in Figure 11-2.

Other current assets shown in Figure 11-1 are inventories of raw materials and finished goods. Raw materials are changed into finished goods through the production process. Between these two stages, they are called goods in process. (See Figure 11-1 again.) The financial manager seeks to reduce excess inventories at all three stages. There may be a conflict with the production manager because production may be simplified by keeping large inventories of inputs.

Suppose that the sales manager of the Wonder Mattress Corporation forecasts a 10 per cent increase in sales during the next year. The production manager bases estimates of raw materials needs on this sales forecast. Now, suppose further that the purchasing agent can receive a 12 per cent discount if the order of raw materials is increased by 20 per cent rather than by 10 per cent. Should the financial manager approve this use of funds to earn the additional discount?

This depends on whether the Wonder Mattress Corporation could use the additional funds tied up in raw materials more profitably elsewhere. If the raw material is perishable or if storing it would be costly, the larger order would probably not be approved. On the other hand, if the price of raw materials is expected to rise, the manager would probably approve the larger order. The value of a systems approach to decision making is clear here. Better decisions are made when finance, marketing, and production are viewed in terms of their overall goal of helping to increase the firm's profit.

Still another use of working capital involves the current asset "prepaid expenses". You can buy a three-year fire insurance policy on your home, for example. Paying the three-year premium in one lump sum means that you are prepaying your insurance coverage. The same is true for a firm.

Opportunity costs

A financial manager carefully evaluates the option to pay insurance premiums "in advance". The choice depends on the other uses which could be made of those funds. Prepaying expenses is wise when the savings exceed the opportunity costs. **Opportunity costs are costs of losing the option to use the funds in another way.** Let's assume that for the Sanford Ice Cream Company the savings from paying a lump sum for three years of fire insurance coverage, rather than paying on a year-to-year basis, amounts to $100. Assume further that Sanford's comptroller could have earned $150 of interest on the prepaid part of the expense during the last two years of the policy life. Clearly, the comptroller would not prepay in this case. The current (and ex-

Dun & Bradstreet

Please note whether name, business and street address correspond with your inquiry.

BUSINESS INFORMATION REPORT

BASE REPORT

SIC	D-U-N-S	© DUN & BRADSTREET	STARTED	RATING
59 43	20-155-5868	CD 34 JAN 28 197-		
59 47	STANDARD SUPPLY	STATIONERY & GIFTS	1947	DD1
	CHARMAINE'S GIFTS			

165 CHINA ST
OURTOWN ON
TEL 519 729-1141

S P E C I M E N

CHARMAINE B. MALVIN)
LLOYD T. MALVIN JR) PARTNERS

SUMMARY

PAYMENTS	DISC PPT
SALES	$119,519
WORTH	$40,096
EMPLOYS	3
RECORD	CLEAR
CONDITION	STRONG
TREND	UP

PAYMENTS

HC	OWE	P DUE	TERMS			
1100	300		2-10-30	DEC 11 197-	Disc	SOLD Over 3 yrs
500	300		2-10-30		Disc Ppt	Over 3 yrs
300			30		Ppt	1 yr

FINANCE On Jan 7 197- L.T. Malvin Jr., partner, submitted statement Dec 31 197-:

Cash	$ 8,524	Accts Pay	$ 702
Mdse	15,214		
Current	23,739	Current	702
Fixt & Equip	5,020		
R E	10,600	NET WORTH	40,096
Prepaid	1,439		
Total Assets	40,798	Total	40,798

REAL ESTATE TITLE VALUE MTGE
Business property Partners $10,600 Clear
197- sales $119,519. Gross profit $55,238. Net profit $7,879. Fire insurance on merchandise and fixtures $20,000, on building $10,000.
Signed Jan 7 197- for STANDARD SUPPLY, by Lloyd T. Malvin Jr, partner.
-----0-----
During interview Malvin stated sales up 5% in 197-, and profit up by 2% during same year. These comments have been confirmed.

BANKING Non-borrowing account. Balances average medium four figures. Relations satisfactory.

HISTORY Style registered Jan 5 1950 by partners. Charmaine's Gifts is unregistered, used for advertising.
C.B. MALVIN, born 1903, married. 1938-1947 employed as bookkeeper-stenographer by Public Works Department. Purchased this business in 1947. Her husband, Lloyd Malvin, is employed by City Fire Department.
L.T. MALVIN JR, born 1927, married. Attended Speed Business College 1945-1946. Served Canadian Army 1946-1949. Employed by mother prior to 1950. Then became partner.

OPERATION Retails stationery (60%), gifts, school supplies and greeting cards (40%). Terms 100% cash. Employs three, including the partners. LOCATION: Owns one-storey block building providing 6,200 square feet, in normal condition. Housekeeping good.
10-23 (14 66)

THIS REPORT MAY NOT BE REPRODUCED IN WHOLE OR IN PART IN ANY FORM OR MANNER WHATEVER.
It is furnished by DUN & BRADSTREET of Canada, Limited in STRICT CONFIDENCE at your request under your subscription agreement for your exclusive use as a basis for credit, insurance, marketing and other business decisions and for no other purpose. These prohibitions are for your own protection—your attorney will confirm the seriousness of this warning. Apprise DUN & BRADSTREET promptly of any question about the accuracy of information. DUN & BRADSTREET of Canada, Limited does not guarantee the correctness of this report and shall not be liable for any loss or injury caused by the neglect or other act or failure to act on the part of said company and/or its agents in procuring, collecting or communicating any information. C-516 L (3-73)

Figure 11-2. A credit report. (Reprinted by permission of Dun & Bradstreet, Canada, Ltd.)

pected) interest rate is a major factor in making all such financial decisions because it determines what unused dollars can "earn".

A manager should have an overview of the flow of working capital such as we saw in Figure 11-1. He or she must understand this flow and the timing of it from one use to another. If the firm has a good sales forecast and a good collection policy, it can achieve the goal of providing enough working capital but not too much.

Long-term Capital

Long-term capital

Long-term capital is the firm's investment in fixed assets. Such capital is committed for at least one year (usually much longer), and it requires a different perspective. The amounts are larger and the risk is greater. A bad mistake could cause the firm to fail.

Long-term capital is invested in land, buildings, heavy machinery, and other fixed assets which, to a large extent, determine the direction in which the firm is going. When RCA went into the computer business, it directed a large part of its long-term

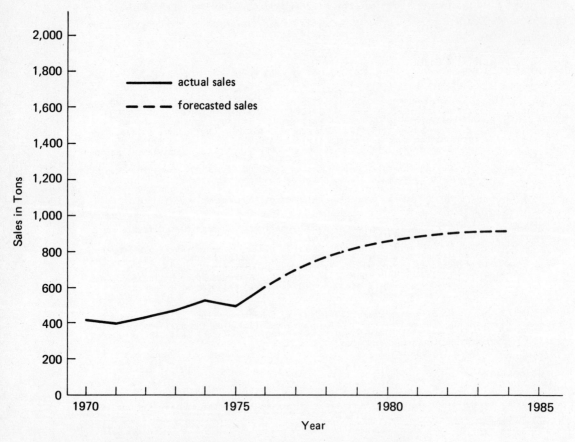

Figure 11-3. A sales forecast for the Rustic Nail Company

capital into assets that could not be sold without loss. RCA management assumed that such an investment would be profitable. This decision proved to be wrong, and RCA later sold its computer holdings at a loss of millions of dollars. Ford had a similar experience with the Edsel. Auto makers in the mid-1970s had to decide how much production capacity to convert to compact and subcompact cars. Decisions like these involve many millions of dollars in long-term investments.

Financial managers must take a long-run view of the firm's operations. They use some of the accounting devices referred to in Chapter 9. **One is the capital budget. This projects the expected need for fixed assets for a period of five to ten years.** This long planning period means that budget makers must use every scrap of information about the long-range plans and expectations of the firm. They must take the environment into account, too. This includes technological events outside the firm and changes in consumer needs and tastes which might require changes in plant machinery and equipment.

Capital budget

Technology is especially hard to predict. It is difficult to say when a competitor might come up with a product or a process which makes your own obsolete. For firms which sell a standard product such planning may depend only on a forecast of sales. The Rustic Nail Company (see Figure 11-3) prepared its sales forecast primarily from government projections for the home construction industry in its market area.

Some steel producers have plants which are nearly obsolete because of recent developments in production technology. Yet these firms compete with foreign steel producers whose plants, in many cases, are newer and use the latest equipment and techniques. The capital budgeting problem in this case is to find sources of funds to enable the firm to up-date plant and equipment.

Consider the great technological change which has occurred in the airline industry. The propeller-driven airplane has been replaced by jets. An airline must face the possibility that even recently bought planes will soon become obsolete. If they do, a decision must be made whether or not to replace them and, if so, with what. Some airlines did not place orders for the Boeing 747 for some time after it was introduced. Fuel costs have since caused some airlines to reduce or stop their use.

Each decision depends on the expected payoff from this type of capital investment. Such decisions call for expert planning in the use of long-term capital. In some cases, firms will avoid the risk of long-term asset purchase by resorting to leasing.

The Alternative of Leasing

One way of avoiding the need for long-term financing for land, buildings, or equipment is to lease them. **A lease is an agreement to grant use of an asset for a period of time in return for stated regular payments.** Leasing such assets has several advantages over the choice of borrowing funds for their purchase. First, it reduces the outstanding debt of the firm. Secondly, leased equipment may be replaced with more modern equipment without the losses which result from replacement of owned equipment. Thirdly, it is often a tax advantage to lease. The entire lease payment is tax deductible.

Lease

The decision to lease, however, is not always so obvious. There are advantages to outright ownership. Often the cash payments on a lease are considerably higher than the equivalent financed purchase payments. Also, there are often restrictions on the way a firm might use or modify leased assets. Such restrictions don't apply to owned equipment.

A bank or a manufacturer often leases computer equipment instead of buying it. Such leasing can be viewed as a "source of long-term funds" rather than as an alternative to borrowing.

Evaluating and Selecting Sources of Funds

Once a financial manager determines the needs for short- and long-term funds, the question is, "who will provide them?" For some firms there is little choice. For most, however, there is some choice of sources of funds. The basic choice is between debt financing and equity financing.

Debt financing

Debt financing is the use of borrowed funds. This could mean a major corporation issuing bonds or it could mean a barber shop borrowing $1,000 for 60 days from a local bank.

Equity financing

Equity financing means the provision of funds by the owners themselves. This could involve issuing stock or using retained profits. Canadian non-financial corporations since 1970 have used somewhat more equity financing than debt financing. Less than one-fourth of the equity capital was gained through stock issues.

Criteria for Evaluation

Maturity

There are several important features of financing which help a firm decide between the use of debt and equity. These are shown in Table 11-1. **Maturity is the factor of time of repayment. When a debt matures, it must be paid.** If funds are internal, they need not be repaid at all. If they are borrowed, the date of maturity (due date) may vary.

Equity and debt financing also differ in the way they affect the claims on assets and earnings. To issue bonds means that the new bondholders will get the designated interest payment before stockholders get any dividends. They have a prior claim on income. Bondholders also come first in the event that the firm goes out of business. They are paid off out of the proceeds of the sale of the firm's assets before stockholders receive anything.

Still another factor in the choice of debt or equity capital relates to control of the corporation. If a firm issues more common stock and this stock is bought by newcomers to the firm, the original common stockholders may lose control over the election of the firm's board of directors. They might lose some influence over policy decisions. A bond sale would not run such a risk for the controlling shareholders.

Of course, the main reason businesses borrow in the first place is that they feel they can earn a higher return on borrowed dollars than the cost or interest they must pay to their lenders. To improve earnings by borrowing is called leverage. Financial markets also play a role when a firm seeks new sources of capital. Sometimes there is a lot of money available to lend and sometimes there is not. The final selection is often a

Table 11-1. Five crucial differences between debt and equity financing

Criterion	Debt	Equity
Maturity	Has a due date.	No due date.
Claim on income	The lender has a prior claim on earnings of the firm over the owners of the firm. This claim is fixed in amount.	The owners have no prior claim on earnings of the firm. They have a residual claim.
Claim on assets	The lender has a prior claim on assets.	Preferred stockholders have a claim prior to the residual (common) stockholders but after those of creditors.
Right to voice in management	Issuance of bonds or borrowing money from other types of lenders does not have any direct effect upon the control of the firm.	Issuance of additional common stock may result in dilution of control.
Income tax effect	Interest on bonds is tax deductible.	Dividends to stockholders are not tax deductible.

compromise between what management would like most and what suppliers of capital are willing to give.

A large firm that pays its bills on time and has a high current ratio is in the best position for selecting from among financing sources. Small firms may overcome the advantage of larger firms by building a good credit record and practising good money management.

WHAT DO YOU THINK?

Should Interest on Home Mortgages Be Deductible for Federal Income Tax Purposes?

The purchase of a home is the single largest investment a typical household will make. Most Canadians want to own their own home; however, unlike Americans, Canadians cannot (as of 1979) deduct interest paid on a home mortgage from their income. On the other hand, the Canadian government has instituted a Registered Home Ownership Savings Plan under which $1,000 per year can be deducted from income if it is being saved in preparation for buying a home. Which system is most beneficial to the individual? What will happen to interest rates and house prices if a bill goes through the Commons making house payments tax deductible? WHAT DO YOU THINK?

Sources of Short-term Funds

What are the sources of short-term funds? "Short-term" here means a period of one year or less.

Most firms need to borrow short-term funds regularly for many reasons. The most common reason is to meet working capital needs. Of course, the expected level of working capital needs may be largely financed from long-term sources, while temporary liquidity is provided by short-term sources. (See Figure 11-4.) The Klutz Cor-

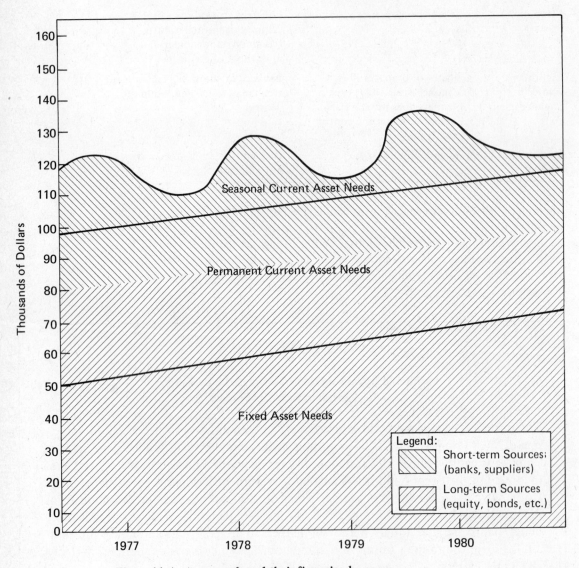

Figure 11-4. Asset needs and their financing by source

poration normally uses $10,000 in working capital (divided in equal parts among cash, accounts receivable, and inventory). Suppose there is a seasonal need to expand this to $15,000. This puts the firm in the market for $5,000 of temporary short-term credit. These funds are likely to be obtained either from trade creditors (open-book account), from bank loans, or from secured loans made by a variety of lenders. Let's examine these three sources.

Trade Credit

Trade credit or "open-book account" differs from other types of short-term credit because no financial institution is directly involved. It is simply credit extended by sellers to buyers. To the seller trade credit means accounts receivable. To the buyer it means accounts payable. When one firm (manufacturer, wholesaler, retailer) buys materials or merchandise from another, the transaction is handled in open-book accounts. There are no complex credit papers. The buyer records a new account payable. The seller makes an entry showing a new account receivable. Nearly 90 per cent of sales are handled this way.

    In effect, the seller "lends" the buyer money for the time between receipt of the goods and payment for them. Without this type of credit, many firms could not survive. The same type of credit exists between the consumer and the retailer. When you "charge" the purchase of a TV on your account, the seller is really lending you money for a while. Instead of calling this an open-book account, most people call it a "charge account".

    Most trade credit involves cash discounts for early payments. Invoice terms of "2/10, net/30" mean that the buyer can deduct 2 per cent of the invoice price if paid within 10 days; otherwise it is due in 30 days. In other words, the buyer is "giving up" the use of the seller's money for 20 days for which the buyer earns a 2 per cent discount. This is equal to an annual rate of 36 per cent. Thus, many firms will often borrow from their banks in order to take advantage of cash discounts.

Trade credit

Chartered Banks

The bank, as we saw in Chapter 10, accepts time and demand deposits and lends a part of these funds to businesses for their short-term commercial needs. Banks are a popular credit source among smaller business borrowers.

    Depending on the current balance between the demand and supply of commercial credit, bankers will adjust standards for lending. When money is short, bankers are likely to become more careful about those to whom they lend money. In any case, a bank will always check a new borrower's past credit record and ability to manage. The banker will screen loan applications also on a basis of the current ratio of the firm as well as on some of the other key credit ratios. The bank loan officer may obtain a credit report such as the one described earlier.

    A banker expects that the loan will be repaid normally out of seasonal declines of inventories and accounts receivable held by the borrower. The bank and borrower must agree on four principal terms of a commercial loan: (1) the general nature of the

> ## THINK ABOUT IT!
>
> **Borrowing to Save Money**
>
>
>
> When this business manager bought supplies from Acme Paper Company, they sent a bill for $1,000 with terms of 2/10, net/30. He had the choice of paying $980 by June 10 or the full amount by July 1.
>
> Since he had no spare cash, he went to the bank and talked to the vice-president. She offered him a loan of $980 for 20 days at an annual interest rate of 10 per cent. Total interest cost was only $5.44. (20/360 × 10% × $980 = $5.44) He saved $14.56 as follows by borrowing from the bank to pay the bill early:
>
>
>
> | | |
> |---|---:|
> | Bill saving: | $20.00 |
> | − Bank interest | 5.44 |
> | Net Saving | $14.56 |

arrangement; (2) the interest rate; (3) the quantity and type of security (if any); and (4) when the loan will be repaid. Firms which have a continuing need for funds from the bank often use a line of credit or revolving loans. **When a line of credit is set, the bank stands ready to lend up to this amount to the borrower with some restrictions.**

Line of credit

Revolving credit

A revolving credit agreement, on the other hand, is a very formal and specific agreement which guarantees funds for a period of time with strict rules limiting the borrower.

Secured Loans

Secured loans

Many commercial loans to smaller firms and to firms with lower credit ratings are "secured" loans. Here, the lender is protected by a pledge of the borrower's assets. This also may be done by firms which have reasonably good credit records and wish to borrow unusually large sums or want favourable interest rates. Items pledged as security for loans may include accounts receivable, inventories, equipment, or stocks.

A special kind of secured financing is called floor planning. An auto dealer who gets a shipment of new cars signs a note to a bank or other financing agency for the amount due. Title passes to the lender who pays the bill. A trust receipt serves as a substitute for the actual asset. The bank holds the trust receipt for the cars until they are sold and the loans are paid. Chartered banks are also the major institutions involved in the use of the other credit instruments described in Chapter 10—promissory notes, drafts, acceptances, and certified cheques.   **Floor planning**

Other lending institutions used by firms for short-term financing include factors, commercial finance companies, and sales finance companies.

Sources of Long-term Funds

The sources available to corporations for long-term funds are much more numerous than the sources available to partnerships and sole proprietors. In any case, there is some choice between internal (equity) and external (debt) financing.

Sources for Corporations

Corporate long-term capital is available in the public market by means of stock issues and bond issues. As we have seen, there are several types of stocks and several types of bonds. Issuing preferred stock is a source of growth funds for corporations with stable earnings. Common stock is more likely to be issued when good growth is expected but earnings are considered to be unpredictable. The "new" common stockholder joins the "old" common stockholder on an equal footing. Depending on the relative size of "old" and "new" common stockholdings, the "old" shareholders may risk loss of, or reduction in, their control over the firm's affairs.

A corporation which needs additional debt financing for the long run may issue bonds. Firms have an implied right to borrow as long as it is done for the company's benefit.

A major decision in a bond issue is selecting a method to pay off bondholders. The most attractive way for a strong, growing firm is by debt replacement—relying on the ability of the firm to exchange maturing bonds for new bonds or for stock. **In other cases, firms set up a sinking-fund to retire maturing bonds. This means putting aside money each year from profits to pay them off.** This is more likely to be done in a firm which is not expanding or which might find this investor protection feature the only way to attract bond investors. Also, using a sinking-fund will bring the interest rate on the bonds down.   **Sinking-fund**

A profitable firm has the option of using profits to pay dividends or to "plough back" into operations. This permits fixed asset growth without the use of money markets. It is a form of equity financing.

Common shareholders may or may not favour such financing, depending on their investment goals. Investors who view their shares as "growth stock" expect it to appreciate over the long run. They don't demand immediate dividends. Others invest for immediate income. They prefer to receive regular (and large) dividend cheques rather than have profits stay in the business.

Conflict often exists among shareholders concerning the "dividend vs. retained earnings" policy. Managers and shareholders may conflict on this point, too, unless managers are also shareholders. The availability of large amounts of capital is usually more attractive to managers than it is to shareholders. Retained earnings (a form of equity financing) has no maturity date and does not dilute managerial control as common stock does. Like common stock, retained earnings places no prior claim on income or assets. Financial managers are attracted to retained earnings as a source because they are not subject to the evaluation of the marketplace in their use of such funds.

Another outside source which has been available for long-term funds in recent years is insurance companies. Such companies have huge reserves to invest. They make these funds available, especially to large corporations, for long-term expansion at rates similar to, or somewhat lower than, those paid to bondholders.

Sources for Non-corporate Firms

Sole proprietorships depend on the personal funding of the owner-operator for equity financing. Partnerships have the same kind of limited source of funds except that there are two or more partners who may contribute fixed capital.

All forms of ownership may generate new funds internally if the firm succeeds. The amount available, in part, depends on the amount of retained earnings. In addi-

---

TWO POINTS OF VIEW

Stockholders Often Disagree

Stockholder A:
"Right, we will have to build a new production plant in the western region in the next couple of years. But, the important thing is to finance it by issuing a new block of common stock. We've already got a pretty heavy load of bonds outstanding, and retained earnings just have not been large enough. Besides, if we put all of our earnings into expansion, we won't be able to pay a decent dividend."

Stockholder B:
"You're all wrong, buddy. Plant expansion should be financed by issuing $5 million worth of debenture bonds. That way, the chances are that common stockholders like you and me will be able to gain greater returns in the long run. The profits from the new plant will far exceed the cost of servicing the debt. This multiplies our earning power. It gives us leverage. Besides, to issue additional stock will dilute our control of the corporation. Let's use somebody else's money to grow. Maybe current dividends will be smaller than usual. I can wait."

tion, the amount charged for depreciation of assets is "available" for current or long-term financing. Such equity capital is available for an indefinite period of time and has a subordinate (after creditors) claim on the firm's earnings and assets. It can lead to dilution of control of the present owners if a new partner is added.

PROTECTING THE FIRM'S RESOURCES

Owning and using resources leads to risk of many kinds. Risk is the possibility of loss. To protect these resources—to deal with these risks—is the third major task of a financial manager.

Kinds of Risks

There are risks in everything and the degree of risk may vary greatly. In lending money, we risk loss. In buying things, we risk the possibility of defective merchandise. In running a factory, we risk liability for accidents to employees or visitors. In owning buildings and cars, we run risks of fire, vandalism, and theft. We risk that secret processes or designs will be discovered and risk the death of officers. All of these threaten the firm's resources and must be dealt with if the firm is to survive and prosper.

Self-Insurance and Risk Avoidance

There are three basic approaches to dealing with risk (see Figure 11-5). They can be used in combination. One is to assume it yourself. The second is to avoid or reduce it. The third approach is to shift the risk to others.

    Many firms practise self-insurance of certain types. This means assuming your own risk and preparing for loss. If, for example, a large chain of shoe stores sets aside regularly a certain amount to cover the possibility of fire in one of its outlets, it is practising self-insurance. The idea is that, with a very large operation, some fire damage is bound to happen. Instead of paying insurance firms, the self-insurer pays itself. *Self-insurance*

    Such a practise, of course, makes a firm very conscious of ways to avoid fires so as not to have to spend its reserved funds. It will install sprinklers, inspect heating systems, prevent unnecessary smoking in the stores, and take other steps to minimize risk. There are many other types of operational precautions a firm can take to reduce or avoid risks in different phases of its business.

    Mechanized cash control systems help to protect against theft, as do basic cash audit procedures and regulations for writing cheques and making cash purchases. Related procedures provide for systematic purchasing (sometimes on a sealed-bid basis) to avoid losses because of favouritism in buying or commercial bribery. Usually, more than one signature is required for approval of purchases over a certain amount. Careful inspection of both quantity and quality of goods received also plays an important part in resource control.

1. **Risk assumption:** Self-insurance against fire, theft, etc.

2. **Risk avoidance and/or reduction:** Sound business management enables the firm to avoid unnecessary risk (i.e., credit policies which grant credit only to persons with good credit records) and/or reduce risk (i.e., safety training to teach employees safe work habits).

3. **Shifting risk:** Shifting the risk to another party such as an insurance company.

Figure 11-5. How a business manages risk

Firms which deal in new ideas and processes must maintain secrecy. This includes careful personnel screening and constant development of employee loyalty. Also included are normal security precautions such as checking visitors to plants. Careful patent protection is another means of protecting this kind of resource from being copied by competitors.

CAREERS IN FINANCE

People who hope to rise to a position of president or chief executive officer of a large corporation might be wise to consider a career in finance. Many chief executives in this country started in financial management.

Positions in finance include credit manager, financial analyst, and chief financial officer (comptroller or treasurer). Credit managers supervise the granting of credit to customers and establish criteria and procedures for evaluating credit applicants. Cash-flow managers oversee the cash position of the firm to assure liquidity and avoid loss of earnings on idle cash balances. They choose short-term investments. Financial analysts or project analysts study a firm's major capital investment decisions such as new plants and new ventures so as to maximize return on them. The chief financial officer supervises all of these activities and plans for security issues and major long-term borrowing. He or she helps set dividend policy.

All of these positions, with the possible exception of credit manager, usually require a university degree with some specialization in accounting or financial management. Persons who want to advance in the financial field must possess skills and aptitude for mathematical computation. They often start as financial management trainees. These people are shifted around various financial departments of the firm and often to various geographic locations. Information is available from the Canadian Bankers Association, Box 282, Toronto Dominion Centre, Toronto, Ontario M5K 1K2, and from the Financial Executives Institute Canada, Suite 409, 45 Sheppard Avenue East, Willowdale, Ontario M2N 5W9.

There is always a danger that changes in market conditions will hurt a firm. A retailer may find some goods hard to move. To avoid losses, he or she may use one of several legal devices which allow the return of merchandise to the producer or wholesaler. We call this buying on consignment. To protect against sudden rises in the cost of needed supplies, a firm might have a policy of stocking up large quantities when the market is down.

Such protective measures complement the role of insurance, which is our next topic.

Shifting Risk and the Use of Insurance

Insurance companies assume the risks of their policyholders for a price. A firm pays a premium for a policy which pays if it sustains certain types of losses. The policy specifies the types of risks which are covered, the amount of coverage, and the premiums.

Pure risk

The insurance company is a professional risk taker, but it takes only certain types of risks. Insurance is available only for pure risks. **Pure risks are those which offer only a chance of loss. There is no chance of gain.** Examples are risk of fire and risk of death.

Speculative risk

Speculative risks are "gambles" in which there is possible gain as well as loss. This can happen at the race track or whenever someone goes into business, where the quality of a new product or the quality of management may lead to profit or loss. Insurance is not available to deal with such speculative risks.

The federal and provincial governments are also in the insurance business. Their involvement is evident in such diverse programs as unemployment insurance, workmen's compensation, social insurance, crop insurance, and automobile insurance.

Firms deal with insurance companies for the same reasons that individuals do. Most homeowners, for example, carry fire insurance rather than bear the entire burden of the risk themselves. The same is true of firms. Insurance companies combine the risks of many policyholders—firms or individuals—into a group.

The Law of Large Numbers

Insurance firms study the past to see how many people die each year at age 50 or age 60 or at any age. They develop mortality tables from these facts. Mortality tables are used to predict the number of policyholders who will die in a given year. **This prediction depends on the law of large numbers. In other words, if the insurance firm has a large number of policyholders, it can pretty well predict how many of them will die in a year from the mortality tables.**

Law of large numbers

The same principle applies to other insured risks such as risks of fire or theft. Past experience and the law of large numbers give insurance firms a fair idea of how much they will have to pay out in claims. They set their premiums at a level which will allow them to cover expected claims as well as to cover operating costs and profit.

Of course, an insurer attempts to avoid writing insurance for a group if the peril (danger) insured against would damage all members in the group at the same time. For example, a fire insurance firm would not concentrate all its coverage in one section of a city. A major fire there could affect too many policyholders and could ruin the insurance firm.

The risk from the insurance firm's point of view relates to how accurately it can predict total losses within a group of policyholders. If the probable range of losses is great, the risk is great.

EXTRAORDINARY FINANCING ARRANGEMENTS

Sometimes, special conditions lead a firm to make extraordinary financing arrangements. These include devices for external expansion as well as the processes of recapitalization and reorganization.

Merger

External expansion can be achieved through merger, amalgamation, or the creation of a holding company. **In a merger one firm (the absorbing firm) keeps its identity, and another firm is absorbed and loses its identity.**

In an amalgamation both firms lose their identity. A new corporation is formed, and shareholders of both former firms receive stock in the new corporation in exchange for their old shares.

A holding company owns the stock of one or more other corporations and controls them. When the holding company is itself an operating firm, it is known as an operating holding company. Control can be exercised without majority ownership by means of proxies. (See Figure 11-6.)

Amalgamation

Holding company

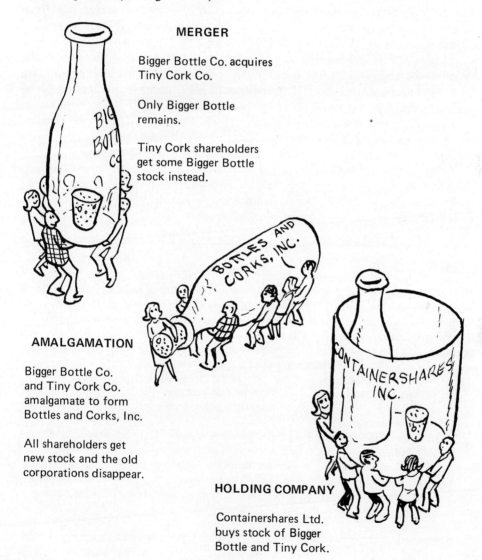

MERGER

Bigger Bottle Co. acquires Tiny Cork Co.

Only Bigger Bottle remains.

Tiny Cork shareholders get some Bigger Bottle stock instead.

AMALGAMATION

Bigger Bottle Co. and Tiny Cork Co. amalgamate to form Bottles and Corks, Inc.

All shareholders get new stock and the old corporations disappear.

HOLDING COMPANY

Containershares Ltd. buys stock of Bigger Bottle and Tiny Cork.

Original firms continue to operate.

Figure 11-6. Methods of expanding a firm

**Recapitaliza-
tion**

Recapitalization occurs when a firm changes its capital structure to meet changing conditions. It does not raise more capital. It may involve replacing a high-yield preferred stock with a lower-yield preferred stock or floating a new bond issue to replace a maturing one. Sometimes a stock split is used to attract investors. Recapitalization often requires that the corporation amend its charter or receive permission from a provincial securities commission.

Reorganization

Reorganization is an involuntary process. It occurs when a firm is in very serious financial trouble and the court steps in to protect creditors. Sometimes creditors force an out-of-court reorganization.

Firms go into debt (both long- and short-term) in the hope that they will be able to pay the interest and principal out of earnings. If this doesn't work out over the period of indebtedness, the firm is in trouble. A firm which cannot meet its maturing financial obligations is insolvent. **If, in addition, its liabilities are greater than its assets, the firm is bankrupt. Such a firm is said to be in bankruptcy.**

Bankruptcy

Under voluntary bankruptcy, a person or firm files a petition in court claiming inability to pay debts because the debts exceed available assets. This petition asserts willingness to make all assets available to creditors under court supervision.

Under involuntary bankruptcy, a person or firm's creditors seek to have a debtor declared bankrupt by proving that the debtor committed one or more acts of bankruptcy as defined in the law. Once a defendant is declared bankrupt by the court, the procedure is the same as it was in voluntary bankruptcy.

SUMMARY AND LOOK AHEAD

The financial manager's job is complex and requires a variety of talents. One must be aware of present and future needs for working capital and fixed capital. Capital budgeting helps in performing this task. The financial executive must also evaluate and select sources of funds. He or she thinks in terms of maturity, claims against assets and income, and control of the firm when deciding how to get funds. For short-term funds, trade credit and banks are the major sources. Stocks, bonds, and retained earnings are the sources of most long-term funds.

To protect the firm's resources, a financial manager must think of ways to avoid, reduce, or shift risk. Insurance companies play a major role in risk shifting for firms as well as for individuals.

In addition to the ordinary long-term financing sources—retained earnings and stock and bond issues—extraordinary means for long-term financing are sometimes needed. These include external expansion devices, recapitalization and reorganization.

In the next chapter we will begin to explore how the human resource is dealt with by studying personnel. We will also see how a firm maximizes its use of people and the institutions involved in such efforts.

KEY CONCEPTS

Amalgamation A method of combining corporate firms by which both original firms lose their identity and all old stock is exchanged for stock of the new, combined firm.

Bankruptcy A condition of insolvency (inability to meet current debt) combined with liabilities greater than assets.

Capital budget A projection of expected needs for fixed assets over a 5 to 10 year period.

Debt financing Obtaining funds by going into debt (e.g., issuing bonds).

Equity financing Financing by selling additional shares of stock or obtaining additional funds from a proprietor or partner.

Floor planning A method by which sellers of consumer durables finance those products so that they do not have to pay for them until they have sold them.

Holding company A corporation owning all the common stock of other firms so that the operation of those firms is co-ordinated and controlled.

Law of large numbers The fact that the likelihood of occurrence of a peril within a large group of policyholders may be estimated fairly closely.

Lease Agreement to grant use of an asset for a period of time in return for a periodic payment.

Line of credit A stipulated volume of credit made available to a person or firm by a lending institution.

Liquidity The quickness and ease with which an asset can be turned into cash.

Long-term (fixed) capital Assets committed for more than one year or funds used for the purchase of such assets.

Maturity The date on which a debt obligation becomes due.

Merger A combination of formerly independent business firms under the name of one of the firms.

Opportunity cost Cost of losing the option to use funds in another way.

Pure risk Opportunity for financial loss without opportunity for financial gain.

INCIDENT:

Mrs. Winthrop Barnes

Mrs. Winthrop Barnes has been very successful in her small real estate sales firm. She has been earning a net profit of 20 per cent on her investment for the last three years. She sees the opportunity for adding a branch sales office in a nearby smaller city. Profit opportunities there, she believes, are at least as good as in her present operation. She plans to sell $30,000 worth of bonds paying 14 per cent interest to finance the expansion.

Questions:
1. Would you advise Mrs. Barnes to sell the bonds?
2. What are the main reasons for and against the sale?
3. What are her alternatives?

> # INCIDENT:
>
> ## WANDALUST Bath Oil
>
> The WANDALUST Bath Oil Company, Inc., has a running battle going. Ever since Leroy Coleman became sales manager, he has been pressing for approval of credit sales to the Laroo Unisex Perfumery Chain in and around Toronto. Bob Blunt, the comptroller, has refused to sell to Laroo on credit because of an experience five years ago. At that time the owner of the Laroo Chain was convicted of income tax evasion and served a two-year term in jail. Many of his suppliers at that time had to wait more than six months to collect on the Laroo account.
>
> Questions:
> 1. What are the arguments supporting Blunt?
> 2. What are the arguments supporting Coleman?
> 3. How do you suggest the conflict be settled?

Recapitalization A change in capital structure to reflect changing conditions internal or external to the firm.

Reorganization An involuntary process by which a court steps in to protect creditors from loss due to poor financial conditions in a firm.

Revolving credit Chartered bank credit under a formal agreement. It guarantees the borrower funds for a period of time under strict rules.

Secured loans Chartered bank loans in which assets of the borrower are pledged as security.

Self-insurance Assuming your own risk and preparing for loss.

Sinking-fund A fund set aside over the lifetime of bonds for their retirement.

Speculative risk Risk in which there is a chance for gain as well as a chance for loss.

Trade credit Credit extended a firm by its suppliers.

Working capital Current assets used in business operations or funds employed for the acquisition of current assets.

QUESTIONS FOR DISCUSSION

1. Briefly explain why it is important to distinguish short-term financing from long-term financing.
2. What specific uses are made of working capital?
3. What are some examples of fixed capital needs? Give an example of a firm with great fixed capital requirements and another with very small fixed capital requirements.

4. What factors may affect the supply of credit to a small manufacturer?
5. How do debt and equity financing differ in terms of their maturity? Their claims on income?
6. What is another name for "open-book accounts"? Are they a source of long-term or short-term credit? Explain.
7. Going back to Chapter 10, describe the principal types of stocks, common and preferred, which are used by corporations.
8. Describe two of the operational precautions a firm might apply in order to protect its resources.
9. Why can some firms use self-insurance while others can't?
10. What is the law of large numbers? How does it relate to the size of insurance premiums?
11. How does a firm become bankrupt? Discuss.

ROCKETOY COMPANY VI

After 15 years of growth as a corporation and a successful move from Toronto to Windsor the Rocketoy Company is financially sound. Its credit rating is good, and an analysis of its financial statements by a large national CA firm indicates that it is a very good prospect for continued success if growth funds can be acquired. Presently there is common stockholders' equity worth $4 million. All of it is held by the original stockholders except for Richard Talley, who sold his holdings to Terrence Phillips. Terrence now owns 65 per cent of the outstanding stock. Bonds are outstanding in the amount of $1 million. Half of this amount is maturing (will become due) within a year.

Including the retirement of maturing bonds and capital expansion needs as found in the capital budget, Rocketoy needed $6 million in long-term capital funds in the following two-year period. Pam Carter, the vice-president of finance, still held 5 per cent of the outstanding stock at the time. She proposed issuing $2 million in 9 per cent bonds maturing in 20 years and selling 40,000 shares of stock at $100 per share. This would double the number of shares outstanding. Her brother, the president, is concerned about this proposal because he does not have personal funds sufficient to maintain majority ownership under such a plan. But he does see the advantages of the plan, too. Joe Phillips, Mike Shultz, and Julia Rabinovitz could easily purchase 10,000 additional shares each. Investment bankers have advised that up to $4 million in 9 per cent bonds could be sold at par.

Questions:
1. Assuming the expansion needs are genuine and you are Terrence Phillips, would you oppose Pam's plan? Why or why not?
2. What are the advantages of Pam's plan?

3. What are the advantages of selling $4 million in bonds?
4. What compromises could you suggest?
5. What information about the toy market would you want before reaching a decision?

12 Personnel Management

SECTION THREE BUSINESS DECISIONS

OBJECTIVES: After reading this chapter, you should be able to:

1. Compare a firm's human asset with its non-human assets.
2. Explain the meaning of human resource management.
3. List and discuss the tasks of personnel management and the personnel department.
4. Differentiate among the three types of interviews used in the selection process.
5. Identify two important controversies regarding the use of selection tests.
6. Draw a chart which shows how a job applicant becomes an employee.
7. Compare job-skill training with management development.
8. Compare the merit rating system with the management by objectives approach to appraising employee performance.
9. Discuss the importance of compensation to a firm, its employees, and its publics.
10. List and discuss three ways by which a worker's pay can be determined.
11. Give three examples of employers' compensation philosophies.
12. List and discuss three ways by which employees can be terminated.

KEY CONCEPTS: In reading the chapter, look for and understand these terms:

HUMAN RESOURCE
PERSONNEL ADMINISTRATION
PERSONNEL MANAGEMENT
PERSONNEL DEPARTMENT
MANAGEMENT INVENTORY
JOB ANALYSIS
JOB DESCRIPTION
JOB SPECIFICATION
RECRUITING
PRELIMINARY
   EMPLOYMENT INTERVIEW
JOB APPLICATION FORM
SELECTION TESTS
IN-DEPTH INTERVIEW
BACKGROUND
   INVESTIGATION
FINAL SELECTION
   INTERVIEW
EMPLOYEE ORIENTATION

JOB-SKILL TRAINING
MANAGEMENT DEVELOPMENT
PERFORMANCE APPRAISAL
   SYSTEM
MERIT RATING SYSTEM
SENIORITY
MANAGEMENT BY
   OBJECTIVES (MBO)
WAGE AND SALARY
   ADMINISTRATION
PIECE RATE
INCENTIVE PAY
WAGE
SALARY
PROMOTION
RESIGNATION
EXIT INTERVIEW
DISMISSAL
DISCHARGE

Any firm's success, in the final analysis, depends most on the quality of the people who work for it—its personnel or human resource. This includes workers and managers at all levels in the firm—from top to bottom.

Top management, of course, is responsible for staffing the firm with good personnel. Staffing is a vital function of management. Its importance, unfortunately, is sometimes underestimated. Personnel administration is concerned with setting broad company policy regarding the firm's management of its human resource.

Top management delegates to the firm's personnel manager the task of implementing its human resource policies at the firm's "operations" level. Usually, this involves setting up a personnel department. This staff department is accountable for building and keeping a good work force.

In this chapter, we will discuss the firm's approach to human resource management. In practice, all managers participate in managing this vital resource. But most of the activities of human resource management are delegated to the personnel manager. Our main focus, therefore, is on the nature of personnel management activities.

HUMAN RESOURCE MANAGEMENT

A firm's most important resource, or asset, is its human resource. **The human resource is the personnel who staff the firm—its workers and managers.** It includes maintenance workers, salespeople, assembly-line workers, typists, and managers at all levels.

Human resource

A firm's workers and managers make its non-human resources, or assets, productive. Without good personnel, the best-equipped plant will not function properly. A well-financed firm will not make a profit if its workers and managers are incompetent.

If you look at a company's balance sheet, you will find cash, accounts receivable, and inventories listed as current assets. Machinery, equipment, furniture, buildings, and land are listed as fixed assets. But there is no specific accounting for the human asset. Yet, a firm does invest money in recruiting and training employees. These employees are "earning assets". Without them, the firm would fold up.

The human resource, of course, must be managed in any type of organization. All managers participate in managing this resource. Assembly-line bosses, for example, help to train their subordinates and to motivate them to good performance. Production managers, in turn, help to train and to motivate their assembly-line bosses.

Human resource management, therefore, is concerned with the management of a firm's most important resource—its personnel. Wherever there are managers and workers, human resource management activities are involved. These activities are discussed in this chapter.

PERSONNEL ADMINISTRATION

The broadest possible view of a firm's human resource is that of personnel administration. **Personnel administration is top management's effort to develop an overall effective work force which includes managers and workers at all levels in the firm.** Thus,

Personnel administration

personnel administration is concerned with human resource management on a grand scale.

A firm's top management sets the tone for the firm's approach to its human resource management. The importance that top management attaches to the human resource plays a large role in determining how it will be managed.

PERSONNEL MANAGEMENT AND THE PERSONNEL DEPARTMENT

Personnel management

Top management's policies regarding human resource management are carried out through the practise of good personnel management. **Personnel management consists of recruiting, selecting, training, developing, compensating, terminating, and motivating employees to good performance.**

As we said earlier, the management of personnel is the job of every manager. But as firms grow, they create specific departments to help managers manage their personnel. Such a department is a personnel department. Top management delegates to the personnel manager the task of carrying out its human resource policies at the firm's "operations" level. **A personnel department is a staff department which is headed by a personnel manager. It advises and helps line managers to manage their personnel by performing specialized activities which are assigned to it.**

Personnel department

The role played by the personnel department in a firm depends on the authority granted to it by top management. As a staff department, it advises and helps line managers to recruit, select, train, and motivate workers. Thus, final decisions about personnel matters are made by the line managers. But personnel departments in some firms have functional authority. In these firms the personnel department actually makes decisions regarding the hiring of new workers, granting promotions, and so forth.

In large organizations, the personnel function employs many people. Many of

---

A CONTEMPORARY ISSUE

Human Asset Accounting

Although most managers recognize the value of company personnel, very few of them try to put a value on the firm's human asset. Consider a firm whose balance sheet lists a typewriter as an asset. Shouldn't the typist also be listed as an asset? Without a typist, a typewriter is useless.

A major problem, however, is how to value the human asset. What, for example, would be the "acquisition cost" for a firm's personnel? What about the replacement cost? What about "depreciating" the human asset?

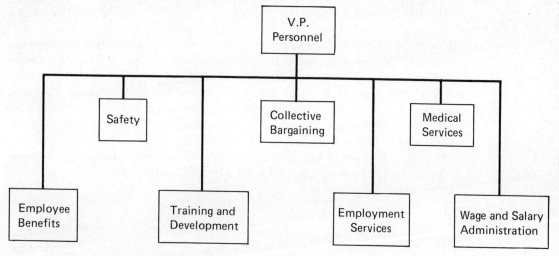

Figure 12-1. Organization of the personnel function

these are experts in a certain field such as safety, collective bargaining, or employee testing. Figure 12-1 indicates the functions which are usually performed by a personnel department in a large manufacturing firm.

In the discussion which follows, we look at the work of the personnel manager and the personnel department. This work can be divided as follows:

1. Determining human resource needs.
2. Searching for and recruiting applicants to fill those needs.
3. Selecting applicants for employment.
4. Training and developing personnel.
5. Appraising employee performance.
6. Compensating employees.
7. Promoting employees.
8. Terminating employees.
9. Providing personnel services.
10. Performing other personnel activities.

DETERMINING HUMAN RESOURCE NEEDS

For a new firm, the best place to start in order to determine human resource needs is the statement of company objectives. This helps in preparing an organization chart, which provides a skeleton of the firm's needs for workers and managers.

In the case of established firms, some workers and managers are already on the payroll. These firms also must continually study their need for personnel. Old employees retire, leave to take jobs with other employers and so on. They must be replaced.

Forecasting Human Resource Needs

There are two basic ways to determine a firm's human resource needs. One approach involves no advance planning. The need is recognized as urgent only when a present employee leaves or a new job opening must be filled.

The other approach is to anticipate needs before they become urgent. This takes careful planning and forecasting of human resource needs. It involves estimating the number and types of positions which will be opening up and making a time schedule of when they will probably open up.

Ideally, a firm should take its sales forecast and "back into" production scheduling and financial forecasting. This helps it get a better fix on the exact number and types of jobs to be filled.

The Management Inventory

Management inventory

A firm's management inventory is a list of present managerial positions along with the names and personal data (age, experience, length of service, etc.) of the persons holding them. This inventory helps top managers determine the reserve of managerial talent on hand for promotion to higher positions. It also helps in making projections regarding future need for managers.

Specifying Human Resource Needs

Job analysis

Job analysis involves defining the jobs which must be done if a firm is to reach its objectives. Each job is studied to determine what work it involves and what qualifications are needed by the persons who will fill them.

Job description

A job description is prepared from the job analysis. **A job description outlines the nature of a given job.** Such things as how each job relates to other jobs; the specific duties involved; and the tools, machinery, and supplies needed are stated in writing. (See Figure 12-2.)

Job specification

A job specification is prepared from the job description. **A job specification states the personal qualifications needed of the person who is to fill each job—education, skills, experience, etc.**

Management inventories, job descriptions, and job specifications help make a firm's human resource needs more concrete. They give the personnel manager a clearer idea of what types of employees are needed. They permit more effective evaluation of people who apply for jobs. Thus, they help make the best match of specific workers with specific jobs.

SEARCHING FOR AND RECRUITING APPLICANTS

A firm which waits for recruits to come to it has to be satisfied with whatever shows up. This is, at best, a poor approach. A firm should be an active recruiter.

GENERAL SALARIED JOB DESCRIPTION

TITLE ___Clerk – General I_____ CODE _____

LOCATION _____ DATE _____

PREPARED BY _____

PRIMARY PURPOSE OF JOB

Computes the number of parts to be supplied by specific vendors and codes line item information for machine preparation of releases. Checks releases and posts a variety of information to record cards.

DETAILS:	% of Time
1. Computes manually the number of parts to be supplied by vendors by interpreting part usage, transit time, float requirements, production program, vendor deviations, and minor adjustment for preparation of material releases; prepares draft form of the release.	
2. Prepares vendor release transmittals which instructs an electronic machine how to prepare material release by assigning schedule code numbers to line items from a pre-established schedule code structure and all necessary data for key-punching.	
3. Checks machine-prepared releases to ensure accuracy by computing total authorization.	
4. Posts inventories, adjustments, vendor deviations, and the like to Material Record Cards from source documents.	
5. Assists in stripping and mailing releases.	
6. Performs related duties as required.	

Nature and Purpose of Contacts Required in Performance of Duties

Has contact with work associates only.

Consequence of Errors in Performance of Duties

Errors in manual calculations or assigning schedule codes could result in excess or short inventories at assembly plants.

Nature of Controls or Restraints

Work is spot-checked by Schedule Manager and Auditors.

Working Conditions

Works in office area.

Figure 12-2. A job description

The Search Task

The nature of the search task depends on the types of jobs being filled and the general conditions in the job market. The more skills needed, the more complex is the search process. The supply of, and the demand for, people possessing certain skills also affects the search.

Deciding where to search is an important first step. Sources of recruits include present and past employees, their friends and relatives, public and private employment agencies, vocational and technical schools, universities, and many others.

The extent of the search depends largely on company policy. A firm with a policy of not promoting from within will conduct an extensive external search. But excluding present employees in the early search may have a bad effect on their morale. The personnel department must remember that building and maintaining employee morale are basic tasks. The various personnel activities must be viewed as interrelated. The advantages of external search must be balanced against those of internal search.

As a college student, you know that many firms recruit on college campuses. You also probably know that there seem to be "ups and downs" in this recruiting activity, depending on general economic conditions. You should pay close attention to the Appendix in which we discuss how to prepare student data sheets, resumés, letters of inquiry, and how to interview.

Recruiting should be an ongoing process for most medium- and large-size firms.

Recruiting **Recruiting is the task of attracting potential employees to the firm.** "Beating the bushes" to find recruits when the firm is in desperate need for new employees should be avoided. The recruiter, of necessity, will have to be less selective in choosing. It is much wiser to view recruiting activity as a continuous process.

There are many advantages of recruiting for the long range. First, it forces a firm to take stock of its present work force and to plan for future human resource needs. Secondly, it helps to eliminate the chance that new employees will be selected from among a small number of applicants. Thirdly, a firm with a good recruiting program may be able to gain an edge over competitors. It will be able to hire people who are not available to "panic" recruiters.

Recruiting Managers

There are two sources of management recruits: (1) promotion from within the present managerial and non-managerial ranks; and/or (2) going outside the firm. Promoting from within boosts employee morale. Recruiting outsiders, however, may bring needed "new blood" into the firm.

SELECTING APPLICANTS FOR EMPLOYMENT

If the personnel department succeeds in getting applicants, its next task is to select the best ones. This takes a great deal of preparation.

> ## CAN YOU EXPLAIN THIS?
>
> ### Front-end Bonuses
>
> Some firms, especially those having serious financial and other management problems, have a hard time recruiting good managerial talent. In some cases, these firms might offer experienced executives "front-end bonuses".
>   These bonuses are not dependent on how profitable the firms become under their new managers. They are payments to executives who "sign-up"— much like the bonuses which many professional athletes get. They are recruiting devices. Why are they used? CAN YOU EXPLAIN THIS?

In the discussion which follows, we examine the preliminary employment interview first. This is the first step in campus interviewing. If a prospective employee searches out prospective employers without the benefit of a campus interview, that person usually completes a job application form first. You should interview with campus recruiters when they visit your campus.

The Preliminary Employment Interview

The preliminary employment interview is the first time that the employer and the applicant meet together. The employer is usually represented by an interviewer who informs the applicant of job openings. The applicant has an opportunity to ask questions and to discuss skills, job interests, education, etc. If a match seems to exist between the firm's needs and the prospect's abilities, the applicant is asked to fill out a job application form.

Preliminary employment interview

The Job Application Form

The job application form is prepared by an employer and filled in by a job applicant. The applicant provides job-related information which helps the reviewer determine if the applicant has the needed training, education, and so on, for the job. Many applicants fill out these forms carelessly because they think the forms are unimportant. This often is the reason that a person is turned down. The application form is really important.
  All provinces have passed laws prohibiting employers from discriminating against prospective employees on the basis of race, colour, religion, national origin, or sex. They have also passed legislation designed to ensure equal pay for men and women performing comparable work. Anti-discrimination laws also make it illegal to require

Job application form

an applicant to supply data regarding religious preference, colour, race, sex, or nationality. Many firms which formerly required applicants to attach photos to the job application form have dropped the practice.

Selection Tests

Selection tests

Recruits who make it through the preliminary employment interview and whose application forms are acceptable are then given selection tests. **Selection tests are used to measure an applicant's potential to perform the job for which he or she is being considered.** These tests include intelligence tests, aptitude tests, and performance tests.

Intelligence tests measure general verbal ability and specific abilities such as reasoning. Aptitude tests measure ability such as mechanical aptitude or clerical aptitude. Performance tests measure skill in a given type of work such as typing. These tests help to determine if applicants match up to job requirements. But the tests themselves are far from perfect.

For example, it is easy to measure a typist's ability to type by giving a typing test. But not all jobs require such specific skills. Testing is more successful in spotting an applicant's shortcomings than in picking guaranteed successes. It is, for example, hard to test for job motivation. But that plays a large part in a person's success or failure on the job. Selection tests should be used to supplement judgment and any other information which is available on an applicant. Tests should not be used as a substitute for them.

The use of selection tests is controversial today. Some people believe that the tests are biased. Intelligence tests, for example, have been attacked because of their white,

---

WHAT WOULD YOU DO?

The Swinging Banker

Bob Jones is a 30-year-old applicant for a job opening at a bank in Calgary, Alberta. He is applying for the job of loan officer. During routine reference checks, the bank's personnel manager learned that Bob had some rather "radical" political beliefs while he was a student in college. He also was arrested one time for public drunkenness after a fraternity party.

Other than that, the investigation turned up very "favourable findings". The only "bad comment" came from one of Bob's current neighbours: "He is an unmarried young man who does an awful lot of partying until the wee hours of the morning".

Suppose you are the person who decides whether or not Bob gets the job. WHAT WOULD YOU DO?

middle-class cultural bias. If an organization requires all applicants to take an intelligence test, those who do not come from a white, middle-class background may find themselves at a disadvantage, not because they are unintelligent, but because they are not familiar with the middle-class way of life.

Critics argue that selection tests should measure a person's ability to do the job or measure the person's ability to be trained to do it. Other people argue that a job involves more than simply the ability to do a given task. For example, a recruit's ability to get along with others is important on the job and should be considered. To stay within the law, selection tests should be carefully designed to measure ability to do the specific job for which an applicant is being considered.

The In-depth Interview

An applicant who passes the selection tests is usually scheduled for an in-depth interview. **An in-depth interview is one conducted by trained specialists from the personnel department to shed light on the applicant's motivation, ability to work with others, ability to communicate, etc.**

In-depth interview

The Background Investigation

After in-depth interviewing, an applicant's references are checked. **In a background investigation, the applicant's past employers (if any), neighbours, former teachers, etc., are questioned about their knowledge of the applicant's job performance, character, background, etc.**

Background investigation

There is a lot of controversy here. An employer has the right to look into an applicant's suitability for a job, but it is easy to go overboard and invade the applicant's privacy. There also are problems related to misinformation or falsehoods. Employers might be sued by former employees for giving false or misleading information on reference forms. Some employers now refuse to fill them out.

The Final Selection Interview

If the results of the background investigation are satisfactory, the applicant is likely to be called in for the final selection interview. **In the final selection interview, all company personnel who have interviewed the prospect are present, along with the manager under whom the applicant will work. The manager is the person who makes the decision whether or not to hire the applicant.** Any hiring decision, however, is contingent on passing the company's physical (and perhaps, psychological) exam.

Final selection interview

In recent years, there has been some controversy regarding these exams. As long as the exam pertains to job-related requirements, there is no problem. But when a firm sets standards that are not job-related, trouble arises. There is no excuse for denying a physically handicapped person a job if the handicap does not hurt job performance. The same is true of the mentally retarded.

Recruiting, interviewing, screening, preparing job specifications, and other personnel activities require care to avoid illegal discrimination based on an applicant's race, colour, religion, sex, age, or nationality. An employer who wants to hire a female fashion model would not be breaking the law by screening out a male applicant, since employers are allowed to set "bona fide" occupational qualifications. But these must be related to the job being filled. Thus, an employer cannot refuse to hire a female applicant on the grounds that the job is thought to be too physically demanding or because it involves travel with men.

The entire process described above is shown in Figure 12-3.

Selecting Managers

Applicants for management positions are screened in somewhat the same way as other employees. Depending on the level of the position, however, greater or less emphasis is placed on the applicant's human or "people skills" versus technical or "how-to-do-it" skills. Personality tests often are given to applicants for jobs that require a lot of "people skills". Such a test measures ability to "get along" and to work with others.

TRAINING AND DEVELOPING EMPLOYEES

Up until this point, the main goal of the personnel activities we have discussed is to reject or accept the applicant. Early rejection of unfit applicants saves the firm the expense of interviewing and testing them. It also minimizes the applicant's anxiety while waiting for the final decision.

Employee Orientation

Employee orientation

Employee orientation involves introducing the new employee to the job and to the firm. It reflects the fact that the person is no longer an applicant but is an employee.

An applicant gets some orientation to the firm and the job. But it only touches the surface. As an employee, the orientation is much more formal and complete.

The new employee is told about the firm's history, its products, and its operation. Company policies and rules are explained, as are company-sponsored employee services. To acquaint the new employee with the firm, a tour of the plant is made, co-workers and the boss are introduced, and the new employee's questions are answered.

Job-skill Training

Job-skill training

Job-skill training teaches employees specific job skills. For simple jobs, on-the-job training often is used to teach new employees their jobs. They learn by doing. It is also used to teach experienced employees new job skills.

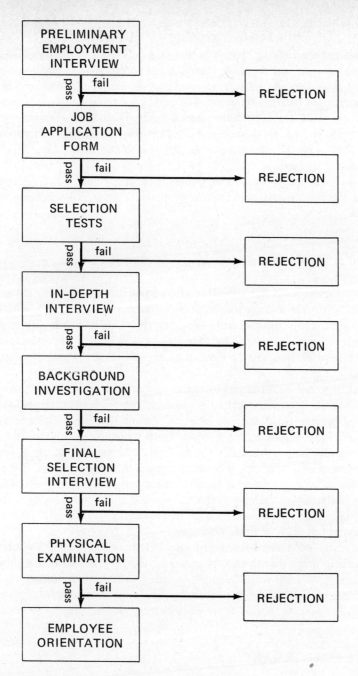

Figure 12-3. How an applicant becomes an employee

If a higher level of skill is required or if on-the-job training is too dangerous, training may be done away from the job. The employee goes through a company-sponsored training program either at the plant or in the company's training school. In this "vestibule" training the employee is trained on the same machines which are used in the plant, but the training is done in a classroom. When the training is completed, the employee works under the close supervision of an experienced employee.

Training is an ongoing process. New employees must be trained, and those with longer service often must be retrained. The rapid pace of technological change means that employee skills become obsolete much faster. Furthermore, employee expectations about jobs and careers have been rising. Many employees want not only a job but a career. They want the opportunity to "grow with the firm". Employee training, therefore, is a crucial task.

Canadian firms spend millions of dollars each year training and retraining their workers in the hope that it will improve employee morale and will keep their work force intact.

Management Development

Management development *Management development refers to efforts to sharpen managerial skills.* These skills are more general in nature and are much harder to teach than specific job skills. Managerial skills are basically "people skills" and conceptual skills. But managers also must have enough technical knowledge to understand the work they supervise. Many techniques have been created to help develop managerial talent. Most are group learning techniques.

In sensitivity training, managers meet together daily over a period of several weeks. The group's leader does not structure relationships among members or tell them what to do. They become frustrated and blow off steam. Thus, they come to know one another quite well. Relationships become more structured and group ties become stronger. This develops a manager's ability to see the meaning of his or her actions from the point of view of others. It makes a manager sensitive to others' feelings.

Performance appraisal training trains a manager to rate subordinate performance more objectively. Other training includes the case study, role-playing, and various simulation techniques. Managers are able to project themselves into situations even though they are not really in those situations.

How a manager is trained varies with the skills and experience brought to the job. A young management trainee may be rotated through a sequence of different jobs. This develops familiarity with the total company operation. A manager being considered for promotion may be given special assignments to see how he or she measures up to the higher position. It also shows whether more training is needed.

APPRAISING EMPLOYEE PERFORMANCE

Performance appraisal system The personnel department also helps to develop a formal performance appraisal system. This enables a supervisor to rate a subordinate's performance on the job. *A good performance appraisal system provides a basis for measuring an employee's contribution to the firm.* It reduces employee suspicion that promotions are based on favouritism rather than performance. These systems must be understood by employees and the supervisors who rate their performance. There are several types.

Traditional Employee Performance Appraisal Systems

Traditional employee performance appraisal systems require a manager to appraise a subordinate's work habits and personal traits. **A widely used approach is the merit rating system. Each employee's job performance is appraised every six months or every year. Initiative, dependability, ability to work with others, judgment, etc., are appraised by each employee's supervisor.** Employees with "good" ratings are given raises and offered promotions when they open up.

Merit-rating system

In practice, however, these traits and characteristics are hard to rate objectively. Two extreme supervisor reactions are to give all subordinates "good marks" or to use performance rating to "dress down" or to "get even" with those they dislike.

The result, in many cases, is that merit raises tend to be automatic for favoured employees. Because of this, many firms rely on an employee's seniority for pay raises

---

TWO POINTS OF VIEW

Employee Evaluation

Supervisor A:
"As a supervisor, I have to evaluate my subordinates every six months. The people over in the personnel department keep tabs on these evaluations for use when promotions open up and for granting pay raises. Honestly, I think it's a lot of bull. If I rate any of my subordinates low, it's a bad reflection on me. After all, my job is to inspire them to good performance. For example, if I check the box that says 'poor attitude toward job', in effect I'm admitting that I'm failing as a manager.

"On the other hand, if I give 'top grades' to my subordinates, I look good to them and to my boss. It sort of reminds me of school and the lousy grading system. You can bet that all of my subordinates always get good marks."

Supervisor B:
"As a supervisor, I want to evaluate my subordinates every six months. I have an obligation to my subordinates to rate their performances realistically and objectively. They want to know where they stand and how well they are progressing on the job. I also have an obligation to my superiors to make the best use of my subordinates.

"As a manager, I must work through others. The more that I can help them develop into good employees, the more valuable they are to me, to the company, and to themselves."

Seniority   and promotions. **Seniority refers to an employee's length of service. The longer that service, the more seniority an employee has.** Labour unions prefer that raises and promotions be awarded on the basis of seniority. Although managers generally are opposed to this, they encourage it when they do not try to rate employee performance. Some firms, however, are turning to a new approach.

TWO POINTS OF VIEW

Management by Objectives (MBO)

Supervisor A:
"This MBO is for the birds. Why should I let my subordinates help to set the goals they are supposed to accomplish? This mutual goal-setting lets them tell me what they should be trying to accomplish. But my job is to tell them what to accomplish.

"What is worse is that MBO lets workers help decide what acceptable levels of performance are. You can bet that it results in lower employee performance.

"Periodic meetings with each worker to discuss his or her progress towards reaching goals is another requirement of MBO which wastes both my time and my workers' time. Then we're supposed to meet again at the end of the period during which the worker is supposed to have accomplished the mutually agreed-upon goals. The big joke is that the worker and I are supposed to evaluate his or her performance and set new goals for the next period. Thus, not only does a worker help to set goals, a worker also helps evaluate his or her own performance. It's a never-ending cycle of giving away authority to make decisions."

Supervisor B:
"MBO is the best possible way to motivate subordinates. By letting them participate in setting their goals, they know what is expected of them. By letting them participate in evaluating their performance, they know how their performance will be evaluated. In other words, MBO improves boss-worker communication. It is at the heart of participative management. I think it shows workers that their ideas and opinions count.

"MBO helps each worker to better understand how his or her performance is related to the firm's accomplishment of its objectives. It also puts an end to the notion that promotions and pay raises are based on favouritism. Each worker knows what goals he or she is supposed to be working towards and how his or her performance will be evaluated. Supervisors who believe that MBO is no good assume that workers are lazy, unconcerned about accomplishing goals, and indifferent about performance evaluation."

Management by Objectives

This new approach is called management by objectives (MBO). **In the MBO approach, boss and worker get together to set objectives for the worker. If those objectives are accomplished, the worker is considered to have performed well.** It's the result that counts!

The MBO approach does away with some of the subjective elements in a merit rating system. The boss does not have to evaluate workers on their dependability, co-operativeness, ability to get along with others, or whether or not they make good use of their time. But MBO requires mutual trust and respect between boss and worker. It also requires that they be able to communicate effectively with each other and have faith in each other's abilities.

Management by objectives (MBO)

COMPENSATING EMPLOYEES

Wage and salary administration is the process of developing and implementing a sound and fair method of compensating employees. It involves setting pay ranges for all jobs in the firm and setting a specific wage or salary for each employee. The system used must be fair and equitable. Wage and salary administration is one of the toughest tasks faced by the personnel department.

Wage and salary administration

Importance of Compensation to Employees

Pay is the main factor which determines the standard of living for most Canadian workers. For most Canadian families, the husband's and/or wife's salary is their only source of income. If it were suddenly cut off, a family would have a hard time paying its bills. Its buying power would be greatly reduced and, possibly, eliminated entirely. Thus, pay is important in an economic sense. But pay is also important in a "psychological" sense. Many workers measure their importance to their employers in terms of their pay.

Importance of Compensation to Employers

Wages and salaries are a cost of doing business. Although wage and salary administrators want to hold down this cost, they know that low wages and salaries usually lead to low employee productivity. The firm will be able to hire only low-quality workers. They will tend not to improve, and many will look for a chance to leave for higher-paying jobs.

Low pay does not always mean more profit to the firm. If it did, then most of the goods produced in the world would be produced in the so-called "low-wage" countries. We can't look only at the cost of labour. We must also consider its productivity. It is better to pay a worker $6 per hour for turning out 100 widgets than it is to pay a worker $3 per hour for turning out 30 widgets.

An employer is wise to view pay as an incentive to greater effort, not as a cost which should be reduced.

Importance of Compensation to a Firm's Publics

A firm's approach to wage and salary administration can also affect its various publics. Take the case of a large firm which builds a plant in a small town. If the firm pays above the going rate in that town for certain types of workers, it might attract some workers away from other employers. It might raise the prevailing wage level in the town for those types of jobs. If the local labour supply cannot meet the new plant's needs, workers might move to the town from other areas. An "employment boom" might occur in the town.

Determining the Basis for Payment

A worker's pay is based on (1) output produced; (2) time spent on the job; or (3) a combination of both.

Piece rate   **Some workers are paid a piece rate. Each worker is paid a certain rate for each acceptable unit of output produced.** A worker who sews material together to make a pair of coveralls could be paid a piece rate. This method, of course, can be used only when each worker's labour can be identified with specific units of output.

Incentive pay   **To encourage greater worker productivity, some firms offer incentive pay. For those units produced by a worker above the normal output per day (the quota), the piece rate is increased.** Suppose the quota is 100 pairs per day. A worker gets 25 cents for each pair sewed together. Meeting quota means a daily payment of $25. If a bonus (incentive payment) of 5 cents were paid for each pair produced above quota, a worker who produced 120 pairs would receive $31 for that day's work (100 pairs × 25 cents + 20 pairs × 30 cents or $25 + $6 = $31).

Wage   A worker paid on the basis of time spent on the job gets paid by the hour, week, or month. **A worker paid on an hourly rated basis receives a wage.** Thus, a worker who gets $5 per hour and works 40 hours a week gets $200 in wages for that week. Hourly workers can get overtime pay by working extra hours.

Salary   **A worker paid a fixed amount on a weekly, bi-weekly, or monthly basis receives a salary.** Salaried workers usually work for a fixed amount of pay per year. Most white-collar jobs are salaried jobs. Salaried workers usually do not get overtime pay.

A salesperson might be paid a base salary plus a commission for sales made. This is a combination of time spent on the job and output produced. If a salesperson gets a bonus for selling more than the quota, the pay plan provides incentive.

A Philosophy of Compensation

Now we know what the bases are for determining employees' wages and salaries. But what determines how much a particular employee will earn? This is a tough question.

Let's start with the firm's overall philosophy of compensation. A firm can base its wage and salary plan on its ability to pay. The more profitable the firm, the more it can afford to pay its workers. This gives employees some stake in the firm's profitability. Workers accept this as long as profits are increasing. They tend to reject it when profits are falling, since it would mean less pay. But some employees take pay cuts to avoid being laid off in periods of business decline.

Another approach is the supply and demand approach. The supply of and demand for labour determines its price. When labour is in short supply, wages will be higher than when it is in excess supply, given the same level of demand for it.

Still another approach is to view pay in terms of its ability to deliver a standard of living. Employees are paid a wage or salary which enables them to enjoy a "good" standard of living.

Finally, a firm can base pay on employee productivity. This is hard to put into practise when employee productivity cannot be measured accurately. The piece rate system, in effect, bases pay on productivity.

Depending on its philosophy of compensation, a firm's level of employee compensation might be at the going rate, above it, or below it with respect to its competitors. In general, the more that a firm's top management believes that the human resource is its most important resource, the more it is willing to pay its workers—the more it is willing to invest in that asset. Of course, other factors are present. The presence or absence of a union and the general state of the economy are two examples. Unions are discussed in the next chapter.

Compensation for Particular Job Classifications

Top management's compensation philosophy must be translated into pay ranges for specific types of jobs. How much pay should typists, receptionists, accountants, salespersons, and so forth, receive? Important factors here are the prestige and status attached to different jobs; the desirability of the work; and the amount of skill, experience, or education needed to perform a given job.

Different firms take different approaches. Some firms set pay ranges for different jobs on a strictly judgment basis. At the other extreme, some firms develop detailed rating systems or point systems. For example, the important components in each job (experience, education, skills, etc.) are identified, and points are assigned to each component. These are then totalled and jobs with the highest points are paid the most and those with the fewest points are paid the least. Alternatively, jobs might be classified on the basis of how difficult they are to perform. Workers in more difficult jobs would receive more pay then those in easier jobs.

Compensation for Particular Employees

Finally, wage and salary administrators must decide how much pay each individual worker should receive. If a pipefitter should get a wage in the range of $9 to $12 per

> # WHAT DO YOU THINK?
>
> ### The Civil Service vs. Private Enterprise:
> ### Is There a Difference?
>
> Many people see major differences between the employment practises of government and business. The government uses the Civil Service System, which often results in rigid pay schemes and many levels of bureaucracy. The private sector does not use the Civil Service System, yet many of the same characteristics are apparent. In both cases specific qualifications are set out for jobs, and experts in the administration of personnel have developed. Perhaps the only major difference between employment practises in public and private firms is the higher job security evident in government jobs. However, even this is changing as government payrolls are cut due to lack of revenue. WHAT DO YOU THINK?

hour, how much should pipefitter Eugene Bunch be paid? Important factors here are anti-discrimination laws, employee productivity, seniority, and an employee's mobility—ability and willingness to change jobs. Of course, the presence or absence of a union will influence this decision also.

Compensating Managers

An owner-manager's compensation is tied to the firm's profit. But with professional managers, this is not always the case. Yet, if managers are to be well-motivated, they must be fairly compensated.

Many of the same factors that determine a worker's pay affect how much a manager will receive. The firm's profitability and the competition among executives for certain positions influence the amount of pay. But there are no union wage scales to serve as guidelines.

Fringe benefits are very important to managers in high income tax brackets. Deferred income plans enable them to receive part of their compensation after retirement. This shelters some earnings from the high tax rates they pay in their working years. A firm tries to "lock in" good executives by offering many benefits which hinge on their staying with the firm. Liberal retirement benefits are an example.

PROMOTING EMPLOYEES

Promotion   **A promotion means moving up to a higher position in the firm, usually one which involves more pay and more challenge.** Promotion is a way of compensating an employee for good performance in the previous job or position.

You have probably heard jokes about the employee who got a promotion but no pay increase. "I got promoted, a fancy title, more duties, but no more pay." This underscores the psychological benefit of promotion to an employee. It is a clear form of recognition for good performance. It moves the employee up the firm's job ladder. In most cases, a promotion is much more visible to others than is a salary increase.

As we saw earlier, there are many problems in rating employee performance. Some methods are more objective than others. But employees do want to be rated. The important thing is that the system used is fair, is understood by employees, and is consistently and objectively used. Performance in the present job should be the basic factor in determining an employee's promotability to a higher job.

TERMINATING EMPLOYEES

Eventually, every employee will leave the company's service. This may come about by death, retirement, voluntary resignation, or dismissal.

Retirement

Many firms have retirement plans. Employees whose service has been good over their working years get compensation from the firm during their non-working or retirement years. It is a type of deferred compensation. But when is an employee "ready" for retirement?

Some employees are ready much earlier than others. But most retirement plans are based on the employee's age. This forces some workers who perhaps should retire

---

WHAT DO YOU THINK?

The Peter Principle*

Generally, managers believe that an employee who performs well in a lower job is ready for promotion to a higher job. Following this logic, therefore, a person might end up in a job which is beyond his or her capabilities. While the person performed well in the lower job (and this was the basis for promotion), performance in the next-higher job may be poor. The person, in effect, is "trapped" in that job. Job performance is not good because the employee is incompetent in that job.

What, then, should determine a person's promotability? WHAT DO YOU THINK?

*See Laurence J. Peter and Raymond Hull, *The Peter Principle* (New York: William Morrow, 1969).

earlier to stay on the job. It forces others to leave perhaps before they would like to and even though they are still good workers. This is shortsighted. A compromise is to grant year-to-year extensions to productive employees who still want to work but who have reached retirement age.

Some workers want early retirement and their employers grant it. They also help the employee to adjust to this new life style by providing various types of counselling services.

Voluntary Resignation

Resignation

Resignation occurs when an employee voluntarily leaves the employer's service. There are many reasons for employee resignation. Some employees want to leave to take a job with another employer. In fact, a lot of firms try to hire away good employees from other firms. Some employees quit in order to dramatize a point of difference with higher-ups.

A wise employer does not want to hold on to an employee who can improve his or her position with another firm because of the poor effect on morale. But it is a good practise to conduct an exit interview with an employee who is quitting the firm. **The purpose of an exit interview is to determine the reasons why an employee is leaving.** Perhaps the work environment could be changed to discourage others from leaving.

Exit interview

Dismissal

Dismissal

Dismissal is an involuntary temporary or permanent type of separation of the employee. Some employees are temporarily laid off when business is slack. Many auto workers are laid off at the end of a model year when plants are being retooled for the new models. Seniority usually determines the order in which they are laid off and the order in which they are called back. Of course, a layoff can become permanent if laid-off workers are not called back.

Discharge

Discharge

Discharge is a permanent type of involuntary separation due to a permanent layoff or outright firing of an employee. A firm might permanently lay off workers in a plant when it is closed down. An employee might be fired because of an inability to do the job or serious violations of work rules.

PROVIDING PERSONNEL SERVICES

Modern managers know that good employer-employee relations depend on more than "a fair day's work for a fair day's pay". These other-than-pay incentives come under

A CONTEMPORARY ISSUE

Job Safety and Health

Most provinces have passed Industrial Safety Acts which are designed to protect the health and safety of workers. These laws govern such things as sanitation, ventilation, dangerous machinery, plant design, etc. However, there are problems in setting "acceptable standards", and workers, employers, and labour unions sometimes disagree on how stiff the standards should be.

There are other problems as well. Suppose a steelworker has to be shifted from one job to another job because the present job has exposed the worker to the risk of cancer from breathing oven fumes. Should the worker be guaranteed that he or she will not lose seniority and/or not be transferred to a lower-paying job?

CAREERS IN PERSONNEL

Both in private business and in government there will be continued growth in demand for people in the field of personnel. Much of this stems from new laws which place stricter requirements on hiring practises and health and safety conditions in firms.

Besides labour relations, which we will discuss in Chapter 13, personnel job areas include training and development, employee benefits, safety and health, records management, and employment processes. A college education is often required to reach the level of manager in any of these personnel areas. Certainly, some college training, preferably in business or public administration, is needed.

Personnel workers do recruiting, interviewing and testing of applicants, check application forms, contact references, prepare and administer training programs, inspect health and safety conditions, plan and conduct recreational activities, provide personnel counselling, explain company fringe benefits, and keep the many records relating to employee performance.

Nearly all personnel positions require good human relations skills. Safety and health jobs often require technical or professional training in industrial engineering or nursing. Information is available from the Council of Canadian Personnel Associations, 2221 Yonge Street, Toronto, Ontario M4S 2B4.

personnel services. Examples are company-sponsored recreational programs, cafeterias, credit unions, and group medical insurance. Often, these are referred to simply as "fringe benefits".

Personnel services help to build and keep a high level of employee morale. This reduces absenteeism and labour turnover. Absenteeism means that a worker fails to report to work. Labour turnover means employees leaving the firm. High absenteeism and labour turnover lower employee productivity.

In recent years, many Canadian firms have become very concerned about employee productivity. This is important to the firm, its workers, and society. When productivity declines, it means rising costs and less ability to compete with other producers, mainly foreign competitors. To the employee, it means little hope for good raises. To society, it means more inflation. When overall wage increases outstrip productivity gains in the economy, the result is increased prices for the goods people buy.

Employee safety and health programs are two personnel services which help to reduce absenteeism and turnover. Although all managers are concerned with health and safety matters, the personnel department must create and implement the company-wide program. This helps to reduce lost time due to accidents and illness. It raises productivity and boosts morale by making jobs safer and healthier.

A good safety and health program helps to reduce job accidents. An accident is an unwanted interruption in work. The safety program helps to eliminate the cause of accidents and the injuries which often result from them. Accidents are caused by unsafe working conditions and/or careless and unsafe activities of employees. Once the causes of accidents are known, measures can be taken to reduce or eliminate them. Likewise, by studying employee activities, steps can be taken to eliminate unsafe ones.

PERFORMING OTHER PERSONNEL ACTIVITIES

Personnel departments often provide other types of services as well. There will always be circumstances which affect the firm which are beyond management's direct control. During periods when business is slack, management must decide what to do with employees whose services are no longer needed. Perhaps one department will have to cut its work force while another may need more workers.

The personnel department might help the two department managers to shift workers among departments. This protects the firm's investment in the work force and maintains employee morale. The personnel department might help in a retraining effort if that is needed. It might also help to develop a plan for sharing the available work among employees.

Often, employers who have to cut back their permanent work forces try to minimize the problems of workers who are being discharged. If a plant closing is being considered, the employer advises workers in advance. The personnel department might help these workers find jobs with other firms. It might also handle the job of paying them severance pay.

The personnel department plays an important role in formulating employee discipline policy and in explaining it to workers. Although most workers abide by the rules, some workers do break them. Disciplinary action is administered by the worker's supervisor. For first offences, it usually means an oral reprimand in private. The more serious the offence and the greater the number of prior offences, the stiffer the penalty.

SUMMARY AND LOOK AHEAD

A firm's most important resource is its human resource—its personnel. Top management is responsible for putting together and keeping intact a productive work force which includes managers and workers. This is the task of personnel administration.

Top management delegates to the personnel manager the task of carrying out its human resource philosophy at the firm's operations level. Usually, this means setting up a personnel department headed by a personnel manager. This is a staff position created to advise and assist line managers in managing their personnel.

The personnel department's main activities are (1) determining human resource needs; (2) searching for and recruiting applicants; (3) selecting applicants for employment; (4) training and developing employees; (5) appraising employee performance; (6) compensating employees; (7) promoting employees; (8) terminating employees; (9) providing personnel services; and (10) performing other personnel activities. (See Table 12-1.) All of these activities are interrelated and are vital aspects of human resource management.

In our next chapter, we look at labour relations. The viewpoint for personnel management is the employer-individual employee relationship. In labour relations, the viewpoint is the employer-union relationship. The focus is on the employees as a group—as members of one or more labour unions. Although some personnel managers handle labour relations matters, many firms have labour relations specialists to deal with their employees' unions.

KEY CONCEPTS

Background investigation Checking up on a job applicant's references to help in assessing his or her suitability to the job. Conducted by the firm's personnel department.

Discharge A permanent type of involuntary separation due to permanent layoff or firing of an employee.

Dismissal A temporary or permanent type of involuntary separation of an employee.

Employee orientation The process of introducing a new employee to the job and to the firm.

Exit interview An interview conducted with an employee who is leaving the firm. Purpose is to find out the reasons for leaving.

Final selection interview The last in a series of interviews of a prospective new employee. Comes after the in-depth interview.

Human resource The personnel who staff a firm. Includes workers and managers.

Table 12-1. The personnel department's activities, objectives, and procedures

Activities	Objectives	Procedures
1. Determining human resource needs	To specify the firm's need for applicants.	(a) Study company objectives (b) Study organization chart (c) Forecast human resource needs (d) Develop management inventory (e) Perform job analysis (f) Prepare job description (g) Prepare job specification
2. Searching for and recruiting applicants	To attract applicants to the firm.	(a) Specify sources of recruits (b) Recruiting
3. Selecting applicants for employment	To select the most desirable applicants.	(a) Conduct preliminary interview (b) Prepare job application form (c) Administer selection tests (d) Conduct in-depth interview (e) Conduct background investigation (f) Conduct final selection interview (g) Schedule physical examination
4. Training and developing employees	To build and maintain a productive work force.	(a) Handle job orientation (b) Perform job-skill training (c) Aid in management development
5. Appraising employee performance	To rate employee performance objectively.	Develop performance appraisal system
6. Compensating employees	To develop a fair and equitable system for paying employees.	Wage and salary administration
7. Promoting employees	To reward productive employees in order to utilize the human resource more effectively.	Developing promotion policies
8. Providing personnel services	To build and enhance employee morale.	(a) Provide fringe benefits (b) Employee safety and health
9. Performing other personnel activities	To advise and assist line managers in coping with special personnel problems.	(a) Formulate employee discipline policy (b) Retrain employees (c) Share work among employees (d) Counsel employees

Incentive pay Payments made to a worker who exceeds his or her quota under a piece rate system. In general, the purpose is to encourage and reward increased employee productivity.

In-depth interview The interview given to an applicant who passes the selection tests. Comes after the preliminary employment interview.

Job analysis   Defining the jobs which must be done if a firm is to reach its goals.

Job application form   Prepared by an employer to be completed by a job applicant. The applicant gives a written summary of his or her education, experience, skills, etc.

Job description   Prepared from the job analysis. Lists and describes the requirements of a job.

Job-skill training   Training an employee how to do a job. Main purpose is to teach specific job skills.

Job specification   Prepared from the job description. States the personal qualifications (education, skill, experience, etc.) needed of the person who is to fill each job.

Management by objectives (MBO)   An approach to evaluating employee performance. Boss and worker set objectives for the worker. If they are achieved, the worker has performed well. Purpose is to make performance appraisal more objective.

Management development   Training present and potential managers to improve their administrative abilities. Focuses mainly on developing conceptual and human relations skills.

Management inventory   A list of present managerial positions in a firm and the names and personal data of the persons who hold those positions.

Merit rating system   An approach to evaluating employee performance. Performance is appraised periodically by each worker's supervisor. Performance criteria are the worker's initiative, dependability, ability to work with others, etc.

Performance appraisal system   A system used by management to measure and evaluate employee performance on the job. The merit rating system is an example.

Personnel administration   Refers to top management's efforts to develop good company personnel including both managers and workers. The broadest possible view of the firm's human resource. Top management, through personnel administration, sets the tone for the firm's approach to human resource management.

Personnel department   A staff department headed by a personnel manager who advises and assists line managers in managing their personnel. Performs personnel tasks such as recruiting and training.

Personnel management   Charged with the task of implementing top management's policies regarding human resource management. It is the job of the personnel manager who heads the personnel department. More generally, it refers to the duty of all managers to manage their personnel.

Piece rate   A method of paying workers. Each worker is paid a certain rate for each acceptable unit of output produced.

Preliminary employment interview   A job applicant's first interview with a prospective employer.

Promotion   Moving up to a higher position on a firm's job ladder. Usually involves more pay and more challenge.

Recruiting   The task of attracting potential employees to a firm.

Resignation   Occurs when an employee voluntarily leaves an employer's service. By tendering one's resignation, the employee announces his or her intention to leave.

Salary Fixed compensation to an employee who is regularly paid on a weekly, bi-weekly, or monthly basis.

Selection tests Tests given to job applicants whose preliminary employment interviews and job application forms were satisfactory. Purpose is to select the best applicants and to reject the others. Include aptitude tests, intelligence tests, and performance tests.

Seniority An employee's length of service. The longer that service, the greater the seniority. It can be computed on the basis of length of service with the firm, length of service in a particular job or department, or length of membership in a labour union.

Wage Compensation to a worker who is paid by the hour.

Wage and salary administration The process of developing and implementing a sound and fair method of compensating employees.

INCIDENT:

The Larson Corporation

The Larson Corporation recently began hiring minority group members as part of its efforts to help solve some of society's more pressing problems. Larson management, after many hours of meetings and thoughtful analysis, believed that this would benefit both the firm and the low-income people who would be hired under the program. Larson is a large mining company with operations in Manitoba and Saskatchewan. The new program was launched in one of its northern Manitoba plants because that area is home to many native Indians who did not hold steady jobs.

In the past, Larson hired very few of these people. Many of Larson's jobs require a great deal of skill, and there always were enough trained people to fill job vacancies. As a result, Larson's management lacked experience in dealing with the type of worker this program was designed to help.

But Larson and its employees want to make this new program work. The personnel manager at the plant, Leslie Parks, was instructed to begin actively recruiting minority group members for new job openings. Ms. Parks had to develop a training program for new recruits and integrate them into their new environment.

Initially, the new recruits would be sought for those jobs which required very few skills. In this way, a training program could be developed which would take only a short time to complete. Thus, a new worker could be placed on the job in as short a time as possible. The reasoning here was that this type of employee would become discouraged by a long training program which might take months to complete. Management was afraid that the dropout rate would be too high.

Ms. Parks is now involved in setting up a recruitment and training program.

Questions:
1. Why do you think the firm decided to hire from the ranks of the disadvantaged unemployed?
2. What type of training program would you develop?
3. What kinds of problems might Ms. Parks encounter?
4. What might Ms. Parks do to increase the likelihood that the new program will be successful?

INCIDENT:

Reliable Motors Company

Adam Cranford is the personnel manager of Reliable Motors Company, a large car dealership in Ontario. Over the years, Reliable has taken pride in the training given to its mechanics. The firm plays this up in its advertising. It claims that most competing dealers hire almost anyone, put them in uniforms, and call them auto mechanics. Mechanics at Reliable, however, must go through a tough training program before they can begin servicing cars. But in recent months, labour turnover at Reliable has been very high.

Mr. Cranford knows that car dealers in the area have been under fire from customers to give better service after the sale. As a result, well-trained mechanics get above-average pay.

After conducting many exit interviews with mechanics leaving Reliable, Mr. Cranford knows that they are going to work at other dealerships in the area. He also learned that very few of the mechanics leaving were dissatisfied at Reliable. Some even said that they had second thoughts about leaving, but they just couldn't pass up the chance to make more money. Mr. Cranford believes that the other dealers can pay more because they do not spend money on training. What their mechanics learn, they learn on the job.

Questions:
1. Do you think that Reliable's mechanics have an obligation to remain with the company after they have gone through the training program? Why or why not?
2. Compare Mr. Cranford's philosophy about the human resource to that of the other car dealers.
3. What should Mr. Cranford do about the situation?

QUESTIONS FOR DISCUSSION

1. When a firm buys a new typewriter, it expects that its value will diminish with the passage of time. It becomes "used up". Is the same true of a new management trainee who recently began working for the firm? Explain.
2. Do you agree that a firm's most important asset is its human resource? Why or why not?
3. Does "personnel administration" mean the same thing as "personnel department"? Explain.
4. Why might a firm create a personnel department?
5. What is a management inventory? What purpose does it serve?
6. Discuss the problems in developing and administering selection tests.
7. What should be covered in an employee orientation program?
8. What are the major similarities and differences between job-skill training and management development?
9. Why is it important to have a good performance appraisal system? Discuss the firm's and the worker's view.
10. What is involved in wage and salary administration?
11. Would it be possible to pay all employees a piece rate? Why or why not?
12. Applicants for managerial positions are usually given more "subjective" tests than are applicants for non-managerial jobs. Why is this? Are there any potential dangers? Explain.
13. Should a firm retrain employees whose skills have become obsolete if it can hire persons already possessing the needed skills? Discuss.
14. Do you think that it is fair for a firm to have a policy of not promoting from within? Explain.
15. Do you think that it is proper for a retailing firm to require job applicants to submit to a polygraph (lie detector) test as a condition of employment? Discuss. Is it fair to require all salespeople to take a polygraph test every six months in order to keep their jobs? Discuss.

13 Labour Relations

OBJECTIVES: After reading this chapter, you should be able to:

1. Explain how the Industrial Revolution changed the "world of work" for workers and led to efforts among workers to form unions.
2. Illustrate the general nature of labour-management relations prior to the 1930s.
3. Give examples of how the legal environment prior to the 1930s was hostile to unionization.
4. List the major federal labour laws and discuss the major provisions of each.
5. Discuss the provincial labour board's role in certifying a union as the exclusive bargaining agent for a firm's employees.
6. Show how a local union and a national union are related to each other.
7. Distinguish between craft and industrial unions.
8. Give examples of political, social, and economic objectives of unions.
9. List several important reasons why workers join unions.
10. Explain how employees and employers collectively bargain through union and management representatives.
11. Cite specific issues which might lead to labour-management conflict.
12. List and discuss labour and management's "weapons" in dealing with conflict.
13. Appraise the future prospects for the union movement in Canada.

KEY CONCEPTS: In reading the chapter, look for and understand these terms:

LABOUR UNION
COLLECTIVE BARGAINING
LABOUR CONTRACT
BLACKLISTS
UNFAIR LISTS
THE CONCILIATION ACT
THE INDUSTRIAL DISPUTES
   INVESTIGATION ACT
PRIVY COUNCIL ORDER 1003
THE BRITISH NORTH AMERICA ACT
THE CANADA LABOUR CODE
BARGAINABLE ISSUES
CRAFT UNIONS
INDUSTRIAL UNIONS
LOCAL UNION
INDEPENDENT LOCAL UNION
NATIONAL UNION

INTERNATIONAL UNION
CLOSED SHOP
UNION SHOP
AGENCY SHOP
OPEN SHOP
GUARANTEED ANNUAL WAGE
COLA CLAUSE
CONCILIATION
MEDIATION
VOLUNTARY ARBITRATION
COMPULSORY ARBITRATION
STRIKE
PICKETING
BOYCOTT
LOCKOUT
INJUNCTION
GRIEVANCE PROCEDURES

The relationship between employer and employees is crucial in a firm. When the relationship is between the employer and the employees as individuals, it is called employee relations or personnel relations. In many firms, however, the employees belong to labour unions. The relationship between the employer and the employees' union(s) is called labour relations. In this relationship, a third party, the union, comes between the employer and the employees. No longer does the employer deal with an employee solely as an individual.

In labour relations, management representatives bargain with union representatives over wages and working conditions for unionized employees. In some firms, the personnel manager does the bargaining for the employer. But as unions have become more powerful, labour relations specialists are increasingly found in unionized firms.

We begin this chapter with a brief look at the history of the union movement in Canada and the major laws which deal with unions. We will also look at why workers join unions, the nature of labour-management relations, and the future of unions in Canada.

WHY UNIONS BEGAN

Individualism is a basic characteristic of capitalism. When we talk about labour unions, however, we imply collective action. **A labour union is an organization of employees formed for the purpose of dealing collectively with their employers in order to further the interests of those employees.**

Labour union

The dealing which occurs between the employer and the union is called collective bargaining. **Collective bargaining is the process of negotiating a labour agreement between union representatives and employer representatives.**

Collective bargaining

The agreement negotiated between employer and union representatives is called a labour contract. **A labour contract sets forth the terms and conditions under which union members will offer their labour services to an employer.**

Labour contract

To explain the apparent conflict between capitalism's emphasis on individualism and unionism's emphasis on collective action, let's briefly study the history of unionism in Canada.

The Work Environment

Before the Industrial Revolution, many workers were skilled artisans. Production was organized around the domestic system. Shoemakers, for example, worked in small shops located in their homes. They bought materials needed for making shoes, made the shoes, and sold them. They were more like independent business owners than workers. They worked for themselves.

The Industrial Revolution changed this. The domestic system was largely replaced by production in factories. Many formerly independent artisans became employees who worked for firms owned by others. Workers lost control over buying, producing, and selling activities. These became activities of organized businesses. The only thing over which workers could have any control was in offering their services to employers. Thus, workers began to organize in order to control that supply collectively.

Furthermore, many skilled artisans became machine operators in big factories. Mechanization (substituting machines for human labour) was also seen as a threat by many workers. To deal with this threat to their security, workers began to organize unions. The "fear" of being replaced by machines still is a fact of life for many workers in our highly mechanized and automated production system.

Because wages paid to employees are a cost of doing business, employers want to hold down both the number of employees they hire and the wages they pay to them. This gives rise to the workers' concern about job scarcity, low wages, and poor working conditions. Labour responded to these concerns by organizing unions for collective action.

Thus, the Industrial Revolution brought rapid changes in the "world of work". But the human problems of this revolution, especially for workers, were not quickly solved. Although organization was a way for workers to deal with their changed world, workers were going to have an uphill battle to have their right to form labour unions recognized.

Blacklists

Employers and workers sympathetic to unions battled with each other for many years. For example, employers circulated blacklists among themselves. **Blacklists contained the names of workers who were known to be in favour of unions.** These blacklisted workers were refused employment.

THEN AND NOW
Child Labour

During the nineteenth century young children were hired to work in factories at very low wages and under extremely poor working conditions. Because most of them were denied the chance to attend school, they remained unskilled. Improvements came slowly.

The founding conference of the Canadian Labour Union in 1873 supported a resolution to prohibit the employment of children under 10 in manufacturing establishments where machinery was used. In 1886 the Ontario Factory Act was passed. It prohibited the employment of boys under 12 and girls under 14, and set a limit of 60 hours of work per week for women and children.

During the twentieth century major advances were made. At present, the federal government and all provinces have passed child labour laws. The Canada Labour Code allows the employment of individuals under the age of 17 only if (a) they are not required to be in attendance at school under the laws of their province of residence, and (b) the work is unlikely to endanger their health or safety. In addition, no one under 17 is permitted to work during the hours from 11:00 P.M. to 6:00 A.M.

> **THEN AND NOW**
>
> **Equal Pay for Equal Work for Women**
>
> Women, like children, worked for a fraction of men's wages and held only unskilled jobs during the Industrial Revolution. For example, in 1882 male members of the Brotherhood of Telegraphers of the United States and Canada were paid approximately $35 per month, while female operators earned about 1/7 that amount. During a strike of the brotherhood, one of the union's proposals was equal pay for men and women.
>
> Other unions were also seeking equal pay for women. In 1882 the Toronto Trades and Labour Council proposed labour legislation which included an "equal pay for equal work, regardless of sex" provision.
>
> Although gradual improvements in working conditions and opportunities for women have occurred during the twentieth century, it was not until the mid-1950s that serious efforts were made to ensure equal pay for equal work.
>
> The Canada Labour Code prohibits an employer from maintaining different wages for men and women employed in the same industrial establishment and who are doing the same or similar work. Differences in pay between men and women *can* be maintained if they are based on factors other than sex.

Labour unions circulated their lists also—"unfair lists". **Unfair lists contained the names of employers whom unions considered unfair to workers because these employers would not hire union members.**

Unfair lists

The Development of Canadian Labour Unions

The earliest evidence of labour unions comes from the Maritime provinces early in the nineteenth century. Generally, these unions were composed of individuals with a specific craft (e.g., printers, shoemakers, barrelmakers). Most of these unions were small and had only limited success. However, they laid the foundation for the rapid increase in union activity which occurred during the late nineteenth and early twentieth centuries.

A succession of labour organizations sprang up and just as quickly faded away during the years 1840–70. In 1873 the first national labour organization was formed—the Canadian Labour Union. By 1886 the Knights of Labour (a United States-based union) had over 10,000 members in Canada. The Canadian Labour movement began to mature with the formation of the Trades and Labour Congress in 1886. The TLC's purpose was to unite all labour organizations and to work for the passage of laws which would ensure the well-being of the working class.

The growth of labour unions began in earnest early in the twentieth century as the concept of organized labour gradually came to be accepted. Within the ranks of labour, various disputes arose which resulted in numerous splits in labour's ranks. For example, there was concern that United States-based unions would have a detrimental effect on Canadian unions. The Canadian Federation of Labour was formed in 1908 to promote national (Canadian) unions instead of U.S. unions. These and other disputes (such as how communists in the movement should be handled) often led to the creation of rival union organizations which competed for membership. By 1956 these disputes had been largely resolved, and the two largest unions—the Trades and Labour Congress and the Canadian Congress of Labour—merged to form the Canadian Labour Congress. This brought approximately 80 per cent of all unionized workers into one organization.

The Legal Environment

There were political and legal barriers to collective bargaining until well into the twentieth century. Courts held that unions were conspiracies in restraint of trade. Employers viewed their employees' efforts to unionize as an attempt to deprive the employers of their private property. The employment contract was between the individual worker and the employer—not between the employer and employees as a group. The balance of bargaining power was in favour of the employer.

The employer-employee relationship became much less direct as firms grew in size. Managers were, themselves, employees. Hired managers dealt with other employees. Communication among owners, managers, and workers became more formalized. Big business had the upper hand. Because of mounting public concern, laws were passed to place the worker on a more even footing with the employer.

In 1900 government concern for labour disputes resulted in the passage of the Conciliation Act. **The act was designed to assist in the settlement of labour disputes through voluntary conciliation and was a first step in creating an environment more favourable to labour. A more comprehensive law, the 1907 Industrial Disputes Investigation Act, provided for compulsory investigation of labour disputes by a government-appointed board before a strike was allowed.** However, this act was later found to violate a fundamental provision of the British North America Act (see below).

The current positive environment for labour did not come into being until 1943 when Privy Council Order 1003 was issued which (1) recognized the right of employees to collectively bargain, (2) prohibited unfair labour practices on the part of management, (3) established a labour board to certify bargaining authority, and (4) prohibited strikes and lockouts except in the course of negotiating collective agreements. Thus, approximately 45 years of dealings between labour, management, and government were required before the labour movement achieved its fundamental goal of having the right to collectively bargain.

The British North America Act, passed in 1867, has also affected labour legislation. **This act allocated certain activities to the federal government (e.g., labour legislation for certain companies operating inter-provincially) and others to individual provinces (labour relations regulations in general).** Thus, labour legislation emanates from both the federal and the provincial governments.

SOME IMPORTANT DATES IN CANADIAN LABOUR HISTORY

1827	First union formed; shoemakers in Quebec City
1840–1870	Many new unions formed; influence of U.S. and British unions felt
1871	Formation of Toronto Trades Assembly; composed of five craft unions; went out of existence a few years later
1873	Canadian Labour Union formed; objective was to unite unions across Canada
1879	First coal miners' union in North America formed in Nova Scotia
1881	The U.S.-based Knights of Labour enter Canada
1883	Canadian Labour Congress formed; lasted until 1886
1886	Canadian Trades and Labour Congress formed; later became known as the Trades and Labour Congress of Canada (TLC)
1902	Knights of Labour expelled from TLC [K of L became affiliate of AFL]
1902	Expelled unions form the National Trades and Labour Congress (became the Canadian Federation of Labour [CFL] in 1908); purpose was to promote national unions instead of international ones
1902–1920	Rapid growth of union membership in both major unions (TLC and CFL)
1919	One Big Union formed; organized in opposition to the TLC
1919	Winnipeg General Strike
1921	Canadian Brotherhood of Railway Employees (CBRE) expelled from TLC
1921	Confédération des Travailleurs catholiques du Canada (CTCC) organized by the Roman Catholic clergy in Quebec; goal was to keep French-Canadian workers from being unduly influenced by English-speaking and American trade unions
1927	All-Canadian Congress of Labour (ACCL) formed; objective was to achieve independence of the Canadian labour movement from foreign control; made up of One Big Union, the CFL, and the CBRE
1939	TLC expels industrial unions; Canadian Congress of Industrial Organization (CIO) committee formed
1940	ACCL and the Canadian CIO Committee unite to form the Canadian Congress of Labour
1956	TLC and CCL merge to form the Canadian Labour Congress; remnants of One Big Union joined new organization
1960	CTCC dropped association with Roman Catholic Church and chose a new name—Confédération des Syndicate Nationaux (CSN); in English, was called the Confederation of National Trade Unions (CNTU)

FEDERAL LEGISLATION—THE CANADA LABOUR CODE

Canada labour code

The Canada Labour Code is a comprehensive piece of legislation which applies to the personnel practices of firms operating under the legislative authority of parliament. The Code is composed of four major sections as follows:

Fair Employment Practises

This section prohibits an employer from either refusing employment on the basis of a person's race or religion or using an employment agency which discriminates against people on the basis of their race or religion. These prohibitions apply to trade unions as well, but not to non-profit charitable and philanthropic organizations. Any individual who feels a violation has occurred may make a complaint in writing to the Department of Labour. The allegation will then be investigated and, if necessary, an Industrial Inquiry Commission will be appointed to make a recommendation in the case. (As of 1978, Fair Employment Practises are covered by the Canadian Human Rights Act.)

Standard Hours, Wages, Vacations, and Holidays

This section deals with a wide variety of "mechanical" issues such as standard hours of work (eight-hour day and 40-hour week), maximum hours of work per week (48), overtime pay (at least 1½ times regular pay), minimum wages, equal wages for men and women doing the same jobs, vacations, general holidays, and maternity leave. The specific provisions are changed frequently to take into account changes in the economic and social structure of Canada, but their basic goal is to have consistent treatment of employees in these areas.

Safety of Employees

This section requires that every person carrying on a federal work project do so in a way which will not endanger the health or safety of any employee. It also requires that safety procedures and techniques be implemented to reduce the risk of employment injury. This section requires employees to exercise care to ensure their own safety; however, even if it can be shown that the employee did not do this, compensation must still be paid. This section also makes provisions for a safety officer whose overall duty it is to assure that the provisions of the act are being fulfilled. The safety officer has the right to enter any federal project "at any reasonable time".

Canada Industrial Relations Regulations

This is the final major section of the Canada Labour Code and deals with all matters related to collective bargaining. It is subdivided into seven divisions, as follows:

- Division I—gives employees the right to join a trade union and gives employers the right to join an employers' association.
- Division II—establishes the Canada Labour Relations Board whose role is to make decisions on a number of important issues (e.g., certification of trade unions).
- Division III—stipulates the procedures required to acquire or terminate bargaining rights.
- Division IV—indicates the rules and regulations which must be adhered to during bargaining; also presents guidelines for the content and interpretation of collective agreements.
- Division V—states the requirement that a conciliation officer must be appointed by the Minister of Labour if the parties in the dispute cannot reach a collective agreement.
- Division VI—stipulates the conditions under which strikes and lockouts are permitted.
- Division VII—a general conclusion indicating methods which might be used to promote "industrial peace".

PROVINCIAL LABOUR LEGISLATION

Each province has also enacted legislation to deal with the personnel practises covered in the Canada Labour Code. These laws vary across provinces and are frequently revised; however, their basic approach and substance is the same as that in the Canada Labour Code. Certain provinces may exceed the minimum Code requirements on some issues (e.g., minimum wage).

Each province also has a labour relations act. To give an indication of what these acts cover, the Manitoba Labour Relations Act is briefly described below.

The Manitoba Labour Relations Act

The Manitoba Labour Relations Act is a comprehensive document dealing with the conduct of labour relations in the province. It is divided into six sections as follows:

Part I—labour practises and rights. This section states some fundamental rights of labour and management (e.g., every employee has the right to be a member of a union, and every employer has the right to be a member of an employers' organization). Neither employers nor unions may discriminate against union members simply because they are union members. Other basic union-management issues are also addressed in this section. Overall, the section is designed to show what unfair labour practises are.

Part II—certification and bargaining rights. This section describes the detailed procedures which must be followed before a union can be certified as a bargaining agent. Included are activities such as who has the right to apply for certification, how the appropriateness of the unit is determined, the certification vote, and the effect of certification.

Part III–collective bargaining and collective agreements. The actual practise of collective bargaining is described in this section. The major points are as follows:

1. The union can require management to commence bargaining within 10 days of its being certified.
2. Conciliation and mediation are available if labour or management request it to help in the development of a contract.
3. Management is required to deduct union dues from employees' wages and remit these dues to the union.
4. Employees who oppose unions on religious grounds may stipulate that their union dues be given to charitable organizations.
5. Management must notify the union at least 90 days in advance if it plans to implement technological changes which will affect the security of a significant number of employees.

Part IV–lockouts and strikes. Unauthorized lockouts and strikes are prohibited in this section. Both management and labour are expected to exert a reasonable effort to reach an agreement before engaging in these tactics (at least 90 days must pass after the union has been certified before a strike is allowed).

Part V–conciliation, mediation, and arbitration boards. The three basic forms of third-party intervention and the administrative structure in which they operate are discussed in this section. The purpose of these boards is to help labour and management settle differences they encounter in developing a new contract or interpreting an existing one. Each person serving on one of these boards must take an oath of impartiality and confidentiality. The boards must write reports of their activities, and these reports must be made available to labour and management. For conciliation and mediation, the expense of the proceedings is paid by the province; for arbitration, the expenses are split by labour and management.

Part VI–general. The concluding section considers a number of other labour relations issues not treated in earlier sections. These include the function and powers of industrial commissions, prosecution of employers or unions for offences under the act, penalties for offences, liability for damages, and the date of commencement of the act.

THE ORGANIZING DRIVE

A union might try to organize a firm's workers when it is trying to break into new geographical areas, when some workers in a firm are members and it wants to cover other workers, or when it is attempting to outdo a rival union. Thus, in some cases, a union might try to organize workers for purposes other than helping a group of employees to help themselves.

Management often becomes aware of union organizing effort through the grapevine. These "rumblings" may set off a counter-effort by management to slow the drive. Management must know, however, what it can legally do. A do-nothing approach is rare today. Employers can exercise the right of free speech to present their side of the story to the workers.

Suppose that a union is trying to organize employees of a Manitoba company. If it can show that at least 50 per cent of the employees are members of the union, it can apply to the Manitoba Labour Board (MLB) for certification as the bargaining agent for the employees.

A problem may arise regarding the right of different types of workers to join or not join the union. For example, supervisors may or may not be included in a bargaining unit along with non-management workers. The MLB has final authority in determining the appropriateness of the bargaining unit. Professional and non-professional employees are generally not included in the same bargaining unit unless a majority of the professional employees wish to be included.

Once the MLB has determined that the unit is appropriate, it may order a certification vote. If a majority of those voting are in favour of the union, it is certified as the sole bargaining agent for the unit.

THE COLLECTIVE-BARGAINING PROCESS

An employer must bargain with a union which is certified by the appropriate labour board. The employer does not have to grant the union everything it wants, but both parties must bargain in good faith on bargainable issues. **Bargainable issues are aspects of the work or job environment which are subject to collective bargaining between union and management representatives.**

Bargainable issues

The scope of these bargainable issues has been broadened over the years. A non-union worker's wage and fringe benefits are set by the employer, except to the extent that the worker can bargain for himself or herself. Those of a union member are set through the collective-bargaining process. The union bargains for the employee. The employer's right to move a plant to another location may be subject to collective bargaining. Because the area of managerial discretion is narrowed when a union is present, non-union companies usually try to provide good working conditions to discourage employee interest in unions.

The bargaining process may take many forms, depending on industry practice, the degree of hostility between union and management, and the nature of the union's demands. Both deal face to face at the bargaining table. Management negotiators usually want to give less to union negotiators than they want. The union negotiators usually ask for more than they will finally settle for. If both parties come to terms, the agreement is voted on by the union's rank-and-file members. If they ratify the agreement, it becomes a binding contract.

TYPES OF UNIONS

Like business firms, unions also have organization structures. In fact, some workers are as far removed from the top officials in their national organizations as they are removed from the top managers of the firms for which they work.

Craft unions

The two basic types of unions are craft and industrial unions. **Craft unions are organized by crafts or trades—plumbers, barbers, airline pilots, etc. Craft unions restrict membership to workers with specific skills.** In many cases members of craft unions work for several different employers during the course of a year. For example, many construction workers are hired by their employers at union halls. When the particular job for which they are hired is finished, these workers return to the hall to be hired by another employer.

Craft unions have a lot of power over the supply of skilled workers. This is because they have apprenticeship programs. For example, a person who wants to become a member of a plumber's union will have to go through a training program. He or she starts out as an apprentice. After the training, the person is qualified as a "journeyman" plumber.

Industrial unions

Industrial unions are organized according to industries—steel, auto, clothing, etc. Industrial unions include semi-skilled and unskilled workers. For example, all production workers in an automobile plant would belong to the same union regardless of the specific job they do. Industrial union members typically work for a particular employer for a much longer period of time than the craft union member. But an industrial union does have a lot of "say" regarding pay and personnel practises within unionized firms.

Whether unions are craft or industrial, they are organized on a local, national, or international basis.

Local union

The local union is the basic union organization. A local of a craft union is made up of artisans in a local area. A local of an industrial union is made up of workers in a given industry in a local area.

National union

A national union is one which has members across Canada. These members belong to locals which are affiliated with the national union. There are many national unions in Canada, including the Canadian Union of Public Employees, the National Railway Union, and the Canadian Airline Pilots Union.

International union

A union which has members in more than one country is called an international union. The prime example is the United Automobile Workers, an international union made up of locals in the United States and Canada.

Independent local union

An independent local union is one which is not formally affiliated with any labour organization. It conducts negotiations with management at a local level, and the collective agreement is binding at that location only. The University of Manitoba Faculty Association is an independent local union. Membership in local unions in 1977 was 2.7 per cent of total union membership.

Table 13-1 shows the ten largest unions in Canada by membership.

UNION OBJECTIVES

The long struggle by unions to be recognized in their right to exist has resulted in their general acceptance today—at least in acceptance of the idea that workers have a right to form unions. Union objectives are political, economic, and social.

Table 13-1. The ten largest unions in Canada, 1977

Union	Membership
1. Canadian Union of Public Employees	228,687
2. United Steelworkers of America	193,340
3. Public Service Alliance of Canada	159,499
4. United Automobile, Aerospace and Agricultural Implement Workers of America	130,000
5. National Union of Provincial Government Employees	101,131
6. United Brotherhood of Carpenters and Joiners of America	89,010
7. International Brotherhood of Teamsters, Chauffeurs, Warehousemen and Helpers of America	86,603
8. Quebec Teachers' Corporation	85,000
9. International Brotherhood of Electrical Workers	63,914
10. Ontario Public Service Employees Union	63,340

Source: Labour Canada, *Labour Organizations in Canada, 1976-1977,* Minister of Supply and Services, Canada, 1977. Reproduced by permission of the Minister of Supply and Services Canada.

Political Objectives

During the nineteenth century union objectives were basically political, since unions were fighting for recognition of their right to exist. This involved them in the political process. Once this right was recognized, unions shifted their emphasis to that of advancing the economic interests of their members. But unions do have political goals.

Unlike most Western European nations, the union movement in Canada is a minority movement. Only about 38 per cent of our non-farm workers are unionized. (See Table 13-2.) Nevertheless, politicians speak of the "labour vote". There are two dimensions here: (1) the organized (unionized) and (2) the unorganized (non-unionized). The ability of labour leaders, however, to deliver the labour vote is questionable. Most Canadians probably vote on issues other than the candidates' pro- or anti-labour views.

But organized labour does try to speak for the labour force. Although there is no party called the "Labour party" in Canada, the NDP is very supportive of labour's

Table 13-2. Characteristics of national and international union membership, 1977

Total Union Membership	3,100,000
Percentage in Canadia Labour Congress	68.7
Percentage in independent or unaffiliated unions	23.7
Percentage in unions with headquarters in the U.S.	49.0
Percentage of total labour force in unions	31.0
Percentage of non-agricultural workers in unions	38.2

Source: Labour Canada, *Labour Organizations in Canada, 1976-1977,* Minister of Supply and Services, Canada, 1977. Reproduced by permission of the Minister of Supply and Services Canada.

goals. Union leaders attempt, with varying degrees of success, to have legislation passed which is favourable to workers. This is particularly true for the minimum wage and hours of work issues.

Economic Objectives

Union leaders are elected by union members. Thus, those leaders seek to satisfy their members' economic needs.

Improved Standard of Living

Unions seek to improve their members' standard of living. In the past this meant getting higher wages. The pay envelope, however, is no longer the only concern of union members. Many want more leisure time. They want to work fewer hours but make more pay per hour. The entire package of fringe benefits is part of the improved standard of living which unions seek. Working conditions, pensions, paid vacations, and so on are important bargaining issues. Inflation in recent years has led union members to demand that labour contracts protect their standard of living. We will discuss this later in the chapter.

Security Objectives

The growing security consciousness of Canadian workers is reflected in union goals. The seniority provision in most contracts spells out the worker's rights when layoffs, transfers, and promotions occur. Employees are ranked in terms of length of service. Those with longer service get better treatment.

Much conflict exists regarding seniority. Two specific examples of conflict relate to women and minority groups. They typically have less seniority and are the first to be laid off and the last to move up to higher jobs. These workers tend to oppose the tradition of seniority.

Union security is another issue. The greatest security is found in the closed shop. **Closed shop** **In a closed shop an employer can hire only union members.** For example, a plumbing or electrical contractor who hires workers through a union hiring hall can hire only union members.

Union shop **In a union shop, an employer may hire non-union workers even if the employer's present employees are unionized.** New workers, however, must join the union within a stipulated period of time (usually 30 days).

Agency shop **In an agency shop, all employees for whom the union bargains must pay dues, but they need not join the union.** This is a compromise between the union shop and the open shop, and is called the Rand Formula after the individual who proposed it.

Open shop **In an open shop, an employer may hire either union and/or non-union labour. Employees need not join or pay dues to a union in an open shop.**

Another security issue is job security, especially in highly automated industries. The guaranteed annual wage reflects the worker's concern about job security. **The**

TWO POINTS OF VIEW

The Right to Work

Labour:
"Sure, a worker in a union plant has to belong to the union. The majority rules, and the majority of the workers who voted did vote to organize our union. Under the law, two persons doing the same job must be paid the same wage and receive the same fringe benefits. It would be unfair for a person who doesn't pay union dues to enjoy the same benefits won by a union which is supported by members' dues. Freeloaders should have to pay their fair share.

"To say that unions are unnecessary is plain hogwash. Do away with the union and see how employee-oriented most employers are. Business firms exist to make a profit. The less they pay to their workers, the more they keep as profit. The claim that union leaders have no incentive to do a good job is a lot of XXX! No law requires any firm's employees to form a union. If they weren't benefiting from the union, the members could have it decertified as their exclusive bargaining agent."

Management:
"A worker should not be forced to join a union in order to hold a job in my company. Granted, employees do have a right to form a union, but an employee should have a right *not* to join. Yet, if a union is certified as the exclusive bargaining agent for my production workers, even though not all of them voted to unionize, all of them have to pay dues or else be fired. They don't have freedom of choice.

"Unions have done a lot for the worker in the past, but they are not needed today. Social insurance, workmen's compensation, minimum wage laws, job safety laws, and enlightened management take away the need for unions.

"When you think about it, a customer who isn't satisfied with our product can stop buying it. But if one of our unionized workers isn't satisfied with the union, he or she must still pay dues. What reason do union leaders have to do a good job?

"Finally, when bargaining time rolls around for one of our suppliers and its union, we often build up our inventory of that supplier's product. This is done in case there is a strike at that plant. A strike would cut off our supply source. The end result, of course, is higher prices to consumers to make up for the cost of carrying extra inventory."

Guaranteed annual wage guaranteed annual wage is a provision in a labour contract which maintains the workers' income level during a year. Most labour contracts which provide for this guarantee the worker a minimum amount of work during the contract period. This lends stability to the worker's employment. Some contracts provide for early retirement, lengthy vacations, and sabbatical leaves for employees.

The security of unions themselves is also in question. Our most organized industries—steel, auto, transportation, etc.—are almost totally unionized and are rapidly automating. Prospects for new members are not attractive. Furthermore, the stronghold of unions—blue collar labour—is diminishing as a percentage of our labour force. Thus, unions have sought new areas for growth, such as government workers and white-collar workers.

Working Conditions

Suppose that a long strike has been settled. Nevertheless, some locals of the national union remain on strike. The national contract covers major issues such as wage rates and fringe benefits. But specific provisions on rest periods, sanitary facilities, washup periods, and so on vary from plant to plant. These disputes over working conditions are worked out by the local union and the specific plant.

Social Objectives

In the past, much of the emphasis of collective bargaining was on economic issues such as those discussed above. Recently, however, more and more unions have been stressing improvements in the non-economic aspects of workers' jobs. The most popular issue here appears to be something called the "quality of working life". This term denotes the concern that many jobs (particularly blue-collar ones) are boring and depressing to those who perform them. Improvements in these jobs are being experimented with by a number of companies.

WHY DO WORKERS JOIN UNIONS?

Let's assume that workers are not required to join unions in order to get or keep their jobs. Why, then, would they join?

First, there is strength in numbers. An individual worker's demands may receive little attention from management. Those same demands, when expressed by an organization of workers, are likely to receive a lot more attention. The individual worker's threat to strike, for example, would cause little interruption of the work flow in a plant. A collective strike, however, could easily cripple the work flow.

Secondly, union members are represented in collective bargaining by professional negotiators. The employer is also represented by professional negotiators. The outcome is likely to be better for each worker than if each negotiated by himself or herself.

A third reason is the feeling of power workers get from union membership. Although the employer-employee negotiations are handled by professionals, the workers at least have a chance to vote for union officers and also have veto power over any "settlement" which is reached regarding wages, fringe benefits, and working conditions.

Finally, many workers believe that union membership is necessary to keep employers interested in, and concerned with, the well-being of their workers as human beings. We will discuss this in greater detail later in the chapter.

SOURCES OF LABOUR-MANAGEMENT CONFLICT

In Canada, labour and management bargain within a framework in which they share certain basic common beliefs. This is unlike negotiations in some other countries in which the bargaining takes on an air of class warfare—the "working class" is pitted against the "capitalists". In Canada today the vast majority of unions accept the capitalist system. It is within that framework that labour and management bargain.

There are, of course, some basic differences in outlook between labour and management. We will look at several sources of conflict and discuss each from a "labour viewpoint" and a "management viewpoint".

WHAT DO YOU THINK?

Do Unions Benefit Society?

Some Canadians still feel uneasy about labour unions. Much of this is probably because a lot of the publicity unions get is bad. Strikes are discussed in terms of lost production time; unions often are blamed for inflation; and cases of union corruption get a lot of news coverage.

Some people, even some union members, think that unions are too powerful. They want tighter government control of unions. Many of them want to deny unemployment benefits to workers who are on strike.

Emotions run high when we talk about unions. But unions have played a major role in improving our society. Few people would want a return to the conditions of labour during the nineteenth century. Unions help make democracy work. Labour has a voice in the political process without dividing society into warring classes. Unions have helped create our large middle-income group of consumers without which many firms would have a lot fewer customers. Do unions benefit society? WHAT DO YOU THINK?

The Profit Issue

Both labour and management accept the idea that a firm must make a profit to stay in business. But there is some difference of opinion as to how profits should be distributed. Unions, for example, keep a keen eye on company profits. High profits, to some, mean that workers are not getting enough pay and benefits. This leads to bigger wage and benefit demands in collective bargaining.

The Jobs Issue

Some workers think that a firm's major goal should be to provide jobs for workers. Recent high levels of unemployment have made many workers very concerned with the prospect of being laid off. They want job security.

Our federal government also looks to private firms as the major source of jobs for people. It is committed to a policy of full employment. When jobs in private industry cannot absorb all those people who are willing and able to work, many people want government to step in as the "employer of last resort". In other words, many workers believe they have a right to a job.

In some countries (e.g., Japan) a worker enters into a sort of long-term unwritten contract with his or her employer. In return for good service, the worker is more or less guaranteed lifetime employment. Some Canadian workers would like to have a similar guarantee. The strategy most evident in Canada, however, is for workers to bargain for shorter work weeks and extended vacations. The purpose is to spread the available work around to more workers. But workers want more pay per hour to keep their same take-home pay. Some managers argue that pushed-up wages lead to higher costs, which are passed on to consumers in the form of higher prices.

The "Right-to-Manage" Issue

As the scope of bargainable issues grows, management's authority to make decisions on its own decreases. Over time, unions have sought to increase the scope of bargainable issues. The most controversial issue here appears to be compulsory overtime. Many unions want to make overtime work voluntary, but management contends that compulsory overtime is necessary for the efficient scheduling of plant operations. To date, the management position seems strongest. For example, the protracted Griffin Steel strike in Winnipeg in 1977 had compulsory overtime as one of its main issues. In that case, the provincial government refused to introduce legislation banning compulsory overtime.

The Seniority Issue

Unions are often critical of wage and salary plans which are not based on seniority. Labour argues that any other system is subjective. We discussed this issue in Chapter

12. But management resists basing all pay and promotion decisions on seniority on the grounds that it ignores employee productivity.

The Inflation Issue

A growing number of labour contracts include cost-of-living allowance (cola) provisions. This clause means that, during the period of time covered by a labour contract, wage hikes will be granted on the basis of changes in the cost of living. Some are tied to increases in the consumer price index. This helps to ensure that the worker's standard of living will not deteriorate due to increases in the cost of living. The worker is sheltered from inflation. This helps to avoid demands for big "catch-up wage hikes" when a current contract expires and a new one is being negotiated. In some cases, unions want to bargain for shorter pacts and to reopen negotiations when their contracts do not protect their members adequately from inflation.

Cola clause

CAN YOU SETTLE THE ISSUE HERE?

Job Enlargement and Job Enrichment

One reason for Canadian workers' high productivity is technological progress. But there are problems. This progress, for example, makes some jobs repetitive and boring due to overspecialization of labour. Many workers today are not satisfied with these types of jobs. Management wants to improve jobs in order to increase employee productivity, and workers and their unions want more satisfying jobs.

One approach is job enlargement—adding more tasks to a job. This makes the job less routine and less boring.

Another approach is job enrichment—improving the quality of the job by providing opportunities for advancement and recognizing employees who do good work.

In some cases, these efforts have raised employee productivity. This benefits management. In many cases, however, these efforts lead to increases in production costs. Unions often criticize these efforts on the grounds that they are merely attempts to get more work out of workers without really improving their job satisfaction.

Managers, therefore, often are caught in a dilemma. They want to achieve efficient production from a cost and quality standpoint, but they also want to provide jobs which satisfy the demands of modern employees and their unions. CAN YOU SETTLE THE ISSUE HERE?

In general, management acceptance of the cola clause is growing, but most employers would prefer to have an upper limit on the cost-of-living benefits they might have to pay during a contract period. In fact, bargaining over the formula to use for making cost-of-living allowances can be tough.

The Productivity Issue

The productivity issue has become very important in recent years. Productivity means the ability to produce. In measuring productivity, we compare output in relation to input. The output is goods and services and the inputs are labour and capital. If we can get more output from the same input or a lesser input, productivity has increased.

The problem, however, is determining how much of the increase in output is due to labour and how much is due to capital. Labour and management often conflict over this. Suppose a worker doubles his or her output per worker-hour. The worker wants to be paid more. But management argues that at least part of the increase in the worker's productivity is due to the fact that the worker works with a new and better machine. At any rate, if all of the increase in output is paid to the worker in higher wages, management would have little or no incentive to invest money in new plant and equipment. The result, according to management, is rising costs, less profit, and eventually, a business failure because of inability to compete.

DEVELOPING THE COLLECTIVE BARGAINING AGREEMENT

When an old contract expires, or when a newly certified union wishes to develop its first collective agreement, labour and management representatives must sit down at the bargaining table and work out an agreement which will be acceptable to both parties. The issues discussed above (seniority, productivity, inflation, etc.) are normally the key points of contention. Generally speaking, management wishes to give as little as possible, while the union wants to get as much as possible. It is not surprising, therefore, that the two parties may find it very difficult to reach an agreement. Various procedures are available to help the parties resolve their differences so that a strike is not necessary.

Conciliation Conciliation may be used. **Conciliation is a process in which a neutral third party is called in to prevent negotiations from breaking down.** If negotiations do break off, the conciliator tries to get the two parties back to the bargaining table. The conciliator, however, has no authority over either party and cannot impose a solution on them.

Mediation Mediation goes a step further. **In mediation the neutral third party's task is to suggest a possible compromise.** The mediator tries to persuade the parties to settle the dispute. Like the conciliator, the mediator has no authority over either party to the dispute.

Voluntary arbitration **In voluntary arbitration the neutral third party hears both sides of the dispute and settles the issue.** The two parties decide voluntarily to submit the dispute to arbitration. Both parties are bound by the settlement.

Finally, **compulsory arbitration may be compelled by federal or provincial law in certain cases.** The arbitrator's settlement is binding. Compulsory arbitration is rare and is used only when essential public services are involved.

Compulsory arbitration

Either or both of the parties may feel uncomfortable with the idea of third parties getting involved in the bargaining process, and they may try other tactics.

Labour's Weapons

Labour's main weapons are the strike, the picket, and the boycott. (See Table 13-3.) **A strike is a temporary withdrawal of all or some employees from the employer's service.**

Strike

Table 13-3. Types of strikes, pickets, and boycotts

STRIKES:	
1. *Primary strike*	The employer's workers withdraw from their jobs for their direct and immediate benefit.
2. *Secondary strike*	B Company's union strikes to force B to bring pressure on C Company because the union has a gripe with C.
3. *Sympathy strike*	A strike called by one union primarily for the benefit of another union.
4. *National general strike*	All the workers in the nation strike.
5. *General strike*	All or most of the workers in a particular industry go out.
6. *Sitdown strike*	The workers cease working but do not leave their place of employment.
7. *Slow-down strike*	The workers "slow down" rather than cease working altogether. This type of strike is most effective in mass production industries.
8. *Partial strike*	Only part of the work force strikes, but those who go out are strategically selected to place the employer in a difficult position.
9. *Wildcat strike*	Some union members go out even though the union did not authorize a strike.
PICKETS:	
1. *Primary picket*	The employer's workers walk around the building or place of employment with placards informing other workers and the public that the company is unfair to labour.
2. *Secondary picket*	Several employees of X Company (with whom the union has a gripe) picket Y Company, a customer of X Company, in order to induce Y to cease buying from X Company.
BOYCOTTS:	
1. *Primary boycott*	The workers refuse to do business with (buy from) their employer.
2. *Secondary boycott*	The boycotters cause third parties to the labour dispute to refrain from dealing with the employer of the boycotters.

The presumption is that they will return when their demands are met or a compromise is worked out. The strike is the union's ultimate weapon. It ordinarily will not be used, however, unless the union has the financial resources to ride it out.

Picketing

Picketing means that persons (pickets) form a picket line and walk around a plant or office building with placards (signs) informing other workers and the general public that the employer is unfair to labour. Strikes are usually accompanied by picketing, but picketing may take place without a strike.

In general, picketing is protected under the right of free speech as long as it does not include any fraud, violence, and/or intimidation. An effective picket may keep other employees who belong to different unions from entering a plant. If a picket line around a plant is honoured by truck drivers, the picketed firm finds itself without deliveries.

Boycott

In a boycott a union tries to get people to refuse to deal with the boycotted firm. There are primary and secondary boycotts. Suppose that the employees of Company Y are involved in a dispute. They might send circulars to Y's customers and suppliers asking them not to do business with Y. This secondary boycot is generally legal if the circulars are not fraudulent.

CAREERS IN LABOUR RELATIONS

There are three separate approaches to careers in labour relations. One is through the business firm, one through the labour union itself, and the third through independent organizations (sometimes government) serving as mediators between the two. In any case, labour relations careers focus primarily on the interface between a union and a firm. Basically this means negotiating, living with, and interpreting the provisions of a labour contract.

From the union side, careers start as a union member who may be nominated as shop steward by his or her co-workers. The shop steward deals with day-to-day employee problems on behalf of the union. In larger problems, including contract negotiation, the union business agent assumes responsibility. The business agent also promotes the union and recruits new members.

On the management side, supervisors deal with shop stewards. The industrial relations manager deals with the business agent. The industrial relations manager usually has a college degree in labour economics or management and often is promoted from lower positions in the personnel department.

Qualifications for labour relations work include a good knowledge of labour law, human relations skills, and unusual strength of personality. Information is available from the Canadian Labour Congress, 2841 Riverside Drive, Ottawa, Ontario K1V 8X7, and from the federal Department of Labour in Ottawa.

Management's Weapons

For many years management used the lockout to counter labour's threat to strike or to organize. **In a lockout employees are denied access to the plant until they accept the employer's terms of employment.** This weapon is now used mainly as a defensive weapon once a strike is called.

Lockout

Today the layoff is more effective. A general strike against steel-makers that lasts long enough to deplete their inventories leads to layoffs in steel-using industries. Although auto workers who are laid off claim they are "behind" the steelworkers, a lengthy steel strike brings hardship to the auto workers. This may lead to indirect pressure on the steelworkers to reach an agreement.

The injunction is a court order forbidding union members from carrying on certain activities such as intimidating workers or impeding other operations of the company. Injunctions have been used frequently in the past, but recently they appear to be used only in extreme cases (such as when workers are actually damaging plant and equipment).

Injunction

Employers' associations are important in industries with many small firms and one large union which represents all workers. Member firms in an industry might contribute to a strike insurance fund which is used to help members who are struck. They are similar in purpose to strike insurance funds built up by unions.

Firms or industries with labour problems often publicize their side in newspapers and other media to gain public support. Unions also do this. The strength of public opinion sometimes leads to new laws.

LIVING WITH THE COLLECTIVE-BARGAINING AGREEMENT

At one time most employers believed that all unions were bad. This is not the case today. A firm may benefit from the presence of a union. The threat of unionization has made many managers improve their management practices. Furthermore, employer-employee communication often is improved when a union represents the workers.

The labour contract enables management and labour to co-exist. The rights of each are stated. But no contract can cover every situation in which trouble might arise. Also, problems sometimes arise over the interpretation of the contract.

A grievance is something which causes a worker to complain. Not all complaints, however, are grievances—only those complaints which relate to alleged violations of the labour contract or the law are grievances. To deal with grievances, labour contracts include grievance procedures. **Grievance procedures spell out the sequence of steps a grieved employee should follow in seeking to correct the cause of the grievance.** These procedures are set up to reduce the chance that employee gripes will cause a breakdown in labour-management relations.

Grievance procedures

Suppose Homer Anderson's supervisor tells him to do a task which Homer does not think is part of his job. He might take the matter up with his boss or with his union steward (a union representative in the plant), who would then take it to Homer's boss. If the issue is not settled at this stage, it goes to a higher union official, who takes it to a higher manager. If the issue is still not resolved, the union can ask the firm's industrial relations manager to meet with the grievance committee in an attempt to solve the problem. If agreement is still impossible, the matter is submitted to binding arbitration, that is, a neutral third party is called in to consider both sides and to make a decision which is binding on both parties.

FACTORS AFFECTING UNIONIZATION

Why are some industries more unionized than others and why are unions stronger in some areas than in others? Industries in which a few firms dominate are easier to organize because the workers are more concentrated than when many firms exist. Thus, the auto industry is more unionized than the retailing industry. Furthermore, industry-wide bargaining is more likely where there are few firms.

Another factor is the nature of the production process. It is not as profitable to organize highly automated industries which employ only a few workers as it is to organize industries which employ many workers. To make organization a paying proposition, a union must expect the dues from new members to more than offset the costs of organizing them. This is generally the case, except when a union is trying to break into an area or industry where unions are weak. Thus, the organizing effort in unorganized industries, such as banks, is progressing.

Workers hired out of union halls identify very little with their employers. Collective bargaining here is quite different from that in which a union represents workers who work for one employer over a long period. Management knows the workers' needs better in the latter case. Union and management can work together to build a healthy relationship. This also explains why different unions seek different goals. A worker hired out of a hiring hall is more concerned with job security than one who has a long record of employment with one firm.

In general, the more skilled workers are, the greater their bargaining strength. Thus, workers in the building trades earn high wages. They are skilled and their skills have not become obsolete. Many other skilled workers, such as typesetters, have found their skills outmoded by machines.

THE FUTURE OF UNIONISM

The union movement began in a period when the excesses of early capitalism placed the average worker at the mercy of the employer. For decades unions sought to bargain at arm's-length with employers. This goal was met through legislation and the collective-bargaining process. But what is in store for unions in the future?

Public Attitude

Many people considered unionism to be a worthwhile cause up through the 1930s. But some people became critical of some unions towards the end of the 1940s and have continued to be critical up to now. These people argue that some unions have become too powerful and disrupt the economic system when they do go on strike. Nevertheless, white-collar unions continue to grow rapidly as individuals join them in an attempt to get improved economic benefits.

Youth Looks at Unions

Some young people view unions the same way that they view big businesses. They believe that big union organizations are just as impersonal and "distant" as some big businesses. These young people, even if they are union members, do not get deeply involved in union affairs. This difference is apparent in the contrast between the dedicated, older union members who regularly attend union meetings and some of their young counterparts who pay their dues and do not attend.

White-collar Workers Look at Unions

The proportion of blue-collar jobs is declining, while the proportion of white-collar jobs is increasing. White-collar workers now account for roughly half of our labour force. It is often assumed that they identify closely with management and are hesitant to join unions—particularly professional employees. However, unions are having increasing success in organizing white-collar workers. Examples are teachers, university professors, professional athletes, and government workers.

THEN AND NOW

University Professors and Collective Bargaining

For many years, university professors resisted unionization on the grounds that the "adversary system" of labour-management relations was inconsistent with the "community of scholars" concept at the university.

Recently, however, professors in Quebec and Manitoba have unionized and others are likely to follow suit. The increased sympathy for unions has come from a number of areas, including inflation and faculty perception that university administrators are not responsive to faculty wishes.

Women Look at Unions

For years the working woman worked to supplement her family's income. Today, many more women are career-oriented. They do not take jobs only between children. In fact, a growing percentage of the female work force is made up of working mothers. In the decade 1966–76 alone, the female participation rate in the labour force increased from 32 to 45 per cent. Career-oriented women view work differently from the way their earlier counterparts did. Along with the sexual revolution there are new ideas about marriage and family. For many women a career rather than the family has become their primary concern in life. As a result, organized labour is making strong efforts to appeal to women.

SUMMARY AND LOOK AHEAD

The modern labour union in Canada evolved from the early craft unions. This was not a smooth evolution. The earliest attempts of workers to form unions were frustrated by court rulings. Labour unions were considered conspiracies in restraint of trade.

Gradually, public sympathy shifted in favour of unions, and laws were passed to guarantee the worker the right to join the union. Once a union is certified, union and management must bargain in "good faith" on bargainable issues. The broadening scope of bargainable issues has narrowed the area of managerial discretion over the years.

Unlike unions in the days prior to World War II, today's unions do not have to fight for the right to exist. Their goals reflect this. Modern unions seek a broad variety of economic, political, and social goals.

We looked at the reasons why workers join unions and the sources of labour-management conflict. Then we discussed how labour and management bargain collectively to negotiate a labour contract and how the parties "live" with the contract. Grievance procedures are important here.

Labour's main weapons are the strike, the picket, and the boycott. Management also has its weapons—the lockout, the layoff, the injunction, and appeals to the public.

The union movement has always been a minority movement in Canada. In recent years, unions have sought to increase their membership by growing in new directions, such as organizing government employees and other white-collar workers. This is necessary because of the declining percentage of the labour force which is in blue-collar jobs.

In our next chapter, we will study computers and quantitative analysis in business. Few developments can rival the impact that computers have had on business.

KEY CONCEPTS

Agency shop A type of union security. All employees for whom the union bargains must pay dues but need not join.

Bargainable issues Aspects of the work or job environment which are subject to collective bargaining between union representatives and management representatives. Examples are wage and fringe benefits.

Blacklists Lists circulated among employers containing the names of workers who are known to be in favour of unions. Blacklisted workers were denied employment. The use of blacklists is now illegal.

Boycotts In labour relations a boycott is a union's attempt to get people to refuse to deal with boycotted firm. Also used by other groups to get people to unite against another person, business, non-business organization, or country and to agree not to buy from, sell to, or associate with that person, organization, or country.

British North America Act 1867 act which divides legislative powers between the federal and provincial governments.

Canada Labour Code A personnel practises law which prohibits discrimination and stipulates working conditions.

Closed shop A type of union security. Only members of the union can be hired by an employer.

Cola clause A provision in a labour contract that wage hikes will be granted on the basis of changes in the cost of living.

Collective bargaining The process of negotiating a labour contract between union representatives and employer representatives. The parties bargain in good faith on bargainable issues.

Compulsory arbitration If management and labour cannot agree on the terms of a new contract or the settlement of a grievance, the issue must be submitted to a neutral third party who will impose a binding settlement.

Conciliation A method of helping labour and management settle disagreements which arise. A neutral third party tries to prevent negotiations from breaking down. If negotiations break off, the third party (conciliator) tries to get the disputants back to the bargaining table.

Conciliation Act A 1900 law which assists in the settlement of labour disputes through voluntary conciliation.

Craft unions Unions organized by crafts or trades—plumbers, carpenters, machinists, etc. Membership is restricted to workers with specific skills.

Grievance procedure Included in most labour contracts. Spells out the sequence of steps an aggrieved employee should follow in seeking to correct the cause of the grievance. A grievance is a complaint about an alleged violation of a labour contract or the law as it applies to a worker.

Guaranteed annual wage A provision in a labour contract which maintains the worker's income at some agreed-on level during a year.

Independent local union A union which is not formally affiliated with any other labour organization.

Industrial Disputes Investigation Act A 1907 law which requires the use of conciliation before a strike is allowed.

Industrial unions Unions organized according to industries—steel, auto, clothing, etc. Include semi-skilled and unskilled workers.

Injunction Issued by a court. A mandatory injunction requires performance of a specific act. A prohibitory injunction orders the defendant to refrain from certain acts. In labour relations, an injunction granted to an employer orders employees not to strike or to return to work.

International union A union with members in more than one country.

Labour contract The agreement negotiated between employer and union representatives which sets forth the terms and conditions under which union members work.

Labour union An organization of employees formed for the purpose of dealing collectively with their employers in order to further the interests of those employees.

Local union The basic unit of organization in a union. Made up of workers in a local area. Is part of a national or international union.

Lockout Employees are locked out or denied access to their place of employment until they accept the employer's terms of employment.

Mediation A method designed to help labour and management iron out their differences. A neutral party suggests a possible compromise of the dispute. The mediator's suggestions are not binding on the parties.

National union The organization set up to bring all the local unions of a particular craft or industry together for bargaining purposes.

Open shop An employer may hire either union and/or non-union workers. Employees need not join or pay dues to a union.

INCIDENT:

Gordon McGhee

Thirty years ago, Gordon McGhee established the McGhee Manufacturing Company. Mr. McGhee has always been "anti-union". According to him, unions only cause trouble. He once told a friend, "A union cannot raise the worker's pay; only the employer can do that. All the union can do is take some of the worker's pay away in the form of dues on the promise that he or she will receive something in return."

Until recently, Mr. McGhee has had little "trouble" with unions. However, there is now quite an effort being made at his plant to unionize the workers. Mr. McGhee knows those employees who want the union and he called them in for a "talk". He told them, "I'll go out of business before I'll let you ruin the company by bringing in a union."

Questions:
1. Do you agree with Mr. McGhee's statement in paragraph one? Why or why not?
2. Reread Mr. McGhee's last statement. Describe the nature of the "talk" you would recommend him to have with his employees.

Picketing In labour, persons (pickets) form a picket line and walk around a plant or office building with signs informing other workers and the general public that the employer is unfair to labour. In general, a means of communicating with others.

INCIDENT:

Operation Uplift

Union leaders in a large city, the mayor, the school board president, the heads of several community action groups, and the leaders of the city's business community are planning "Operation Uplift". Its main purpose is to reduce unemployment among teen-agers and women in that city.

But there are some conflicts. The union leaders and the business leaders "had it out" on the issue of the minimum wage. The businesspeople want to start a "grass-roots" effort to have the minimum wage considerably lower for teen-agers than for adult workers. The union leaders flatly rejected these ideas.

The school board president wants career programs for students who are not going on to college. The head of an Indian group said this would deny some students the preparation they need if they should later decide to go to college. The same person also said that this approach trains people for low-level jobs from which it is hard to advance.

The mayor and the business leaders argued over taxes. The mayor wants the business community to support a special municipal levy on business firms in order to create city government jobs. The business leaders say the tax would only lead to higher prices for their goods and services, lower their sales, and eventually lead to layoffs of employees.

The group representing women wants better enforcement of anti-discrimination laws. But one union leader said that these laws irritate male workers with a lot of seniority.

"Operation Uplift" has been in the planning stage for about six months, but no one feels that there has been enough progress towards a compromise to put the plan to work.

Questions:
1. What do you think of the proposal to have a lower minimum wage for teen-agers? Explain.
2. Do you think that the criticism of career education is valid? Why or why not?
3. What do you think of the municipal levy proposal on business firms to finance city government jobs? Explain your reasoning.
4. Prepare a plan of action for reaching a compromise among the people who are planning "Operation Uplift".

Privy Council Order 1003 Recognized the right of employees to form labour unions and to bargain collectively with management.

Strike A temporary withdrawal of all or some employees from an employer's service.

Unfair lists Lists circulated among workers containing the names of employers whom unions consider unfair to workers because these employers refuse to hire union members.

Union shop A type of union security. An employer whose employees are unionized may hire non-union workers, but they must eventually join the union (usually within 30 days of being hired) or else be fired.

Voluntary arbitration A method of settling a dispute between labour and management. A neutral third party decides how the dispute is to be settled. The two parties decide voluntarily to arbitrate the dispute. The parties are bound by the agreement.

QUESTIONS FOR DISCUSSION

1. Discuss the Industrial Revolution from the viewpoint of the typical worker during that period.
2. Prior to the 1930s, how did employers resist their employees' efforts to form unions?
3. Discuss the provisions of the Canada Labour Code.
4. Is the scope of "managerial discretion" narrowed when a firm's workers form a union? Explain.
5. Distinguish between a craft union and an industrial union.
6. Define (a) local union; (b) national union.
7. Does a union have any responsibility to persons or groups other than the members of that union? Discuss.
8. Do the employees of a firm whose management is "employee-oriented" have a need for a union? Why or why not?
9. How would you explain the fact that the majority of Canadian workers do not belong to a union?
10. List and discuss two major economic objectives of modern unions.
11. In the light of the declining proportion of our labour force which is engaged in blue-collar jobs, what are unions doing to increase their membership?
12. List and discuss four issues that might lead to labour-management conflict.
13. List and discuss the "weapons" of labour and management.

ROCKETOY COMPANY VII

Rocketoy's work force had expanded to a total of 300 workers. In spite of its size, Rocketoy's management was able to preserve the good employer-employee relationship that had existed ever since the company was founded. Furthermore, Rocketoy had a profit-sharing plan which enabled its employees to share in the firm's profit. According to Terrence, the plan gives each worker a real stake in the company. He believed this was crucial in getting them to put forth their best efforts.

Joe Phillips was still vice-president of production. Terrence had noticed "people problems" between Joe and ten young production workers who had recently been hired. Terrence asked these ten workers to come to his office to discuss the situation. He wanted to take steps to solve the problem between the workers and "Uncle Joe".

The ten workers showed up in Terrence's office and it became apparent to Terrence that the "people problems" between Joe and his workers were a lot more serious than Terrence had thought. They accused Joe of treating his subordinates like children.

Charlotte Ethridge was the most outspoken. She accused Joe of playing the "father role". Charlotte cited an example. Two months ago, a promotion opened up. The former supervisor of the loading dock retired and the obvious choice for the job was Pauline Williams. She had the most seniority in the entire production department, and she wanted the job. Pauline was a good worker and had never had any trouble with Joe or anyone else at the plant.

But Joe passed over Pauline and gave the promotion to Gordon Hooker. Gordon had less seniority than Pauline, but Joe believed he would be a better choice for the job because the loading dock supervisor sometimes has to "be tough" to keep things moving.

After the meeting, Terrence thought the situation over. He knew Joe was not ready to retire. Terrence also felt that he owed him a lot. Without his help, Rocketoy might never have gotten started.

But Terrence did understand the source of the workers' complaints. He recalled Joe's discussing with him the reasons he promoted Gordon Hooker over Pauline Williams. "Terry, I know I'm out of step with some of the newer thinking. If Pauline were a man, I'd have given her the promotion. She really deserved it. But—she is a 28-year-old woman. Can you imagine the problems I'd have with her as supervisor on the loading dock?"

At first Terrence decided to let Joe run his department as he saw fit. But during the next three months Charlotte "stirred up" several of the production workers. At first it was mainly Charlotte and the other nine workers who had visited Terrence's office. But it soon spread. Complaints about Joe began multiplying rapidly.

Before long, union organizers tried to unionize Rocketoy's production workers. The Pauline Williams "incident" was a big issue. Rocketoy's workers were the only ones among the big toy manufacturers who were not unionized.

Questions:
1. Discuss the meaning of a "good employer-employee relationship".
2. Do you think that a profit-sharing plan for employees helps to make employees feel that they have an ownership interest in the company for which they work? Explain.
3. Do you think that Terrence should have called the ten workers in for a talk? Explain.
4. Could Terrence have done anything to prevent the problem as outlined by Charlotte Ethridge? Discuss.
5. Once the problem involving Pauline Williams was out in the open, what should Terrence have done about it? Explain.
6. How successful do you think the union organizers were in unionizing Rocketoy's production workers? Discuss.
7. If you were Terrence, what would you have done to try to keep Rocketoy a non-union company. Explain.

14
The Computer and Other Tools

OBJECTIVES: After reading this chapter, you should be able to:

1. Describe one important way that a computer enters your life.
2. Illustrate the complementary relationship between people and computers in doing a job.
3. Explain what makes a system automated.
4. Compare manual, mechanical, and electronic data processing systems.
5. Draw a diagram of a computer system.
6. Recognize common input-output devices for computers.
7. Explain the function of a language such as COBOL.
8. Draw a simple flow chart.
9. Contrast several ways that people react to computers.
10. Discuss some of the jobs available which are related to computer operation.
11. Compute an arithmetic mean and median.
12. Prepare a break-even chart.

KEY CONCEPTS: In reading the chapter, look for and understand these terms:

COMPUTER	COBOL
AUTOMATION	DOCUMENTATION
COMPUTER PROGRAM	QUANTITATIVE TOOLS
DATA PROCESSING	STATISTICS
PUNCHED CARD	ARITHMETIC MEAN
HARDWARE	MEDIAN
CENTRAL PROCESSING UNIT (CPU)	MODE
OUTSIDE MEMORIES	FREQUENCY DISTRIBUTION
INPUT-OUTPUT DEVICE	SAMPLE
ON-LINE SYSTEM	BREAK-EVEN ANALYSIS
TIME SHARING	OPERATIONS RESEARCH (OR)
SOFTWARE	MODEL BUILDING
FORTRAN	LINEAR PROGRAMMING

14 THE COMPUTER AND OTHER TOOLS

In this chapter we'll first discuss what a computer is and what it can do. We will examine major components of a computer and the programming and other things needed to make it work properly. We'll also study some of the practical problems encountered in using computers. Finally, we will turn to statistics and mathematical techniques used by business firms. Topics include break-even analysis and linear programming.

WHAT IS A COMPUTER?

A **computer** or, more exactly, a computer system is an electronic machine capable of storing huge amounts of data and performing mathematical calculations very quickly. It is also called an electronic data processing system. Computers are important parts of many thousands of business firms.

What a Computer Does

Computers play a big role in your everyday life. Stop and think about it! They do some of the little everyday things like figuring your bank balance or the size of your family's bill at Sears. A computer might even prepare your term grade report. Computers do thousands of repetitive operations like these. They perform very efficiently

---

YOU BE THE JUDGE!

The Computer's Personal Touch

Have you ever received a personal letter from a computer? What about that big sweepstakes letter you got from your favourite magazine publisher? Wasn't the letter informing you of the contest addressed to you personally? Didn't the salutation on the letter say "Dear (your name)"? Wasn't your name mentioned at several places in the letter? How nice of such a big company to take the time to be so friendly and personal! Think of what it had to pay for typing all those personal letters!

Of course, you know the letters were not from a computer. They were from the publisher. But the publisher had help from a computer-controlled typing or printing system. Hundreds of thousands of letters come out looking like personal letters because of the computer. Such "personal communication" should flatter you. But is it really personal communication or just another piece of "junk" which is typical of the computer's ability to impersonalize company-customer relations? YOU BE THE JUDGE!

for a big institution like a corporation or a university. They store a huge mass of information and make thousands of routine calculations.

Suppose you use your Visa card to buy gas at a service station. The attendant uses it to print your account number before you sign the charge card. This card ends up being processed by a computer. The amount of your purchase is added to your Visa card bill. Because of the great speed and accuracy of computers, thousands of transactions like your purchase of gasoline, can be quickly and accurately processed.

But a computer can do a lot more than routine data processing. Think of its role in the space program! Without computers, moon landings and weather satellites would

---

WHAT DO YOU THINK?

What Is the UPC?

You've probably seen something like this many times and wondered what it is. It's really a universal product code (UPC). Under those lines are two 5-digit numbers. This code is like a fingerprint to the laser-beam scanner of a computerized checkout system.

In an automated checkout system based on the UPC, a computer and its memory are tied in to the checkout stands. The computer memory has stored in it the prices of all the products sold by the supermarket. At each checkout is an electronic terminal, a scanner, and a display screen.

As each package passes over the scanner, the laser light "reads" the UPC symbol and sends the information to the supermarket's computer. The computer, within a fraction of a second, displays the product's name and price on the display screen and also prints that information on an itemized receipt for the shopper. The computer also keeps track of how many units of each product are in inventory because it deducts from inventory each item sold.

Whether or not a particular supermarket will install such a system depends on the cost savings it makes possible, the cost of the system, and the expected response of its customers. But some people claim that consumers want to see the price stamped on each product they buy. If so, a lot of the benefit of the system is lost. Meanwhile, however, variety stores and some other retailers are installing similar systems. Will they work? WHAT DO YOU THINK?

have been impossible. In fact, much of the progress in science and technology has depended on computers. Computers touch your life in at least these two ways—as a go-between for large institutions and their clients and as an instrument which speeds up technological and economic progress.

Businesses use computers to prepare payrolls, to analyze past-due accounts receivable, to prepare and mail out bills, to keep social security and tax records, to keep track of inventories, and to simplify reordering of goods. There are many other jobs, too, which a typical firm can find for computers to do—if the firm can afford the computers.

All businesses, as we have seen in Chapters 6 through 13, need to collect and use all kinds of information. They need facts about their resources—people, materials, machines, and money—which are their inputs. The facts about these inputs include cost, quantity, and source data for the past, present, and future. Firms also need facts about their own processes—rates of input usage, costs, efficiency of use, and work stoppages. They need facts about their outputs—their rate of flow in the distribution system, their quality, and expected future demand.

Factual needs also include data about competitors and the environment—competitors' sales and growth, customer characteristics and habits, and government regulations and tax decisions.

All of these bits and pieces of information tell the story of today's and tomorrow's business success. A successful firm needs to be able to gather, store, combine, and use this mass of data at a reasonable cost—a cost which is lower than the benefit it brings. If a computer is well-designed and well-used, it can do this.

Computers can't predict the future, but they can help to make the future manageable, given complete and accurate data and a good understanding of the causes of success. Informed decisions are almost always better than those which are uninformed.

Master or Servant?

Some fear that computers are "taking over" civilization. The truth is that, although the computer can perform millions of simple computations in a very short period of time or solve amazing problems, it is still only a servant. Some say it is not even a servant, but only a tool in the hands of people.

Whether we call it a servant or a tool, it still is a great multiplier of human power. It does some jobs which people could never do. It can work tough math problems at fantastic speeds without a mistake.

At the same time, there are some tasks the computer will never do and which only people can perform. People can set values of things and can create things. A computer cannot judge a beauty contest or write real poetry. The relationship between people and computers is complementary—they can work very well together.

BASIC COMPUTER-RELATED IDEAS

Before discussing how businesses use computers, let's discuss two important ideas related to the use of computers. These include automation and data processing.

Computers and Automation

Automation

In earlier sections of our book we used the term "automation". An activity or process is automated when it is possible to set its controls in advance so that it can work a long time without human attention.

Some automated processes are fairly simple and don't require computers. An example is a household heating and air-conditioning system. The thermostat permits the system to operate without much human interference.

A petroleum refinery is a more complex system. There are many points in the refining process at which information must be fed continuously into a computer. The information relates to things such as the rate of flow, temperature, and so on. The refinery's central computer has been programmed so that it uses this information to control the refining process.

Programming, as we will see in greater detail later in this chapter, is the process of telling a computer what to do. **A computer program is a detailed set of instructions in a special computer language.**

Computer program

The computer automatically makes certain computations and relays instructions to machinery in the factory. It does this in accordance with the program fed into it at an earlier time. Thus, valves are opened and closed and temperatures are raised and lowered automatically.

Computers and Data Processing

Data processing

As we saw in Chapter 9, all businesses need to accumulate, store, manipulate, interpret, and report data. This is called data processing. It includes financial and non-financial data. Governments and other non-business institutions need to process data, too. These data processing needs vary a lot because of the great differences in the type and the size of data flows among these institutions. In all cases, however, there is a need to keep accurate tallies of all those numbers which are important to these institutions. It could be counting hospital admissions or adding cash collected by a grocery store or figuring the net profit of Labatt's Breweries.

The size of the data flow (need) and the financial resources of the institution (ability to pay) determine the scale and complexity of its data processing system. Not all firms need, nor can they afford, computers.

On a very small scale—such as would be found in a small rural gasoline service station—no machines at all may be involved. Keller's Service Station in Bush, Saskatchewan, uses only a pencil and a loose-leaf notebook to record sales and expenses.

An early type of data processing system was the keysort card system. The keysort card has a system of notches and holes at the edges. (See upper part of Figure 14-1.) It permits easy retrieval of collected information according to certain classification codes. For example, by using a simple sorting needle, a clerk can quickly remove from a file all accounts with balances of more than $2,000 or those of customers living in a certain part of the city. Use of keysort card systems is decreasing as their cost increases relative to the cost of mini-computers.

14 THE COMPUTER AND OTHER TOOLS

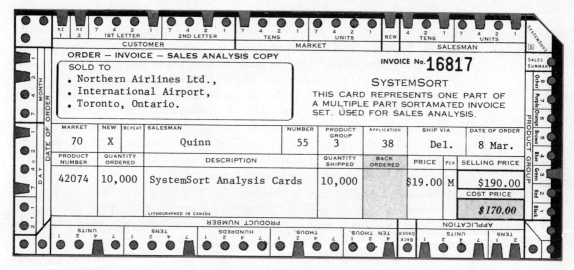

KEYSORT CARD

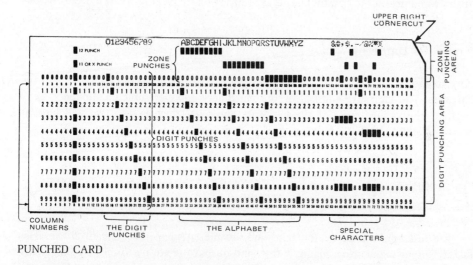

PUNCHED CARD

Figure 14.1 (Illustrations courtesy of Systems Equipment Limited, Winnipeg, Canada, and International Business Machines Corporation.)

A second type of mechanical system uses punched cards. These cards have space for 80 characters across and 12 positions in each of the 80 "columns". (See Figure 14-1.) Numbers and letters are punched on a card by a keypunch machine. This machine works like a typewriter. But in addition to printing, it makes small holes in the card. The vertical position of a hole in a column indicates a number. Two vertical holes in a column indicate a letter. Notice the top of the lower card in Figure 14-1. A keypunch usually prints the letters across the top of the card.

Punched card

SECTION THREE BUSINESS DECISIONS

Data on punched cards can be processed in many ways. A sorting machine reads the holes in one column of a deck of cards and sends each card to a special slot in the machine. Each slot represents a certain number or letter punched on the card. All cards sent to a particular slot have the same hole punched in a particular column.

A collating machine organizes or collates separate decks of cards into one combined deck. Accounting machines also do data processing jobs. A worker in a furniture store's billing department can take a deck of punched cards which represent customer credit purchases and print bills for those customers with an accounting machine. This is a mechanical data processing system. This kind of system is rapidly becoming obsolete, although some of its parts (keypunches, sorters, etc.) are still widely used to support the more modern electronic data processing systems.

Computers are also known as electronic data processing (EDP) systems. They are generally much more complex and powerful than mechanical systems. EDP systems are used by nearly all large companies today and by growing numbers of middle-size firms.

EDP systems differ from mechanical systems in many ways. One of the main differences is that EDP systems allow instructions to be "fed" into the system rather than "wired in" on the control boards as found in mechanical systems.

Computer Hardware

Hardware

Any discussion of computers usually falls under two headings: the hardware and the software. **The hardware consists of the machinery and electronic components.** Let's examine the various parts of the hardware and what they can do. We will look at software later.

The tasks performed by the hardware are presented in the diagram above. The tasks of inputting, storing, manipulating, and outputting are performed by four kinds of parts or components: (1) input devices; (2) central processing units; (3) outside memories; and (4) output devices. (See Figure 14-2.)

Central processing unit (CPU)

The heart of any computer is its central processing unit (CPU). **The CPU includes an internal memory for storing data, an arithmetic unit for performing calculations, a logic unit for comparing values and helping to "make decisions", and a control unit which actually operates the computer and sends instructions for controlling all of the other components.**

Outside memories

A computer's internal memory can be added to by outside memories. **Outside memories are separate systems for storing information, such as magnetic discs, magnetic tape, decks of punched cards, and punched paper tape.** (See Figure 14-3.) Information can be recorded on any of these devices and fed back into the computer itself

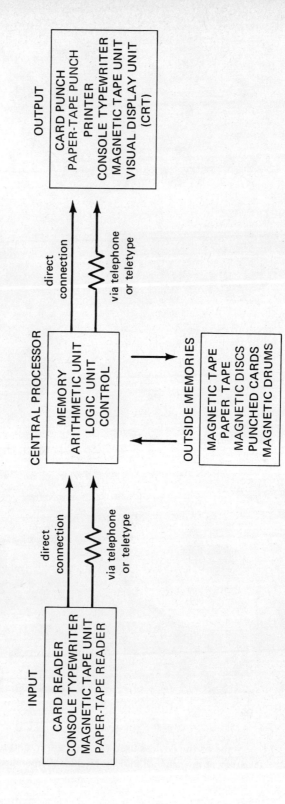

Figure 14-2. Hardware components of a computer system

MAGNETIC DISC

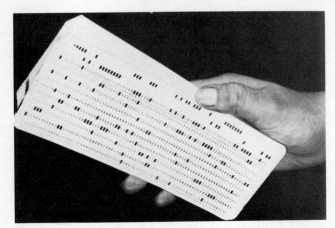

PUNCHED CARDS

MAGNETIC TAPE

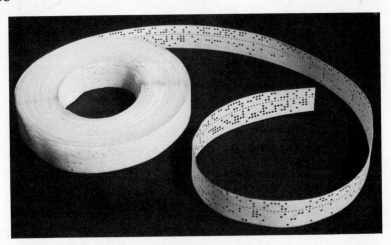

PAPER TAPE

Figure 14-3. Some of the devices used to store, enter, and retrieve data from computer systems. (Courtesy of International Business Machines Corporation.)

at any time. Outside memory devices are usually used for information which is not needed constantly in the operation of the computer system. More frequently used data are stored in the internal memory.

The input and output (I-O) devices are the means of getting information in and out of the computer. Information may be put into a computer by means of a card reader, a paper-tape reader, or a magnetic tape unit. Sometimes, information is directly fed in by means of a typewriter. Typewriters and magnetic tape input units also serve as output devices. Automatic card punching units, paper-tape punching units, visual display units similar to a TV screen (CRTs) and a variety of high-speed line printers also can serve as output devices. Some line printers can print out hundreds of lines of information per minute.

Input-output device

One important modern idea involves the use of remote input-output devices. The input or output unit may be far away from the central processing unit. When the operating needs of a business require that information be fed or retrieved from the CPU by someone located at some distance, it is possible to do so by means of telephone lines. **A computer system which employees use constantly to get operating facts, such as parts availability or accounts receivable balances, is called an on-line system.** When data are collected for a period of time before being processed, this is called batch processing.

On-line system

The use of remote I-O devices also facilitates an on-line process known as time sharing. This involves connecting several devices to the same computer so that several different firms or different users in the same firm can use it at the same time. The computer's high-speed calculation makes this possible. A computer can "handle" several jobs at the same time because the CPU operates much faster than any of the input or output devices. Several departments of a firm can use the same system at the same time. The inventory clerk in a warehouse can be typing in data on a newly arrived

Time sharing

THEN AND NOW

Computers and Small Business

At one time, the use of computers was limited to large firms. The hardware was very expensive, and the computer makers concentrated on selling them to big firms.

Over the years, however, the cost of computers has declined. Small firms can now afford to buy "minicomputers". They vary greatly in sophistication and price, but a small firm can buy a minicomputer which can be fully programmed for under $25,000 or lease it for only a few hundred dollars a month. If they are used properly, minicomputers enable small firms to do a better job of inventory management, cost analysis, sales analysis, and budgeting. Clearly, the use of computers is no longer restricted to big firms.

> ## THEN AND NOW
>
> ### Computer Hardware and Tomorrow's Market Research
>
> A few years ago, many marketing researchers used computers primarily to tally the findings of surveys. In fact, mechanical tabulating systems were often enough.
>
> Market research experts of today are predicting radical new computer hardware applications to their trade. Among these are a CRT display device and computerized questionnaire for carrying out telephone surveys; optical scanners for tabulating sales data at supermarkets and other retail outlets; self-administered questionnaires completed by shoppers at consoles set up in shopping malls; and large computer-accessible data tapes containing demographics and product usage for studies of market segments. Computers even help to make large numbers of automated phone calls for marketing and marketing research purposes.
>
> Many of these new systems have already undergone field testing. What will the computer mean for the future of marketing research?

shipment of bolts while the payroll department is feeding in a paper tape containing information needed to print this week's pay cheques. There is no noticeable interruption of either input. The CPU is seeing to it that all the incoming numbers and instructions go where they belong.

Time sharing also makes it possible for many small firms to use computers at a moderate cost. They pay only for the time during which they are "connected" to the central computer. A small business, for example, may make a time-sharing contract with a local computer services company. The small business's only investment is in the cost of the teletype or other I-O device.

The Arc Trucking Company operates a fleet of twenty trucks in Alberta. It must maintain an accurate file of scheduled truck usage, including full information on dates, times of shipment, cargo sizes and types, and so on. The central office needs to be able to tell a prospective customer if and when a particular shipment can be made. By means of a teletype or visual display unit an employee can "call up" the computer at ABC Computer Services Ltd. and get an immediate report on what will be available in the next two weeks. This information is stored in a special part of the memory of the computer. The computer's CPU "handles the traffic" of Arc Trucking's computer needs as well as the needs of many other firms. The fee that Arc Trucking pays is less than the cost of setting up its own small computer.

Of course when the time-sharing arrangement was originally made, Arc Trucking needed the help of programmers and others from ABC to learn how to do it right.

Computer Software

Computer software consists of those things which complement the hardware, such as computer languages and programming. Software is just as important as hardware. Hardware can do nothing until a person sends it instructions. A set of instructions is called a program. Writing such instructions is called programming. These instructions can reach the computer's control unit only if they are in a language which the computer understands. The kind of language a computer understands varies even among models of the same computer manufacturer. New software developments, however, make it possible to write instructions for the computer in a language which is nearly like human language. Some of these languages are FORTRAN, BASIC, COBOL, RPG, and PL1.

Software

FORTRAN is a widely used language with many applications. It is used in relatively simple tasks and also in complex mathematical computations. BASIC and PL1 are simpler languages and are easier to learn than FORTRAN. They have, however, fewer applications.

FORTRAN

COBOL is a specialized language which is extremely close to English and is used for business data processing. RPG is a specialized language for small computers to write business reports. RPG stands for "report program generator".

COBOL

A business student can make good use of all of these languages. They are rather simple to learn. With the proper educational tools you can teach yourself BASIC or PL1 in a few days and FORTRAN or COBOL in a few weeks.

The BASIC program in Figure 14-4 tells a computer to do a series of steps in the order of the program line number at the left of each line. Line 10 says to read the first number in the first DATA line. This is the number 30 in this case. Line 20, in effect, tells the program when to stop adding. It does this by picking a "dummy" number, −99.99, which is placed at the end of the DATA list. Line 20 says: "When you come to the dummy number (−99.99), skip to line 50." Line 30 is the actual adding process. It tells the computer to set up a counter, T, and to add the next value of X in the DATA line to this counter. Line 40 says to start the cycle at line 10 again. The computer continues the cycle of lines 10 to 40 until it reaches the dummy number, −99.99. Next, line 50 says to print the words "THE SUM IS" and the final value of T, which is 220. Line 70 tells the computer to stop the program.

```
10   READ X
20   IF X = −99.99 THEN 50
30   LET T = T + X
40   GO TO 10
50   PRINT "THE SUM IS"; T
60   DATA 30, 25, 63, 91, 11, −99.99
70   END
```

Figure 14.4. A BASIC program for finding a sum

This may seem like an awful lot of trouble for a simple addition like this. It is. But if we had hundreds of numbers and much more complex mathematics than addition, it would be worth the trouble.

Documentation

Documentation **Documentation means preparing a chart or diagram to illustrate the steps in a program and how they relate to the job a computer is doing.** To help people use a program, to explain, or to modify a program, documentation is needed. Programs are documented in the form of a block diagram or flow chart. (See Figure 14-5.) A block diagram uses certain standard symbols indicating specific kinds of operations or steps in a program. Figure 14-5 is a block diagram of a program to print pay cheques from time cards and wage rate information.

HOW ARE COMPUTER SYSTEMS CREATED?

A sad fact about the typical use of computers in businesses is a lack of sufficient planning before installing the systems. This is caused by a variety of things. Consider this typical sequence of events leading up to the installation of a computer system at the Ajax Bolt Company.

Ajax is a middle-size manufacturer with an outdated, partially manual system of processing data. The information systems in the production, marketing, and finance departments are all somewhat different. The system at Ajax' newly acquired subsidiary is completely different. There is a serious need for a modern computer-based information system.

The subject came up once before at a board of directors meeting. The need began to be realized, though, when Ajax' president, Sam Black, visited a competitor's plant. He saw the fancy computer room over there—with all the blinking lights and spinning reels of tape. Of course, he also noticed that there was a lower volume of paperwork. Ajax' president also learned that the competitor's profits had improved since the system was installed.

The next time the computer firm's sales representatives visited Ajax, Sam listened closely to their strong sales pitch. They presented what seemed to be a very good application of the computer company's hardware and software to Ajax' needs. Being a smart businessman, Sam arranged for three computer firms to make presentations to his staff. After study and comparison of the proposals (which were all similar), they picked a medium-priced computer made by the firm with the best service reputation of the three.

Ajax' need for a computer system was clear. But Sam Black did not take the best approach. Ideally, Ajax should have started by hiring an independent systems consultant or a permanent systems analyst. This person would have talked to all department heads to develop a complete set of concepts of the firm's needs. Next, with the analyst's help, Ajax would have examined the computers made by a larger number of

14 THE COMPUTER AND OTHER TOOLS

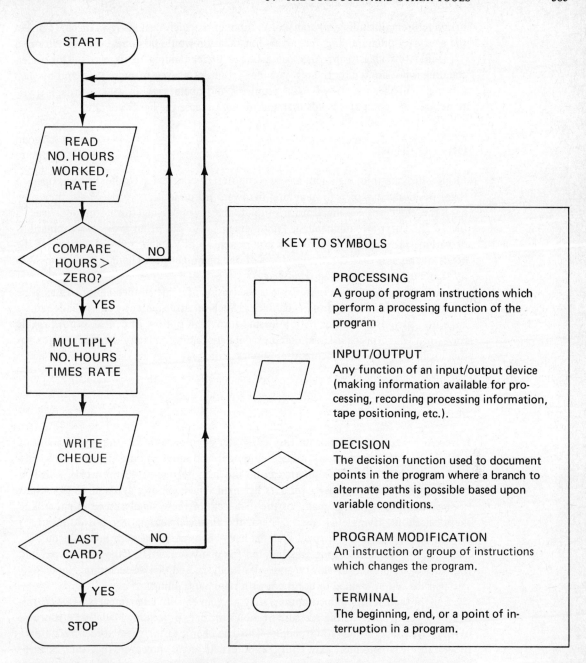

Figure 14-5. A form of documentation known as a block diagram or flow chart. *This describes a program which reads hours worked and hourly rate from a record, skips records of workers who have not worked during the week, computes total wages earned for the week, and writes the pay cheque. Standard flow chart symbols are also explained.*

manufacturers, including a variety of components which best fit Ajax' needs. Finally, the necessary programming, testing, and installation would be done.

Under this ideal plan, Ajax has a better understanding of its needs. This understanding is not limited by the ready-made systems of a given manufacturer and by such a firm's "outsider's view" of Ajax' problems and operations. In other words, it's wise to define your computer needs first and to buy or lease the hardware later.

Other Options

Firms which are thinking about a new computer, of course, have other options open to them besides the choice from among hardware manufacturers. First of all, they now have the choice of leasing from manufacturers or of leasing from computer leasing firms. The latter are independent "middlemen" who can often provide combinations of various manufacturers' hardware components which might more closely meet the needs of the user than can the products of one manufacturer. There is often a possibility that a user can save money that way.

Another option which is available is the leasing of a line connected into an existing large system with time-sharing capabilities. Such an arrangement can be made with a large user such as a bank or with a leasing firm. This option, of course, is not feasible for a firm which needs a large, complex data processing capacity. Another option for the small firm is to let a computer service firm take over its data processing needs entirely.

Minicomputers

In recent years the revolution in tiny computer circuits has led to the development of a whole new set of business "minicomputers". The minicomputer, which costs only 1/40 as much as a large computer, can do much more than 1/40 the work. This low cost of models made by many vendors has made the computer available to thousands of smaller firms. Although many of the managers of these smaller firms might still be skeptical about having their own system, the financial reasons for not doing so are rapidly disappearing. These small systems have been marketed with broad accounting packages and specialized programs tailored for many industries. They can do payroll, general ledger, billing, inventory and sales analysis; and clerks, secretaries, and floor salespersons can be trained to use them with minimal training.

For the larger firm, minicomputers may have an equally large impact. Larger firms, in recent years, have felt the pressure of mounting data processing loads and a shortage of trained computer-operating people. This has been causing them to centralize data processing in larger, more powerful computers. It seems, however, that this, in turn, has often led to conflict between those actually *using* the systems and those *controlling* them at data centers. The falling costs of minicomputers has started to reverse the trend towards centralization of data processing. Firms which wish to implement decentralized management philosophies, then, can avoid the frustration created by large centralized computers.

Minicomputers can also be linked to large computer systems so that maximum flexibility can be achieved. The mini can serve both as a form of intelligent terminal and as an independent data processor.

How People React to Computers

Human reactions to the computer range from worship to outright fear. Most people who know computers reject these extremes. Rather, they learn that the computer is a marvellous tool. They find that people and computers can bring their different abilities together and that this combined power can be used very effectively.

Fear of the computer takes different forms. Some people fear it because they feel it can bring about mass unemployment. Others fear it because it brings change in the firm—new ways of doing things. Still others fear the computer for another reason. They know that the government and many private agencies have stored huge quantities

---

GUIDELINES

Selecting a Computer System*

Hardware evaluation:
- What are the costs involved? The rent, lease, or purchase price as well as the costs of operation and maintenance must be considered.
- Can it perform to your satisfaction? This includes work volume and speed.
- Is it compatible with the system already in operation?
- Is it expandable to meet your growth expectations?
- Can your staff be trained to run it?

Software evaluation:
- Do packaged programs fit your needs and staff expertise?
- Are programs documented adequately?
- Are the operating system, compilers, etc., reliable?

Vendor evaluation
- What are the maintenance capabilities of the vendor?
- Does the local office of the vendor provide programming and systems support?
- What support will vendor provide during conversion?
- What is the record of the vendor as to past performance, consistency during negotiations, etc.?

*Adapted from Barry Render and Ralph M. Stair, Jr., "How to Evaluate Computer Systems," *Louisiana Business Survey* (October, 1977), pp. 14-15.

(a) 3800 Line printer, (b) 3525 Card punch, (c) System/370 model 145 CPU, (d) 3330 Direct access storage device, (e) 3505 Card reader, (f) 3420 Tape drive, (g) 3211 Printer.
(Photo Courtesy of International Business Machines Corporation.)

of personal data about private citizens and feel that this is a violation of their right to privacy. This is a special problem because of the prevalence of electronic spying. Another fear is that people will become so dependent on complex control systems that a small human error could produce chaos.

TWO POINTS OF VIEW

Converting to a Computer System

Gregory Snare has been head bookkeeper for 30 years at Wing Fanbelts Ltd. He feels that the company does not need a computer system. He is quoted as saying, "We can't expect to keep our customers or employees happy if they feel they are dealing with a computer. They want the feeling that human beings are taking care of their accounts and their payrolls. Besides, half of my clerical staff would quit. They feel it would only be a matter of weeks before they would be fired!"

Dave Delaney, the company's new vice-president, disagrees strongly. "We have grown 300 per cent in the last five years. The volume of business demands computerization of our accounting. Bills are often late and we've had several payroll errors recently. If we convert to a computer system, all of these problems will disappear. The people in Snare's department have no reason to fear for their jobs. We'll retrain them to work with the computer. The XYZ Computer Company will do all of the training and installation. Our business will grow much faster and we'll have a more modern image!"

Who has the more realistic point of view? Could there be still another viewpoint?

Sometimes people (even managers) feel that a computer will solve all their problems like magic. This can cause as many problems for a firm as fear of the computer. The truth is that managers must plan very carefully. They must get accurate data and a "debugged" program (one in which all the problems have been worked out) before they can count on using a computer's output. Someone invented the term GIGO (garbage in–garbage out) to describe how much the computer depends on reliable human input. A chimpanzee is unlikely to be able to count his toes with a computer's help.

Management Effects

As far as the management structure is concerned, it is hard to predict the long-range effects of computer installations. Some predict there will be less need for staff executives and that line executives, aided by computers, will assume staff functions. Depart-

---

CAREERS WITH COMPUTERS

Computer-related jobs cover a wide range of skills and educational requirements. There are many people working as operating personnel, programmers, and systems analysts. Operating personnel such as keypunch operators, data typists, console operators, and high-speed printer operators do the job of processing computer data. The number of keypunch jobs will probably not grow much because of new ways of getting data into computers, but console and auxiliary equipment operators' positions are expected to grow rapidly. Such workers usually need a high school education, and some experience.

Programmers are in great demand. College degrees are not usually required, except in scientific applications of computers. College-level courses in data processing and accounting are helpful to prepare for the job. Many firms train their own employees for promotion to programming jobs.

Systems analysts are people who recommend the kinds of data processing equipment which should be used. They prepare instructions for programmers, too, and help business managers learn how to use data processing systems efficiently. Some programmers are promoted to jobs as systems analysts. Some systems analysts are recruited from the business, accounting, and science departments of colleges and the computer education programs at community colleges. The demand for systems analysts is expected to grow very rapidly in the years to come. For additional information write to the Data Processing Management Association of Toronto, Box 116, Station F, Toronto, Ontario M4Y 2L4, or the Canadian Association of Data Processing Service Organizations, 299 Waverly Street, Ottawa, Ontario, Canada K2P 0W1.

SECTION THREE BUSINESS DECISIONS

ments assuming major responsibility for computer operations (often the accounting department) may gain power as the computer becomes more important to the firm.

There are widely differing views of what effect computers are having or will have on "middle management". Some middle managers say that computers have increased their workload except for the controlling function. Others predict that computers will eventually make it possible to eliminate most middle management positions. Suppose, for example, that the Nutt Corporation has several small departments. The department heads ordinarily make decisions only about the levels of inputs and outputs in their departments. The firm might, with the help of a new computer, develop programs to make such decisions routinely. Whatever else these middle managers did might be assigned to line executives either above or below them. How widely such substitutions of computers for managers could occur is hard to predict at this time.

While computers have grown to be a vital part of business activity, a parallel growth has occurred in the use of mathematics and statistical tools. The fact that these two things have grown at the same time is not a coincidence. Many of the mathematical and statistical tools depend on computers for their practical application. **We refer to this whole set of mathematical and statistical applications to business as quantitative tools.**

Quantitative tools

SOME QUANTITATIVE TOOLS FOR MANAGEMENT DECISION MAKING

The use of quantitative tools by managers is increasing. Whereas managers in the past often relied only on their own judgment, modern managers strengthen their judgment by collecting and organizing data to support it. We will examine some basic statistical concepts and some examples of quantitative tools used today.

Statistics for Business

Statistics

Managers have dealt with numerical data in their decision making for many years. These numerical data and methods of summarizing them are called statistics. Data may represent internal facts, such as number of units sold, or external facts, such as the population of the provinces in which a firm does business.

It is often helpful to summarize numerical data by using special kinds of averages. For example, we may wish to refer to average family income in Canada or to the average number of years of school completed. An average is a summary figure which describes the facts we are studying. There are three principal types of averages: (1) the arithmetic mean, (2) the median, and (3) the mode.

Arithmetic mean

The arithmetic mean is an average computed by first adding numbers, finding a total, and then dividing that total by the number of numbers which were added together. Look at Table 14-1. It is a list of the ages of seven employees in the receiving department of a factory. It shows that the sum of their ages is 210 years. We can compute the arithmetic mean of their ages—their average age—by simply dividing 210

Table 14-1. Ages of employees in the receiving department

Employee	Age
Harold	30
Janet	28
Gordon	38
Clyde	27
Susan	27
Richard	34
Thomas	26
Total	210 years

years by 7. The answer, of course, is 30 years. This is the most common form of average.

Another measure of an average is called a median. It means the middle number when numbers are listed in rank or array—from smallest to largest or vice-versa. To find the median of the ages of the employees in Table 14-1, we first rank ages in an array, starting with the youngest. The list becomes 26, 27, 27, 28, 30, 34, 38. The middle number is 28 years (Janet's age). There are three people older than Janet and three who are younger.

To find a median of an even number of numbers, we still rank the numbers and then take an arithmetic mean of the two "middle" numbers only. If we added Paul, aged 29, to our list the median would be 28.5, which is the arithmetic mean of the two middle numbers—28 and 29.

A third average is called a mode. It is the most common or frequent number in a list. In our example, only two people are the same age. Clyde and Susan are both 27, so this is the mode. Their age, 27, is the modal age.

These types of "averages" are different ways of making a summary measurement of a characteristic of a group. Which one is best depends on the use to which the measurement is put and how the "raw" data are distributed.

Suppose we collected statistics on family incomes in your city and organized them into five groups. The statistics might appear in grouped form as shown in Table 14-2.

Median

Mode

Table 14-2. Frequency distribution of family incomes

Annual family income	Number of families
$ 0– 2,999.99	420
3,000– 5,999.99	652
6,000– 8,999.99	809
9,000–11,999.99	602
12,000 and over	248
Total	2,731

Frequency distribution

A frequency distribution shows how many things fall within different ranges or intervals of values. Table 14-2 is a frequency distribution. It shows the number of families in each income category. In this case the "things" are families and the ranges of values are income brackets: 420 families had less than $3,000 income. We determine the income distribution of families without actually listing the income of each. This frequency distribution tells us more about income than any simple "average" could. Distributions and averages are examples of descriptive statistics. Managers often use them in preparing reports which describe their business or their markets in some way.

Sample

There is also wide use of the statistics of sampling. A sample is a part of a larger whole called a universe, or population. Political analysts, for example, base their projections of winners in elections by studying a relatively small number of voters (the sample). The time, costs, and effort involved in interviewing every voter (the universe) would be too great. By interviewing a sample of voters, a pollster can make a good estimate of the election results.

Businesses also use sampling. Suppose a manufacturer of light bulbs wishes to guarantee that its bulbs will last a certain number of hours. The company might find, based on a study of a sample, that the average bulb life is 200 hours. It would be unrealistic to base the guarantee on a study of all the light bulbs it produces. Its entire inventory would have to be "burned out". So it tests a sample of these bulbs.

A TV program sponsor uses a "rating" to decide whether to keep a particular program. A TV rating firm cannot check all viewers in the country. It contacts a sample of them and then estimates the national audience from the sample.

A sample is a part of a larger group and represents that group. Some samples are drawn in a way that you can estimate certain things about the larger group with a given degree of confidence. Other samples are not drawn according to strict mathematical rules but still try to approximate the characteristics of the larger groups. They don't provide a measurable degree of confidence in their accuracy, but they are cheaper to get.

Break-even Analysis

Break-even analysis

A useful management tool in both production planning and pricing of products is break-even analysis. Break-even analysis demonstrates the profitability of various levels of production. The break-even point shows at which level total costs are exactly equal to total revenue. As you can see in Figure 14-6, the number of units produced is measured on the horizontal scale, and dollar costs and revenues are measured on the vertical scale. The Sales Revenue line starts at the zero point (lower left corner). Since the product sells for $100, this line moves up $1,000 each time it moves to the right by 10 units. If we make and sell 20 units, we get $2,000 in revenue.

Costs are of two kinds: fixed and variable. Fixed costs occur whether we produce zero or 20 or 1,000 units. These costs are often called "overhead" costs. They include depreciation on plant, insurance, and other costs which do not vary with the level of production. In this case fixed costs are $1,000.

Variable costs depend on the number of units produced and sold. These might

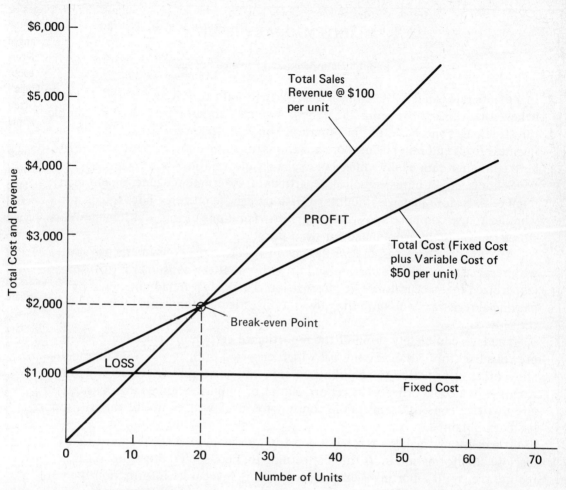

Figure 14-6. A break-even chart

include raw materials, labor, and other costs which go into each unit produced. For each unit we produce and sell, it costs us $50 more—in addition to fixed costs. The effect of variable cost is represented by another line sloping up from the $1,000 mark on the left vertical axis. This line, since it starts from the $1,000 fixed-cost level, also measures total cost at various levels of production. If 20 units are made, the total cost is $2,000, consisting of $1,000 in fixed costs and $1,000 (20 × $50) in variable costs. If 25 are made, the total cost is $2,250.

At the level of 20 units of production, the cost and revenue lines cross. This is the break-even point—the point in production level beyond which the firm begins to make a profit. For each additional unit made and sold the firm realizes an increase of $50 in profits. This is so because unit revenue minus unit variable cost ($100 − $50) equals $50.

A break-even chart can help a plant manager decide several things. It can help a

A CONTEMPORARY ISSUE

Computer Data Banks and Personal Privacy

In a democratic society the citizens cherish their right to privacy. Many people believe that this right is being threatened. A single computer system, for example, can store pages and pages of "information" on each and every person in Canada. Business firms and non-business organizations, including government agencies, have computer data banks which can exchange information with each other.

Schools, credit bureaus, police departments, government agencies, and many retail outlets store massive amounts of data on private citizens. The issue is not whether to use computers to store data, but what kinds of information should be stored on whom, by whom, and for what purposes.

All provinces have passed laws designed to protect the consumer's privacy with respect to credit information and to ensure that the consumer is not jeopardized by inaccurate credit information. These laws are administered by provincial registrars who have the power to investigate complaints by consumers.

The laws completely prohibit the reporting of certain information (such as previous bankruptcies of the subject which occurred prior to a certain date) and allow other information to be published only under certain conditions (e.g., in response to a court order). The report cannot be published unless the consumer has given written consent or unless the consumer is given written notice that the report has been obtained.

If a consumer makes a written request to an agency, the agency must then disclose all information in its files concerning that consumer. Most provinces require the credit reporting agency to correct any inaccurate information the consumer finds in the file.

---

manager decide whether to install expensive new machines which would change the production cost structure. It can help in setting prices or help to decide whether to buy or to lease a plant. Retailers could also use a break-even chart to make similar decisions.

Operations Research

In recent years many new mathematical tools have been developed. Many were originally developed to analyze and solve military problems. The army's problems in assigning personnel are a lot like the problems found in business. Assigning aircraft is similar to a business assigning salespeople or distributing inventory among warehouse locations.

Determining how many units of a product to keep in its warehouses is a tough problem for a company selling hundreds of products around the world. Mathematical means are available to help with such a problem. Others are suited for scheduling operations and for making other common decisions. These various quantitative techniques as a group are called operations research (OR). Sometimes the term "management science" is used to mean the same thing.

Operations research (OR)

Computers enable managers to apply these techniques. Otherwise, the mathematical calculations might take years to perform. Several examples of the use of operations research in business follow.

Model Building

Sometimes the best way to understand and solve a problem is to create a simplified model of it. An organization chart, for example, is a model of the organizational relationships in a firm. It is a very simple model. It does not show all the formal and informal relationships which exist.

An architect builds a model of a bridge to evaluate certain structural ideas. It is subjected to tests of strength. If the results are favourable, the architect feels more confident in the design. The same is true in business. Managers, however, cannot usually build that kind of model themselves. They may, however, call on a specialist to build a mathematical model or simulation of a problem situation. **Such model building represents the way in which several important factors (variables) work together to explain how other factors such as profits or sales vary.**

Model building

The equation $S = bY + cV - dR$ could be a simple model explaining variation in sales of Blanche's restaurant. (See Figure 14-7.) S stands for total annual sales of the firm. Y stands for total projected income of Edmonton, Alberta, where Blanche's is located. V means number of expected visitors to Edmonton, and R is the number of rainy days expected in the year. The letters b, c, and d are called co-efficients. They are values to be multiplied times the values of Y, V, and R, respectively. It would then be possible to forecast sales with some degree of assurance, if we could plug in good estimates of Y, V, and R. The co-efficients must be estimated from past testing of this

Expected sales of Blanche's restaurant = b (Next year's income of Edmonton, Alberta) + c (Predicted number of visitors to the city) − d (Predicted number of rainy days)

$$S = bY + cV - dR$$

Figure 14-7.

model. A model can help in other things besides forecasting. Making a model can help the manager of Blanche's understand better the operations of the firm and the market.

A model is tested by seeing how it "fits" the real facts (variables) in the past. If it seems to work, it can be used to predict what might happen if the firm takes a certain action. A marketing manager who wants to introduce a new product might introduce this as a new variable in the model to see what this does to the other variables, including profit levels. There are, of course, no guarantees of accuracy.

By using model building, the decision maker can see what will happen if a particular course of action is chosen. The likely outcome of a particular decision can be predicted by plugging it into a model of the organization—without the "real" decision being made. Without a model, this could be determined only by actually making the decision.

Linear Programming

Linear programming

Managers must make informed decisions about how resources will be allocated. **Linear programming is a mathematical tool used to allocate resources in the "best way" so that a desired objective is maximized or minimized.** This desired objective may be the greatest profit, the least cost, or another "best" result for the firm.

Linear programming is widely used in problems such as determining the best (optimum) inventory level or plant location. A manager who must select a location for a new plant would find this technique helpful. It could allow him or her to minimize the total cost of getting raw materials into the plant, getting finished goods to warehouses, and moving them to customers. Actually, these are only a few of the many variables which influence plant location decisions. A linear programming technique considers all variables in order to arrive at a "best" location. It can be applied in a situation when there are limited resources (time and money) and a value (profit or cost) which is to be maximized or minimized. The limitations are called constraints. An example is given in Figure 14-8.

SUMMARY AND LOOK AHEAD

How the special talents of computers fit in with unique human talents is gradually being learned as people and the computer work together solving business problems. Computer hardware and software together comprise a computer system which can serve a firm in many ways.

We have now reviewed how computers work in information storage and retrieval and in data manipulation and organization. Computers, together with people who know techniques of operations research, can greatly improve management ability. We have described the tools of statistics, break-even analysis, and several types of operations research tools. They fit in well with the use of modern computers.

In the next chapter we focus on the special problems related to running a small business. We will look at the risks and the opportunities, the advantages, and the pitfalls of running your own business. We'll also see the role played by federal and provincial agencies set up to aid small business.

KEY CONCEPTS

Arithmetic mean A kind of "average" computed by adding a group of values and dividing by the number of values.

Automation An activity or process is said to be automated when it is possible to set its controls in advance and permit it to function normally for extended periods of time without human interference.

Break-even analysis A technique for estimating the relationships among volume of operation, costs, and revenue. Costs are divided into fixed and variable parts. When revenues cover total costs, the firm breaks even.

Central processing unit (CPU) The device which performs the actual calculations and logic in a computer system.

COBOL A specialized language which is extremely close to English and is used for business data processing.

Computer An electronic data processing system which stores great amounts of data and manipulates data quickly and accurately.

The Purple Advertising Agency has $1,000 to spend and 10 working days to complete the job for the Creeky Hinge Company's ad campaign. They want to produce the largest amount of sales possible. Past experience shows that the average newspaper ad produces $3,000 in sales; the average radio ad, $4,000.

It costs $33.33 each to run newspaper ads and $100 each to run radio ads. It takes a day to prepare a newspaper ad and half a day for a radio ad. Purple can only work on one ad at a time. The question is how many newspaper ads to run and how many radio ads? The problem must be solved in terms of producing maximum sales.

In mathematical terms, the agency must maximize the expression $S = 4,000R + 3,000N$, where R equals the number of radio ads and N equals the number of newspaper ads. S equals expected sales.

The constraints or limits to the values of N and R in the solution are in the form of linear inequalities:

$$33.33N + 100R \leq 1,000$$
$$N + .5R \leq 10$$

The upper inequality says that the $1,000 must be spent at the rate of $33.33 per newspaper ad and $100 per radio ad. The lower inequality says the 10 days must be used at a rate of one day per newspaper ad and half a day per radio ad.

Solving these as you would a set of simultaneous equations gives values of $N = 6$ and $R = 8$. This combination will give maximum sales revenue in the amount of $50,000.

Figure 14-8. **Linear programming and the advertising budget**

Computer program A detailed list of instructions to a computer in a special computer language.

Data processing The accumulation, storage, sorting, interpretation, and reporting of facts, mostly in number form.

Documentation Explanation or illustration of a computer program.

FORTRAN Short for "Formula Translator". FORTRAN is a science-oriented computer language.

Frequency distribution A means of describing a large group of values. Intervals or ranges of values are established first. Then the number of values falling within each interval is computed and listed next to the interval.

Hardware The electronic and mechanical components of a data processing system.

Input-Output device Any piece of equipment which allows information to be fed into a computer or permits the computer to make information available to its user (e.g., a typewriter).

Linear programming A quantitative technique used to allocate resources in the best way so that some desired objective is maximized or minimized. Desired objectives might be greatest profit or least cost.

Median The middle value of those ranked in order of size.

Mode The most frequent value in an array.

Model building A model is a simplified version of a set of relationships. By simulating the real-world operations of the firm through model building, the decision maker can see what will happen if he or she chooses a particular course of action.

On-line system A computer system which permits direct and continuous access by employees to stored information.

Operations Research (OR) Various complex quantitative techniques used by managers to assist them in decision making.

Outside memories Means of storing data other than in the CPU itself.

INCIDENT:

Wanda Viller Ltd.

The cost accountant at Wanda Viller Ltd. wants to demonstrate the effect of growing production volume on profits. Total overhead cost amounts to $100,000. Each additional unit of production costs $10. The product, a stainless steel bun warmer, sells for $20 each.

Questions:
1. Could the accountant make use of a break-even chart? Explain.
2. Draw such a chart and estimate profits if Wanda Viller Ltd. makes and sells 12,000 units.

Punched card A paper card which is used in computers to store and transmit information by means of holes which have been perforated in a special arrangement through the card.

Quantitative tools Methods of using mathematics and statistics to help solve problems and understand facts.

Sample A part of a larger group which represents the larger group.

Software The programs, languages, and routines used in electronic data processing. Software complements hardware.

Statistics Can refer to data themselves or the science of manipulation, interpretation, and summarization of the data.

Time sharing A method whereby a number of users may utilize the facilities of a single computer at one time.

QUESTIONS FOR DISCUSSION

1. Besides the trips to the moon, what dramatic modern accomplishments do you think never would have happened without computers?
2. Review the experiences you had yesterday and try to determine which of these were in some way influenced by the existence of computers.

INCIDENT:

Kelley Enterprises

Jeremy Baltz was hired to help install and run the new computer installation at Kelley Enterprises. He faced a genuine challenge. The small toy manufacturing firm had been using strictly manual accounting, record-keeping, and planning systems. The manufacturing plants were located in Kingston and London, Ontario. The computer, however, was located at the administrative and sales headquarters in Kitchener, Ontario.

Many of the plant supervisors and office staff were opposed to the idea of the computer. They feared their jobs might be lost. Others feared loss of communication between the plants and headquarters.

Questions:
1. What do you think led Kelley Enterprises to install the system in the first place?
2. What kinds of functions might the computer system perform?
3. Describe some of the most serious problems which the firm might anticipate in connection with the installation and give some suggestions as to how these problems might be minimized.

3. Do you think that computers will someday dominate us? Why or why not?
4. Name the three major hardware components of a computer. How are they interrelated?
5. What is a computer language? Name two common ones.
6. What kinds of business processes make good use of the rapid repetitive capabilities of a computer?
7. What is a computer program?
8. Check the business magazines in your library to find a report of an operations research application of a computer in a real situation. Prepare a brief summary of the report.
9. Can you "invent" a business problem which linear programming could solve? What are you maximizing or minimizing?
10. Fixed costs are $100 and variable costs are $1 per unit. How many units must be sold at $2 to break even?

Section Four

Special Types of Business

You can lose your perspective by thinking of business firms only as national corporations. Firms come in all sizes and dimensions. For many firms there is a need to "think small" or a need to "think international".

We will adjust our sight in Chapter 15 to take a look at businesses which are quite small and which have many problems peculiar to small business.

In Chapter 16 we will look at businesses which have an international perspective. There we examine the special challenges and opportunities involved in conducting business across national borders.

15
Small Business

15 SMALL BUSINESS

OBJECTIVES: After reading this chapter, you should be able to:

1. Explain, in your own words, the meaning of a "small-business firm".
2. List and discuss three ways a person might become a small-business owner.
3. Give two or more reasons why small firms can compete effectively with larger competitors.
4. Develop a "self-test" to find out whether you have the basic requirements to be a small-business owner.
5. Compare the benefits and burdens of entrepreneurship.
6. List and discuss the first steps in starting your own business.
7. Explain, in your own words, the nature of a franchise operation.
8. List and discuss the benefits of franchising to the two parties to a franchising agreement.
9. Develop criteria by which a person can assess the potential franchising offers as a method of doing business.
10. Compare the views of critics and supporters of government aid to small business.
11. Describe the kinds of government assistance which are available to small business in Canada.
12. Identify the challenges to survival faced by small firms.
13. Explain, in your own words, why a small firm's owner may want to seek growth of his or her firm.
14. Identify two or more ways by which a small firm might grow in size.

KEY CONCEPTS: In reading the chapter, look for and understand these terms:

ENTREPRENEUR
BETTER BUSINESS BUREAU (BBB)
CHAMBER OF COMMERCE
ECONOMIC DEVELOPMENT COUNCIL
FRANCHISER
FRANCHISEE
FRANCHISING AGREEMENT
SMALL BUSINESS LOANS ACT
FEDERAL BUSINESS DEVELOPMENT BANK (FBDB)
COUNSELLING ASSISTANCE TO SMALL ENTERPRISE (CASE)

Although our discussion on production, marketing, finance, and so on in earlier chapters applies to all firms regardless of size, there are several aspects of small business which merit separate treatment. If you have ever considered going into business for yourself or if you already have done so, you will be especially interested in this chapter.

Most people who go into business for themselves start out in small businesses. A small business is a firm which is small in relation to its competitors. "Smallness" relates to such things as number of employees, dollar sales volume, funds invested in the firm, and so on. As we will see, it's hard to provide a definition of "small business" which would be meaningful and satisfactory in all cases.

We also discuss how and why a person might start a small business. Then we look at the pros and cons of going into business for yourself. After that, we take a close look at franchising, a way of doing business which has enabled many people to go into business for themselves. Finally, we examine the various ways that provincial and federal government agencies have become involved with small business.

In the end, a small-business owner's success (or lack of it) is the result of good (or poor) management. As you will see, it's still possible to make "big money" by going into business for yourself. On the other hand, it's just as possible to lose everything you have invested.

WHAT IS A SMALL BUSINESS?

In Chapter 3 we saw that there are thousands of sole proprietorships, partnerships, and corporations in Canada. Many of these firms are small businesses, but "small" is a relative term. For example, you might consider Home Oil to be a "small" company in comparison with Imperial Oil, but compared with many other business firms, Home Oil is very large.

At present, the federal government defines a small business as one where gross revenue does not exceed $1.5 million for any fiscal year. However, there is no single definition of small business which is entirely satisfactory. Let's say, then, that a small business is one which can be started with a relatively modest investment of funds by the owner or owners. Of course, what is "relatively modest" is just that—it is relative to the industry in which the firm exists. Some industries, however, are made up almost entirely of small firms. Examples are barber shops, beauty parlours, sandwich shops, and so on. The concept of relative size in its industry has little real meaning in these cases.

BECOMING A SMALL-BUSINESS OWNER

Entrepreneur

A small-business owner is an entrepreneur. **An entrepreneur is a person who assumes risk by organizing and managing a business in the hope of making a profit.** A person becomes a small-business owner in one of three ways: (1) by taking over the family's business; (2) by buying out an existing firm; or (3) by starting a new firm. Each way has its own set of problems and opportunities.

Taking Over the Family Business

Not too long ago, it was common for a father to train his son to take over the family's business someday. Usually, these firms were very small and employed only family members. The oldest son was expected to keep the business going after the father stepped down. This meant the father made the choice of occupation for his son. A grocer's son would become a grocer. A barber's son would become a barber.

During the twentieth century, this method of becoming one's own boss lost much of its appeal. Apprenticeship under the father generally has been replaced by formalized education.

Nevertheless, every year many firms are taken over by relatives of the former owners. In many cases the person taking over a firm is the spouse or one of the children. But this often is not planned in advance and makes it very hard for the person taking over, especially when the former owner is not there to help the new owner.

Buying Out an Existing Firm

Many people go into business by buying out an existing firm. In many cases an agreement can be reached whereby the seller helps the new owner to learn the business from the ground up. It is good practise to have a written contract outlining the duties of the former owner after the new owner takes over.

Sometimes, buying out an existing firm means buying it from a surviving spouse. This presents a different problem, since the owner is not there to help the new owner to get oriented to the firm. In many of these cases the surviving spouse has neither the desire nor the ability to continue the firm. This fact points out the need for a plan to continue the firm after the owner's death. However, many small-business owners do not ever prepare such a plan.

A person might go into business with the intention of selling out after the firm becomes a going concern. Some companies are formed to buy out small firms which have good growth potential. Often, these firms give financial and managerial help to promising new firms and offer to buy them out. In some cases they "go public". This means that they become corporations and sell shares of stock to the public. Many investors want to buy stock in new ventures.

Starting a New Firm

In the previous examples the business owner takes over an existing business. It has an established customer base and is a going concern. A person who starts a new firm, however, must build a going concern.

A new firm is not troubled with many of the problems which accompany the takeover of a firm which has been in business for some time. There are no dissatisfied customers, no fixed plant or store location and layout, and no bad debts for a new firm. The owner has the opportunity to build the firm from the ground up.

> **CAN YOU EXPLAIN THIS?**
>
> **Profit Expectations and Being Your Own Boss**
>
> Relatively few small new firms are successful enough to make a profit in their early years to match their owners' expectations. This, of course, often proves to be a big disappointment to these entrepreneurs.
>
> Some entrepreneurs react by lowering their expectations. Often, this takes away their motivation to excel. Some others, however, react by trying harder to "make it". There are still others who "throw in the towel" and admit failure. What accounts for these different reactions? CAN YOU EXPLAIN THIS?

Many young people today are turned off by the thought of working for somebody else. They want personal and direct involvement which can be best achieved when they are their own bosses. Going into business may reflect a search for identify. In cities where universities are located there are small firms which were started by students who recognized profit opportunities. Examples are small clothing stores which cater to young adults, swimming pool maintenance firms, and home or apartment maintenance services.

Some universities now offer courses in entrepreneurship or in starting a new business firm. While some students don't have the money to start a business right after graduation, many want to do it after they gain the necessary funds and experience.

OPPORTUNITIES IN SMALL BUSINESS

There are countless opportunities for small firms to serve ultimate consumers and industrial users. Opportunity exists in manufacturing, agriculture, retailing, wholesaling, and even in the extractive industries. Each year many new firms are opened to exploit opportunity in those fields. Many start out as small firms.

The big question is: "How does a would-be entrepreneur learn how to spot opportunity?" Is it partly a matter of being sensitive to the environment. One of your authors recalls a success story of a former student. This was just before the widespread popularity of pre-washed denim jeans. As a part-time salesclerk in a "jeans shop", the student recognized a growing preference for pre-washed jeans among young people. He wanted to try "something on his own", so he bought 100 pairs of jeans from his employer. He had them washed by a commercial laundry so that they "looked old". He gave ten pairs free to college students to wear. They were asked to take orders from people who wanted the "new look" in jeans. The cost of the jeans which were given away, the laundering costs for 100 pairs, and the commissions he paid to his "sales force" amounted to $845. His revenue on the whole deal was $1,500, leaving a profit of $655. He now owns a very successful clothing store which caters to young men and women.

We could discuss many such "success" stories. What they teach is that opportunity always exists but hardly ever "knocks at your door". In a competitive economy you can be sure that it never knocks twice! Opportunity does not come to you. You must discover it!

Some people say that opportunity has all but dried up for small businesses. They argue that only large corporations can afford to hire the talent needed to spot trends and to capitalize on them. But "bigger" does not necessarily mean "better" as far as business is concerned. A large discount chain store can't give its customers the personal attention a small clothing store can give. A big manufacturer with many products can't give any one of them the attention a small firm can give to its one or two products. A small firm is usually more adaptable than a large firm. It often can react to change a lot faster.

SHOULD YOU GO INTO BUSINESS FOR YOURSELF?

Let's assume for a minute that you want to become your own boss. You have a rich uncle who will lend you the money to get started. If that's not enough to tempt you, you also can assume that you have a product or service which will definitely lead to a good profit if you can succeed as a manager. Should you go into business for yourself?

Given these assumptions (and they may be very unrealistic), you still must evaluate a third vital input—you, the entrepreneur. Before you make your decision, you must consider your goals in life. Then you must determine whether you have what it takes to reach those goals.

The Basic Requirements

Ask yourself the following questions to see whether you have the basic requirements for starting your own business:

1. Are you afraid of risk?
2. Are you unable to put off enjoying the "good life" today because you are afraid you won't be here tomorrow?
3. Are you overly security conscious?
4. Do you have trouble getting along with people?
5. Do you lose interest in things which don't work out as quickly or as well as you thought they would?
6. Are you a thinker and not a "doer"?
7. Are you a "doer" and not a thinker?
8. Are you easily frustrated?
9. Do you have trouble coping in situations which require quick judgments?
10. Do you "cave in" under stress?
11. Does your family make unreasonable demands on your time?

12. Are you emotionally unstable?
13. Are you unable to learn from your mistakes?
14. Are you "too good" to do manual labour?

If you answered "yes" to several of these questions, you probably are not ready to start your own business. In any case, this sort of thinking is very important in deciding whether you measure up to the job of being your own boss. If you do "measure up", then you can begin to weigh the benefits and the burdens of starting your own firm.

The Benefits of Entrepreneurship

Perhaps the best thing about being your own boss is the sense of independence you feel. You get a great deal of personal satisfaction from being directly involved in guiding your firm's growth. It is also possible to make a sizeable personal fortune. You not only draw a salary, but you also own the firm, the value of which may increase many-fold over the years.

Owning your business also is good for your ego. You are respected by others because you are not a "cog in a wheel". You are "the wheel"! For many people, achieving their true potential means the same thing as becoming their own boss. Clearly, it gives you personal, economic, and social benefits.

The Burdens of Entrepreneurship

Being in business for yourself, however, requires your full attention. You usually will not leave the office or shop at 5:00 P.M. Nor do you leave "job problems" at the office or shop. They follow you home as "business problems" and "business homework". This means you have less time for your family.

A person with very limited abilities may be able to hold a job in a large company by doing just enough to get by. Maybe others will cover up for his or her shortcomings. There is, however, no one to "carry" you when you are in business for yourself.

While you may not have to report to a boss, you often have to bend over backwards for your customers. You have to contend with creditors, employees, suppliers, government, and others. In short, you are never completely independent.

When you work for somebody else, you may be paid a salary based on a normal work week of 40 hours. If you work 50 hours, your pay will be greater (if you get paid on the basis of the number of hours worked). If you are a manager, your salary is based on a "normal work week", even if it is usually more than 40 hours. As your own boss, however, you might work for many months and not be able to take a penny out of the business. Your profits may have to be reinvested to meet short-term demands for cash or for long-term growth. Thus, you may not be able to draw a salary during the period it takes to get the firm to become a truly "going concern". Many small business owners fail to anticipate this when they start out. It is a major cause of new business failures.

STARTING A SMALL BUSINESS—THE PRELIMINARIES

It's a sad fact that many people who want to start their own firms do not do so because they just don't know "how to do it". Some people think that there is so much red tape involved that you need a team of lawyers to start a firm. At the other extreme, some people "open shop" without even checking to see if they need permits or licences. Both views are wrong.

Starting a business always requires careful planning of preliminary details. A firm does not come into being by itself. It takes planning. In the following discussion, we will outline some of the steps in getting started. It is very general because the laws of different cities and provinces differ on how to start a business.

Licences and Permits

Before you start operations, you must have the required licences and permits. Which ones you will need depends on provincial and city laws. In many cities you will need a certificate of occupancy which certifies that your type of business is permitted in your location. For example, you could not operate a pet shop in a location zoned exclusively for single-family residences, nor could you set up a steel foundry in a location zoned "light commercial".

In most provinces, you must get an occupational licence to engage in business or professional activity. These licences are usually available from the city in which the business is carried on. The cost depends on the type of business activity which is engaged in.

If your business deals with food, you will probably need a local food permit. These are usually issued by city health departments. If you plan to sell liquor or tobacco, you will also need a special permit. Requirements in this area vary considerably across provinces. If you wish to go into business for yourself selling door to door, you will have to have a local vendor's permit.

The best advice on licences and permits is to check with the local chamber of commerce or the provincial or city government. You may need a greater number of licences than you might expect!

Sales Taxes

Most provinces collect retail sales taxes. In these provinces you must register with the department of finance in order to comply with the law. This permits you to collect the tax for the province.

Employer Taxes

If you hire employees, you must withhold federal and provincial income taxes as well as social insurance taxes from your employees' wages. You must also pay your own share of their social insurance taxes. You'll also pay taxes for unemployment insurance (1.4 times the employee's contribution).

Workmen's Compensation Insurance

If you hire employees, you must carry workmen's compensation insurance. This covers employees who are injured or killed on the job. The premium is a percentage of your estimated payroll.

Information Sources

In starting your new firm, you may need more information on many aspects of your business. Of course, if you are in a large city, you will have access to more sources than if you are in a rural area or a small town. The following discussion covers some of the basic sources.

Better Business Bureau (BBB)

Many cities have a Better Business Bureau (BBB). **A BBB is a non-profit organization of business firms which join together to help protect consumers and businesses from unfair business practises. Businesses "police" themselves through the workings of the BBB.** Suppose you have doubts about buying from a particular supplier. You can call the local BBB to ask if any complaints have been filed against that supplier. If you have trouble with a supplier, you can file a complaint against that firm. It's a good idea to join the BBB.

Chamber of Commerce

Many cities also have a chamber of commerce. **The Chamber of Commerce is a national organization of local chambers of commerce in cities and provinces throughout Canada. Its purpose is to improve and protect the free-enterprise system.** A local chamber of commerce is a useful source of data on business conditions in that area. Active involvement in the local chamber can put you in contact with potential customers and suppliers.

Economic development council

Some cities have an economic development council. **An economic development council is an organization of business firms and local government officials. It seeks to further the economic development of the area in which it is located.** It is a good source of data on the local economy. In many areas local governments co-operate with local action groups to aid small businesses. A check of local libraries may help you to find other information you might need in starting your business.

Business Insurance

Before you start operations, you should have business insurance. Your selection of an insurance agent is an important step. The agent can give advice on the types of coverage you'll need in your line of business. But be sure to shop around for the best combination of price and coverage.

FRANCHISING

Franchising is not a form of ownership like a sole proprietorship or a partnership. It is a method of doing business. During the 1960s, franchising became a very popular way of going into business for many people. Familiar examples are Holiday Inn, Pizza Hut, McDonald's, Midas Muffler, Radio Shack, and Canadian Tire.

There are two parties in franchising. Each has certain duties or obligations and each receives certain benefits. **The franchiser is the firm which licences other firms to sell its products or services.** The McDonald's Corporation is a franchiser.

Franchiser

The party that is licensed by the franchiser is called the franchisee. **A franchisee has an exclusive right to sell the franchiser's product or service in his or her specified territory. Each franchisee is an independent business owner.** The McDonald's Corporation licences independent franchisees to make and sell McDonald's hamburgers. Each franchisee pays an initial fee and yearly payments to the McDonald's Corporation for the right to use the McDonald's trade name and to receive financial and managerial assistance from the franchiser.

Franchisee

The franchiser and the franchisees are related to each other through the franchising agreement. **A franchising agreement is a contract between a franchiser and its franchisees which spells out the rights and obligations of each party.**

Franchising agreement

Franchising and the Franchisee

Franchising can be profitable to both the franchiser and the franchisee. Let's discuss it from the franchisee's point of view.

THINK ABOUT IT!

Cultivating Outside Expertise

Starting your own business takes a lot of planning. Running your business also takes a lot of know-how. Unlike most larger firms, however, you will not have staffs of experts in your firm to turn to on tax problems, insurance problems, legal problems, and financial problems. In fact, lack of expertise in these areas is one of the major reasons why some entrepreneurs fail.

Remember, you are not out "to do it all without any outside help". Often, whether you can get that "outside help" will be the main factor in your success as an entrepreneur. At the minimum, you should know an accountant, a banker, an insurance agent, and a lawyer to whom you can turn for advice in setting up your firm and guiding it through its early life. Take the time to cultivate this type of assistance from "outsiders". It can make the difference between success and failure. THINK ABOUT IT!

Recognition

A person who wants to go into business in the fast-growing services industry (such as the fast-foods industry) will run into many obstacles. Perhaps the most important one is becoming known. A new Joe's Hamburger Joint just doesn't have the instant recognition that Joe Jones would get as a McDonald's franchisee.

Standardized Appearance and Operation

A franchiser enjoys widespread consumer recognition because the units are all basically alike. A Midas Muffler shop in Winnipeg is very similar in appearance and operation to one in Toronto. The franchiser usually provides the franchisee with a blueprint for constructing a building which will be just like all other franchised outlets.

The franchiser also insists on standardized operation of all outlets. These are spelled out in the franchiser's operations manual and are backed up with standardized forms and control procedures so that all outlets look and operate alike. This is important in a society where people are highly mobile. A newcomer to a town feels a lot safer about buying a McDonald's hamburger than he or she might feel about eating one at Joe's Hamburger Joint. A traveler is more relaxed about staying at a Holiday Inn than he or she might be about staying at the Three Pines Motel.

Management Training and Assistance

A major reason why many small firms fail is the owner's lack of management skills. Many franchisers operate training schools where franchisees learn business skills like record keeping, buying, selling, and how to build good customer relations. Ongoing training is also important. Many franchisers send representatives to give their franchisees advice and assistance. Franchisees with special problems can turn to the franchiser for help. Thus, franchisees are not left entirely on their own in managing their businesses.

Economies in Buying

A franchiser either makes or buys ingredients, supplies, parts, and so on in large volume. These are resold to franchisees at lower prices than they would pay if each of them made or bought them on their own. In the past some franchisers required franchisees to buy their supplies from them. Even if a franchisee could get a better deal from another supplier, he or she had to buy from the franchiser to keep the franchise. This is no longer legal. A franchisee who can get a better price without sacrificing quality can buy from any supplier.

Financial Assistance

A franchisee can get financial assistance from the franchiser to go into business. Usually, a franchisee puts up a certain percentage of the cost of land, building, equipment, and initial promotion. The rest is financed by the franchiser, who is paid back

> **CAN YOU SETTLE THE ISSUE HERE?**
>
> **How Independent Is a Franchisee?**
>
> David Darren has been a franchisee of a large franchiser in the fast-foods field for the past three years. David's business has been very successful. He has had a good working relationship with the district representative of the franchiser, Rachel Petersen. David always consulted Rachel on important business matters and followed her advice on running his business.
>
> Last week, David had lunch with his nephew, Sam. Sam is a college student majoring in business. He wants to go into business for himself after he finishes school.
>
> When David told Sam about the terrific opportunities in franchising, Sam reacted rather negatively. "But, Dave, you're not your own boss. You're pretty much the same as a store manager—just like the person who manages a Loblaws or a Safeway. The only differences are that franchisers make you think you're boss and make you risk your own money. You really work for them—you're an employee, not an independent businessman." David was somewhat upset by Sam's assessment. CAN YOU SETTLE THE ISSUE HERE?

out of revenues earned by the franchisee. The franchising agreement spells out the amount of financing the franchiser will provide and the terms of repayment. The franchiser also provides working capital by selling to franchisees on account.

In some cases the two parties agree on a joint-venture arrangement. The franchisee does not pay back the money put up by the franchiser. Instead, the franchiser becomes a part owner of the franchisee's business.

Finally, a franchisee may find local banks more willing to grant loans than if he or she were completely on their own. Bankers know that a reputable franchiser will license only dependable franchisees and will help them to be successful.

Promotional Assistance

Franchisers usually supply their franchisees with various types of promotional aids. This includes in-store displays, advertising mats for use in local newspaper advertising, radio scripts, publicity releases, and many others. Franchisers also help them to develop their promotional programs.

Franchising and the Franchiser

The franchiser, of course, also stands to benefit from the franchising agreement.

Recognition

A franchisee benefits from being able to use the franchiser's name and products at his or her location. The franchiser benefits by expanding the area over which the trade name is known. A franchiser can achieve national and, perhaps, international recognition much faster than if he or she were to do it alone. This increases the value of the franchise to both parties.

Promotion

A local franchisee pays a lower rate for newspaper advertising than a national franchiser. By sharing the cost of advertising, the franchiser and the franchisee both benefit. This is called co-operative advertising. Also, by using local radio and TV advertising in his or her franchise areas rather than blanket network coverage, the franchiser may avoid wasted coverage—advertising in areas which don't have a franchisee. There are benefits from "localizing" promotion to suit customer tastes in a given area or to tie in with local events. Furthermore, a franchisee can promote the business as being locally owned. This may give the franchisee a competitive advantage over chain-store operations in some areas.

Franchisee Payments

The franchising agreement sets out the amount and type of payments which the franchisee will make to the franchiser. Sometimes the franchisee pays a royalty based on monthly or annual sales. In some cases the fee is fixed at a certain amount and is payable monthly or annually. Often the fee is determined on the basis of the market size in the franchisee's territory. In still other cases a combination of these methods is used. Less frequently, the franchiser gets only a one-time payment from the franchisee. Also, keep in mind that franchisees increase the funds available to a franchiser for expansion purposes through their payments of fees.

Motivation

Some large chain stores have trouble recruiting and developing well-motivated store managers. These hired managers are not independent business owners. Franchisees are their own bosses. Their profits belong to them. A franchisee is, therefore, more likely to accept long hours and hard work than a hired manager.

Attention to Detail

The headquarters of a chain-store operation must keep payroll, tax, and other records on all of its units. It must be concerned with local laws regarding sales taxes, licences, permits, and so on. In a franchise operation, keeping records and complying with local laws are the job of local, independent franchisees.

Franchising and You

Do you have a future in franchising? The answer depends on your willingness to work, your ability to find a good franchise opportunity, and your ability to buy into the operation. Many independent business owners have been very successful as franchisees.

There are, however, some possible drawbacks. Franchising has become a "get-rich-quick" scheme for some "fast" operators. These fast-talking promoters will try to develop a franchise operation around practically anything. If you are thinking of becoming a franchisee, check into the franchiser's reputation for honesty and record of past performance.

Also, carefully read and take time to understand any proposed franchising agreement. Some make promises which the franchiser cannot fulfil. Don't be in a rush to "sign up before someone else does". Look for clauses which might permit the franchiser to buy you out at his or her discretion. If necessary, consult a lawyer to help you to understand the proposed agreement.

Look out for oversaturation (too many firms) in that particular type of operation in your area. This is why location is so important. Show-business people and gimmicky promotions are poor substitutes for facts about market potential in a given location. Some of the best-promoted franchises have failed because they were poorly conceived.

Fortunately, the "franchising fever" of the 1960s has died down. But, as in any type of investment, franchising has its risks and rewards. On the balance, however, it still offers a lot of promise to would-be-entrepreneurs. But don't be sold on the rewards without considering the risks. Be careful! Ask for the names and addresses of current franchisees. Talk to them. They can give you more objective insight. But don't be all-trusting here, either. They might be looking for someone to sell out to!

GOVERNMENT AND SMALL BUSINESS

Throughout our history the small-business owner has been admired and respected. The entrepreneurial spirit is most closely associated with the small-business owner's dedication to thrift and hard work. The owner realizes the dream that anybody can become their own boss if they can get enough money to start a firm and keep it going through hard work and know-how.

During this century, however, several basic changes took place in our economy, and many Canadians lost sight of the dream of going into business for themselves. For many, "making a living" meant going to work for someone else. Growing aversion to risk and the high failure rate among small businesses led many Canadians to reject the role of the entrepreneur.

Recently governments at both the federal and provincial level have become more involved in programs to assist small business. These programs and agencies have given substance to the belief that small business should be encouraged. They are also partly intended to ensure that the economy is not exclusively controlled by large firms.

Critics of government assistance to small business say that the government should

---

WHAT DO YOU THINK?

Is the Entrepreneurial Spirit Declining in Canada?

A lot has been said during recent decades about the decline of the entrepreneurial spirit in Canada. Some people believe that it has not only declined but is "deathly ill". Others see a revival of the entrepreneurial spirit during very recent years. They argue that it is "alive and well". WHAT DO YOU THINK?

---

not equate preserving competition with preserving the number of competitors. They suggest that if a small firm cannot survive without the government's help, it should fail. They say that although everyone has a right to go into business, no one has a right to expect the government to help him or her to start a business or to stay in business. The Canadian taxpayer has little to gain, they argue, by subsidizing business.

The argument that preserving competition does not mean preserving the number of competitors is rejected by most small-business owners. They believe that big firms want to reduce the number of competitors by setting their prices so low that small firms cannot make a profit. A firm which sells 100 different products can sell one of them at a price which only covers cost. It can make its profits on the other 99. A small firm which sells only one product can't do this. Yet, when the small firms are run out of business, the large firm can raise its price back to a high level. In other words, small businesses feel that competition laws are not enough.

Government Financial Assistance

Unlike the United States, where small-business assistance is federally co-ordinated through the Small Business Administration, in Canada, separate federal and provincial organizations exist to help small-business owners. These are examined below. (Additional information is provided in Chapter 19 in the section entitled "Incentive Programs".)

Federal Government Assistance

Small Business Loans Act

Federal Business Development Bank

Federal assistance is available in both the financial and management areas. **The Small Business Loans Act helps proprietors get loans to purchase and/or up-date fixed assets and for improvement of facilities in general.** The government places a ceiling on the amount which will be loaned ($50,000) as well as a time limit on the loan (10 years maximum). **The Federal Business Development Bank also encourages loans to small business by charging increasingly higher interest rates as the amount borrowed increases.** This encourages small businesses, with their relatively small needs, to receive the lowest interest rates. The Income Tax Act allows all Canadian-controlled private corporations (many of which are small) to pay a low tax rate on the first $150,000 income per year.

As far as management aid is concerned, firms with less than 75 employees are eligible to use the Counselling Assistance to Small Enterprises (CASE) services. The CASE program is designed to help owner-managers more effectively manage their business in all areas, not simply the financial end of the organization.

Counselling Assistance to Small Enterprises

Provincial Assistance

In addition to federal aid, each province has passed legislation or formed agencies to aid small business. This aid varies widely from province to province, but the aim is always to encourage small-business activity. Some examples of provincial aid are as follows:

CAREERS IN SMALL BUSINESS: THE STORY OF PHIL COVER

Ever since he was 14 years old, Phil Cover wanted to run his own business. Maybe it was his experience as a stock boy in Mr. Palmer's successful food store. Maybe it was the great affection he felt for his Uncle William, who owned a small insurance agency.

In any case, the idea continued to grow as Phil entered college and enrolled in the general business curriculum. He began to appreciate how complicated a business could be and how specialized many of the positions in large businesses are as he studied accounting, marketing, and finance. Although he enjoyed these subject areas, Phil wanted to see how they all fit together. He always thought in terms of applying what he learned to a small business.

The specific choice came the summer before he graduated. He was looking for a summer job in the small city where he grew up and realized that there was no private employment agency available. It all "clicked"! He saw an opportunity for profit. He still had that motivation to "do his own thing". His general business studies taught him a lot about how an employment agency serves job seekers and the business community. He even visited an agency in a distant city to study its operation.

Fortunately for Phil, his cash savings and training were sufficient to qualify him for a provincial business development loan. The agency opened soon after his graduation and almost made a profit the first year. Phil had a lot going for him. Not many people do it quite that way, but there is no "one" way to start your own business.

1. *The Enterprise Development Group of Manitoba* is designed to specifically assist the 83 per cent of all firms in the province which have less than 50 employees. Help is provided in areas such as small-enterprise development, existing enterprise improvement, human resource management, and marketing.
2. *The Alberta Opportunity Company* is designed to facilitate economic growth by establishing new businesses in the province. A wide variety of profit-oriented businesses are eligible for both financial and management counselling assistance.
3. *The Industry Development Branch of Saskatchewan* encourages new business formation in the manufacturing, service, distribution, and processing industries. The Development Branch does this by locating and evaluating development opportunities and providing information on the province's economic climate.

SURVIVAL OF THE SMALL FIRM

Each year many new and old firms go out of business because they cannot meet the competition. Most of these failures are small businesses. (See Figure 15-1.)

The Struggle to Survive

Small firms are often thought to be at a disadvantage in the struggle to survive. Larger firms use banks as a source of short-term funds. These firms usually make 90-day loans. They can borrow from other lenders, such as insurance companies, for longer-term capital needs. They also can issue more stocks and bonds. Small firms, for the most part, depend mainly on their local bankers for borrowed capital.

A small firm's loan may take as long as ten years to repay. This, by itself, makes the small firm's loan risky. Thus, small firms pay higher interest rates than larger firms. Their loan requests are more closely checked to see if they can meet monthly loan payments. Whereas a larger firm can shop around for a low interest rate, a small firm usually has to deal with a local banker. This is why the small business owner's relationship with his or her banker is so important.

Government loans are vital to small firms' survival. But some small-business owners argue that these have a negative side. Because they are easier to get than loans from banks, some business owners tend to borrow too much. This may overburden the firm with loan repayments and lead to a strain on working capital.

Challenges to Survival

Clearly, a small firm faces many challenges to its survival. A major one is lack of funds. Too many firms are started on a shoestring. Often, the owner realizes too late that more funds are needed to stay in business.

A firm which starts off with too much money tied up in fixed assets runs into a shortage of working capital. Too much money tied up in plant and equipment leaves little working capital to finance accounts receivable, pay off trade creditors, and maintain adequate inventories. But if a firm can't manage these, it will go out of business.

Small firms also find it hard to estimate and control expenses because they lack the advanced cost control methods and accounting procedures of larger firms. Also,

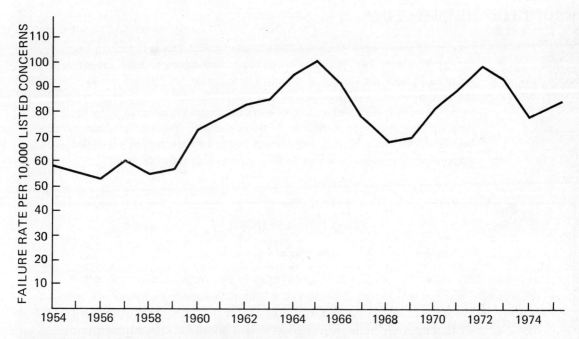

BUSINESS FAILURES include those businesses that ceased operations following assignment, receiving order, or proposal; ceased with loss to creditors after such actions as execution, or foreclosure; voluntarily withdrew leaving unpaid obligations; were involved in court actions such as receivership, or arrangement; or voluntarily compromised with creditors.

LIABILITIES as used in this booklet, refers to current liabilities, and has a special meaning; it includes all accounts and notes payable and all obligations, whether in secured form or not, known to be held by banks, officers, affiliated companies, supplying companies, or the Government. It does not include long-term, publicly-held obligations. Offsetting assets are not taken into account.

TOTAL LISTED CONCERNS represents the total number of business enterprises listed in the Dun & Bradstreet Reference Book. This Book includes manufacturers, wholesalers, retailers, building contractors, and certain types of commercial service including public utilities, water carriers, motor carriers, and airlines. This count by no means covers all the business enterprises of the country. Specific types of business not listed are: financial enterprises including banks, and mortgage, loan and investment companies; insurance and real estate companies; railroads; terminals; amusements; and many small one-man services. Neither the professions nor farmers are included.

Figure 15-1. Business failures. (Reprinted by permission from the *Canadian Business Failure Record*, 1976, Copyright © 1978 by Dun & Bradstreet Canada Ltd.)

small firms ordinarily don't have dependable data on the market potential for their products and services. Too often a "sure thing" turns into a disaster in terms of sales. In other cases the owner may be content with a sales volume which is well below potential.

Sometimes the owner pays himself or herself too much salary. Living too high on the firm's profits robs it of funds needed for growth. This is why self-discipline and a proper outlook are so vital to small-business success.

GROWTH OF THE SMALL FIRM

Owners of many small firms tend to "think small". This is especially true when an entrepreneur goes into business "to escape" from having to work for someone else. Many small firms are started by people whose primary motivation was to escape from the "rat race". Many of them don't really want their own firms to go through "growth pains". After all, they had enough of working for growth-oriented larger firms. This is especially the case when entrepreneurs are past middle age when they start out in their own firms. To many, starting and running their own firms amounts to little more than a hobby or something they "have to do" to prove that they can do it.

TWO POINTS OF VIEW

The Fear of Failure

Some people believe that business failures represent a tremendous waste of resources. Buildings and stores become vacant, bank loans are not paid off, employees are laid off, tax collections go down, and the entrepreneurial spirit dies. Some of these people believe that government should control the entry of entrepreneurs into business. For example, before a person could open a clothing store, he or she would have to apply for and receive permission from the local government. This permission would be granted only if there is a demonstrated need for this proposed business firm in the community. This, they say, would at least increase the chances of success and reduce the social and personal costs due to business failures.

On the other hand, some people believe that government control of entry into business would destroy the free-enterprise system. These people are especially upset with the notion that a business failure is a disgrace. Sure, nobody goes into business to fail. But a failure, according to them, is not totally bad. Entrepreneurs can and do learn from their failures. A business failure is not necessarily a personal failure. If everybody were afraid to fail, we would not enjoy the high standard of living we have in Canada.

There is nothing wrong with wanting to remain small. But there are many reasons why small-business owners should consider the benefits of a planned strategy of growth.

Why Seek Growth?

A firm which offers only one service or product is in deep trouble if demand for its product or service falls off sharply. It may have to close down. All its eggs are in one basket.

Growth may enable a firm to achieve the benefits of specialization and economies of scale. There are always fixed costs connected with operating a firm, and by expanding production and sales these fixed costs may be spread out over a greater volume of output. Putting on two shifts of workers, for example, may enable the plant to operate more efficiently. This may help the small firm to be more competitive with larger firms. But the wisdom of any approach to growth depends mainly on the owner's skill and vision.

A small firm which seeks to grow may provide greater motivation to employees. New chances for promotion arise. This may reduce labour turnover. Aggressive and growth-oriented employees don't want to work for a firm which is not growing.

Of course, there is the chance for greater profits from growth. Because business opportunity is dynamic, a firm should also strive to be dynamic.

Many small firms are started by people who want to sell out to someone else eventually, maybe to a larger, growth-oriented corporation. An aggressive strategy of growth may make this possible at an earlier date. The owner may also realize a big profit from selling such a firm.

A firm which doesn't grow may also find it harder to borrow money. The greater the competition among borrowers for loanable funds, the greater the disadvantage in being a "standstill" operator.

Growth Strategies

A small firm can grow in many ways. It can expand its present business. A small gift shop in a downtown location, for example, might expand by opening a branch in a suburban shopping centre.

Some small firms don't grow simply because they lack direction. Maybe the owner spends too much time running the firm and not enough time thinking about growth. Many small firms have been launched into aggressive growth strategies, including new lines of business, by bringing in some "new blood".

Merger is another way to grow. By joining together, two small firms might enjoy economic benefits and be able to exploit opportunity which neither could by itself.

Seeking new customers is always a way to grow. Two often overlooked sources for small firms are government contracts and export sales.

Of course, there are other avenues to growth. The approach which is best for a firm depends on its resource strengths and weaknesses and on its environment.

Survival of the Fittest

Even with government help most small firms find it hard to survive. "Survival of the fittest" is the rule of the game. Unfortunately, not all entrepreneurs realize the need for good management or possess enough management skills. To survive, an entrepreneur must constantly be alert to new opportunities and be careful about spending. Creativity, determination, careful planning, and a willingness to work are the keys to survival.

SUMMARY AND LOOK AHEAD

The entrepreneurial spirit is "alive and well" in modern Canada. Many Canadians, including a growing number of young people, want to start their own firms. The opportunities are there for those who can spot opportunity and exploit it.

But it takes work, know-how, and a determination to succeed in the face of chilling statistics on failures of new, small firms. Lack of funds and management know-how are the most often cited causes of failure.

There are possible benefits and definite headaches involved in going into business for yourself. The best approach is to be realistic in assessing them. Don't let the "dream" of becoming your own boss turn into a "nightmare".

We presented a summary of the first steps in starting a firm. Local requirements may differ, so you must study them carefully.

Franchising is still a growth industry. It provides good opportunities for those who want to start their own businesses, but it does not guarantee success.

The government helps small business in many ways by giving financial and managerial help. These agencies can help you to get started and to stay in business.

Small firms must face the challenges of survival and growth. Recognizing the challenges to survival is an important first step in developing the ability to survive. The same is true about growth.

In our next chapter we will look at international business. Here, too, there are challenges and opportunities for small, medium-size, and large firms.

KEY CONCEPTS

Better Business Bureau (BBB)   A BBB is a non-profit organization of business firms which join together to help protect consumers and businesses from unfair business practices.

Chamber of Commerce   A national organization whose main purpose is to improve and protect our free-enterprise system. Many cities have local chambers which belong to the national organization.

Counselling Assistance to Small Enterprise   A government program designed to help small owners more effectively manage their business.

Economic development council   An organization of business firms and local government officials who join together to further the economic development of the area.

> ## INCIDENT:
>
> ### Cubley Roofing Contracting Company
>
> Dolores Cubley's husband, Fred, started in business 20 years ago as the Cubley Roofing Contracting Company. Last month Fred had a severe heart attack and died. It came as a complete shock to Dolores, since Fred had always enjoyed very good health.
>
> From the outset Fred ran the firm by himself. His 10 employees never participated in managing the business. Neither did Dolores. She was content to leave "business matters" up to Fred.
>
> Dolores really wants to keep the business going. She believes that is what Fred would want her to do. But she is worried about her ability to do so. She also got an attractive offer one week ago from one of Cubley Roofing Contracting Company's competitors to buy out the firm. The proceeds of the proposed sale plus Fred's life insurance proceeds would enable Dolores to lead a reasonably comfortable life.
>
> Questions:
> 1. How might Fred have avoided placing his wife in this predicament?
> 2. Why would a competitor want to buy the firm? What would the buyer receive?
> 3. What advice would you give to Dolores? Why?

Entrepreneur A person who assumes the risk of organizing and managing a business in the hope of making a profit.

Federal Business Development Bank A federal agency which loans money at low interest rates to small business.

Franchisee A person or firm which is licenced by a franchiser to sell its products or services in a specific territory under a franchising agreement.

Franchiser A person or firm which licences franchisees to sell its products or services in specified territories under a franchising agreement.

Franchising agreement The contract between a franchiser and his or her franchisees.

Small Business Loans Act A federal law which facilitates loans to small business.

QUESTIONS FOR DISCUSSION

1. What are some of the problems in defining "small business"?
2. What are the three ways a person might become a small-business owner? Discuss each.

3. What are the basic requirements for becoming a successful entrepreneur?
4. What are the benefits and burdens of entrepreneurship? Discuss each.
5. Briefly list and discuss the preliminaries of starting a small business.
6. Explain how a franchising agreement works.
7. List and discuss the advantages of franchising to an entrepreneur.
8. How does a franchiser benefit from a franchising agreement?
9. Discuss the types of federal and provincial assistance which are available to small business.
10. List five key reasons why a small business may have trouble surviving.
11. What are the potential benefits of a planned strategy of growth for a small business?

INCIDENT:

Janitoil

Roger Bond has always been very interested in business. In college now, Roger is an active member of several business and professional organizations.

In his third year at university, Roger and two other student entrepreneurs (Kay Dobbs and Randy Pitts) formed a small business. They recognized an opportunity for a firm which could perform household cleaning chores for working couples who lived in apartment complexes.

The firm is called Janitoil. The advertising slogan is "Don't Toil. Call Janitoil". From a rather modest beginning, the firm now employs 15 part-time student employees. Business is booming!

Roger, Kay, and Randy are now in their fourth year at university. They are thinking about making this a full-time business after graduation. They are considering franchising. They want to license entrepreneurs on other college campuses in Canada to offer the same service.

Questions:
1. Why do you think franchising is appealing to Roger, Kay, and Randy?
2. What advice would you give them at this point in time with respect to the attractiveness of the franchising opportunity they see for Janitoil?
3. What kinds of "problems" will the three entrepreneurs face if they decide to go into franchising?
4. Would you be willing to become a Janitoil franchisee? Why or why not?

16
International Business

OBJECTIVES: After reading this chapter, you should be able to:

1. Give an example of why two nations would trade because of the principle of absolute advantage.
2. Give an example of why two nations would trade because of the principle of comparative advantage.
3. Differentiate between natural and human-made barriers to international trade.
4. List and discuss the human-made barriers to international trade.
5. List and discuss the arguments used to justify tariffs and present counter-arguments to those justifications.
6. Distinguish between a country's balance of trade and its balance of payments.
7. Explain how Canada's balance of payments could be unfavourable while its balance of trade is favourable.
8. Give examples of how the Canadian government aids Canadian firms in conducting international business.
9. Give reasons why firms import and export goods.
10. Compare the different degrees of commitment a firm might have to international business.
11. Compare a multinational firm's view of opportunity with that of a purely domestic firm.
12. Discuss the extent of foreign ownership of Canadian business and the steps which are being taken in response to this situation.

KEY CONCEPTS: In reading the chapter, look for and understand these terms:

ABSOLUTE ADVANTAGE
COMPARATIVE ADVANTAGE
TRADE BARRIERS
TARIFFS
EMBARGOES
NON-TARIFF BARRIERS
COUNTERVAILING DUTIES
BALANCE OF TRADE
BALANCE OF PAYMENTS
DEVALUATION
EXCHANGE CONTROL
ISOLATIONISM
REGIONAL TRADING BLOC
TAX CONTROL
EXPROPRIATION
NATIONALIZATION
CONFISCATION
INTERNATIONAL TRADE
STATE TRADING COMPANY
COMBINATION EXPORT
   MANAGER
PIGGYBACK EXPORTING
FOREIGN ASSEMBLY
CONTRACT MANUFACTURING
LICENCING
JOINT VENTURE
MULTINATIONAL COMPANY
CARTEL
FOREIGN DIRECT
   INVESTMENT
PORTFOLIO INVESTMENT
EXTRATERRITORIALITY

Up to now we've discussed business mainly in terms of Canadian firms operating in Canada. In this chapter we look at international business. Business activity is becoming more international in character. Some of our large corporations sell more abroad than they sell in the "home" market. Many of them have set up manufacturing operations overseas.

The days of political and economic non-involvement with other nations are gone. Canadian-made goods are exported to almost every nation in the world. They are even found in many of the communist countries. Canadian-owned manufacturing plants are also located in many foreign countries.

But foreign firms also export goods to us. Familiar examples are Japanese television sets, Colombian coffee, West German cars, and French perfumes. A less familiar example is Russian tractors.

Foreign firms also have manufacturing plants in Canada and invest heavily in "Canadian" business. There are many examples of foreign ownership of Canadian businesses.

International business creates new types of business opportunity, stimulates international contact, and leads to new business practices and new business challenges. Ours is truly an exciting age of international business. Let's begin with a discussion of the reasons why nations trade with each other.

WHY NATIONS TRADE

In general, greater specialization makes possible greater output. But specialization requires exchange or trade. The greater the market, the greater the opportunity to exchange and specialize.

There is no economic reason why specialization and exchange must be limited to the people within a country. Broadening the scope of the market makes greater exchange and specialization possible. This is one reason why nations buy from, and sell to, each other. Let's examine some basic principles of trade which explain why nations specialize in certain kinds of goods.

The Principle of Absolute Advantage

A country enjoys an absolute advantage in producing a good when either (1) it is the only country which can provide it, or (2) it can produce it at lower cost than any other country. If a good can be produced only in Switzerland, any other countries which want it must trade with Switzerland. Lake Winnipeg goldeye is produced only in Canada, so other countries who want this product must buy it from us. If a good can be produced at a lower cost in France than in any other country, the other countries must either trade with France to get it or pay the higher cost of producing it themselves.

Absolute advantage

If all nations followed this principle, each would specialize in producing the goods in which it enjoyed an absolute advantage. Each would import all others it wished to have. Let's assume that Japan and Canada are the only countries in the world. If Japan can produce steel more cheaply than Canada, Japan has an absolute advantage in steel production. We would import steel from Japan. Suppose Canada has an absolute advantage in producing grain. According to the theory, we should specialize in producing grain and import Japanese steel. The Japanese should specialize in producing steel and import our grain. Both countries would be better off.

The Principle of Comparative Advantage

Suppose the president of a large firm can type faster than his or her secretary. The president, therefore, enjoys an absolute advantage over the secretary in typing ability. But should the president spend time typing? Although he or she is a better typist than the secretary, the president is, more importantly, a much better decision maker than the secretary. The president is wiser to attend to the duties of president rather than to type. The choice here should not be based only on "absolute advantage".

The same is true of countries. Suppose a country can produce everything its people consume more efficiently than all other countries. It can still benefit from trade. Decisions made to cope with the economic problem are made on the basis of the best use of our resources. While the company president may be making "good" use of limited time by typing, he or she is not making the "best" use of that time. "Comparative advantage" rather than "absolute advantage" should guide decision making here. **Comparative advantage means that a country should specialize in producing those goods in which it has the greatest comparative advantage or the least comparative disadvantage in relation to other countries.**

Comparative advantage

WHAT DO YOU THINK?

Textile Industry Assistance

The textile industry in Canada, located primarily in Quebec, employs many workers. The costs of production of the Canadian textile industry are higher than in some other countries of the world. Some people argue that because the textile industry is important to Canada, and because it employs so many people, tariff barriers should be maintained to keep cheaper foreign imports out and to protect the Canadian industry.

Other people say that if the industry is not competitive internationally it should not be protected by artificial measures. They argue that if protective measures were removed and adjustment assistance given to the companies and workers, the country would be better off in the long run. WHAT DO YOU THINK? Should textile tariffs be raised, or should we buy more from others who can produce textiles at lower cost?

The products a given country will produce depend on many factors, such as presence of natural resources, cost of labour and capital, and nearness to markets. International trade enables each nation to use its scarce resources more economically. Its people can enjoy a higher standard of living. Since our imports and exports account for about 25 per cent of our GNP, trade is obviously important to us.

Our exports create jobs for people who produce goods and services and get them to overseas markets. This includes manufacturers, transportation firms, banks, and insurance firms. The same is true of our imports. Some people complain that importing textiles "robs" Canadian textile workers of jobs. But it also creates jobs for people involved in the job of importing the textiles.

BARRIERS TO INTERNATIONAL TRADE

International trade among people in free economies is carried out mainly by private business firms in those countries. But government actions and policies can affect the

---

THINK ABOUT IT!

The Principle of Comparative Advantage

Assume that skilled labour is the only scarce factor of production in Canada and Japan. A Japanese worker can make 16 radios or 4 TVs per day. A Canadian worker can make 4 radios or 2 TVs per day. Thus, the Japanese worker is 4 times as efficient in making radios and 2 times as efficient in making TVs than the Canadian worker.

	Radio	*TV*	*Radio-TV ratio of advantage*
Japanese worker	16	4	4:1
Canadian worker	4	2	2:1

Without trade the Japanese have to give up 4 radios to get 1 more TV. The Canadians have to give up 1 TV to get 2 more radios.

Trade can take place and be profitable for both Japanese and Canadian consumers at any ratio of exchange between 4:1 and 2:1. The ratio of exchange is determined largely by the relative bargaining strength of the trading partners. Suppose, therefore, that Japan specializes in radios and Canada specializes in TVs. They settle on a radio-TV ratio of 3:1. Thus, the Japanese give up 3 radios to get a TV. The Canadians give up 1 TV to get 3 radios. This is the principle of comparative advantage. THINK ABOUT IT!

> **CAN YOU SETTLE THE ISSUE HERE?**
>
> **Imports and Domestic Unemployment
> in the United States**
>
> Canada is not the only country where international trade is an issue. In the Trade Expansion Act of 1962 the U.S. Congress first approved trade adjustment assistance benefits for American workers who could prove that increased imports were caused "in major part" by a tariff concession and that these imports were the major cause of unemployment.
>
> In the Trade Act of 1974 the U.S. Congress liberalized the government assistance program for workers idled by foreign imports. Workers who qualify under the 1974 law can get weekly benefits of 70 per cent of former pay, up to $176 a week. The benefits last for 52 weeks for most workers but 78 weeks for workers in training programs and for those who are 62 years old or more. Regular unemployment compensation paid by the states is figured in the maximum.
>
> Union workers sometimes refer to this adjustment assistance as "burial assistance". They argue that this does not save jobs. Many of them want Congress to authorize import quotas and to impose heavy taxes on multinational firms. The multinational firms, of course, disagree. CAN YOU SETTLE THE ISSUE HERE?

willingness of firms to trade. So can natural things like the distance between countries. For example, even if a good can be produced more cheaply in Country X than in Country Y, the cost of shipping the good to Y might wipe out the cost advantage.

Trade barriers

Trade barriers are natural and "created" obstacles which restrict trade among countries. Technology has helped us to reduce many of the natural barriers. Because of the jet airplane, distance between countries is measured in hours of flying time. This is very important for products which spoil rapidly. It is also important for products which are expensive to transport. The big problems, however, are the "created" barriers. Let's discuss these.

Tariffs

Tariffs

Tariffs are duties or taxes which a government puts on goods imported into or exported from a country. Governments rarely impose tariffs on exports, because they generally favour exporting goods to other countries. However, Canada does have a duty on oil exported to the United States.

Purposes of Tariffs

Tariffs serve two main purposes—revenue and/or protection. A revenue tariff raises money for the government which imposes it. The purpose is not primarily to reduce imports of the good on which the tariff is imposed. Revenue tariffs were important during our early history when we had very little domestic industry. By taxing goods coming in from abroad, our government raised revenue. Of course, the final effect is that people pay more for the goods they import since the foreign sellers add the tariff to their selling prices.

The purpose of a protective tariff is to discourage imports rather than to raise revenue. Our government has a comprehensive set of tariffs designed to protect manufacturers and encourage their establishment in Canada. Tariffs have been an important instrument of government economic policy since Confederation. A protective tariff also leads to higher prices of the imported good. Table 16-1 summarizes the major arguments used to "justify" tariffs.

Types of Tariffs

There are three ways of setting tariffs: (1) ad valorem; (2) specific duty; and (3) combination ad valorem and specific duty.

Table 16-1. Arguments for tariff and non-tariff barriers

Argument	Reasoning
1. The infant industry argument	1. Tariffs are needed to protect new domestic industries from established foreign competitors. Once imposed, they are hard to remove. The industry which "grows up" under such a tariff tends to need it in adulthood.
2. The home industry argument	2. Canadian markets should be reserved for Canadian industries regardless of their maturity levels. Canadian consumers pay higher prices, and these tariffs are easily matched by other nations. The principle of comparative advantage is completely overlooked.
3. The cheap wage argument	3. Keep "cheap foreign labour" from taking Canadian jobs. When labour is a large cost of producing a product, firms in low-wage countries can sell it in Canada at lower prices than Canadian manufacturers. In periods of high domestic unemployment, this argument influences policy. Keeping such products out of Canada denies potential foreign buyers the Canadian dollars with which to buy our products.
4. The national security argument	4. Certain skills, natural resources, and industries are judged to be vital to national security. Many such industries are protected. Examples of protected industries in Canada include airlines, railroads, banking, radio and television, and others.

An ad valorem tariff is one levied as a percentage of the imported good's value. It is used mainly for manufactured goods. A specific duty is one levied on an imported good based on its weight and its volume. It is used mainly for raw materials and bulk commodities. The duty is figured on the basis of pounds, gallons, tons, and so on. In a combination ad valorem and specific duty, both types of tariffs are imposed on an imported good.

Embargoes

Embargoes

Embargoes prohibit the import and/or export of certain goods into or out of a country. This may be done for health purposes (the embargo on the import of certain kinds of animals), for military purposes (the embargo on the export of Candu nuclear reactors to certain countries), for moral purposes (the embargo on the import of heroin), or for economic reasons (the embargo of American grain which was advocated by some people during that country's grain shortage in 1973). Embargoes are also used for political purposes such as trade embargoes against Rhodesia in the mid and late 1970s. In 1971 the United States lifted its 21-year embargo on trade with the People's Republic of China. Many countries also have embargoes on trade with South Africa because of their disapproval of its racial policies.

Non-tariff Barriers (NTB's)

Non-tariff barriers

There are a variety of measures which a government can take to restrict imports into a country. **The measures to restrict trade, other than tariffs, are referred to as non-tariff barriers. These can include things such as bilingual labelling requirements, package size or characteristics, product safety requirements, requirements for import licences, inspection of perishable products and emission standards, to name a few.**

Countervailing duties

If a nation introduces tariff or non-tariff barriers which effectively reduce imports, or give it an export advantage, other affected trading nations are likely to retaliate by introducing their own countervailing measures. **Countervailing duties are tariffs or non-tariff barriers introduced by a nation to offset an artificial advantage gained by another nation as a result of trade barriers.**

The Balance of Trade Problem

An export from one nation is an import to another nation. International trade is a two-way street. Some governments, however, see it as a means of gaining an economic "edge" over other countries. National pride and mutual distrust cause many nations to view trade as desirable only if their exports are greater than their imports.

Balance of trade

A nation's balance of trade is the difference between the money values of its exports and its imports. If it exports more than it imports, its balance of trade is considered favourable. If it imports more than it exports, its balance of trade is unfavourable.

At one time nations used gold for settling trade imbalances. A country which imported more than it exported (an unfavourable balance of trade) paid for its excess imports by shipping gold to the creditor nations. Because a nation's gold supply was considered a measure of its "strength", a government would restrict imports and grant tax breaks to firms that exported goods. This enabled it to hold its gold.

Canada has enjoyed a favourable balance of trade for many years. The only recent year merchandise imports exceeded exports was 1975. To understand how this affects international business, let's discuss the balance of payments. (See Figure 16-1.)

The Balance of Payments Problem

A nation with a favourable balance of trade can have an unfavourable balance of payments. **A country's balance of payments is the difference between its receipts of foreign money and the outflows of its own money due to imports, exports, investments, government grants and aid, and military and tourist spending abroad.** For Canada to have a favourable balance of payments for a given year, the following would have to be true. The total of our exports, foreign tourist spending in Canada, and foreign investments in Canada must be greater than the total of our imports, Canadian tourist spending overseas, payments on foreign debt, and the investments made by Canadians abroad.

Balance of payments

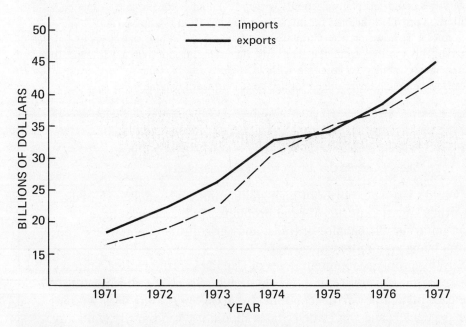

Figure 16-1. Canadian imports and exports of merchandise. (Source: Statistics Canada, *Quarterly Estimates of the Canadian Balance of International Payments*, cat. 67-001, Third Quarter, 1977. Reproduced by permission of the Minister of Supply and Services Canada.)

The Canadian balance of payments was negative in five of the seven years between 1971 and 1977. In every year but 1975 the balance of trade was positive.

The Rate of Exchange

Assume that a French importer buys goods from a Canadian firm. If the Canadian firm wants payment in Canadian dollars, the French importer has to exchange francs for dollars at a bank in France. How many francs are needed to buy a dollar (or vice-versa) depends on the rate of exchange between francs and dollars. Even if the Canadian firm accepted payment in francs, it could exchange the francs for dollars. Our exchange rate is important to our imports and exports. If the exchange rate decreases, the value of our dollar falls in relation to other currencies—our exports will become less expensive to other countries, and their products (our imports) will become more expensive for us to buy. If the exchange rate rises, the reverse situation occurs.

If a country has a persistent balance of payments deficit there will be downwards pressure on its exchange rate. This will eventually bring about a devaluation on its currency in relation to that of other countries.

The Role of Gold

Until 1971 gold played an important role in settlement of the international balance of payments. The price of gold was established at $35 per ounce (U.S.), and foreign countries could settle any accounts with the United States in gold if they chose. This situation provided the base for international exchange transactions.

In 1971, however, the United States had a $6.3 billion deficit in its balance of trade. This resulted from the long-term increasing productivity of foreign firms and their greater ability to compete with U.S. firms for customers. Furthermore, Americans were buying lower-priced foreign goods in record volume.

WHAT DO YOU THINK?

Should Canada Use Agripower in World Markets?

Canada is the world's largest exporter of high-grade milling wheat. The United States is the world's largest exporter of food products. Both countries have recently had an unfavourable balance of trade, and agricultural exports have kept the balance from being even more unfavourable.

To deal with this problem, representatives of Canada and the United States have been discussing the possibility of a "wheat cartel". Since the oil-exporting countries have used oil as a powerful economic "weapon" in trading with Western countries, perhaps Canada and the United States should use food in the same way. WHAT DO YOU THINK?

Because of the United States' shrinking gold supply, some foreigners began to question its ability to continue to convert foreign-held dollars into gold. Many of them tried to convert their dollar holdings into gold. These "runs on the dollar" got so severe that, in 1971, the United States stopped paying gold in exchange for dollars held by foreigners. An embargo was placed on the shipment of gold.

The U.S. dollar was officially devalued in 1972 from $35.00 per ounce of gold to $38.00 per ounce, and in 1973, from $38.00 to $42.22 per ounce. Before the 1972 devaluation, the U.S. dollar was worth 1/35 of an ounce of gold. The 1973 devaluation meant that the dollar was worth 1/42 of an ounce of gold. **Devaluation, therefore, means reducing the value of a currency in relation to gold or in relation to some other standard.**

Devaluation

The Payment Problem

Because buyer and seller are in different countries, international transactions sometimes present a payment problem. Methods of payment vary widely. While cash payment appeals to the seller, the buyer may look at it as seller suspicion of the buyer's creditworthiness. It also ties up the buyer's working capital if payment must accompany an order for goods.

With open account, the exporter ships the goods to the buyer, and the exporter's commercial invoice indicates the buyer's liability to pay. This is risky unless the exporter is sure of the buyer's willingness and ability to pay.

The most common method of payment is with drafts, or bills of exchange. Drafts provide documentary evidence of financial obligation. A draft drawn on, and accepted by, an importer is a trade acceptance. A draft drawn on, and accepted by, a bank is a bank acceptance. A commercial letter of credit is issued by the importer's bank showing that the bank accepts the draft drawn on it by the exporter. Of course, the importer must apply for and be granted this credit to be issued a letter of credit.

There also are other methods of payment. For example, an exporter who ships materials from Canada to be used in a construction project for the Canadian government may be paid directly by the Canadian government. This is practically the same as a domestic sale by the firm.

In some cases exporters accept bonds issued by the importer's government. This is likely to occur when that government is short on foreign exchange.

In still other cases, barter transactions are made. The exporter may be paid in the currency of the importer's country, but the exporter may be required to spend it on goods made in that country. Outright barter deals are becoming more important. Importer and exporter agree to swap goods for goods, and there is no currency involved.

Because of the growing importance of international business, many foreign banks have banking connections in Canada. Some foreign banks operate affiliates in Canada. Foreign bank assets in Canada in December 1976 were about $2.4 billion. A branch office's main job is to keep in contact with banking developments in Canada and to promote goodwill. Affiliates do not accept deposits from Canadians, but they do compete for wholesale and commercial lending, operate foreign exchange, and perform other services for firms in their home countries and in other countries. In one case (the Mercantile Bank of Canada) a foreign bank has control of a chartered bank of Canada.

Canadian banks also do the same. The large Canadian banks have foreign branches in many countries. They also operate foreign departments in Canada to finance imports and exports. Canadian banks also enter into correspondent relationships with foreign banks. An overseas correspondent bank of a U.S. bank handles its transactions in that country. Of course, the foreign bank's U.S. correspondent handles that bank's international dealings in the United States.

Exchange Control

Exchange control

A Canadian firm which opens a branch overseas wants the branch's profits to go to its Canadian owners. If the branch is in West Germany, West Germans pay for the goods in marks. But the Canadian owners want dollars. If West Germany is short on dollars, it might stop the branch from sending the profit to Canada. This is done by limiting the amount of marks which can be converted to dollars. It is called exchange control. **Exchange control means government control over access to its country's currency by foreigners.**

Isolationism

Isolationism

Isolationism is the tendency of a country (or a group of countries) to limit social and economic contact with other countries. Isolationism decreases the amount of foreign trade which occurs. Because foreign trade is so important we are opposed to any tendencies towards isolationism.

Regional trading bloc

In some areas, isolationism has given way to regionalism. **A regional trading bloc is a group of countries which agree to eliminate restrictions which limit trade among member nations.** Often, this is at the expense of "outsider" countries. The members want to trade with each other, not with non-members.

In 1968 the European Economic Community (EEC), or the European "Common Market", was formed by Belgium, France, Italy, Luxembourg, the Netherlands, and West Germany. Since then, several other countries have joined. The member countries have moved towards closer economic, and limited political, integration. They strive to eliminate trade barriers for their mutual advantage. Other regional trading blocs are the European Free Trade Area (EFTA), the Latin American Free Trade Association (LAFTA), and the Central American Common Market (CACM).

Canada, as a member of the British Commonwealth, had the commonwealth preferential tariff. When Britain entered the European Economic Community we lost our preferential status in trade with her. Canada is very dependent upon the United States as a trading partner. More than half of our trade is with the United States. We have been actively trying to increase trade with Western Europe to offset this situation. We sought a "special trading status" with the EEC.

Tax Control

Foreign-based firms are reluctant to invest in countries which practise tax control. **A country practises tax control when it uses its tax authority to control foreign investments in that country.** Under-developed countries, for example, need revenues but have no tax base at home. Taxing foreign-owned firms is an easy out. In extreme cases tax control can lead to virtual control of these firms by a country's government.

Tax control

Expropriation

Perhaps the biggest political risk of setting up a plant in some countries is the risk of expropriation. **Expropriation means that the government of a country takes over ownership of a foreign-owned firm located in its country.** The firm that is taken over may then be sold to private citizens of the expropriating country.

Expropriation

Nationalization, however, means that the expropriating government keeps ownership and runs the firm. When Salvador Allende came to power in Chile, the copper mines in that country were nationalized. Some potash mines in Saskatchewan have been nationalized.

Nationalization

A government which expropriates property may or may not compensate the owner. In some cases the owner is paid part or all of the market value of that property. **Confiscation, however, means that the government does not compensate the owner for the expropriated property.** In Saskatchewan the owners were compensated for their mines. In Chile they were not.

Confiscation

REMOVING TRADE BARRIERS

Governments can help to promote trade. Many seek to attract foreign investment by giving informational, financial, and promotional assistance to foreign firms.

Foreign investment is needed to help newly emerging nations to develop. It helps to raise their standard of living.

Many economically advanced countries lack sufficient domestic supplies of certain resources. Japan's domestic growth, for example, is due largely to its government's encouragement, rather than restriction, of international business.

But what about the home government of a firm which invests overseas? A government's attitude towards its country's firms which invests abroad depends largely on the situation and the time. It can involve foreign policy and the national interest.

Political differences with some nations may cause a government to discourage firms from investing in them. Balance of payments problems can lead a government to restrict foreign investment by its country's firms. On balance, the Canadian government encourages and helps Canadian firms to do business overseas.

The federal government's Department of Industry, Trade and Commerce organizes trade missions and operates permanent trade centres in many foreign countries. The department provides information and promotion services to Canadian firms interested

Table 16-2. Some examples of governmental assistance to international business

Institution	Purpose
1. The Export Development Corporation (EDC)	1. This agency was created by the federal government in 1969 after a study indicated that export insurance and financing facilities were more adequate in other countries than in Canada. The EDC finances export credits as well as insuring and guaranteeing such credits provided by suppliers, bonds, or other institutions.
2. Provincial Export Development Agencies	2. A number of provinces have export promotion programs. They will work with firms to develop export opportunities.
3. Trade Commissioner Service	3. The Trade Commissioner Service of the federal Department of Industry, Trade and Commerce will work with firms to identify and develop potential trade opportunities.
4. The International Bank for Reconstruction and Development (World Bank)	4. The World Bank began operations in 1946 to advance the economic development of member nations by making loans to them. These loans are made either directly by the World Bank using its own funds, or indirectly by the World Bank borrowing from member countries.
5. The International Monetary Fund (IMF)	5. The IMF began operations in 1947 to promote trade among member countries by eliminating trade barriers and promoting financial co-operation among them. It enables members to cope better with balance of payments problems. Thus, if firms in Peru wish to buy from Canadian firms but lacks enough Canadian dollars, Peru can borrow Canadian dollars from the IMF. It pays back the loan in gold or the currency it receives through its dealing with other countries.
6. The General Agreement on Tariffs and Trade (GATT)	6. GATT was negotiated in 1947 by member nations to improve trade relations through reductions and elimination of tariffs. GATT has resulted in tariff reductions on thousands of products. The latest round of GATT negotiations was completed in 1979.
7. The International Development Association (IDA)	7. The IDA began operations in 1960 and is affiliated with the World Bank. It makes loans to private businesses and to member countries of the World Bank. In addition to the IDA, there are similar organizations which make loans to governments and firms in certain country groupings. The Inter-American Development Association, for example, is for countries belonging to the Organization of American States.
8. The International Finance Corporation (IFC)	8. The IFC began operations in 1956 and is also affiliated with the World Bank. It makes loans to private businesses when they cannot obtain loans from more conventional sources.

in overseas business. Table 16-2 discusses several ways governments aid international business.

Many agencies and groups promote international business. Many provincial and municipal government programs seek to boost exports. Provincial industrial development commissions and mayors often go to the United States and overseas to lure foreign firms to their areas.

Private efforts are also important. The Canadian Chamber of Commerce provides information and advice to firms interested in selling overseas. Private firms and organizations also participate in and sponsor trade shows. When banks, insurance firms, transportation firms, ad agencies, accounting firms, and marketing research firms help and promote international business, they bring new business to themselves.

STATE TRADING

International trade involves the exchange of goods between one country and other countries. Our main concern is the international operations of business firms. But some foreign trading does not directly involve privately owned business firms. Government agencies carry out this type of trading.

International trade

For example, the Canadian Wheat Board negotiates wheat sales. The Organization of Petroleum Exporting Countries (OPEC) establishes prices. In many countries and for many products, government agencies are important. The trend is towards more and more direct government involvement in foreign trade. When foreign aid is involved, the government is entirely responsible for the international flow of goods.

Essentially the same occurs when the U.S. government negotiates sales for American firms to overseas buyers. The U.S. Department of Defense actively promotes sales

---

WHAT DO YOU THINK?

Government Subsidies and Competition for World Markets

Japan's computer manufacturers get government subsidies to make them more competitive both in competing for Japanese customers and in competing for customers in other countries. In 1975, for example, the Japanese government provided about $200 million in subsidies to Japanese-based computer firms. In that year it also removed quotas on the import of computers.

Japan is the world's second-biggest market for computers. Foreign-based firms have about 45 per cent of the Japanese computer market. But Japanese firms expect to increase their market share in Japan and world-wide as well.

Should the Canadian government directly subsidize Canadian-based firms to help them compete in world markets? WHAT DO YOU THINK?

of American weapons to some foreign countries. The high cost of imported oil has stimulated these efforts in order to improve their balance of payments position.

The volume of trade with the communist nations has increased in recent years. Since the Soviets view foreign trade as an instrument of foreign policy, politics plays a large role in their trade with the non-communist world. Trade with those countries is handled through the USSR's state trading companies.

State trading company

A state trading company is a government-owned operation which handles trade with foreign countries and/or firms. A Canadian firm selling to the People's Republic of China sells to one of seven state trading companies. These companies do the actual buying for the Chinese. The Canadian firm does not deal directly with the people who will finally use the good.

EXPORTING AND IMPORTING

We've used the terms "exports" and "imports" in our discussion of trade. Imports are things which enter a country from other countries. Exports are things which go from one country to other countries.

Why Firms Export Goods

Exporting is a special kind of "selling". Exported goods cross over national borders. But these borders are political borders. They are not related to the nature of business activity. Some mass production industries have to produce in large volumes to get the cost per unit down to a low level. If the home market is too small to absorb this output, these firms look to other countries for additional customers. Many Canadian

---

THEN AND NOW

The Soviet Union and Consumer Goods

For many years the Soviet Union showed a general lack of interest in consumer goods production. They concentrated on building heavy industry and a first-rate military establishment. Some Soviet citizens complained about the lack of attention to consumer goods and the poor quality of what was available.

The Soviet Union is now eager to expand its exports of consumer goods. Soviet-made watches, television sets, cameras, cars, radios, camping equipment, pianos, and other types of consumer goods are sold in the Western European nations, Canada, and the United States.

The incentive for this is the Soviet desire to bolster its prestige in the world and to earn foreign exchange.

firms do not have sufficiently large-scale plants to produce at a competitive cost relative to other industries, because the Canadian domestic market is too small. This is particularly true in relation to the United states, our largest trading partner. The United States has a domestic market roughly ten times as large as Canada's.

In other industries such as mining the majority of our products are exported. These companies must export because the markets in Canada cannot absorb their level of production. Often exporting is undertaken only after domestic markets are satisfied and overhead costs have already been met. Japan, although a very large exporter, exports only a percentage of the goods which are domestically consumed. This means that margins can be higher in export markets. If the product and the selling approach have been developed and tested in the home market, moving into another country may require little additional preparation expense.

The demand for many products is seasonal. Some firms shift their off-season production into foreign markets where the product is in season. This may lower production cost due to better production scheduling.

Finally, a firm might find it more profitable to expand its market coverage to foreign countries than to develop new products for sale at home. Its skills may be put to best use by producing and selling its traditional product rather than by risking the development of new products.

Why Firms Import Goods

Importing is buying. Canadian firms import bananas and coffee because they are not available in Canada. Sometimes domestic sources must be supplemented by foreign sources. Thus, Canada imports large quantities of crude petroleum for use in the eastern provinces.

Prices of foreign goods are often lower than like goods produced at home. This may be due to lower labour costs overseas. Some cars, steel, textiles, and electronics equipment are imported into Canada. Remember, however, that low-cost labour is a bargain only when it is productive. A Canadian worker who can produce two units of output per hour is more productive than a foreign worker who produces only one unit of output per hour. As long as the Canadian worker's wage is less than double that of the foreign worker, the Canadian worker is more productive in real terms.

Importing goods from foreign producers may lead to exporting goods to them. This accounts for a lot of foreign commerce. This is the type of trading the Soviet Union prefers.

In other cases imported goods have prestige value. Some Canadians, for example, are willing to pay extra for imported wines or perfumes.

TYPES OF INTERNATIONAL BUSINESS

Involvement in international business can range all the way from unintentional exporting to setting up complete branch operations in one or more foreign countries. Let's examine these varying degrees of commitment to international business.

Unintentional Exporting

Many firms' products are exported without their knowledge. For example, a supplier of a part used by Versatile Manufacturing in making farm tractors might be unaware that the part ends up on a tractor used by a farmer in the United States.

Many firms have resident buyers in foreign countries. These buyers buy goods in those countries and send them to their employers. Thus, a Canadian firm might be selling to a resident buyer for an Italian firm without knowing it.

Unsolicited Exporting

Sometimes, a firm might get an unsolicited order from an overseas buyer. Unlike the examples above, the firm is aware that the customer is overseas. Often, however, the firm may not be interested in selling overseas. Many firms, especially small ones, have a mistaken notion that selling overseas involves too much "red tape". Various Canadian government agencies and programs, as we have seen, help these firms to do business overseas. Canadian federal and provincial governments help small firms to get involved in exporting.

Intentional Exporting

Intentional exporting means that the exporter is committed to selling abroad. But there are degrees of commitment. At one extreme are firms which consider their export business to be secondary to their domestic sales. Such a firm might accept orders from overseas buyers but not seek them. It might have a small department which accepts orders from foreign buyers. At the other extreme, a firm might have a large, well-staffed, well-financed division which seeks export sales.

An intentional exporter must decide how to handle its export business. In direct exporting the firm handles the export task for itself. In indirect exporting, outside specialists handle the export task for the firm. Which approach is best depends on such factors as the company's size, its export volume, the number of foreign countries involved, the investment required to support the operation, the profit potential, the risk present, and the desires of the overseas buyers. If the firm exports many products, it may go direct with some and indirect with the others.

Combination export manager

Some firms are afraid to get involved in exporting because they don't know how to do it. **The combination export manager can help. This middleman represents several exporters and handles all the work involved in moving their goods overseas.**

Many firms, especially small and medium-size ones, co-operate with each other in their exporting. Several firms and the federal government have brought small producers together to form export consortia to bid on large-scale international contracts.

Piggyback exporting

Another type of co-operation is piggyback exporting. **In piggyback exporting, one firm (the carrier) uses its overseas distribution network to sell non-competitive products of one or more other firms (riders).**

Overseas Manufacturing

Overseas manufacturing involves a greater commitment to international business than exporting does. Let's discuss several types of overseas manufacturing.

Foreign Assembly

One example of overseas manufacturing is foreign assembly. **In foreign assembly the parent firm exports parts overseas where they are assembled into a finished product by its overseas subsidiary or a licencee.** A Canadian tractor manufacturer, for example, might export parts overseas for assembly into a finished tractor. This may be wise when the tariff on the parts is much lower than the tariff on an assembled tractor. The overseas assembler might be a subsidiary of the Canadian firm. Usually, the Canadian firm would also have a sales subsidiary in that country. Perhaps the overseas assembler is a foreign-owned firm which is licenced to assemble the tractor. The sales task here is handled by the licencee. Sony televisions and Honda motorcycles are "foreign goods" which are assembled in the United States for sale there.

Foreign assembly

Contract Manufacturing

Another example of overseas manufacturing is contract manufacturing. **In contract manufacturing the firm which wants to do business in a foreign country enters into a contract with a firm there to produce the product.** The overseas company, however, does not handle the sales task. This arrangement is popular with big consumer goods firms such as Procter and Gamble. Proctor and Gamble might contract out the production work to a foreign firm, but the company takes on the marketing task.

Contract manufacturing

Licencing

The main difference between contract manufacturing and licencing is that a licencor-licencee relationship usually extends over a longer period of time. **In a licencing arrangement, licencees are licenced to manufacture and market products in their countries. The licencor gets an agreed-upon percentage of the licencee's sales revenues.** Many Canadian firms have licencing agreements with firms in the United States.

Licensing

Joint Venture

A joint venture is similar to the licencing arrangement. **But a joint venture involves some ownership and control of the foreign firm by the firm wanting to do business there.** Suppose a Canadian firm forms a joint venture with a Japanese firm. The Canadian firm would have a partial ownership in the Japanese firm and would also have some say in managing it. Because of this, some governments discourage joint ventures between domestic and foreign firms.

Joint venture

Foreign Manufacturing and Marketing

Foreign manufacturing and foreign marketing represent the greatest commitment to international business. The firm owns and controls the overseas plant and the marketing of its products. It may build the plant from the ground up or buy out a firm already in business there. This brings us to the multinational company.

THE MULTINATIONAL COMPANY

Multinational company

A multinational company is a firm which is based in one country (the parent country) and has production and marketing activities spread in one or more foreign (host) countries. The greater the number of these host countries, the more "multinational" it is. Such a firm truly becomes a global enterprise.

We tend to think of the United States as the home country of most international corporations. Firms such as International Business Machines (IBM), Procter and Gamble, Coca-Cola, and F. W. Woolworth Company are American-owned firms. In fact, they are global firms. They look at the world as their base of operations. Many such firms sell more overseas than they sell in the United States.

The United States is not the parent country for all multinational firms. For example, Switzerland's Nestlé does more than 90 per cent of its business outside of Switzerland. Royal Dutch Shell and Unilever Corporation do more than 80 per cent of their business in host countries. France's Michelin Tire Company and Canada's Massey Ferguson Ltd. and INCO Ltd. are other examples of multinational corporations.

The multinational company has become more visible and controversial since the end of World War II. Some people believe they are not subject to enough social control. These people question the "allegiance" of a firm headquartered in one country but having operations in scores of other countries. Some people think they have too much economic and political power.

THE "WHY" OF MULTINATIONAL BUSINESS

Modern communications and transportation have shrunk the world's size. Satellites in space can bring us live televised news from foreign nations. Jet travel means we are only hours away from foreign cities. Within seconds you can place a telephone call to someone on the other side of the globe. People of different nations know more about each other and each other's wants than anyone could have known at any other time in history. It is not strange, therefore, that many firms consider the whole world when they think about market opportunity.

One key fact helps to explain why firms engage in multinational business—the world's resources are unequally distributed among the nations. Some have lower standards of living and an abundance of labour. This tends to make labour less costly than in other countries where labour is scarce. If those workers have the skills needed by a global firm or if they can be taught those skills, the firm has an incentive to locate a plant there. Of course, the productivity of labour must always be considered.

Raw-materials-producing nations have been at a trading disadvantage with the industrialized nations for a long time. They argue that foreign-based firms come in and "take" their natural resources. These are exported to the industrialized nations and manufactured into goods. The "foreigners" get the high-paying manufacturing jobs; their governments get taxes from the firms; and the workers spend their money at home. They also believe that they are in a poor bargaining position against the big firms.

This is why some raw-materials-producing nations have formed cartels. **A cartel is a group of business firms or nations which agrees to operate as a monopoly**. Thus, they regulate prices and production.

Cartel

Raw-materials-producing nations want a better deal in trading. Multinational firms are aware of this. Many firms, rather than transporting raw materials home from a foreign country, build plants to process these materials at the source. Sometimes the nation providing the raw material insists on this arrangement. In Canada, where there is a heavy emphasis on export of primary goods, we see some evidence of discussion about cartels. The potential wheat cartel mentioned previously is one example. The Canadian government also arranged for a cartel of uranium producers in the early 1970s.

FOREIGN INVESTMENT IN CANADA

Canada's manufacturing economy has developed behind a protective tariff barrier. This has made it attractive for foreign firms to establish or acquire subsidiaries in Canada to supply the Canadian market. In this manner foreign firms supply goods to the Canadian market less expensively than by exporting to Canada and paying the tariff on the product. Canada's rich natural resources—minerals and petroleum—also attracted money from abroad. (See Table 16-3.)

Foreign firms which invest in Canada because of the tariff barriers often build small, inefficient plants to serve only the Canadian market. The Canadian market is much smaller (approximately 23 million) than the over 200 million people in the United States or in Western Europe. This cost of production, in addition to the tariff barriers erected by other countries, and the fact that most export allocations are made by head offices of foreign countries, makes it difficult for these firms to compete actively as exporters.

The market size and the tendency for much technological development work done by foreign-owned companies to be conducted outside Canada means that Canadian subsidiaries do not conduct autonomous new process or product development work. This, it has been argued, is another major disadvantage of the high degree of foreign ownership of Canadian corporations. The other side of the argument, of course, is that Canadian firms would have less access to technological developments if it were not for these foreign subsidiaries. (This subject is discussed further in the section on "The Technological Environment" in Chapter 17.)

Table 16-3. Degree of non-resident majority ownership of corporations in Canada as measured by assets, 1967 and 1974

Industry	Assets of foreign-controlled corporations as a percentage of industry assets	
	1967	1974
Agriculture, forestry and fishing	8.2	9.8
Mining		
Metal mining	42.0	55.1
Mineral fuels	81.7	74.0
Other mining	50.0	58.4
Total mining	60.0	63.0
Manufacturing		
Food	35.7	38.8
Beverages	17.6	21.9
Tobacco products	83.6	99.8
Rubber products	92.4	93.7
Leather products	21.9	22.6
Textile mills	49.6	60.2
Knitting mills	18.8	23.5
Clothing	12.0	15.5
Wood industries	25.8	27.0
Furniture industries	15.8	18.5
Paper and allied industries	38.8	43.7
Printing, publishing & allied industries	11.6	11.5
Primary metals	55.6	37.9
Metal fabricating	44.4	38.3
Machinery	71.9	67.6
Transport equipment	86.2	79.6
Electrical products	65.7	65.1
Non-metallic mineral products	47.1	62.4
Petroleum and coal products	99.6	94.4
Chemicals and chemical products	83.0	76.6
Miscellaneous manufacturing	48.7	47.1
Total manufacturing	56.7	56.6
Construction	14.0	12.6
Utilities		
Transportation	–	8.5
Storage	6.0	3.7
Communication	–	0.5
public utilities	7.3	2.4
Total utilities	6.2	4.3
Wholesale trade	28.5	27.8
Retail trade	20.4	18.2
Finance	12.1	10.7
Services	17.3	23.4
Total all industries	26.0	22.1
Total non-financial industries	38.0	32.8

Source: Statistics Canada, Corporation and Labour Unions Returns Act, *Annual Report,* 1967, pp. 50–103, and 1974, pp. 116–17. Reproduced by permission of the Minister of Supply and Services Canada.

Government Studies

There has been concern about the level of foreign direct investment in Canada since the early 1950s. **Foreign direct investment, defined in the *Report of the Royal Commission on Corporate Concentration,* is the transfer of a package of assets from a foreign-domiciled corporation through corporate channels into an enterprise in Canada, either by acquisition of an existing firm or by the creation of a new enterprise, which thereafter becomes a subsidiary of the foreign corporation and subject to its control.** The assets transferred in the package may include capital, a licence to use a brand name, and preferred access to markets and sources of new materials.

Another type of foreign investment, portfolio investment, is purchase of the stocks and bonds of Canadian corporations by non-residents.

Foreign investment in Canada increased dramatically from $8.7 billion in 1950 to $68.6 billion in 1975. Foreign direct investment, measured as the total of equity investments, undistributed retained earnings, and long-term debt owed to the parent firm, has increased from about 45 per cent of total foreign investment in 1950 to almost 60 per cent in 1975.

There have been several studies of foreign investment in Canada. The first was the Gordon Commission Report in 1957. The report pointed out that there were dangers to foreign investment. A major concern was the possibility that if U.S. subsidiaries in Canada were faced with a conflict between U.S. and Canadian positions they would choose to support the U.S. position. A key recommendation, which was subsequently acted upon, was that financial intermediaries should be in Canadian hands.

The Watkins Report in 1968 dealt with extraterritoriality, that is, the application of U.S. laws to the subsidiaries of U.S. companies in other countries. The report examined U.S. government guidelines for U.S. direct investment abroad, the application of U.S. law on subsidiaries trading with communist countries, and the application of U.S. anti-trust law to subsidiaries.

Major recommendations of the Watkins Report were to create a government agency to survey multinational activities in Canada; to compel foreign subsidiaries to disclose more of their activities in Canada; to encourage nationalization of Canadian industry; to subsidize research and development and management education in Canada; to form the Canada Development Corporation; and to forbid the application of foreign laws in Canada.

The 1970 Wahn Committee investigation examined Canada–United States relations. Much of its work was based on that done by the Watkins Study. One recommendation was that over time all foreign-owned firms in Canada should allow for at least 51 per cent of their shares to be owned by Canadians.

The Gray Report in 1976, *Foreign Direct Investment in Canada,* attempted to determine the economic forces that promoted foreign investment and to measure its benefits and costs. The report saw the major benefits of foreign direct investment as access to new technology. It resulted in increased productivity in Canada and the introduction of new and improved products in Canada. The Gray Report recommended a foreign investment review agency to screen new foreign direct investments in Canada to determine their effect.

Foreign Investment Review Agency

The government of Canada established the Foreign Investment Review Agency (FIRA) to screen new foreign direct investment in Canada. Its purpose is to ensure that significant benefits accrue to Canada from new foreign direct investment. It screens:

1. Most acquisitions of control of Canadian businesses by non-Canadians.
2. The establishment of new Canadian businesses by non-Canadians, who either do not already have any business in Canada or do not have any business in Canada to which the new business is or would be related.

Five general criteria are used to assess potential investments to determine whether significant benefits would accrue to Canada. These include:

1. The effect of the level and nature of economic activity in Canada, including employment, processing of resources, reduction of imports and increase in exports, purchasing of Canadian goods, and other such "spillovers".
2. The level of participation by Canadians as managers, shareholders, and directors.
3. The effect on industrial efficiency, technological development, and product innovation and variety.
4. The effect on the competitive behaviour of firms already in the industry.
5. The compatibility of the investment or acquisition with government industrial and economic policies.

After its initial review FIRA will bargain with the applicant to increase the net benefit of the project to Canadians.

SUMMARY AND LOOK AHEAD

Trade among countries broadens the market and permits greater exchange and specialization. The principles of absolute and comparative advantage show why there is economic benefit in trade. National trade enables each country to use its limited resources to the best advantage in raising its people's standard of living.

Despite the advantages of trade, there are many "created" barriers such as tariffs, embargoes, exchange control, isolationism, tax control, and balance of trade and payments problems. There are also natural barriers, such as distance. But progress has been made in removing many of the "created" barriers. As governments recognize the mutual benefits from trade, they want more trade.

In free economies, international trade basically means international business—firms and individuals in different countries whose perspective is the world, rather than an isolated regional area, buying from and selling to each other. There are, however, some forms of state trading even in free economies. In state-controlled economies, trading is handled through state trading companies.

There are many reasons why firms export and import goods, and there are different types of international business involvement. An unintentional exporter is much less committed to international business than is a multinational company. A multinational company is based in a parent country and has production and marketing activities in one or more foreign (host) countries.

In the next three chapters we consider the many environmental factors which can affect business decisions in Canada. These environmental factors have both national and international dimensions.

CAREERS IN INTERNATIONAL BUSINESS

Almost any business-related job you can name could be found in a firm engaged in international business. In many cases these jobs are exactly the same as they would be in a firm which sells only in its domestic market. Many others, however, require additional skills such as knowledge of a foreign language or knowledge of the culture or business practices of a foreign nation.

Some international firms employ interpreters to deal with language-related problems.

You might be interested in the opportunities available to you as an import sales manager for a firm. This job involves keeping in close contact with overseas suppliers and your domestic sales force. Remember, opportunities are not limited to big corporations. If you like to travel, you will more than likely have the opportunity to visit your suppliers' countries from time to time to negotiate contracts with them.

If you want to go into business for yourself, you might consider becoming an export agent. You would handle the export task for your clients for which you would be paid a commission. This takes a thorough knowledge of the laws and regulations governing exporting. A lot of it is detail work. Thus, you must have a knack for handling details.

Many firms that engage in international business hire people whose main job is to keep in close contact with potential buyers. This person would maintain contact with foreign embassies in Canada and to call on foreign visitors and foreign buyers who are in Canada to negotiate supply contracts.

Resident buyers are hired by importers to place orders for them in the country in which the buyer is located.

Because the physical distribution problem often is complex in export selling, many firms turn to the services of foreign-freight forwarders. They prepare documents needed to export goods and handle all the details involved in moving the goods to the buyer. The experience you'll get working for a foreign-freight forwarder will be very valuable to your future in international business.

Canadian banks also have departments of international business. If you are interested in a career in finance and international business, you should investigate opportunities with the chartered banks.

KEY CONCEPTS

Absolute advantage   A country has an absolute advantage in producing a good when it is the only country which can produce that good or when it can produce it at a lower cost than any other country.

Balance of payments   The difference between money flowing into a country and money flowing out of that country as a result of trade and other transactions. An unfavourable balance means more flowing out than flowing in.

Balance of trade   The difference between the money values of a country's imports and exports. An unfavourable balance means that the money value of imports is greater than the money value of exports.

Cartel   A group of firms or countries that agrees to operate as a monopoly. They regulate prices and production.

Combination export manager   A person or firm which represents several exporters and handles the task of exporting their goods. Helps firms with little knowledge in international business and those with relatively small export volume to become involved in international business.

Comparative advantage   According to the principle of comparative advantage, the people of a country will enjoy a higher standard of living if the country specializes in producing those goods in which it has the greatest comparative advantage or the least comparative disadvantage in relation to other countries. A country's natural resources, labour and capital cost, nearness to markets, labour skills, technological skills, and so on determine where its relative or comparative advantage lies.

Confiscation   In international business confiscation occurs when a government expropriates property from a foreign-based firm and there is no compensation to the firm.

Contract manufacturing   A firm in one country contracts with a firm in another country to produce a product intended for sale in that and other countries.

Countervailing duties   Trade barriers set up in retaliation for another country's trade activities.

Devaluation   Occurs when a nation's government reduces the value of its currency in relation to gold or other currencies.

Embargoes   Legal prohibitions on the import and/or export of certain goods into or out of a country.

Exchange control   Government control over access to a country's currency by foreigners.

Expropriation   In international business, expropriation occurs when a government takes over ownership of a foreign-owned subsidiary. The firm may or may not be compensated, and it may be sold to private citizens in the expropriating country.

Extraterritoriality   The application of U.S. laws to the subsidiary of a U.S.-based company in other countries.

Foreign assembly   A parent firm exports parts to a subsidiary or a licencee in a host country for assembly into a finished product.

Foreign direct investment   The transfer of assets from a foreign corporation into an enterprise in Canada.

International trade Flows of goods between or among nations.

Isolationism Non-involvement with other countries. An isolationist country limits its social and economic contact with other countries.

Joint venture A firm in one country which wants to do business in another country might enter into a joint venture with a firm in that foreign country to produce and/or sell a product. There is mutual ownership of the overseas firm.

Licencing In international business, licencing occurs when a licencee is licenced by a licencor to make and sell the licencor's product in the licencee's country.

Multinational company A firm having production and marketing operations spread over several countries. The ultimate commitment to international business. A global enterprise.

Nationalization In international business, nationalization occurs when a government expropriates a foreign-owned firm and the government runs the nationalized firm. The owners of the firm may or may not be compensated.

Non-tariff barriers Measures taken to restrict trade which do not involve the use of tariffs.

Piggyback exporting One firm (the carrier) uses its overseas distribution network to sell non-competitive products made by other firms (riders).

Portfolio investment The purchase of stocks and bonds of Canadian companies by non-residents.

Regional trading bloc A group of nations which reduces or eliminates trade barriers among themselves.

State trading company A government-owned operation which handles a country's trade with other governments or firms in other countries.

Tariffs Duties or taxes which a country's government imposes on goods imported into or exported from that country.

INCIDENT:

McFarlin Company

The McFarlin Company makes farm machinery. During the past three years McFarlin has marketed heavy equipment in a South American country as part of its effort to develop its export potential more fully. The equipment is produced in Canada and is exported to an independent distributor in that country.

Recently the country passed a law which will go into effect in two years. At that time, all heavy equipment, such as that sold by McFarlin, will have to be produced in that country.

Questions:
1. What does this new law mean for the company?
2. What kinds of information would you want before making a decision as to whether or not to begin production operations in that country?

SECTION FOUR   SPECIAL TYPES OF BUSINESS

Tax control   Practised by a country's government when it uses its taxing authority to control foreign investments in that country. Involves applying discriminatory taxes to foreign-owned firms.

Trade barriers   Natural and "created" obstacles which restrict trade among countries. Distance is a natural obstacle. Tariffs are a "created" obstacle.

INCIDENT:

Forest Products Company

Several years ago the Forest Products Company began exporting timber and timber products to industrial buyers in Japan. The company originally sold through an import agent in Osaka, Japan, but has since opened its own foreign sales office in Osaka. Forest Products has had sales increases each year since it began selling abroad. In fact, its foreign sales are growing much faster than its sales in Canada.

Everything was fine until a few months ago. It seems that legislation being supported by a conservation group in Canada would, if passed, raise timber prices. Forest Products fears it might lose Japanese customers to new competitors in Australia.

Meanwhile, a trade association of home-builders is placing the blame for the rising cost of building materials squarely on the shoulders of "companies like Forest Products which deplete Canadian forests and cause home-builders to pay higher prices, which, in the end, cost the homeowner." The firm has been criticized for exporting "goods needed at home".

Forest has received a lot of unfavourable publicity and criticism. In fact, political candidates on the Pacific coast have picked up on the controversy as a campaign issue.

Questions:
1. Why do you think that Forest Products Company began selling its products in Japan?
2. What types of arguments do you think are being made by the conservation lobby and the home-builders? How should the company react to these arguments?
3. Suppose you were one of the political candidates and you decided it would be "good politics" to side with the conservationists and home-builders. What arguments would you offer the voters to sway them to your side?
4. Develop a strategy for the Forest Products Company in the light of these recent developments.

ROCKETOY COMPANY VIII

Last month, Terrence got a letter from a toy distributor in Japan indicating an interest in importing several models of Rocketoy's products. The firm wanted to import the toys and sell them through its own distribution outlets in Japan. Terrence was somewhat surprised by the letter. He had never really considered selling Rocketoy's products abroad. He called in Julia Rabinovitz, the vice-president of marketing, for her thoughts on the matter.

Julia's first reaction was that Rocketoy should be thinking of moving into new market areas. Competition in the Canadian market had become very intense. Selling in foreign countries would be a logical next step in Rocketoy's development.

Terrence decided to talk to the new vice-president of production—Robert Ensminger. (Uncle Joe had resigned a year earlier due to poor health.) Robert was mildly enthusiastic. He said that it made no difference to him where the toys were sold. He had some extra production capacity, and he could produce a "reasonable" number of extra toys without any problems.

Based on Julia's and Robert's reactions, Terrence wrote a letter to the Japanese firm in which he indicated a willingness to enter into further discussion on the proposal.

Within two weeks the Japanese firm replied. It offered to make an initial cash purchase of 20,000 toys and then negotiate for additional orders if the toys sold well in Japan. Terrence, Julia, Robert, and Pam met together to discuss whether or not to accept the proposal. All agreed that this was a golden opportunity to gauge the potential of international marketing for Rocketoy.

Rocketoy, therefore, agreed to supply the initial order of 20,000 toys on the condition that most of the "details" of exporting and importing would be handled by the importing firm. The Japanese firm agreed and a contract was signed between the two parties.

Questions:
1. What do you think prompted the Japanese firm to contact Terrence about the possibility of importing Rocketoy's products?
2. Why do you think Terrence was "surprised" by the proposal?
3. Compare Julia's and Robert's reactions to the proposal.
4. How would you describe Rocketoy's commitment or involvement in international business after signing the contract with the Japanese firm?
5. Suppose that the market reaction to Rocketoy's products in Japan is very enthusiastic. What advice would you give Terrence about further negotiations with the Japanese toy distributor? Explain.

QUESTIONS FOR DISCUSSION

1. Define the principle of absolute advantage and the principle of comparative advantage.
2. Should a country whose business firms can produce everything its people demand at a lower cost than firms in all the other countries engage in international trade? What assumptions are you making when you answer?
3. Define balance of trade and balance of payments.
4. List and discuss three human-made barriers to international trade.
5. List and discuss four arguments used to justify tariffs. If you are against tariffs, how would you respond to each of these arguments?
6. How does devaluation of a country's currency affect exporters of goods in that country?
7. Why do some countries form regional trading blocs? Are there any similarities between trade among member countries of the European Economic Community and trade among the provinces in Canada? What are the differences?
8. Does greater international trade among nations lead to greater interdependence among them? What are the implications of your answer for world peace?
9. Does our government encourage Canadian-based firms to invest abroad? Why or why not?
10. To which country does the management of a large multinational company owe its primary "allegiance"? Discuss.
11. Is it ethical for a Canadian-based company to open a plant in a low-wage country when unemployment in Canada is at a high level? Discuss.
12. Discuss the similarities and differences between the "exporting firm" and the "multinational firm".
13. What is an unintentional exporter? Give two examples.
14. "Investments made by American-based multinational firms in Canada are good for both the United States and Canada." Do you agree? Why or why not?
15. "It is easier for a firm to live up to its social responsibility when its operations are confined to one country than when it engages in international business." Do you agree? Why or why not?
16. Define: (a) joint venture; (b) licencing; (c) contract manufacturing; and (d) state trading.
17. "Canada should limit foreign ownership of firms in this country." Do you agree or disagree?

Section Five

Business Environments and Your Career

Up to this point we have shown why economic systems exist and why the business firms within them come into being. We have also discussed the motivations of businesspeople in a capitalistic system and have described the production, marketing, and financial functions of business. All management decision making takes place in an environment which can make the difference between success and failure.

The next three chapters look at this environment. It includes social, political, economic, ecological, and technological dimensions. All are inter-connected and change constantly.

Business managers make better decisions if they know how the environment affects their decisions. Recognizing trends in the environment helps managers plan business activity and helps them understand customers and employees. Understanding values in the environment also helps managers set standards for their own business behaviour. This is important because in the long run a firm's management must behave in a way which is acceptable to society. What society expects of business usually ends up in the form of laws. Business executives must understand and comply with the laws.

The economy, its health, and its growth rate have a lot to do with business success; technology often determines the success or failure of an economy and the firms in it. A business must know what is happening in the area of technology as it relates to its products and services.

WHAT IS AN ENVIRONMENT?

The environment of an organization consists of all those outside things which come in contact with it and influence it. The influence can be direct or indirect. A carrot seed planted in a garden has an environment. It includes the soil in which it is planted, the temperature of the air around it, and the moisture which is provided. The growth and development of the seed depend on the environmental conditions and the quality of the seed itself.

The success of a business organization, like that of any living thing, depends on two major things: (1) the quality of the inputs, and (2) the quality of the environment in which it is found. We have examined the inputs of a business—human and material. Business managers cannot be successful as decision makers over the long run if they consider only internal factors. We now turn to the other major determinant of success—the environment.[1]

The environment of business can be studied in different ways. It can be looked at as an almost infinite number of different influences. More appropriately, however, these influences are grouped together into a smaller number of categories. One way of grouping environmental influences is into different institutions. These institutions include government, labour, consumers, environmentalists, professionals, education, religion, technology, the law, customs, ideas and beliefs, and many more. The difficulty with the institutional approach to the study of the environment is the large number of different institutions which must be studied and the complexities of the relationships between them. In our approach, institutions and the relationships between them are discussed as part of the social environment.

We have adopted a "factors" model of the environment. In other words, the environment is viewed as being composed of political, economic, social, technological, physical, and ecological factors.

The Environment Is Dynamic

The diagram on the opposite page suggests that the various aspects of the environment of business are dynamic. They are subject to rapid change. Something which seemed irrelevant last year might become vitally important this year. For example, a change in an immigration law might make it easier for a firm to hire skilled foreign workers. This could make a big difference in its decision regarding the kinds of machines it should install or the price it should charge for its product. A rise in the minimum wage might lead a supermarket chain to reduce the size of its labour force. The growth of the women's movement may cause an advertiser to change the content of a series of TV commercials.

---

[1]Since a firm is often a large set of people and things, it could be said to have an internal environment distinct from those things outside which affect its operation. The internal environment might include the attitude of workers towards the firm and towards their own jobs—their morale. It might also refer to physical working conditions, such as lighting, temperature control, space, and so on.

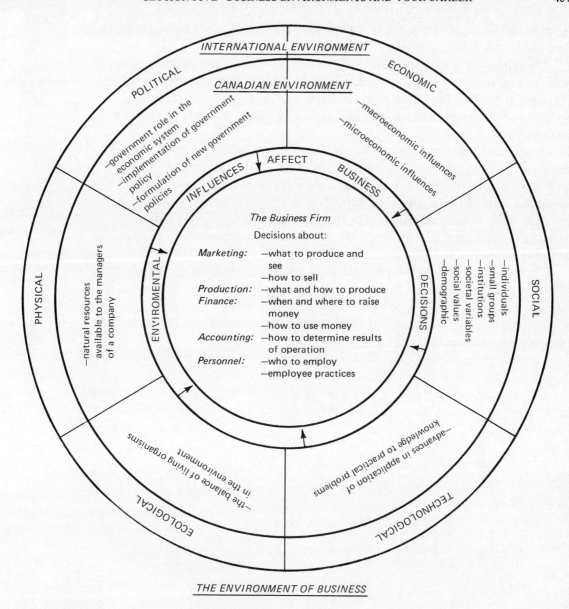

The Environments Are Interdependent

The social, economic, political, and other environments are interdependent. For example, the economic health of a country affects the growth of technology in that country. More money is available for research and development in wealthy countries than in poor countries. Likewise, the ethical and cultural values of a democratic people are reflected in the laws they pass and the kind of government they elect.

Even though the environment is external to a firm, a firm can have some influence on it. For example, business firms work to influence government decisions at federal, provincial, and local levels. Advertising by firms can influence attitudes of members of the consuming public towards products and services.

In the first three chapters of this section you will become acquainted with many of the significant environmental factors which business managers must cope with. You will be presented with a framework for understanding and dealing with the various environmental factors. You will also study many important environmental trends and gain an appreciation of their impact on business decision making. In addition, there will be some discussion about predicting future environmental conditions.

Chapter 17 deals with the physical, ecological, and technological environments of business. It discusses our basic resource endowment, its ecological limits, and how the basic resources can be enhanced by technology.

Chapter 18 describes economic and social environmental variables which business managers must consider.

Chapter 19 describes the political environment of business firms and describes its effects on the decisions of business managers.

The last chapter of the book discusses future career prospects. You will want to consider the impact of environmental trends before you decide which career you will pursue.

17
The Physical, Ecological, and Technological Environments

SECTION FIVE BUSINESS ENVIRONMENTS AND YOUR CAREER

OBJECTIVES: After reading this chapter, you should be able to:

1. Describe how the physical environment can be important to a company.
2. Illustrate how management has improved its ability to cope with the physical environment.
3. Discuss the nature of the ecological environment.
4. Indicate the extent to which business contributes to the pollution problem.
5. Describe how society must make trade-offs when dealing with the pollution problem.
6. Illustrate the importance of technology to our way of life.
7. Discuss what technological progress is.
8. Give examples of how business decisions are influenced by technology.
9. Discuss the two predominant tasks involved in the management of technology.
10. Discuss the relationship between technology and unemployment.
11. Outline "Canada's national interest in technological development."
12. Indicate various forecasting methodologies.

KEY CONCEPTS: In reading the chapter, look for and understand these terms:

PHYSICAL ENVIRONMENT
ECOLOGY
TECHNOLOGICAL ENVIRONMENT
TECHNOLOGY
RESEARCH AND
   DEVELOPMENT
FORECASTS
FORECASTING TECHNIQUES
VISION

INTUITION
EXTRAPOLATION
CORRELATION
SCENARIOS
SIMULATION
CONTENT ANALYSIS
INFORMED JUDGMENT
DELPHI TECHNIQUE

17 THE PHYSICAL, ECOLOGICAL, AND TECHNOLOGICAL ENVIRONMENTS

This chapter deals with the physical, ecological, and technological environments of business. Technology is perhaps the most important contributor to our current standard of living. Technological developments can have a great influence on decisions of business managers. Ecological conditions are, in some instances, disrupted by technological and industrial developments. Maintaining balance in the ecological environment can be an important factor in decisions made by business managers. Physical environmental factors such as the weather and natural resource availability can influence the decisions business managers make. They can also affect the outcomes of actions taken by managers.

THE PHYSICAL ENVIRONMENT

The physical environment is the base from which all economic activity starts. **The physical environment is our natural resource endowment and includes minerals, vegetation, animal, and water resources. The weather, as part of our natural resource endowment, is also an extremely important part of the physical environment of business.**

Physical environment

At first glance, it may be difficult to grasp the importance of the physical environment to business. This is particularly true for individuals who were reared in cities. However, there are many industries in Canada which are not only influenced by, but are dependent upon, the physical environment. These include companies involved in primary industries such as agriculture, fishing, mining, petroleum, and forestry. These industries have been the traditional strength of the Canadian economy, and they still account for considerable employment. Furthermore, they are important exporting industries.

Our dependence upon these industries has led to the observation that Canadians are "hewers of wood and haulers of water". Continued economic prosperity and resolution of problems such as the energy crisis will require that companies continue to develop more efficient ways to utilize the existing resource base. Efforts will also be required to at least maintain and, if possible, expand the size of the resource base. In the case of petroleum and base minerals, this will involve exploration efforts to find new resources. Attempts will be made to bring new agricultural land into production. Efforts must also be made to halt the encroachment of urbanization into highly productive fruit- and vegetable-producing land, such as the Niagara Peninsula and the Okanagan Valley. The forest industries must continue to husband their resources as well.

Other industries are also influenced by another facet of the physical environment—the weather. The classic example is tourism. Hawaii, because of its natural setting and climate, attracts far more Canadian tourists than many parts of Canada. Airline operators are another example of firms which are influenced by the weather. When the weather is bad and prevents flights from taking off or from landing at specified destinations, it interrupts the routine operations of these companies. Farming operations are also influenced by the physical environment. The type of soil will have an effect on what crops can be grown and the yield of these crops. The weather also has an effect on, and will determine what types of, crops can be grown in an area. For example, fruit crops can be grown in the Niagara Peninsula but not in northern Ontario. Variations in rainfall, temperature, and abnormal conditions such as hail can influence the year-to-year crop yield.

Adverse weather conditions over a wide area can result in significant decreases in world grain production. This can result in increases in grain prices. People interested in the grain trade watch with interest for reports of wheat crop predictions in large producing countries such as the United States and the USSR.

Beverage company sales can be affected by weather conditions. For example, hot sunny weather will result in higher sales of beer and soft drinks than will cool and cloudy weather. There has been some concern in recent years that the long-term trend is towards a cooler climate in Canada. This would have serious implications for agriculture in particular.

Secondary Effects

There are many manufacturing and service industries which are influenced by the relative success of the primary industries in coping with and/or exploiting the physical environment. A new mineral or oil discovery can result in expansion of town sites in the vicinity. Supply and service industries benefit from the existence of companies which cope successfully with the physical environment. For example, there is a tremendous amount of activity associated with the development of the tar sands in northern Alberta.

Management and the Physical Environment

In earlier times when people had less understanding of the physical environment, they accepted it as it was and they worshipped various facets of it (e.g., the sun, the wind, and the rain). Phenomena such as solar eclipses, earthquakes, and floods were taken as signs from the gods.

We are now in a position, because of the technology which has developed through the centuries, to better cope with our physical environment. Construction firms are able to divert waterways for electrical generation, flood prevention, and irrigation. By using fertilizers, farmers are able to reduce the adverse effects of low levels of rainfall on crop yields. Chemical herbicides and insecticides make it possible for farmers to cope with pests which threaten their crops.

Weather forecasting techniques also make it possible to anticipate adverse conditions and prepare for them in advance. Sophisticated exploration techniques for oil and metals make it possible to search farther below the surface of the earth for these resources.

The physical environment is still tremendously important to our economy and, to a greater or lesser degree, to various firms in the economy.

Much economic activity is based upon, or influenced by, our natural resource endowment. Changes in the physical environment such as weather conditions can have profound effects on many types of firms. Management cannot control the physical environment. Through use of technology, however, it is possible to predict certain characteristics of the physical environment (such as in mineral exploration) which can lead to better management decision making. It is also possible to anticipate certain changes, such as weather conditions, also enabling more informed decision making.

THE ECOLOGICAL ENVIRONMENT

Ecology is the branch of biology which deals with the relations between living organisms and their environment. The ecological environment of business is important because many people feel that business firms have contributed to pollution of the physical environment. There is particular concern about water and air pollution and about litter and clutter affecting the visual environment.

Ecology

It is true that activities of some business firms have contributed to pollution of the physical environment. There are wastes dumped into inland waters. Examples include the mercury emission controversy in the English-Wabigoon River system in northern Ontario, the discharge of materials by chemical and forest industry companies, and the use of water as a coolant which is returned to its source at a higher-than-normal temperature.

Many examples of air pollution can also be found. One is the emissions in the area surrounding Sudbury. There is also concern in some areas about the aroma of large cattle feed-lot operations.

Products produced by business firms often can be seen littering the countryside and roadsides. Although soft drink manufacturers did not intend their containers to be left after use, they are often held to blame for the problem.

The extent of the overall pollution problem and the contribution to it by business firms must be kept in perspective. Three points are relevant here. First, although there are isolated and individual problems, Canada as a whole does not have a serious pollution problem. We can all think of some specific examples of pollution: the vegetation growth in Lake Okanagan, the Sudbury experience, the pollution of the Great Lakes, and occasional oil spills off our coasts. Certainly these individual situations are undesirable and preventable and remedial action must be taken. However, if we take the broader perspective that we have a vast country and many parts of it are still virtually untouched, we do not have a serious general pollution problem.

Secondly, business is not the only source of pollution. Private citizens and public institutions contribute significantly to the pollution problem. Many municipalities dump untreated sewage into waterways. Some people argue that the city of Montreal would have been better advised to spend money on sewage treatment facilities than on facilities for the 1976 Olympic Games.

Thirdly, the problems associated with pollution are not new. Industrial pollution began with the Industrial Revolution in the 1700s, and the general problem of pollution existed long before that. For centuries, cities around the world have been struggling with the problem of waste disposal. The plagues of the Middle Ages have been attributed in part to poor sanitary conditions in cities. The smog problem in London, England, was mainly a result of householders burning coal to heat their homes. When they switched to smokeless fuel the smog problem dissipated.

WHAT DO YOU THINK?

Private and Public Weather Forecasting

The Atmospheric Environment Service (AES)* of the government of Canada provides weather service to many companies as indicated in the following examples:

A small designer of solar-run buildings requests radiation statistics for Toronto, including mean hours of sunlight, by month, since records were first kept in 1938.

Coca-Cola Ltd. wants to know deviation from normal levels for both temperature and precipitation at different spots across the country so sales curves can be related to local weather conditions.

Imperial Oil Ltd. is routinely supplied with specialized forecasts for its drilling operation. The British Columbia Forestry Service receives forecasts and data related to its needs.

In the summer of 1977, AES was involved in an ambitious company-sponsored program. Five departmental meteorologists were aboard oil rigs in the Beaufort Sea supplying on-site wind and wave forecasts to a subsidiary of Dome Petroleum Ltd.

There are also more than 50 private weather consultants in Canada. Four firms undertake activities which span the full range of meteorological services offered by private industry. They are Weather Consultants of Canada, MEP Company, Acres Consulting Services Ltd., and Weather Engineering Corporation of Canada.

These private companies resent AES activity such as that provided for Dome Petroleum. One comment was that "The government has for years been actively discouraging the growth of the private sector's role in meteorology by increasing the services it offers." WHAT DO YOU THINK? Are weather forecasting services useful to companies? Should this work be done by the private sector or left to the government?

*Based on a July 11, 1977, article from the *Financial Times of Canada*.

While pollution in Canada tends to be found in isolated and individual situations as opposed to being a general problem, it is still important that individual situations be seriously assessed and appropriate action taken. The heightened interest in North America regarding control of the spread of pollution in the 1960s was a positive step forwards. With the oil crisis and economic problems of the 1970s there was some slackening in the move towards pollution control. However, there is still much more interest in, and commitment to, the concept of a clean environment than there was in the pre-1960 period.

Need for Priorities and Trade-Offs

Decisions about which anti-pollution measures to use must take monetary issues into account. It would be possible to control all emissions from public and private sources so that absolutely no pollution would occur. However, there would be a tremendous cost associated with this level of control. Priorities must be established by governments and business regarding what pollution problems to address and how many resources to allocate to pollution control relative to other things.

At a general level, the dilemma for business, and for governments as well, is that emphasis on pollution control in a single industry will result in that industry's output being non-competitive internationally if other countries do not have the same pollution standards. A similar situation exists for individual companies in an industry. If one company buys pollution control equipment and other companies do not, then the innovating company will have costs which are uncompetitively high.

In dealing with the pollution questions raised by the ecological environment, business managers must be aware of the increasing emphasis on pollution control. In

---

WHAT DO YOU THINK?

The Cost of Pollution Control*

A study completed for the federal government's Department of Energy, Mines and Resources estimated the costs for INCO to meet provincial pollution standards at its Sudbury operation. One federal government source said that if the company did meet the standards, economic mining of the Sudbury copper and nickel deposits could end within 20 years. This compared with an expected 40 to 60 year life span under existing conditions. WHAT DO YOU THINK? How should the company go about deciding its position on this issue?

*Based on a September 19, 1977, article in the *Financial Times of Canada*.

general, every attempt should be made to maintain the best possible emission standards. However, the costs of pollution control must be carefully assessed. Business managers must also be aware of the prescribed emission levels laid down by governments. Dealing with the ecological environment is a complex problem for business managers.

Increased concern about the ecological environment has also created numerous opportunities for new products. An entire industry has grown up which sells products to reduce air and water pollution. Devices to process smokestack emissions and waste before it is released into water sources are products for which markets have developed since the early 1960s.

THE TECHNOLOGICAL ENVIRONMENT

Technological environment

Technology

The technological environment includes all applications of knowledge which have an impact upon a business firm. The high standard of living which today's Canadians enjoy depends on our desire and ability to pursue the benefits of technology. Technology has enabled us to move far beyond the products and the life style available from a primitive physical environment. **Technology is the application of knowledge so that people can do entirely new things or do old things in a better way.** It is, in other words, the application of knowledge for practical purposes. Technology created a new "trash compacter" for the household and an "atom smasher" which has revolutionized warfare and energy production. Technology makes it possible for firms to develop new products and processes.

What Is Technological Progress?

Technological progress results in improvements in the state of industry, manufacturing, and commerce. But not all such progress leads to the betterment of humanity. This is one of the most serious challenges facing us today. How do we harness our great technological know-how for human welfare?

In the past, business growth depended on technological development. While this is still true, the progress is partly offset by problems created by technology. These include the harmful effects of obsolescence, waste, ecological disturbances, and the threat of atomic war.

Some Recent Technological Developments

Since 1940 many fantastic discoveries have been made: nuclear power, space exploration, television, computers, the "new biology", and means to control major diseases like polio. Less dramatic, but of major importance to human welfare, are developments in statistical techniques, human psychology, crop yields, contraceptive methods, and exploration of the sea. Businesses must know about new technology and contribute to it in order to survive.

How Does Technology Affect Business Decisions?

Nearly all firms are influenced by new technology. It can affect marketing, production, finance, personnel, and accounting decisions. It presents either an opportunity or a risk. The opportunity for a firm is that if it can develop a successful new product or process it can gain either a sales or cash advantage over its competitors. One risk is that if its competitors develop the technological advantage it can have negative financial consequences for a firm. Another risk is that management will not be able to implement the technological development profitably.

Technology and Management

There is considerable opportunity for business firms in Canada to improve their economic performance by initiating and using technological improvements. For business firms, technology, or the application of knowledge to practical problems, is utilized either to develop new products, to improve existing products, or to improve the process by which a product is produced.

Two different activities are involved in the commercial introduction of technological improvements. The first is the development of the technological improvement, and the second is its implementation. The two activities require very different abilities. The person or firm which is good at one may not be good at the other. Success in developing a new product or process does not guarantee commercial success. In this sense, technological improvements are no panacea for owners and operators of firms. The skill to develop technological improvements must go hand-in-hand with the marketing, finance, personnel, and other management skills required to successfully operate a firm.

Technology and the Need to Plan

Another outgrowth of rapid change in technology is the vital need to plan. If a firm is to succeed in the long run, it must not define its objectives too narrowly. Petroleum producers are wise to think of themselves as "energy companies". They must consider substitutes for oil as sources of energy. Most manufacturers of containers used to be specialists in one kind of container (glass, paper, or metal). Now they produce all three types, mostly because of uncertainty about which type of container will prevail in the long run.

Technological obsolescence means the replacement of a technical product or product feature by a newer, better, or cheaper one. This encourages firms to reduce the risk of obsolescence by introducing their own new product. As we have seen, many firms engage in planned obsolescence. In other words, they plan to introduce new versions of products which will encourage previous buyers to replace their older (but still useful) models.

Patents give a legal protection from theft or imitation of technological ideas. They play an important role in this kind of planning. When a firm wants a patent on a new discovery, it registers it with the Canadian Patent Office. A patent protects that discovery from being copied by competitors for a number of years.

Patent law is highly complex. A large number of patents may be needed to protect a new idea fully. Careful development of patents is one way of planning to meet future product competition.

Technology and Social Problems

Ever since machines became important in the production process, there has been concern that they would take over the work done by humans and that massive unemployment would result. Does automation cause unemployment? The best answer to this question is that automation does cause unemployment in the immediate area where it is applied, but overall, automation creates far more jobs than it takes away. The reasoning behind this conclusion is shown in Figure 17-1.

The use of automation (e.g., in a factory) will generally result in lower costs in producing the product, and hence in increased productivity. This increase in productivity allows the company to sell the product at a lower price. Generally speaking, the lower the price of a product, the greater the demand; increased demand for goods in general results in economic growth. With economic growth comes an increase in demand for workers and an increase in the number of jobs.

The above analysis is an accurate reflection of how automation affects unemployment on a society-wide level, but it may not be accurate in a specific situation. Consider the case of the Weed Drill Company, which is planning to open a highly automated factory with only 25 supervisory and maintenance personnel. The new factory will replace a nearly obsolete factory which presently employs 300 workers. Weed Drill Company must face the problem of the unemployed workers. Economic theory says that improved technology will, in the long run, benefit the whole economy. But this theory gives little comfort to those who will be out of jobs.

The problem is especially tough when a worker in such a case is not easily reemployable. If there is already an oversupply in the worker's specialty, the worker may have to be completely retrained. Sometimes private or government-sponsored retraining programs are available. If the Weed Drill Company has other plants, it may find a place elsewhere for its workers. If it is unionized, the union contracts may have special provisions to protect members who cannot find suitable jobs. In any case, technology can contribute to worker displacement, which, in turn, creates social problems.

We have already discussed another technology-related problem which affects society—pollution. A specific case which relates to energy technology has become a major issue of the day. Rapid depletion of high-grade, low-polluting fuels (natural gas and oil) makes the use of low-grade, high-polluting fuels (coal) more economically feasible. This does not make such a use socially desirable, though. Situations such as these require solutions which are not 100 per cent satisfactory either in terms of ecology or of profit. We will probably design less-polluting techniques for burning coal. In other words, technology will, we hope, overcome the bad side effects of other technology.

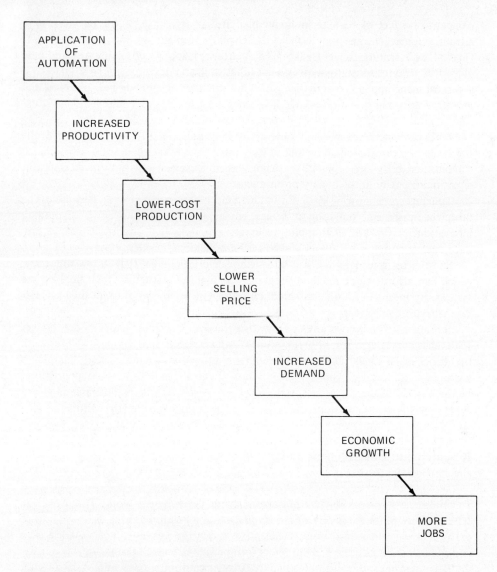

Figure 17-1. Automation and unemployment

Technological advances such as computers are also used to combat social problems. Computers are used to spread job information among various labour markets. They are also used to improve educational systems and to analyze law-enforcement problems.

Research and Development

Firms must be aware of technological change so that their processes, products, and product features will not become obsolete. They must also be ready to counteract the

R&D

competitive effect of such technology. If a firm is aggressive, it will be the first to introduce a new, cheaper way to make its "Gizmo". It might also introduce a "Mark II Gizmo" with features which make a competitor's "Wutzit" obsolete. Any of these objectives requires substantial investment in research and development (R&D). **R&D is a general name applied to activities which are intended to provide new products and processes.** It usually requires a large investment in laboratories, equipment, and scientific talent.

R&D can lengthen a product's life cycle. It can also lead to the quick end of the life cycle of a firm's product or that of its competitors. R&D, then, is the key to much competition today and represents a major class of business activity. It often results in rapid improvement in the quality of products.

Today there is growing criticism of this kind of economic growth. Because of the perceived waste and pollution it brings, some favour a slowing down of product innovation, particularly as it applies to luxury products. Is it important that we produce endless numbers of "bigger and better" products?

Some firms consider R&D as a substitute for price competition. A few large firms which can afford major research programs use them to strengthen their hold on the market. By means of continuous product improvement and innovation, they can prevent entry of competitors.

But a small firm which gives its scientists freedom to explore and a modest budget sometimes comes up with a "breakthrough" which puts it on a competitive basis with the giants in its industry. Dr. E. H. Land's photographic genius made "Polaroid" a name to stand beside "Kodak" in an industry which was nearly a monopoly before World War II. Smaller firms are often forced to merge with others to be able to afford R&D. Sometimes they must purchase patent-licencing rights from larger firms.

Research and Development and the National Interest

Technological advances result from research and development work. This work can take a variety of forms. It may be the inspiration of a practising manager about how to perform a particular task more efficiently. It may be heavily funded work over a period of time by a business, government, or university research team.

It is important that research and development contribute to technological advance in Canada so that our industries can remain internationally competitive.

Because wage rates in Canada are higher than in many other countries, technological leadership is a factor which can give us a competitive advantage over other countries. Technologically advanced products can contribute to the export performance of Canada.

In 1976 expenditures on R&D were $1.932 billion, in current dollars, which amounted to a little under 1 per cent of Gross National Product. This level was lower than previous years. In constant 1971 dollars, R&D expenditures have varied between $1.1 billion and $1.2 billion since 1970. Consequently, the ratio of R&D spending to GNP has declined since 1970 in real terms.

International comparisons show that Canada's R&D effort is below levels in other

Table 17-1. Gross expenditures for R&D in the natural sciences as a percentage of GNP for 10 OECD countries

	Total R&D in all sectors as a per cent of GNP		Total R&D by industry as a per cent of GNP	
	1969	1973	1969	1973
Australia	1.0	1.2	n/a	n/a
Canada	1.3	1.1	0.5	0.4
Denmark	1.0	0.9	n/a	n/a
France	1.9	1.8	1.1	1.0
Germany	1.7	2.0	1.2	1.2
Japan	1.6	1.9	1.0	1.1
Netherlands	2.1	1.9	1.3	1.0
Sweden	1.2	1.6	0.8	1.0
U.K.	2.2	1.9	1.4	1.2
U.S.A.	2.9	2.3	2.0	1.5

N/A not available.
Source: OECD, *Statistical Tables and Notes,* Vol. 5; United Nations, *Yearbook of National Accounts,* Vol. 3; Australia, Project Score.

industrial economies as shown in Table 17-1. This table shows that Canada lagged, both in terms of total R&D and R&D performed by the industrial sector.

Since 1963 expenditures on R&D performed by industry have been constant at about 40 per cent of total R&D spending. The percentage distribution of R&D expenditures by performer is shown in Table 17-2.

Within the private sector, R&D spending is highly concentrated in a relatively few companies and in relatively few sectors. According to the federal Ministry of State for

THE INTERNATIONAL TECHNOLOGICAL ENVIRONMENT

Problems and Opportunities

One of the biggest differences between "have" and "have-not" nations is the technology gap. Advanced transportation and communication are taken for granted in Canada. In many under-developed nations, they are almost non-existent. Managers must evaluate the level of technology in those countries where they expect to operate. This may enable them to adapt their operations to the present technology or attempt to "import" new techniques. But importing these "new ways" sometimes leads to resistance, especially when the local people do not understand them.

Table 17-2. Percentage distribution of R&D expenditures by performer

	1963	1971	1976
Government	42.5	32.2	32.2
Business Enterprise	41.9	39.3	40.7
Universities	15.6	28.5	27.1
Total R&D	100.0	100.0	100.0

Source: Statistics Canada, cat. 13-003. Reproduced by permission of the Minister of Supply and Services Canada.

Science and Technology, 49 companies, constituting 6 per cent of the companies doing R&D in Canada, accounted for over 55 per cent of the total industrial in-house R&D expenditures in 1976. This pattern of a small number of large companies doing the largest share of private sector spending on R&D is not unique to Canada. For example, in 1973 in the United States, companies with more than 25,000 employees accounted for 75 per cent of industrial R&D expenditures.

The high level of foreign ownership of Canadian industry may account for the low level of R&D spending by the private sector. A study conducted by the federal Ministry of State for Science and Technology indicated that large Canadian-owned firms imported less than half the R&D they required. On the other hand, foreign-owned firms tended to concentrate their R&D activities in their home countries. Subsidiaries which performed a significant amount of R&D in Canada relied heavily upon their parent's R&D activities.

Subsidiaries of foreign-owned firms (primarily U.S.) obtain much technology from their parent companies. Domestically owned firms often utilize arrangements whereby they get access to foreign-developed technology through licencing agreements. They pay a fee for the licencing agreement, and often the agreement includes a clause about what markets (export) can be served by the Canadian firm. Exports by U.S. subsidiaries and Canadian-owned firms are also often limited by a smaller scale of operation in this country which results in higher costs than in competitive exporting countries. The extent of exports by U.S. subsidiaries is also often determined by decisions made at the U.S. head offices of these companies.

The argument has been made that Canada's reliance on foreign sources of technology may harm us in the future. Because increasing international competitiveness will in the future be tied to technological advantages, we could be harmed if foreign governments began to limit the export of their technology. The conclusion of the argument is that Canada should become more technologically competitive. One other side of the argument, however, is the fact that we would have to sacrifice some other things in order to apply resources to development of technological leadership.

Forecasting

While it is useful for us to have an understanding of the current environment of business, the most useful information for a business manager is the characteristics of the *future* environment of business which are relevant to the firm.

The most relevant environmental characteristics will vary for different types of firms. For example, a public utility would be very interested in the demographic characteristics of its customers. A retail operation would be very interested in information about competitors' processes and product lines.

In making a decision about allocating resources, the business manager is always making an assumption (implicitly if not explicitly) about what the future environment will be. In business, as in betting on a horse race, the more accurate your assumptions about what will happen (i.e., your forecast), the better your chances are of making money, or the lower your chances of losing it.

Forecasts are assumptions about what the future environment of business will be. A forecast can be based on intuition or on studies which cost hundreds of thousands

Forecasts

WHAT DO YOU THINK?

Need for an Industrial Strategy*

In a speech to the International Federation of Operational Research Societies in June, 1978, Josef Kates, Chairman of the Science Council of Canada, said that there is a trend towards technological protectionism among industrialized countries because of the growing awareness that technology confers certain competitive trade advantages. "This has reached the level of serious policy discussion in the United States, with proposals being considered for a seven-year moratorium on the export of any technology financed by the Government."

Technology, he saw, was one of the few competitive factors currently favouring Western economies. Unless Canada begins to move towards greater technological independence, it will continue to lose its market share in international trading.

The real problem, according to Mr. Kates, lies in Canada's tariff policy, which has created a "fragmentation of manufacturing plants and inhibited the growth of Canadian companies." Extractive industries and branch plant operations have proliferated, but technology-based manufacturing industries have not been encouraged.

He said: "If we do not seize upon some strategy, and quickly, we will continue to lose control of our economic destiny . . . . Ten years from now, our area of choice, our capacity for self-determination and our will to resolve the problem will be further and perhaps irretrievably vitiated." WHAT DO YOU THINK? Do you agree that Canada should adopt a national strategy to funnel more resources into industrial research and development?

*Based on a speech reported in *The Globe and Mail*, Toronto, June 21, 1978.

of dollars. It is never possible to forecast the future with absolute certainty. However, it is possible to gain a better appreciation of what environmental developments might occur in the future. There are a variety of forecasting techniques which can be used. These are discussed below in terms of the physical, ecological, and technological environments. However, they can also be utilized for forecasting the other environments of business as well.

Forecasting techniques

Forecasting techniques are specific methods used for making predictions about the future of business environments. Some of the techniques for environmental forecasting include: vision, intuition, extrapolation, correlation, scenarios, simulation, content analysis, and informed judgments.

Vision

Vision is when a person decides that he or she wants something to happen and goes out and does it. This was essentially the case with the United States' effort to put a man on the moon. In the early 1960s the commitment was made to walk on the moon within a decade, and the vision became a reality in July, 1969.

Intuition

Intuition is a judgment by an individual or a group based on limited information. Brainstorming is another way of describing this method of formulating assumptions about the environment. If you hear someone say that it is their "gut feel" that a certain event will occur, you are listening to a prediction based on intuition. Many decisions are made this way.

Extrapolation

Extrapolation is the projection of past and present circumstances into the future. This could involve projection of past trends. For example, because resource use per capita and pollution have been increasing in the past, it might be assumed that they will continue to increase. Sophisticated extrapolation involves use of time series regression analysis. An example would be the projection by representatives of government that the automobile industry's technological efforts will be able to meet new pollution emission standards. Extrapolation could also involve anticipation of the same conditions in the future which have prevailed in the past.

Correlation

Correlation involves making a prediction about one outcome based on knowledge of other outcomes. For example, the cost of producing a new technologically advanced product can be related to the length of time the product will be in production. Experience curve data compiled by the Boston Consulting Group indicate that for every doubling of the time of production, costs decrease by 20 to 30 per cent. A forecast that inflation will occur as a result of an increase in the money supply is another example of a correlative prediction.

Scenarios

Scenarios are statements of how an event could unfold. One weather scenario is that within the next century our average annual temperature will drop by three degrees. Another is that it will increase by three degrees, and still another is that it will remain the same. The most likely scenario is that the average annual temperature will remain the same, and we can plan on that basis. However, if the possibility is recognized that the other scenarios *could* also happen, contingencies can be developed for them. The time and resources allocated to developing contingencies would have to relate to the probability that an alternative scenario would develop. Scenarios can be developed for a wide variety of things. Examples include: future resource availability; future sources of energy; future levels of pollution; and technology for preventing undesirable atmospheric emissions and weather emergencies such as hail, hurricanes, or tornados.

Simulation involves developing a model of an actual situation and working through the simulation to see what kind of outcome results. Simulation models were used by the "Club of Rome" analysts to predict future high levels of pollution and exhaustion of resources on this planet.

Content analysis is examination of the content of publications for reference to specific items. Content analysis of patent publications or of scientific and trade publications for reference to certain types of technological advances could be used to predict breakthroughs with respect to the development of certain types of products.

Informed judgment is expert opinion on a subject. It is based on a thorough understanding of all the facts in a given situation. It could also be referred to as a projection based upon an analysis of a situation. In most cases, the informed judgment

Simulation

Content analysis

Informed judgment

WHAT DO YOU THINK?

Physical, Ecological, and Technological Forecasts

Many different forecasts can be made. Some may occur; others probably will not. How might you check some of these forecasts?

1. By the late 1990s the natural resources in North America will be so depleted and our environment so polluted that our standard of living will be reduced to that of the late 1800s.
2. By the year 2025 nuclear fission as a source of energy will be an economic reality.
3. Within ten years our dependence upon the United States for technological assistance will make us an economic slave of that country.
4. Cable television companies will replace the current television networks as the prime producers of home entertainment.
5. Solar energy will be economically feasible for individual homes by the year 2000.
6. Technology will solve all pollution problems within the next 30 years.
7. Within the next 50 years the amount of effluent reaching the world's oceans will put them in danger of a pollution crisis.
8. The trend to technological obsolescence of products will be reversed and be replaced with a trend towards higher-quality, longer-lasting, more expensive products.
9. Our weather will become more and more severe.
10. We will deplete our natural oil, mineral, and forest resources by the year 2050.
11. After 1995 we will be unable to further increase agricultural productivity because chemical fertilizers, herbicides, and insecticides will be restricted due to pressure from environmentalists.
12. We will continue indefinitely to find technological solutions to ecological and physical resource shortage problems.

WHAT DO YOU THINK? What forecasts would you accept? What methods would you use to check them?

Delphi technique

of someone is probably the last thing done prior to acceptance of a forecast. **A method of combining the informed judgments of a number of experts on a subject is known as the Delphi technique.** This involves getting opinions on a subject from a group of experts and then going through a process of giving them feed-back about what other experts in the group said and allowing them to reformulate their own opinions.

Different forecasting methods will be used for different purposes by different people. In some cases a variety of methods will be used.

SUMMARY AND LOOK AHEAD

In this chapter we have considered the physical, ecological, and technological environments of business. The physical environment is our natural resource endowment. Technology allows us to develop the physical environment for the benefit of society and to make us less dependent upon it.

The ecological environment has become of greater concern in the last two decades. People have become more concerned with not altering the natural balance of the physical environment for aesthetic reasons and because of concern that the natural physical environment may be less able to serve our needs in the future.

Technology is the application of knowledge to enable people to do entirely new things or to do old things in a better way. Technology has added immeasurably to our standard of living. Management of technology is an important task. The two main aspects of managing technology are developing, or acquiring, the technology and implementing it. Technological development is important to the Canadian national interest. It is the subject of much discussion, since so much technology is imported from the United States.

Environmental forecasting is predicting the future state of all environments which are important to business (including the physical, ecological, and technological environments). Assumptions about what these environments will be in the future are very useful in management decision making. There are a variety of techniques for environmental forecasting.

In the next chapter we look at characteristics of the economic and social environment of business.

KEY CONCEPTS

Content analysis Examination of published material to determine how frequently reference is made to certain concepts or ideas. Can be used to predict technological breakthroughs.

Correlation The relationship between one variable (e.g., the money supply) and one or more other variables (e.g., the rate of inflation). Knowledge of this relationship can be used in predictions.

Delphi technique A method of combining the informed judgments of a number of experts on a subject.

Ecology The branch of biology which deals with the relations between living organisms and their environment.

Extrapolation Projection of past and present circumstances into the future.

Forecasting techniques Specific methods used for making predictions about the future of business environments.

Forecasts Assumptions about what the future environment of business will be like.

Informed judgment Expert opinion on a subject.

Intuition A judgment by a person or group of people based on limited information.

Physical environment Our natural resource endowment including mineral, vegetation, animal, and water resources.

R&D A general name applied to activities which are intended to provide new products and processes.

Scenarios Statements of how a future event could unfold.

Simulation Developing and working through a model of an actual situation.

Technological environment Applications of science which have an impact upon a business firm.

Technology The application of knowledge so that people can do entirely new things or do old things in a better way.

Vision A person deciding that he or she wants something to happen and goes out and does it.

QUESTIONS FOR DISCUSSION

1. What are the physical, ecological and technological environments of business?
2. What steps can a tourist camp operator and a mine manager take to deal with the physical environment?

INCIDENT:

Blakley, Inc.

Blakley, Inc., a manufacturer of washing machines for the home, is the smallest of the major competitors in its market in Canada. Among their competitors are the U.S.-based giants, Maytag, RCA Whirlpool, and General Electric. Rumours persist that a new washer based on "sound waves without water" is nearing perfection in the laboratories of one of the big firms.

Questions:
1. What should Blakley do? Why?
2. Explain how the concept of the product life cycle fits in here.
3. Could any social problem follow from this invention? Explain.

INCIDENT:

Arsenic Storage

In June 1978, a tornado touched down briefly in a small town in southern Manitoba. Fortunately, it did not damage a dilapidated storage shed containing 300 tons of arsenic trioxide—a deadly chemical formerly used for killing rats and insects before it was banned by the federal government a number of years ago. Some of the material had been stored there for 20 years.

There was fear, even prior to the tornado, that the chemical could be dangerous. The one-storey frame-building housing the chemical was old and insecure. At one end of the building the wooden siding could easily be forced open. On hot days fumes could be seen rising from the building.

The company which owned the material did not feel it was responsible for removing the chemical. The company's general manager said that the banned chemical had been in the firm's inventory when he bought the firm and that the change in government regulation was what made it unusable. He indicated that the ideal solution was to transport the chemical and store it in an old missile silo in Idaho. The cost of this, however, was estimated at between $60,000 and $80,000, a very large amount for a small firm.

Just prior to the tornado, the provincial minister of mines, resources and environmental management had indicated in the legislative assembly that the government was planning to see that the arsenic was moved but that it was an extremely complicated situation.

The issue had been under discussion by the village council, the provincial and federal governments, and the company for six years. A company representative said he had a "five-inch stack of letters" from various provincial and federal governments.

A councillor of the town said the villagers were angry because they felt the company and the federal and provincial governments were "first considering who will get stuck with the bill" rather than considering the safety of the community. The townspeople weren't optimistic about the province's plans to have the arsenic removed. "We've heard that so many times before. What we want now is action, not more promises."

Questions:
1. Who is responsible for the situation?
2. Should action be taken? When?
3. What should the company do?
4. What should the provincial government do?
5. What should the federal government do?
6. What should residents of the town do?

3. Does business have a responsibility to not pollute the physical environment?
4. Is business the only contributor to environmental pollution?
5. How would you decide whether to allocate resources to cleaning up weed growth in a lake or to building a new sewage system for a town on the edge of the lake?
6. What are the primary tasks involved in management of technology?
7. Can you give ten examples of technological developments which are important to our way of life?
8. How important is technological development to industry in Canada?
9. Should government support technological development?
10. Does technology contribute to unemployment?
11. Identify five important environmental variables and illustrate how you would forecast each of them for five years from now.
12. What are the differences between the various forecasting methodologies?
13. What role does R&D play in competition in manufacturing?

18
The Economic and Social Environment

18 THE ECONOMIC AND SOCIAL ENVIRONMENT

OBJECTIVES: After reading this chapter, you should be able to:

1. Understand the components and importance of the economic environment of a firm.
2. Discuss the differences between microeconomic and macroeconomic influences on business.
3. Describe the economic theory of the firm and how it is useful in understanding microeconomic influences on the firm.
4. Indicate the various types of macroeconomic environmental influences and how they can affect a firm.
5. Describe various microeconomic and macroeconomic concepts such as demand, supply, productivity, competition, economic growth, income distribution, stagflation, and the balance of payments.
6. Understand the importance of economic forecasting.
7. Understand the components and the importance of the social environment of a firm.
8. Describe the characteristics of the individual, small-group, institutional, and societal variables.
9. Discuss the differences between individual, small-group, institutional, and societal variables.
10. Indicate the differences between "traditional values" and "new values".
11. Explain the differences between the traditional managerial ethic and the professional-managerial ethic.
12. Discuss the concept of social responsibility and indicate the arguments for and against it.

KEY CONCEPTS: In reading this chapter, look for and understand these terms:

ECONOMIC ENVIRONMENT
ECONOMIC THEORY OF THE FIRM
MICROECONOMIC ENVIRONMENT
PRODUCTIVITY
MACROECONOMIC ENVIRONMENT
ECONOMIC DEVELOPMENT
ECONOMIC GROWTH
DISTRIBUTION OF INCOME
PROGRESSIVE INCOME TAX
STAGFLATION
SOCIAL ENVIRONMENT
INDIVIDUAL SOCIAL ENVIRONMENT VARIABLES
SMALL GROUPS
NORM
INSTITUTIONS
POWER
DEMOGRAPHY
VALUES
TRADITIONAL VALUES
NEW VALUES
SOCIAL RESPONSIBILITY OF BUSINESS
TRADITIONAL BUSINESS ETHIC
PROFESSIONAL-MANAGERIAL ETHIC

This chapter contains a discussion of the economic and social environments of business. These two environments have a big influence on business firms. The economic environment determines the prices firms must pay for supplies and the prices which they can charge for their products. There are many microeconomic and macroeconomic factors which influence a firm.

The social environment is composed of all human factors which are external to the firm. The social environment influences marketing, production, and personnel decisions in many ways. The social environment is analyzed in terms of individuals, small groups, institutions, and societal variables.

The final section of the chapter is devoted to a discussion of the responsibilities of Canadian business firms. The main issue is whether or not firms have a social responsibility which goes beyond the simple pursuit of profit.

THE ECONOMIC ENVIRONMENT

Economic environment

The economic environment is composed of external influences which result in changes in prices of inputs used by firms or of products sold by firms. The economic environment must be considered when making decisions about methods of production, location of production, advertising, pricing, volume of output to produce, and selection of inputs. The effect of the economic environment on these decisions is shown in Figure 18-1.

One way of categorizing the multitude of economic factors, and at the same time

Business decisions	Examples of economic environmental considerations
What products to produce	(a) selling prices of products in relation to cost of production
	(b) actions by competitors
Investment in new production process	(a) outlook for economic growth and inflation
	(b) cost of new process in relation to revenues it will generate
	(c) savings in cost of production
Decision about what volume of output to produce	(a) current inventories
	(b) expected growth in demand
	(c) outlook for economic growth and inflation
What price to sell products at	(a) prices charged by competition
	(b) anticipated inflation
	(c) anticipated demand for product
	(d) cost of production

Figure 18-1. Some economic environmental influences on business decision making

relating the discussion to basic economics courses you have taken, is to divide it into microeconomic and macroeconomic influences on business decision making. Microeconomic factors are those which relate directly to the firm's cost of production and revenues from production, while macroeconomic factors are those related to the economy in general.

MICROECONOMIC ENVIRONMENTAL FACTORS

The economic theory of the firm (microeconomics) is that a firm will choose the level of output which results in the maximum profit for the firm. Profit is what remains after costs are deducted from revenues. Figure 18-2 indicates how the relationship between cost and revenue (profit) change as the level of output changes. **Therefore, the microeconomic environment is composed of those factors which directly affect a firm's input and output prices.** These factors include demand and price for a product, supply and price of inputs, productivity of inputs, and competition.

Economic theory of the firm

Microeconomic environment

Demand and Price for a Product

The demand for and price of a product will influence the level and shape of the revenue curve for a firm's product. We discussed demand and price and factors that influenced them in Chapter 2. Some examples will clearly illustrate the effect of changes in demand or price on a firm's revenue.

The traditional example is the buggywhip manufacturer whose market evaporated with the introduction of automobiles. The demand for the product declined and the firm's revenue likewise declined. More recently, the oil-price increase in the mid-1970s resulted in a significant increase in revenue for oil-producing firms.

There are a multitude of factors which can result in changes in demand and/or price for the products of a firm. In many instances a firm's managers have no influence over these external factors.

Supply and Price of Inputs

Just as the demand and price of products influences the *revenue* curve of a firm, the availability of the inputs required by a firm influences its *cost* curve. If, for example, petroleum products such as heating oil and gasoline are a major input factor for a firm, the oil-price increase would have had the effect of raising its costs and lowering profits unless there were a compensating increase in price.

Similarly, if the price of labour increases, the effect is to raise the cost curve of a firm. Again, the managers of firms have little influence over the cost of most inputs they require. Changes in the environment which alter input prices can have a major effect on the profit position of a firm.

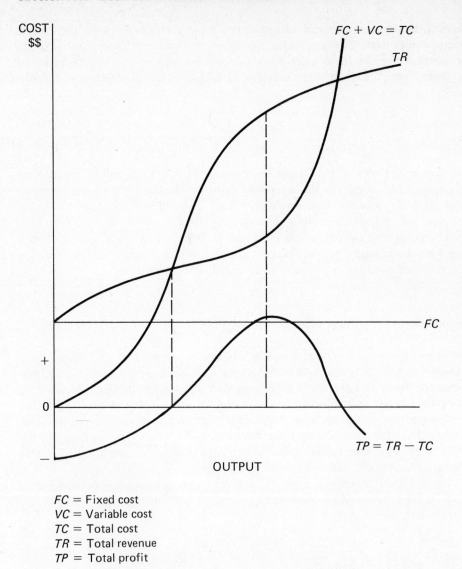

FC = Fixed cost
VC = Variable cost
TC = Total cost
TR = Total revenue
TP = Total profit

Figure 18-2. Theory of the firm

Productivity of Inputs

Productivity

The productivity of inputs is the number of units of output per unit of input. Increases in productivity which result in lower costs per unit of product increase the profit of a firm. For example, when automation makes it possible to produce a given amount of output with less hours of labour, we say that the productivity of labour has been improved. Productivity of any of the factors of production—land, labour, or capital—is normally increased by technological improvements. With any new product there is usually an improvement in productivity as experience is gained with its production. This is referred to as a learning curve. Productivity of all inputs for a new product improve after the product has been in production for a period of time.

A firm must be able to remain competitive with other firms which introduce new technology to improve their productivity. Each firm must therefore make efforts to improve its own productivity.

Competition

Competition is a very real and significant factor in the microeconomic environment of most firms. Competition forces business managers to keep their product prices at a level comparable to those of other firms in the industry. Similarly, competition keeps the price of required inputs up. Competition also forces managers to adopt the most modern management and production techniques in order to keep costs of production down at the lowest level possible. Competition provides excellent discipline in the privately operated portion of our mixed economy. It is harsh discipline, too. The ultimate penalty for not maintaining revenues, or for letting costs rise to the point where operations become unprofitable, is bankruptcy.

Although it is not as direct, competition also provides discipline for publicly owned firms. A government can only afford to provide so much support for an agency whose expenses exceed its revenues.

There are different types of competitive structures in different industries. These create somewhat different competitive environment situations. Three types of market structures are perfect competition, oligopoly, and monopoly. In perfect competition, there are a large number of firms and prices are set by the market. In oligopoly there are a limited number of firms and prices are normally set by a price leader among the firms. In monopoly there is only one firm and it sets its prices so as to maximize profits.

MACROECONOMIC ENVIRONMENTAL FACTORS

The macroeconomic environment is the general economic situation in the country or countries in which a firm operates. It includes factors such as economic development, income distribution, the business cycle, and the balance of payments and exchange rates. These macroeconomic factors are not controllable by business managers, but like many other environmental factors they can have a significant influence on the operation of a firm.

Macroeconomic environment

Economic Development

A nation—even a region within a nation—goes through certain stages of economic development. **Economic development means that a nation grows in terms of the amount its average citizen produces.** Canada has evolved from a simple farm economy to a highly developed, urban, manufacturing and service economy.

Economic development

Some parts of our country are still relatively underdeveloped, largely because of geographic and climatic factors. However, the least-developed parts of Canada are at a more advanced stage of economic development than some of the nations of Asia, Africa, and South America. Some of these nations cannot even feed and clothe their people.

THE INTERNATIONAL ECONOMIC ENVIRONMENT

The economic systems of some countries are simple when compared with the Canadian economy. In many, people live at a subsistence level. Perhaps more important to business people is what the country is doing to advance. Some "backward" countries do not appear to want change. This may be based on religious beliefs which look down on material progress.

Under-developed countries which are rich in natural resources have the potential to develop faster than those which are poor in natural resources. Valuable natural resources attract foreign investment to a country. This brings its people into closer contact with people from more developed nations. This may encourage them to pursue economic growth. But as the world's supply of natural resources dwindles, many countries which export them want to use them to help industrialize their own countries.

The fact that a country is rich in natural resources does not mean that they can be economically exploited. Brazil is rich in natural resources, but they are largely unexploited. One reason is poor transportation facilities. Japan, on the other hand, is rather poor in natural resources. Because of an efficient transportation system and business know-how, the Japanese can import natural resources and convert them into finished goods for export.

The tax structure affects a country's appeal to the multinational firm. Because most of the people in under-developed countries are poor, their governments rely heavily on business taxes. This sometimes discourages foreign business. The same is true of inflation. High rates of inflation are common in many under-developed countries.

In recent years the industrialized nations have had difficult problems with inflation. The members of OPEC (Organization of Petroleum Exporting Countries) have dramatically increased the price of their oil. One reason for this is that inflation in the industrialized nations means that the OPEC countries must pay more for the goods they import. Thus, they want more for the oil they export.

The most common measure of a nation's level of economic development is per capita Gross National Product (GNP). **Economic growth refers to the year-to-year increase in GNP.** If there is rapid economic growth, the prospects for business are normally better than if there is little economic growth.

Economic growth

Canada is rich in resources such as iron, coal, oil, forest products, and nickel. We have relied heavily upon natural resource development for economic growth. Some other countries have few natural resources. They must either import them or do without them. Japan, one of the largest manufacturing nations in the world, has few natural resources. Because of this, Japan has had to follow a different approach to economic development. By importing resources and by developing capital, Japan has overcome the handicap.

Other nations, such as Colombia and Nigeria, have substantial natural wealth but low levels of economic development. These nations can develop their great economic potential if they can bring in more capital and "know-how".

Since our country is rich in natural resources, our firms need not depend on foreign suppliers for most raw materials. This is a strength of the economy.

Income Distribution

The general economic health of a nation and the wealth of its people depends on how the nation's income is distributed among the people and how taxes are levied. The continued economic health of a mature economy like Canada's requires a distribution of income which permits a high level of consumption and investment in business. **Distribution of income means the way in which the people share in the total income of a nation or region.**

Distribution of income

The progressive income tax means taxing people with higher incomes at a higher rate. Jack pays $1,000 on $10,000 income while Jerry pays $1,500 on $12,000 income. A progressive income tax has the effect of increasing total consumption, because poor people spend more of their income than rich people do. Tax policy, however, must avoid being too progressive. This would leave little incentive for

Progressive income tax

YOU BE THE JUDGE!

When Does a Progressive Income Tax Become Too Progressive?

During recent years, some well-known entertainers have left England to live elsewhere. One reason given by some of those leaving is the high income taxes they pay. Some were in the 90 per cent and over tax bracket. When does a progressive income tax become "too progressive"? YOU BE THE JUDGE!

wealthy people to invest. Why take a risk if most of your reward will have to be turned over to the tax collector?

The economic environment is greatly affected by the tax policy and the distribution of income. Business today depends on our large middle class for customers. This group is made up of families in the $10,000 to $20,000 annual income bracket. (See Table 18-1.) Only families with reasonable incomes can buy all the appliances, furnishings, conveniences, and amusements which keep factories and stores open. If our nation's wealth were concentrated in the hands of a few, this vast market and the thousands of firms which serve it would not exist. On the other hand, large individual incomes and wealth accumulations are vital to the environment of business. The opportunity to accumulate a fortune is an incentive for risk taking, which is essential to the capitalist system.

The regional distribution of income is an important issue in Canada. The federal government is trying, through the Department of Regional Economic Expansion (DREE), to balance economic growth in different parts of Canada. In one DREE program, companies establishing plants in "designated areas" are eligible for grants from DREE. This lowers the cost of establishing plants.

The Business Cycle

Although the long-term trend in our economy has been one of economic growth, the level of economic activity and the rate of growth varies over time. The long-term, recurring part of such change is called the business cycle. (See Figure 18-3.) A period

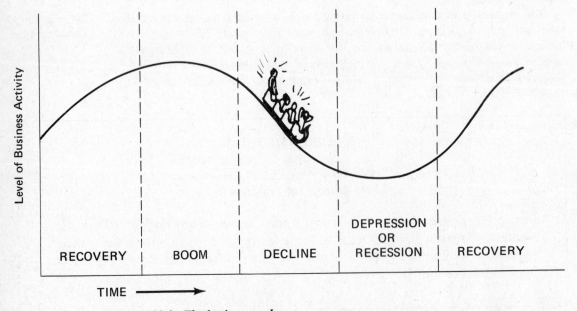

Figure 18-3. The business cycle

Table 18-1. Progressive federal taxation in Canada

Total income ($000)	Number of people	% of people	Taxable income (millions)	% of taxable income	Federal tax paid	% of tax paid	Effective rate of federal tax on taxable income
0– 5	625,462	7.1%	935.1	1.2	26.6	0.19%	2.84
5– 10	3,014,646	34.2	12,519.0	16.18	1,386.7	10.04	11.08
10– 15	2,506,617	28.46	19,934.1	25.75	3,099.4	22.45	15.55
15– 20	1,450,667	16.47	17,200.9	22.22	3,082.8	22.33	17.92
20– 25	622,146	7.06	9,847.1	12.72	1,931.8	13.99	19.62
25– 30	268,048	3.05	5,299.4	6.84	1,108.7	8.03	20.92
30– 40	180,931	2.05	4,581.3	5.92	1,034.7	7.49	22.58
40– 50	61,192	0.69	2,070.8	2.68	520.3	3.77	25.13
50–100	64,930	0.73	3,389.5	4.38	1,006.3	7.29	26.69
100–200	10,176	0.11	1,098.4	1.42	399.2	2.89	36.34
200 & over	1,916	0.02	519.4	0.67	211.9	1.53	40.79
Total	8,806,731	100.0%	77,395.2		15,208.2		

Note 1: This table indicates only federal taxation. In addition to this, provincial taxation would also have to be paid as a percentage of federal tax.

Note 2: Total percentages may not add to 100 because of rounding.

Source: Revenue Canada, Taxation Division, *Taxation Statistics, 1978,* Department of Supply and Services, 1978, p. 12. Reproduced by permission of the Minister of Supply and Services Canada.

of great volume of business activity is called a boom. During boom periods, profits, consumer confidence, capital investment, and employment are high and interest rates are low. But no boom is permanent. For any of a number of reasons, businesses begin to cut back on expansion plans. The economy enters a period of decline. Confidence in the economy fades and consumers cut back on spending. This causes unemployment to rise. If this period is brief, it is called a recession. If it continues so that unemployment and the rate of business failures get very high, it is called a depression. Such a depression last occurred in Canada in the period 1930-39. It led to a rise in the role of government as a stabilizer of the Canadian economy.

As a recession or depression turns into recovery, inventories begin to decline, businesses begin to build factories, households begin to buy cars again, banks expand loans, employees are rehired, and production returns to higher levels. The economy approaches a boom again.

Human psychology, of course, plays a large role in the business cycle. It has a kind of "snowball" effect. During the downturn pessimism is "catching", and because people expect the worst, things actually get worse. Fortunately, the same kind of thing occurs in reverse during recovery. Optimism becomes contagious.

For a small business, the environmental effect of the business cycle can be crucial. Consider the case of Smitty's Restaurant. This restaurant was opened in a neighbourhood made up mostly of factory workers and their families. A recession occurred soon after Smitty's opening. It led to cutbacks at a large production plant and high unemployment in the neighbourhood. Many of Smitty's regular customers lost their jobs and stopped eating out. Other workers, too, who were fearful of being laid off, started to save their money instead of dining out. Smitty's sales soon fell off 30 per cent. The local banks refused to lend him enough to meet current payrolls because they feared he would fail. Within six months after the initial employment cut, Smitty was out of business. If he had opened during a period of recovery, he would have had a much better chance to succeed.

In periods of prosperity prices traditionally rise. We call these price increases inflation. Since the Great Depression of the 1930s, the economic policy followed by the federal government has included a general willingness to permit inflation in order to avoid a recession. **In the 1970s our economy began to suffer from "stagflation". This was a peculiar situation in which we had both a recession and a high rate of inflation.** Economists do not agree on why we experienced stagflation for the first time. It shows that there are still many things that we don't know about managing an economy. We're not sure about the economic effects of welfare programs, the energy shortage, defence cutbacks, and consumer credit expansion, to name a few factors which probably contributed to stagflation.

Stagflation

Inflation affects different groups in our society in different ways. A worker whose union contract provides for increases in wages as the price level increases (an escalator clause) suffers no loss in buying power. The same is true of many businesses which can raise the price of the goods they sell. The groups which suffer are those whose money income doesn't rise, such as those living on fixed-dollar pensions and stable salaries. They find it harder and harder to live as prices rise. Table 18-2 shows how prices have changed in recent years.

The effect of inflation on business varies. Most firms pass on higher costs to their customers. But some, for competitive reasons, cannot raise prices. They must pay inflated prices for supplies and inflated wages to employees. Their profits are "squeezed".

Table 18-2. Consumer price index, 1967–77

Year	Index
1967	86.5
1968	90.0
1969	94.1
1970	97.2
1971	100.0
1972	104.8
1973	112.7
1974	125.0
1975	138.5
1976	148.9
1977	160.8
1978 (August)	177.8

Source: Book of Canada Review [September, 1978], p. S110.

Balance of Payments and Exchange Rates

The balance of payments, the difference (in dollars) of goods and services which our country buys from and sells to other countries, was explained in Chapter 16. The balance of payments affects our exchange rate. If Canadians are buying more abroad than they are selling, we run up a balance of payments deficit. This puts downwards pressure on our currency, and our currency goes down in value relative to the currency in other countries. If our balance of payments is positive, our currency can appreciate relative to that of other countries.

Changes in the exchange rate can have a big effect on companies which buy, sell, or have plants and offices abroad. A decline in our exchange rate will make imported goods more expensive to purchase. It will also reduce the cost of exports. This will aid exporting firms and hurt importing firms.

Another problem created by exchange rate fluctuations is change in the value of assets in different countries. This is particularly significant for a multinational corporation. It must make decisions about what country's currency to hold its cash in. For example, in 1977-78 the Canadian dollar depreciated 35 to 40 per cent relative to the West German mark. If a company had held its cash in marks rather than dollars during this period, it would have been much better off.

THE SOCIAL ENVIRONMENT

The social environment also influences decisions made by business managers. **The social environment is the total of all human variables which are external to the firm.** We will analyze the social environment in terms of individual, small-group, institutional, and societal variables. There are certain environmental variables associated with each of these four groupings of social factors. (See Figure 18-4.)

Social environmental variables must be taken into account in many management decisions (Figure 18-5). These include, as examples, product design, promotion, decisions with social responsibility overtones, personnel and employee relations decisions, plant location, and production process decisions. The social environment is important to many managers in a firm. These include the marketing manager, the personnel manager, the production manager, and the general manager.

Social environment

Individual Variables

Individual social environment variables are those which relate specifically to important individuals in the environment of the firm. There are many important individuals in the firm's environment. These include (1) the customer who decides whether or not to purchase a firm's product, (2) members of the labour force who choose to work or continue to work for a firm, (3) individuals in competitive firms, and (4) government officials who make decisions which affect a firm.

It is important that the management of a firm have as good an understanding as

Individual social environment variables

I. INDIVIDUAL
   A. Consumer Needs and Wants
   B. Consumer Talents and Abilities
   C. Consumer Values
II. SMALL-GROUP
   A. Group Norms
   B. Informal Work Groups
   C. The Role of Groups in the Socialization of Individuals
   D. The Role of the Family
III. INSTITUTIONAL
   A. Business
   B. Government
   C. Labour
   D. Consumers
IV. SOCIETAL
   A. Demographic Variables
   B. Societal Values
   C. Social Responsibility

Figure 18-4. Four types of social environment variables

possible of the characteristics of individuals in the firm's environment. The more intimate the contact with individuals, the greater the importance of having a full understanding of the individual.

The characteristics of individuals which may be of interest include people's needs and wants, their motivation, their values, and their talents and abilities. Students taking, or who have taken, psychology courses will find them of benefit in understanding individuals in a firm's environment.

Customers as individuals. Customers make their decisions to purchase in order to satisfy personal needs or wants. An automobile or a dress may be purchased to satisfy a functional need, but it will often be purchased to satisfy social or ego needs as well. If a firm, through the attractiveness of its product or its advertising, can stimulate individuals to need or want its product, it will increase its sales.

Understanding a customer's motivation to make a purchase is important to firms. Certain types of firms use personal selling to motivate customers to make a purchase, while others use mass advertising. It is important for firms to have an understanding of how customers make purchase decisions and what factors are important to them when they make them. Courses you may take in buyer behaviour will go more deeply into this subject.

Labour force as individuals. People work to satisfy important needs and wants which they have. They agree to contribute their skills and abilities to an employer in return for things which will satisfy their needs. One thing employees receive is money; this allows them to purchase goods and services. In addition, however, a job can satisfy many other needs and wants. Employees can achieve a sense of pride and achievement

| Social environment variables | Management Decisions |||||
	Product design	Promotion	Personnel and employee relations decisions	Plant location	Production process design
Individual Variables (a) motivation of people		The motivation to make a purchase decision affects the way promotions are designed.	People's motivation affects how they should be managed.		Should attempt to take the needs and motivation of people into account in designing jobs.
(b) talents and abilities of people	Products can be designed for people of certain skills and abilities (e.g., tools for professionals).	Design promotions to appeal to people with certain talents.	Will hire people with talents appropriate for jobs.	Will locate plant where it is possible to hire people with talents appropriate for jobs.	Talents and skills of work force can affect design of the production process.
Small-group Variables (a) group norms	People with common skills (e.g., athletes may desire similar products).	Promotion can attempt to convince people that they should use a product because others are using it.	Group norms can affect the decisions of employees.	Community group norms can result in massed opposition to a plant location.	The norms of a work group (e.g., pride in what they produce) can be important in design of a production process.
Institutional Variables (a) institutions other than business	Consumer organizations affect product design.	Consumer organizations affect promotion techniques (e.g., advertising for children).		Conserver and ecological movements can affect plant location and production process.	
Societal Variables (a) societal values and attitudes	Attitudes affect the types of products which appeal to people (e.g., fashion in clothing and entertainment).	Attitudes of potential consumers affect design of promotional campaigns.	Change in values and attitudes affect hiring decisions (e.g., business now hires more women).	Residents may have preferences that firms not locate in some areas.	Must now take greater account of pollution control in designing production processes.
(b) Demographic characteristics (e.g., location, age, sex, occupation, income	Demographics affect the types of products in high demand (e.g., if more babies are born there will be more baby food and other baby products sold).	Promotion must take demographic characteristics into account.	Demographic characteristics of an area affect the supply of labour.	Will locate after consideration of the demographics of the market and the work force.	Availability and type of work force can affect decisions about production processes.

Figure 18-5. Illustration of how social environmental variables can influence management decisions

in their work. People can also satisfy social needs at work through their affiliations with others.

The needs and wants of employees can change over time, as can those of customers. As our level of affluence increases, people develop higher expectations about what they would like to receive from a job, both in terms of financial rewards and working conditions. Failure by a manager to respond to the needs and wants of current or prospective employees can result in labour relations difficulties.

Competitors as individuals. Competitors are a very important part of the environment of a firm. Competitors are also individuals, whether they are managers of competing firms or salespeople from another firm who are attempting to convince a customer to buy their product rather than yours. If you think of competitors as individuals and are able to understand them to the point where you feel you can predict their moves with some confidence, you will improve your ability to compete against them.

Government representatives as individuals. When you deal in the political environment, you do not deal with an abstract entity called "the government". You deal with individuals, whether they be clerks behind a counter, government inspectors, advisors, senior civil servants, or elected officials. Business managers deal with government representatives when existing government policies are implemented or when new government policies are formulated. In both cases, it is wise to remember that government officials are individuals with needs, wants, and feelings.

Government officials are responsible for implementing government policy. It is not wise to belittle or harass them in the performance of their job. Such attitudes, if expressed openly, can create resentment in government officials and cause them to look particularly hard for rule violations by the company.

Government representatives are important in the formulation of new government policies. Government employees make crucial decisions about changes in the policies under which businesses operate. They make them after consideration of all available facts as they understand them. They also make decisions which will satisfy government objectives. The task for the business manager who wants to influence government policy is to first understand the government decision maker as a person in a specific job. The business manager must then frame alternatives which will allow the government official to meet his or her objectives without seriously affecting the business manager's firm. (The interaction between business and government is discussed in detail in the next chapter.)

Small Groups

Small groups have a big impact on people's attitudes and behaviour. An understanding of significant small groups in a firm's environment and of the dynamics of small-group activity is of value to the business manager.

Some small groups of importance to business managers include families, social groups, and recreation groups. Members of these types of small groups are prospective customers, employees, suppliers, and competitors of a firm.

Small groups are composed of members who have some common goal and a structure for achieving it. The real importance of groups is that they have norms to which group members conform. **A norm is a standard way of thinking or behaving which group members feel is reasonable.** The importance of this type of conforming behaviour for business managers can easily be understood.

A group norm to wear a certain style of clothing, or to listen to a certain performer's music, will have positive effects upon sales of these products. Such "fads" as hair length or wearing blue jeans are reflections of our tendency as consumers to conform to group norms.

We can all think of examples of different members of a family, or different people in a community, acquiring major items such as microwave ovens or swimming pools after they had been initially acquired by someone generally respected by other members of the group. Some call it "keeping up with the Joneses"; what it really amounts to is conforming to norms of a group the consumer would like to identify with.

The norms of a group will also dictate whether certain restaurants or cabarets will be frequented. An eating or entertainment establishment which can appeal to a group can become very successful rapidly. If for some reason, however, the appeal of the establishment to the group is lost, sales can decrease rapidly.

The norms of groups of the current employees of a firm, or of groups from which potential employees will be drawn, are important to a firm. It is desirable that current employees' group norms reflect a constructive, satisfied attitude toward their firm. If, on the other hand, current employees' norms reflect dissatisfaction, distrust, and unrest regarding their employer, labour relations difficulties are almost inevitable.

If prospective new employees, such as college students, have a positive attitude towards a firm or industry, it makes it easier for that firm or industry to recruit new employees. If student norms are to shun a prospective employer, it is much more difficult for that firm to recruit the type of people they want. Such situations can occur, resulting in some firms not being allowed to recruit on university campuses at certain times. If, for example, Ajax Corporation has been publicly criticized for polluting recreational waters, a movement to avoid signing up for job interviews with the company may develop.

Company managers should not underestimate the importance of group norms among important groups of consumers and employees. If group norms are favourable to, or can be harnessed by, managers, a business can benefit greatly from sales increases or good labour relations. If norms of important groups are unfavourable, the negative effects on a business can be serious.

Institutional Variables

We live in a society of large organizations and institutions. **Institutions are organizations, or groups of organizations, having an economic, social, educational, religious, or other recognized purpose in society.** Thus, we can talk of labour, business, government, consumers, religions, or charities as examples of institutions. The recognized and established purposes of the institutions are reflected through laws, practises, customs, or traditions. Different institutions have different purposes. They all play important roles in society. Because their objectives differ, there can be conflict between them.

A business manager must be aware of the characteristics, objectives, and views of representatives of different institutions. This is necessary because some of the actions of businesses can be in conflict with the objectives of other institutions. It would be unwise for business managers to take action on certain issues without considering other institutions. For example, the reactions of environmentalists must be taken into account in decisions with pollution implications. Labour's reaction must be considered in decisions about employee salaries. A firm's managers certainly cannot make decisions affecting labour without taking the possible labour reaction into account. Such decisions inevitably lead to labour relations problems.

Business managers also must understand and relate to various government agencies. This relationship is so important that the next chapter is devoted to discussion of it. Business managers must also understand and relate to the media. Many large firms employ people with previous media experience to handle their relations with radio, television, and the press.

Developments in education and research can also affect business decisions. In the previous chapter we saw how important the technological environment could be to a firm.

Power

Business managers must know how much they are able to influence decisions of the various institutions. **Power is the ability to induce others to behave the way you want them to.** The power of some groups is increasing, while that of others is decreasing. The power of the Church, for example, has decreased in the last century. The power of both business and labour has also decreased relative to that of government. Governments have become relatively more powerful and important in economic decision making. It is likely that this trend will continue.

Some people feel there is an elite "establishment" in Canada which possesses disproportionate power and is self-perpetuating. While there is a group of senior executives of major companies and governments who have more power than other Canadians, they do not have anything approaching absolute power. Furthermore, they cannot automatically pass this power on to someone of their choosing, such as their children. Possession of power at any point in time is primarily a function of competence and outstanding job performance.

Another institutional fact of life is that Canadian institutions are getting larger and larger. We have big government, big business, big labour, big universities, and big religious organizations. Large organizations have developed for a variety of reasons, but the primary one is that they are able to perform large tasks. The main concern, however, is not how and why they have developed but that they exist and that Canadian society will continue to be characterized by large organizations. For example, we see efforts by consumer, environmental, and other institutions to become larger and better staffed and financed in order to increase their power.

The fact that there are increasing numbers of large organizations is important for business managers and for you as future managers. In spite of this, there are still many opportunities for smaller organizations. There will always be things which smaller organizations can do better than large ones, such as producing customized products and providing personalized services.

THE INTERNATIONAL SOCIAL ENVIRONMENT

International business requires people of different cultures to interact. Business transactions involve the written and/or spoken language. Unless the parties really understand each other, there can be no basis for business. This problem exists even when people in different countries use the same language. For example, people from England and English Canadians speak the same language but give different shades of meaning to the same words.

Canadian executives overseas should have a basic understanding of the language(s) of the land. Many global firms find it helpful to hire and train local nationals for management jobs. This helps reduce the language gap. Furthermore, many governments demand this.

There are also differences in cultural values. Canadian ideas about getting ahead are rejected in some cultures. Our emphasis on convenience is equated with laziness in some cultures. Although cultural values do change over time, there is cultural resistance to change. For example, Japan accepted a cultural value of fewer children and legalized abortion long before many other cultures accepted family planning. But the Japanese are reluctant to adopt a new alphabet, even though their alphabet contains too many symbols to make use of the typewriter practical.

Tastes also vary. The colours which are considered lucky by one people, for example, often depend on the religious beliefs or superstitions held by that group of people. Western cultures associate black with mourning, whereas Eastern cultures associate white with mourning. This affects a global firm's advertising and packaging decisions.

The religious beliefs of a people influence their ideas of what is ethical behaviour. Our "affluent" society is rejected by devout followers of Buddhism and Hinduism. Employers who favour their relatives in hiring and promotions are frowned on in Canada. Such favoured treatment is part of the religious teaching of Hinduism.

There are other differences in the social environments as well. In extremely class-conscious societies, a person is born into a particular social class and remains there. Ad campaigns built around the idea of "moving up" are not effective, nor are personnel policies which encourage employees to move up the job ladder.

Social customs also vary. In Oriental societies extreme politeness and formality are part of doing business, unlike the more informal approach in Western countries. A Canadian who wants to get down to business without engaging in the proper social behaviour is headed for trouble.

Another implication is the need for detailed planning within organizations, with an emphasis on forecasting the effects of major decisions before they are made. There will also be more emphasis on joint planning by different institutions. An example of joint plans are those undertaken by business and government to improve the Canadian industrial society. When joint decision making is used, the consequences of making bad decisions can be very expensive to firms.

A further implication of working for large organizations is that more people have input into decisions which are made. This is unlikely in small organizations where major decisions are made by one person or a small group of people. Managers in a large organization must understand the process of policy formulation in order to be able to contribute to it effectively.

Societal Variables

In addition to looking at the needs and wants of individuals and the characteristics of small groups and institutions, business managers should also examine the overall social environment of the country. This includes demographic characteristics and the values of Canadians.

Demography

Demographic characteristics of Canadians. **Demography is the branch of anthropology which deals with population statistics, primarily the size and characteristics of the population.**

The total population size and distribution is important to the Canadian business manager. The relatively small size of the Canadian population in comparison with that of the United States, Europe, or Japan gives Canadian business managers a smaller domestic market to serve. Because the domestic market is not as large, plants in Canada cannot be as large as those in other countries. Consequently, these smaller Canadian plants do not have the economies of scale of plants in other countries. This means that Canadian plants have higher costs and are less able to compete on a cost basis with production from plants in other countries.

In addition to the smaller Canadian market, the population is widely dispersed. The transportation costs associated with this population distribution make it even more difficult to have one large plant to serve the entire population.

The major areas of population growth are projected to be Ontario, Alberta, and British Columbia. In absolute numbers, the greatest growth in population will occur in Ontario. In percentage terms, the most significant population growth will occur in the Northwest Territories, British Columbia, and Alberta.

The age distribution of the population is also changing. The average age of Canadians is increasing. People born in the post–World War II baby boom are growing older. This "bulge" of people is moving through the age distribution of the population. This has implications for firms which sell products directed at specific age groups. For example, soft drinks are most heavily consumed by people in the under 25 age category. This age group is not forecasted to increase in size. In order to increase sales, efforts must be directed at encouraging consumption by older persons. A future concern is care of the aged. The proportion of people in the over 65 age group will increase in the future. This presents problems in providing public health services. It will also represent an opportunity for firms which provide services and products for people of this age group.

Another important demographic characteristic of the Canadian population is its bilingual nature. The majority of the population of Canada is English-speaking. However, the majority of Quebecers speak French, and there are also significant pockets of French-speaking people in provinces such as New Brunswick and Manitoba.

These language differences have implications for business managers. Advertising must be different in the two languages. English labels must be translated into French for products sold in Quebec. Firms from other parts of Canada with plants and offices in Quebec must be prepared to conduct operations in that province in French.

The values of Canadians. We have already discussed individuals and their needs, wants, and attitudes. Individuals also have values. **Values are simply what people think is important**. If we could combine the values of all Canadians, we would have a statement of Canadian societal values of the Canadian culture.

There has been considerable discussion in recent years about whether Canadian societal values have been changing. In a book entitled *Canada Has a Future,* prepared for the Hudson Institute of Canada, "traditional" values and "new values" are discussed.

Traditional values are "regard for duty, honour, custom, order, restraint, prudence, loyalty to family, church, and nation and the pursuit of knowledge, technology, and economic growth." New values refer to "spontaneity, self-actualization, sensory awareness, equality, concern for self, humanity, and nature, and indifference or opposition to traditional values."[1]

In the decade of the 1960s there was considerable emphasis on the new values. In the 1970s, particularly after the economic difficulties caused by the energy crisis and possibly as a reaction to the liberal views of the 1960s, there was a resurgence of the traditional values. We saw a definite conservative trend in elections in many of the provinces, including Manitoba, Nova Scotia, British Columbia, Prince Edward Island, Alberta, and Ontario. There was also talk of restraint and financial conservatism at the federal level of government.

The conservative reaction of the 1970s and the return to more traditional values was associated with economic difficulties and uncertainty. Many of the attitudes associated with the "new values" are being incorporated into the decisions of many individuals and firms. As economic conditions improve and the long-term outlook again looks more favourable, there may be a re-emergence of the new values.

The societal values of a country influence the decisions which business managers make. These values are incorporated into business decisions because they influence the criteria which managers apply to the decisions they make.

Values

Traditional values
New values

---

[1]Marie-Josee Drouin and B. Bruce-Briggs, *Canada Has a Future* (Toronto: McClelland and Stewart Ltd.), 1978, p. 233.

Traditional business ethic	Professional-managerial ethic
1. Maximum short-term profits	1. Satisfactory long-term profits; other values are weighted
2. Minimum government control	2. Government-business "partnership"
3. Protectionism	3. Internationalism
4. Stockholder-oriented	4. Serves several masters, including stockholders, customers, citizens, and employees

Figure 18-6. Two opposing ethics in business

THE SOCIAL RESPONSIBILITY OF BUSINESS

Business firms exist to produce economic goods and services which people want. In a really competitive situation business firms must produce useful goods and services or they will not survive. When they produce, businesses also create jobs and income for people. This creates a healthy economic base for the society. Without a healthy economic base, social problems would probably be much more serious than they are at present. The federal and provincial governments would also have much smaller tax revenues with which to attack social problems.

Social responsibility of business

To some people, this means that business should not worry about social problems. An increasing number of people, however, feel that business firms should go beyond the mere pursuit of profit and should be truly socially responsible. **A firm demonstrates social responsibility when, in the process of making decisions, it takes things in addition to profit into consideration.** For example, a firm might decide to start a training program for the hard-core unemployed, even though in the short run the costs of the program will mean less profit for the firm.

There are two opposing points of view about the obligations of business to Canadian society, depending upon whether the "traditional" or the "new" societal values are emphasized. One view is called the traditional ethic. The other is the professional-managerial ethic. Let's examine both of them. (See Figure 18-6.)

The Traditional Business Ethic

Traditional business ethic

In the traditional business ethic, business decisions are based only on how they affect the short-term profit for owners. Profit is measured in the short run. What is "best" for the firm means what provides measurable profit soon.

This position has dominated our business climate for many decades. It is still supported widely by many business managers. Most small- and medium-size firms still feel pretty much this way. This ethic calls for minimum government control of business, and it serves the interest of the business's owners or stockholders almost exclusively. All in all, it is rather conservative and is in sharp contrast to the professional-managerial ethic.

WHAT DO YOU THINK?

Should Business Be Socially Responsible?

Arguments in Favour of Social Responsibility

1. The public supports business by purchasing its output. If the public expects business to be socially responsible, business must be or it will lose customers.
2. If business does not voluntarily behave in a socially responsible way, government legislation will be passed in order to force business to be responsible.
3. Business has a great deal of power. In a democracy, the possession of power means acceptance of responsibility for the way the power is exercised.
4. If a company is socially responsible, it will benefit from good public relations.
5. If business is socially responsible, all companies will benefit because the work force will be more capable, the air and water will not be polluted, and consumers will think highly of the business community.

Arguments against Social Responsibility

1. Business makes its greatest contribution to society by producing goods and services as efficiently as possible. Worrying about whether decisions are socially responsible distracts business managers from their primary goal—making a profit. If profit is not made, the business will go out of existence and many jobs will be lost.
2. The owners of the company (the stockholders) take all the risks, and they must decide if they want company funds spent on socially responsible projects such as hiring the disadvantaged or contributing to charities. If they decide not to do these things, the company should not be forced to do them.
3. When business is socially responsible, the cost of the company's products increases. This is detrimental to consumers and causes inflation.
4. Since business managers are not elected by the people, they have no right to make decisions about social issues. This amounts to them imposing their value system on the public, and in a democracy this is not tolerated.
5. Government agencies have been set up to ensure that various segments of society behave in a responsible way. Business managers therefore do not need to be concerned about whether their actions are socially responsible, since government agencies will make that decision.

The Professional-Managerial Ethic

The professional-managerial ethic[2] has become accepted in recent years by an increasing number of the largest corporations and many smaller firms. **The professional managerial ethic holds that managers represent the interests of stockholders, customers, employees, and the general public.** Decisions are weighed in terms of longer-

Professional-managerial ethic

---

[2]The professional-managerial ethic is described in a booklet published by the Committee for Economic Development, *Social Responsibility of Business Corporations* (New York: Committee for Economic Development, 1971).

range company welfare, not immediate profit. Also, it is assumed that what is good for the employees and the general public is good for the company. For example, a firm which participates voluntarily in training and hiring the disadvantaged may not expect increased profits to result in the next year or two. However, the firm may expect that such activity will "pay off" in the long run. The society gains, and a stable, healthy society is presumed to have long-run beneficial effects on the firm. There are also indirect benefits in the form of good public relations from such activities.

Another important part of this ethic is the belief in a co-operative "partnership" relationship between business and government instead of the traditional hostility. This thinking fosters business participation in solving social problems.

The modern professional-managerial ethic has developed since World War II. It is a product of changing political ideas and of people's acceptance of a greater social consciousness. It is an extremely practical, long-run view. It says a firm is successful when it is accepted by society as a contributor to social welfare. It presumes that stockholders themselves will recognize the wisdom of sacrificing short-run profits for the long-run stability of the society. This kind of thinking led Xerox Corporation to spend half a million dollars a year to release employees with full pay to work at socially oriented tasks in their communities.

It is harder to evaluate managers when values other than the direct economic interests of the firm are brought into the decision-making process. Deciding which non-economic values are to be considered is tough—especially when executives and board members disagree. There is also the added problem of weighing these non-economic values along with the economic ones. Economists and accountants are working on methods of evaluating such activities. These methods fall under the general heading of "social auditing".

THINK ABOUT IT!

A Contemporary View of Social Responsibility

A well-known analyst of the role of business and society offers a "model" for business to follow.* It consists of five guidelines which may keep businesses in harmony with the will of society: (1) We must remember that social responsibility arises from social power. (2) Business must exchange information openly with the general public. (3) Business must weigh every decision in terms of its social cost. (4) Business should pass on to consumers the full cost of including the social goods in its decisions. (5) Businesses are responsible to help solve social problems not of their own making when they have the special abilities to do so.

*Keith Davis, "Five Propositions for Social Responsibility," *Business Horizons* (June, 1975), pp. 19-24.

> ## WHAT WOULD YOU DO?
>
> ### Being Socially Involved
>
> The board of directors of the Wranston Department Store is considering the complaint of two of its major stockholders that the dividends for the last two years have been too small. A review of the firm's operations over the last two years showed that the two largest stores, usually major profit contributors to the corporation, have each made only a 2 per cent return on investment. Further study showed that the major cause of this was their participation in a massive training program for the disadvantaged unemployed.
>
> Wranston's president, Will Carter, was the force behind the employment program. He believes that such social programs are essential in the long run for the big-city stores to survive. He, along with some other board members, believes that reduced short-term earnings are unfortunate, but necessary, in the interest of social stability for the urban environment.
>
> Other directors feel that endangering the corporation's ability to attract investors is too risky. Low profits drive investors away and the basic economic health of Wranston is at stake. The issue is to be resolved at the next board meeting. Assume that you are on the board. WHAT WOULD YOU DO?

The difference between the two ethics is clear. Some say that a capitalist system cannot survive if firms bring non-economic values into their decision making. Others say that if business (especially large corporations) does not assume social responsibilities, the entire society could perish.

SUMMARY AND LOOK AHEAD

This chapter has dealt with the economic and social environments of business. The economic environment includes all the factors which influence the prices a firm must pay for its inputs (land, labour, capital, and management) or the prices which it will charge for its output. We discussed the microeconomic environment of the firm, which includes supply, demand, productivity, and competition.

The macroeconomic environment of business is the general economic situation in the country in which a firm operates. Macroeconomic variables discussed include economic development and growth, income distribution, the business cycle, inflation, and exchange rates.

The social environment is the total of all human variables external to the firm. Individual, small-group, institutional, and societal variables were discussed.

Individuals external to the firm, but relevant to it, include customers, competitors, labour force members, and government representatives. Small groups of importance include informal work groups and the family. Group norms can have an impact upon the success of business operations.

Institutions in society are very important. Business managers often have objectives which conflict with the objectives of other institutions. The relative power of different institutions will determine how much influence they have and to what extent they will be able to impose actions on society.

The values of society were also discussed. A distinction was made between traditional values and new values. Values also influence the ethics which business managers apply in making decisions. The traditional management ethic and the professional-management ethic were discussed.

The question of social responsibility was also raised and the advantages and disadvantages of socially responsible behaviour were noted.

In the next chapter the political environment of business is outlined. Various federal, provincial, and municipal influences on business are indicated.

KEY CONCEPTS

Demography The branch of anthropology which deals with population statistics, primarily the size and characteristics of the population.

Distribution of income The way in which the people share in the total income of a nation or region.

Economic development The growth of a nation in terms of the amount its average citizen produces.

Economic environment The external influences which tend to result in changes in prices of inputs used by firms or of products sold by firms.

Economic growth Year-to-year increases in per capita GNP.

Economic theory of the firm Producing at the level of output which results in the maximum profit for the firm.

Individual social environment variables Factors which relate specifically to important individuals in the environment of the firm.

Institutions Organizations, or groups of organizations, having an economic, social, educational, religious, or other recognized purpose in society.

Macroeconomic environment The general economic situation in the country or countries in which a firm operates.

Microeconomic environment Factors in a firm's environment which directly affect its input and output prices.

New values Refers to spontaneity, self-actualization, sensory awareness, equality, concern for self, humanity, and nature, and indifference or opposition to traditional values.

Norm A standard way of thinking or behaving which group members feel is reasonable.

Power The ability to induce others to behave the way you want them to.

Productivity The number of units of output per unit of input.

Professional-managerial ethic The view that managers represent not only the interests of stockholders, but also of customers, employees, and the general public. Decisions are weighed in terms of long-range company welfare, not only immediate profits.

Progressive income tax Taxing people with higher incomes at a progressively higher rate.

Small groups Composed of members who have some common goal and a structure for achieving it.

Social environment The total of all human variables external to the firm.

Social responsibility of business When making business decisions, managers take factors other than profit into consideration.

Stagflation A situation characterized both by recession (slow economic growth) and a high rate of inflation.

Traditional business ethics Business decisions should be based only on how they affect the short-term profit for owners.

Traditional values Regard for duty, honour, custom, order, restraint, prudence, loyalty to family, church and nation, and the pursuit of knowledge, technology, and economic growth.

Values The things Canadians think are important.

QUESTIONS FOR DISCUSSION

1. What is the economic environment of the firm? How can the economic environment be important to a firm?
2. What type of environmental change would you call a change in ocean conditions which resulted in a shortage of a certain species of fish which subsequently raised the price of this fish to processors and consumers? Discuss.
3. What are the microeconomic variables which can influence a firm?
4. What is the competitive environment of Bell Telephone? Of a provincial telephone utility? Of a travel agency? Of a large mining company?
5. Are macroeconomic or microeconomic environmental factors more important to a firm?
6. What can the consequences be for a firm if one of its competitors makes major improvements in the productivity of labour?
7. What are the implications of stagflation for a Canadian manufacturing firm?
8. What is the difference between individual, small-group, institutional, and societal variables in the social environment of a firm?
9. How can individual social environmental factors affect a firm which manufactures dresses and pantsuits?
10. What is the role of business as an institution in Canadian society? Illustrate how the power of business has changed relative to that of consumers and the government.
11. Should business managers attempt to impose their views on other social institutions?

12. Do you think there will be a shift from traditional values to new values?
13. Is the traditional managerial ethic more appropriate in times of economic recession than the professional-managerial ethic?
14. Can you think of five types of decisions by business managers which are not influenced by either the economic or social environments?
15. How can a person decide whether a given business decision was a socially responsible one?

INCIDENT:

Roberts' Implements Ltd.

Jim Roberts, the president of Roberts Implements Ltd., contemplated his situation in mid-December, 1978. His firm manufactured a specialized line of farm implements—the Roberts cultivator—in a rural Saskatchewan community of approximately 2,000 people. He employed 47 people on a two-shift basis. The cultivator was sold primarily in Manitoba, Saskatchewan, and Alberta.

After increasing dramatically from $320,000 and $15,000 in 1972 to $810,000 and $90,000 in 1977, sales and profits levelled off in 1978 as farm incomes, with which farm implement sales were directly correlated, stabilized. Because of these good years the firm was in very solid financial shape. The outlook for 1979, however, was uncertain. Agricultural experts were divided, although the balance felt that grain prices would remain low for 1979. Furthermore, there was concern about a very dry growing season in 1979 which, if it occurred, would adversely affect yields. There was evidence that inventories of competitive firms were increasing. In addition, there was pending legislation on farm implement warranties which would increase manufacturing costs.

Mr. Roberts was considering whether he should reduce his planned output for the coming year. It would mean cutting back to a work force of 30 men. He anticipated that if he did do this the men laid off would have difficulty finding other jobs, and this would affect the community's image of his firm. On the other hand, if he did not decrease production, there was an opportunity to increase market share if the anticipated market decline did not occur.

Questions:
1. What economic environment factors have influenced Roberts Implements Ltd.?
2. What social environment factors have influenced the firm?
3. How have these environmental factors created problems for Mr. Roberts?
4. After assessing his problems and options, what is your recommendation to Jim Roberts?

INCIDENT:

Jones' Carpets

Jim Hannesson, his wife Doreen, and their three teen-age children lived in Halifax. Jim had worked as an accountant for a plumbing supply firm in the city for a number of years. His last year's pay, including a bonus of $3,400, was $33,248.18. He had recently become aware of an opportunity to purchase a retail carpet business.

The firm, belonging to a long-time friend of the Hannessons, Harold Jones, was apparently in difficulty because personal and family problems were preventing Mr. Jones from devoting appropriate attention to the business. As a last resort, Mr. Jones was considering sale of the firm. Mr. Jones had indicated that he felt the business was worth about $50,000, plus inventory. However, Jim Hannesson had heard the opinion expressed by a mutual friend that the price could probably be bargained down to under $40,000. Jim had $30,000 in savings available for investment and could borrow at least an equivalent amount.

Sales of the carpet store in the previous year had been $200,000. The cost of goods sold was $125,000. Expenses were $25,000 for two full-time salespersons; advertising expenditures of $23,000; building rental and maintenance of $6,000; and administrative expenses of $15,000. Harold Jones was drawing a salary of $2,000 a month for himself.

Jim Hannesson was planning, if he did decide to purchase the firm, to increase advertising expenditures by $15,000. It was his hope that this would increase sales by 50 per cent. His reasoning for this was that the store was located in a growing area of the city. Jim Hannesson also felt that potential sales for a store in the area with adequate management would be over $500,000 within three years. The size of the store would require expansion for sales of over $400,000.

Jim was concerned about how beneficial his past experience would be for operation of a retail store. There had been recent complaints from consumer groups to the Carpet Association concerning the difficulty of distinguishing between different qualities of carpet. Jim did not know what effect, if any, this would have upon the operation of a carpet store.

Jim was also concerned about Harold Jones' alcohol problem. He wondered if it was proper to consider buying a business from a friend who was selling because of this type of problem. He also felt that Mr. Jones would ask to be retained to work for the firm on a part-time sales and advisory basis at $1,000 per month.

Questions:
1. What economic, environmental, and social factors are reflected in the case?
2. What problems do they create for Mr. Jones? For Mr. Hannesson?
3. What would you recommend to Mr. Jones? To Mr. Hannesson?

19 The Political Environment

19 THE POLITICAL ENVIRONMENT

OBJECTIVES: When you finish this chapter, you should be able to:

1. Indicate the three different forms of business-government interaction.
2. Discuss the various levels of government and their responsibilities.
3. Understand and describe the various aspects of an economic decision.
4. Explain how governments can influence business decision making.
5. Indicate the specific tools available to government to influence business decision making.
6. Discuss how government and business interact in the day-to-day administration of government programs.
7. Describe the stages involved in the process of government policy formulation.
8. Indicate the different people who are involved in government policy formulation.
9. Describe the ways that business executives can get involved in changing government policy.

KEY CONCEPTS: In reading this chapter, look for and understand these terms:

DISTRIBUTION OF GOVERNMENT POWER
CO-OPERATIVE FEDERALISM
CROWN CORPORATION
COMPETITION POLICY
ROYALTIES
INCENTIVE PROGRAMS
ADMINISTRATION OF GOVERNMENT PROGRAMS
GOVERNMENT POLICY FORMULATION

The political environment is important to business in Canada. Governments influence many of the economic decisions mentioned in Chapter 1 about what to produce, how to produce, and how to distribute income. Some of these economic decisions are made directly by government. Others are made by private sector firms, but are influenced by government. We begin this chapter with a description of how the government is organized. We then discuss the role of government as a decision maker in the economic system and as an influencer of decisions made by private companies. This role of government in the economic system is one concern of business. There are two other interactions of business and government. One of the interactions occurs when government policies are implemented (e.g., administration of the tax laws or obtaining a licence to do business). Another interaction occurs when new government policy is formulated.

The major part of this chapter deals with the first of these concerns, the role of government in the economic system. The administration of existing government policy and the way government policy is changed are also examined.

THE GOVERNMENT OF CANADA

The political environment of Canadian firms is composed of a number of different governing bodies. These include the Canadian federal government, the provincial and territorial governments, and the nearly 5,000 municipal governments. Foreign governments are also part of the political environment of Canadian firms with operations in other countries, or for firms buying from or selling to firms located in other countries.

The Structure of Government

Distribution of government powers

Canada has a federal structure of government as outlined in the British North America (BNA) Act. Powers are distributed between the national government and provincial governments. **The distribution of government powers means that different levels of government have responsibilities for different matters.** The federal government has legislative jurisdiction over all matters of general or common interest. Provincial governments have jurisdiction over all matters of local or common interest. The specific areas of jurisdiction for federal and provincial governments are indicated in the boxed inserts (on pp. 511-512).

The sharing of responsibilities between federal and provincial governments has been a controversial matter. The tasks outlined in the boxes (pp. 511-512) sometimes overlap as in the case of agriculture, immigration, and old-age pensions. New social, technological, economic, and political developments have created new problems to be dealt with. Some of these problems have included aviation, broadcasting, telecommunications, and energy shortages.

Problems have arisen over provincial control of natural resources. This was a factor in disputes about how mining tax revenues were to be collected and allocated between the federal and provincial levels.

Broadcasting is another area of concern. The Quebec government, in particular, wants provincial control over broadcasting to assure consistency between programming and the cultural aspirations of Quebec citizens. The government of Canada, on the other hand, does not wish to give up control of broadcasting.

FEDERAL GOVERNMENT CONSTITUTIONAL RESPONSIBILITIES

The legislative authority of the parliament includes:

- the amendment of the Constitution of Canada
- the public debt and property
- the regulation of trade and commerce
- unemployment insurance
- the raising of money by any mode or system of taxation
- the borrowing of money on the public credit
- postal service
- the census and statistics
- militia, military and naval services, and defence
- the fixing of and providing for the salaries and allowances of civil and other officers of the government of Canada
- beacons, buoys, lighthouses, and Sable Island
- navigation and shipping
- quarantine and the establishment and maintenance of marine hospitals
- seacoast and inland fisheries
- ferries between a province and any British or foreign country or between two provinces
- currency and coinage, banking, incorporation of banks, and the issue of paper money
- savings banks
- weights and measures
- bills of exchange and promissory notes—interest
- legal tender
- bankruptcy and insolvency
- patents of invention and discovery
- copyrights
- Indians and lands reserved for the Indians
- naturalization and aliens
- marriage and divorce
- the criminal law, except the constitution of courts of criminal jurisdiction, but including the procedures in criminal matters
- the establishment, maintenance, and management of penitentiaries
- agriculture and immigration
- old-age pensions

Source: Reproduced by permission of the Minister of Supply and Services Canada.

PROVINCIAL GOVERNMENT CONSTITUTIONAL RESPONSIBILITIES

The legislature of each province may make laws in relation to the following:

- amendment of the constitution of the province except as regards the lieutenant governor
- direct taxation within the province
- borrowing of money on the credit of the province
- establishment and tenure of provincial offices and appointment and payment of provincial officers
- the management and sale of public lands belonging to the province and of the timber and wood thereon
- the establishment, maintenance, and management of public and reformatory prisons in and for the province
- the establishment, maintenance, and management of hospitals, asylums, charities, and eleemosynary institutions in and for the province, other than marine hospitals
- municipal institutions in the province
- shop, saloon, tavern, auctioneer, and other licences issued for the raising of provincial or municipal revenue
- local works and undertakings other than inter-provincial or international lines of ships, railways, canals, telegraphs, etc., or works which, although wholly situated within one province, are declared by the federal parliament to be for the general advantage either of Canada or of two or more provinces
- the incorporation of companies with provincial objects
- the solemnization of marriage in the province
- property and civil rights in the province
- the administration of justice in the province including the constitution, maintenance, and organization of provincial courts, both of civil and of criminal jurisdiction, including procedure in civil matters in these courts
- the imposition of punishment by fine, penalty, or imprisonment in enforcing any law of the province relating to any of the aforesaid subjects
- generally all matters of a merely local or private nature in the province
- education
- agriculture and immigration

Source: Reproduced by permission of the Minister of Supply and Services Canada.

The BNA Act granted the federal government far more extensive taxing powers than the provincial governments. Since the costs of some provincial responsibilities such as education and health care exceed their revenue-raising capabilities, revenue sharing has been instituted by the federal government. Education expenses, for example, are shared by the federal, provincial, and municipal governments.

The federal government, through the Department of Regional Economic Expansion, also makes funds available in various provinces for economic development. Examples of other programs for which the federal government transfers money to the

SERVICES OFFERED BY LOCAL GOVERNMENTS

- transportation
- protection
- environmental health
- environmental development
- recreation
- community services
- education

provinces include medical care, agriculture, tourism, crop insurance, and pest control. **The techniques and arrangements for dealing with federal-provincial economic relations have come to be known as co-operative federalism.**

The major source of revenue for local governments is taxation of real property. Revenue is also obtained from licences, permits, rents, concessions, franchises, fines, and surplus revenue from municipal enterprises such as transit systems.

Co-operative federalism

The Canadian Parliamentary System

Our federal and provincial governments are characterized by a parliamentary system of government. We have an executive branch of government headed by the Queen, whose representative in Canada is the Governor General. (See Figure 19-1.) The Queen's representative in each of the provinces is the Lieutenant Governor. At the federal level,

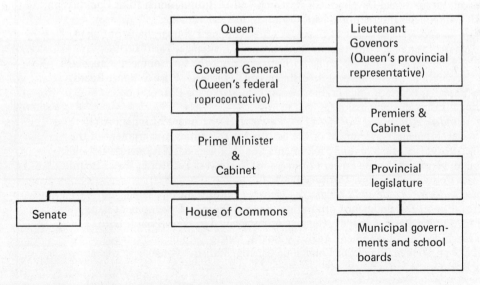

Figure 19-1. Structure of the government system in Canada

the Prime Minister and his Cabinet are formally the Queen's advisors. In the provinces, the provincial premier and the cabinet perform this function.

The Cabinet plays a key role in this system at the federal and provincial levels. It determines what executive actions will be taken by the government. It also places

---

The following are examples of federal government ministries, departments, and agencies reporting through them to Parliament as of January, 1977.

- *Minister of Agriculture* Department of Agriculture, Agricultural Products Board, Agricultural Stabilization Board, Canadian Dairy Commission, Canadian Grain Commission, Canadian Livestock Feed Board, Crop Insurance Corporation, National Farm products Marketing Council.
- *Minister of Communications* Department of Communications, Canadian Radio-Television and Telecommunications Commission, Teleglobe Canada.
- *Minister of Consumer and Corporate Affairs* Department of Consumer and Corporate Affairs, Anti-Inflation Board, Canadian Consumer Council, Copyright Appeal Board, Patent Appeal Board, Restrictive Trade Practices Commission.
- *Minister of Energy, Mines and Resources* Department of Energy, Mines and Resources, Atomic Energy Control Board, Atomic Energy of Canada Limited, Board of Examiners for Dominion Land Surveyors, Canadian Permanent Committee on Geographical Names, Columbia River Treaty Permanent Engineering Board, Eldorado Aviation Limited, Eldorado Nuclear Limited, Energy Supplies Allocation Board, Inter-provincial and Territorial Boundary Commissions, National Energy Board, Oils and Gas Committee, Petro-Canada, Uranium Canada Limited.
- *Secretary of State for External Affairs* Department of External Affairs, Canada International Development Agency, Foreign Claims Commission, International Boundary Commission, International Development Research Centre, International Joint Commission, Roosevelt Campobello International Park Commission.
- *Minister of Finance* Department of Finance, Anti-dumping Tribunal, Bank of Canada, Canada Deposit Insurance Corporation, Department of Insurance, Tariff Board.
- *Minister of Fisheries and the Environment* Department of the Environment, Canadian Saltfish Corporation, Environmental Assessment Panel, Fisheries Price Support Board, Fisheries Research Board of Canada, Freshwater Fish Marketing Corporation, International Fisheries Commissions.
- *Minister of Indian Affairs and Northern Development* Department of Indian Affairs and Northern Development, Government of the Northwest Territories, Government of the Yukon Territory, Historic Sites and Monuments Board of Canada, National Battlefields Commission, Northern Canada Power Commission, Northwest Territories Water Board, Oil and Gas Committee, Yukon Territory Water Board.
- *Minister of Industry, Trade and Commerce* Department of Industry, Trade and Commerce, Export Development Corporation, Federal Business Development Bank Foreign Investment Review Agency, General Adjustment Assistance Board, Loto Canada, Machinery and Equipment Advisory Board, Metric Commission Canada, National Design Council, Standards Council of Canada, Statistics Canada, Textile and Clothing Board.

- *Minister of State (Small Business), Minister of Justice, and Attorney General of Canada* Department of Justice, Anti-Inflation Appeal Tribunal, Law Reform Commission, Tax Review Board.
- *Minister of Labour* Department of Labour, Canada Labour Relations Board, Merchant Seamen Compensation Board.
- *Minister of Manpower and Immigration* Department of Manpower and Immigration, Canada Manpower and Immigration Council, Immigration Appeal Board, Unemployment Insurance Commission.
- *Minister of National Defence* Department of National Defence, Defence Construction (1951) Limited, Defence Research Board, National Emergency Planning Establishment.
- *Minister of National Health and Welfare* Department of Health and Welfare, Advisory Council on the Status of Women, Medical Research Council, National Advisory Council on Fitness and Amateur Sport, National Council of Welfare, Office of the Co-ordinator, Status of Women, Pension Appeals Board.
- *Minister of State (Fitness and Amateur Sport)*
- *Minister of National Revenue* Department of National Revenue (Customs and Excise), Department of National Revenue (Taxation), Office of the Administrator, Anti-Inflation Act.
- *Postmaster General* Post Office Department.
- *Minister of Public Works* Department of Public Works.
- *Minister of Regional Economic Expansion* Department of Regional Economic Expansion, Atlantic Development Council, Canadian Council on Rural Development, Cape Breton Development Corporation, Prairie Farm Rehabilitation Administration, Regional Development Incentives Board.
- *Secretary of State of Canada* Department of the Secretary of State, Canada Council, Canadian Broadcasting Corporation, Canadian Film Development Corporation, National Arts Centre Corporation, National Film Board, National Library of Canada, National Museums of Canada, Public Archives of Canada.
- *Solicitor General* Department of the Solicitor General, Canadian Penitentiary Service, National Parole Board, Royal Canadian Mounted Police, Correctional Investigator.
- *Minister of Supply and Services* Department of Supply and Services, Canadian Arsenals Limited, Canadian Commercial Corporation, Canadian Government Specifications Board, Crown Assets Disposal Corporation, Royal Canadian Mint, Offices of the Custodian of Enemy Property.
- *Minister of Transport* Department of Transport, Air Canada, Canadian National Railways, Canadian Transport Commission, Canadian Wheat Board, Maritime Pollution Claims Fund, National Harbours Board, Northern Transportation Company Limited, Pilotage Authorities (Atlantic, Great Lakes, and Pacific), St. Lawrence Seaway Authority, Seaway International Bridge Corporation Limited.
- *President of Treasury Board* Treasury Board Secretariat.
- *Minister of State for Urban Affairs* Ministry of State for Urban Affairs, Central Mortgage and Housing Corporation, National Capital Commission.
- *Minister of Veterans Affairs* Department of Veterans Affairs, Army Benevolent Fund Board, Bureau of Pensions Advocates, Canadian Pension Commission, Pension Review Board, War Veterans Allowance Board.

Source: Government of Canada, *Canada Year Book,* 1977. Reproduced by permission of the Minister of Supply and Services Canada.

legislative proposals before the House of Commons, Senate, or relevant provincial legislatures.

The unit of local government is usually the municipality, which is normally incorporated as a city, town, village, district, or township. The powers and responsibilities of municipalities are delegated to them by the provincial governments.

Government Departments and Agencies

We cannot simply talk about the federal, provincial, and municipal governments as single units. Each of them is composed of various departments and agencies. Different departments and agencies affect business decision making in very different ways.

Some idea of the number and diversity of federal government departments and agencies is given in the boxed insert on pages 514–515.

Each provincial and territorial government has a number of departments and agencies. A study published by the Ontario Economic Council, for example, identified 292 boards, commissions, and advisory and research bodies in that province. Other provinces also delegate authority to many agencies, boards, and commissions.

Municipal governments, particularly the larger ones, also have numerous departments, agencies, boards, and commissions. City planning commissions, environmental boards, licencing agencies, and zoning commissions are examples.

THE ROLE OF GOVERNMENT IN THE ECONOMIC SYSTEM

The various levels of governments with their many agencies and departments have a definite role in business decision making. This role and the forms it can take are discussed in this section.

A business decision has four elements: the decision maker(s), the options available to the decision maker(s), the criteria by which the options are judged, and the payoffs of the options. A simple example would be the owner of a store, Mr. Lamoureux, who has to choose one of two competing products to sell in his store. His objective is to make as much profit on the product as possible. The criteria he will apply in judging the two brands will be the anticipated profitability of each. Based upon his knowledge of the brand, image, price, quality of product, and sales in other stores, he estimates that he can sell 12,000 units of Product A at an average profit per unit of $2.73. He estimates that he can sell 15,000 units of Product B at an average profit per unit of $1.68. Total profit from Product A would be $32,760. Total profit from Product B would be $25,200. Mr. Lamoureux would therefore choose Product A. (See Figure 19-2.)

The political environment (government) can influence the way this decision is made. It could change the decision maker, the options available to the decision maker, the criteria used to make the decision, or the anticipated payoffs from one or more of the decision options.

The decision maker	The options	The criteria	The anticipated payoffs
Mr. Lamoureux	1. Product A	Profit for the store	$12,000 × 2.73 = $32,760.00
	2. Product B		$15,000 × 1.68 = $25,200.00

Figure 19-2. Four elements of the decision about what product to sell

Governments Can Change the Decision Maker

In this example, Mr. Lamoureux is the only decision maker. He is an entrepreneur. This would be changed if there were a government decision to nationalize the sale of the product. The decision maker in that case would not be Mr. Lamoureux. He would simply be a representative of the government. The authority and responsibility to make the decision shifts from the private sector to the public sector.

WHAT DO YOU THINK?

The Saccharin Ban

In mid-1977 it was announced that products containing saccharin (an artificial sweetener) would have to be removed from retail shelves by September, 1977.

The reason for the ban was because experimental results had linked saccharin with cancer. In order to give retail operators time to comply, the deadline was subsequently extended to December, 1977.

Retailers and food processors were upset by the ban. They claimed that the experimental results did not clearly show saccharin to be a health hazard. They pointed to the fact that saccharin had not been banned in the United States.

Food industry representatives were also concerned about the way in which the ban was enforced. They claimed that government bureaucrats were "overzealous" because in early January, 1978, they began visiting stores and breaking open packages of products containing saccharin, rendering them unsaleable. WHAT DO YOU THINK? Is it proper for a government to control the options available to business decision makers in this way?

Source: Government of Canada, *Report by the Sector Task Force on the Canadian Food & Beverage Industry*, 1978, Appendix C-5, p. 6. Reproduced by permission of the Minister of Supply and Services Canada.

Another way in which the authority and responsibility for making the decision could shift from Mr. Lamoureux would be if a government appointed a board or tribunal to review and approve decisions about what products could be sold. If this type of body was introduced, it would mean that Mr. Lamoureux would no longer be free to decide what product to sell. The government, through its board, would have control over the decision. When the government control over the decision increases, so does the government responsibility for the decision.

Governments Can Change Options

The options currently available to Mr. Lamoureux are to sell either Product A or Product B. Government action could affect these options. This could be done by making it illegal to sell one or the other, or both, of the products; this can certainly happen.

Governments Can Change Decision Criteria

Government actions can change the criteria which will be applied to the making of economic decisions. This occurs when government becomes involved as a decision maker, either through Crown corporation ownership or through operation of a regulatory board or tribunal.

The traditional decision criteria or goals for private economic activity are some combination of profit, growth, and survival. We have assumed in Mr. Lamoureux' case that his primary goal is profitability.

Imagine a situation where Mr. Lamoureux calculated that he could increase his profitability by manufacturing Product A himself rather than purchasing it from someone else. He decides that he will go into a joint venture with an American owned firm to buy out the Canadian-owned manufacturer of Product A. Because a foreign-owned firm is involved in the takeover, the transaction must be approved by the Foreign Investment Review Agency (FIRA).

The criteria which the purchase must satisfy for FIRA are more complex than the original profitability criteria which motivated the transaction. (The criteria which are applied by FIRA were discussed in the chapter on international business.)

Governments Can Affect Payoffs

There are many government policies which can affect the anticipated payoffs from decision options. Taxes, incentive grants, and tariffs are some of these. Government purchasing decisions can also affect the anticipated payoffs.

Let's assume that Product B was imported from France and the federal government decided to impose an import tariff of $1.00 per unit on it. This would have the effect of reducing the profit per unit by $1.00 if the selling price were not raised by

$1.00. This would reduce the potential profit on Product B to $10,200.00. This would make it even less attractive in relation to Product A than it previously had been.

Assume that Mr. Lamoureux has just received an order from the government for 12,000 units of Product B in addition to the initial 15,000 units he had estimated could be sold. Suppose also that Mr. Lamoureux could make the same profit per unit on these 12,000 units that he could on the initial 15,000. He was also aware that the government would not purchase Product A. He has to recalculate the estimated profitability of Product B. Instead of $1.68 × 15,000 units or $25,200.00, it is now $1.68 × 27,000 units or $45,360.00. This would make Product B more profitable than Product A, and Mr. Lamoureux would decide to sell it if he wished to maximize his profits.

In addition to specific policies like the ones mentioned, government policies can have a general influence upon the payoffs of decision options. Two items are the costs of implementing government regulations and collecting statistics on business activity.

Government intervention in the macroeconomy also affects payoffs. The exchange rate can influence costs. In our example, Product B is imported. If the value of the Canadian dollar decreases, the cost of Product B to Canadian firms increases. Interest rate changes which are influenced by the government through the Bank of Canada can also affect the anticipated payoffs of decisions, because, if a firm is borrowing money, interest rate fluctuations affect its interest costs.

GOVERNMENT INFLUENCE IN BUSINESS DECISIONS

There are many specific ways the government directly or indirectly participates in, or influences decisions made by, business firms. These can be grouped into four basic categories. First, government can get directly involved in business decision making by forming corporations to produce products or services. When this is done, government may become a competitor to firms in the private sector. Secondly, government can set up agencies which must give approval before certain business activities are allowed. Thirdly, the government may introduce legislation which prohibits certain business activities. Finally, the government can take actions which will change the payoffs of various business activities. Each of these four categories are discussed below.

Government Corporations

Crown corporations are used by government in Canada to participate directly in business decision making. **A Crown corporation is one which is accountable, through a minister, to Parliament for the conduct of its affairs.** The federal government utilizes three types of Crown corporations: departmental corporations, agency corporations, and proprietary corporations. These were discussed in Chapter 3.

Crown corporations exist at both the provincial and federal government level. In the mid-1970s, over 70 per cent of the electricity in Canada was generated by provincial government utilities. A number of provinces also own the telephone utilities within their borders. In addition, various provincial governments are involved in

Crown corporation

> **WHAT DO YOU THINK?**
>
> What types of economic decisions do the following organizations which are Crown corporations of the federal government, make which could not be made by privately owned firms?
>
> - Air Canada
> - Canadian Broadcasting Corporation
> - Export Development Corporation
> - Federal Business Development Bank
> - Freshwater Fish Marketing Corporation
> - Canadian National Railways
> - Petro-Canada

ownership and operation of other types of businesses. Manitoba, Saskatchewan, and British Columbia have provincial automobile insurance operations. Saskatchewan has taken over operation of much of the potash industry in that province. The Alberta government owns Pacific Western Airlines. Government-owned and government-operated enterprises account for a significant amount of economic activity in Canada.

Government Review Boards

For governments wishing to control business decisions there are alternatives to Crown corporations. Governments can, as we saw earlier, use administrative boards, tribunals, or commissions to screen decisions of private companies before they can be implemented. There are many examples of such boards, including the Foreign Investment Review Agency (FIRA), the Canadian Radio-Television and Telecommunications Commission (CRTC), the Canadian Transport Commission (CTC), the National Energy Board (NEB), and provincial boards.

FIRA reviews proposed acquisitions of Canadian businesses by non-Canadians. A non-Canadian cannot complete a takeover transaction until approval has been received from FIRA.

The CRTC regulates and supervises all aspects of the Canadian broadcasting system. The CRTC issues broadcasting licences and renews licences of existing broadcasting outlets subject to certain conditions. For example, a licence may stipulate the type of programming, the power of the station, or the minutes of commercial messages which can be broadcast. The CRTC also decides upon applications for rate changes submitted to it by federally regulated telecommunications carriers such as Bell Canada.

The Canadian Transport Commission makes decisions about route and rate applications for commercial air and railway companies. For example, before rail service can be expanded or contracted, the changes must be approved by this agency. The same holds true for other transportation activities (e.g., trucking and airlines). Proposed changes in rates must also be approved.

The National Energy Board is responsible for regulating the construction and operation of oil and gas pipelines which are under the jurisdiction of the Canadian government. This includes decisions about routes of pipelines and the size of pipe to be used in the lines. The NEB also decides on the tolls to be charged for transmission by oil and gas pipelines and the levels of export and import of oil and gas. It also issues guidelines on internal company accounting procedures and allowable rates of return.

There are also certain provincial boards which consider and pass judgment on proposed decisions by private companies. One example of such boards is provincial liquor boards or commissions. They must authorize price changes by breweries within provinces. Milk prices charged by farmers, dairies, and supermarkets are also regulated in a number of provinces. Other marketing boards for commodities such as pork, eggs, and vegetables have important roles in establishing prices and/or production levels of producers.

Government Regulation

One effect of government regulations is to limit the options available to business managers. Certain trade practices, products, and manufacturing processes are disallowed. There are many other regulations which company managers must observe. In 1955 the federal government's regulations totalled 2,000 pages; by 1977, this had increased to 12,000 pages. Governments use regulations to protect property rights (e.g., patents and copyrights); to stop abuses of market power (e.g., competition policy); and to set standards for consumers and/or producers in the health, safety, and fairness areas.

Three important areas of regulation are (1) competition policy, (2) consumer protection, and (3) environmental policies.

Competition policy. Canada's competition policy has been the subject of much discussion, with supporters arguing it is necessary for a healthy economy and critics claiming it is not effective. **Competition policy seeks to eliminate restrictive trade practices and thereby stimulate maximum production, distribution, and employment through open competition.**

Competition policy

The guidelines for competition policy are contained in the Combines Investigation Act, a comprehensive document which regulates the practices of Canadian business firms (the act does not apply to labour unions). The act is divided into 49 sections; these deal either with the actual laws or with activities which are necessary to administer them. The act is designed to encourage competition between business firms. If competition exists, all Canadians should benefit from efficient production systems, lower prices, and a healthier economy.

Consumer protection. A number of government programs related to consumer protection have been implemented. Many of them are administered by the federal Depart-

THE COMBINES INVESTIGATION ACT

Section number	Provisions
32	Prohibits conspiracies and combinations which are formed for the purpose of unduly lessening competition in the production, transportation, or storage of goods. Persons convicted may be imprisoned for up to five years or fined up to $1 million or both.
33	Prohibits mergers and monopolies which substantially lessen competition. Individuals who assist in the formation of such a monopoly or merger may be imprisoned for up to two years.
34	Prohibits illegal trade practices. A company may not, for example, cut prices in one region of Canada while selling at a higher price everywhere else if doing this substantially lessens competition. A company may not sell products at "unreasonably low prices" if this substantially lessens competition. (This section does not prohibit credit unions from returning surpluses to their members.)
35	Prohibits giving allowances and rebates to buyers to cover their advertising expenses, unless these allowances are made available on a proportionate basis to other purchasers in competition with the buyer given the rebate.
36	Prohibits misleading advertising. There are many types of misleading advertising which are prohibited, including (1) false statements about the performance of a product, (2) misleading guarantees, (3) pyramid selling, (4) charging the higher price when two prices are marked on an item, and (5) referral selling.
37	Prohibits bait-and-switch selling. No person can advertise a product at a "bargain price" if there is no supply of the product available to the consumer. This is usually done to "bait" prospects into the store and then "switch" them to higher-priced goods. This section also controls the use of contests to sell goods and prohibits the sale of goods at a price higher than that advertised.
38	Prohibits resale price maintenance. No person who produces or supplies a product can attempt to influence upwards, or discourage reduction of, the price of the good in question. It is also illegal for the producer to refuse to supply a product to a reseller simply because the producer believes the reseller will cut the price.

ment of Consumer and Corporate Affairs. The department initiates programs to promote the interests of Canadian consumers. Some of the programs which the department administers are indicated below. The Hazardous Products Act regulates two categories of products. The first category is products which are banned because they are dangerous. Some examples are: toys and other children's articles painted with coatings containing harmful amounts of lead and other chemical compounds; certain highly flammable textile products; and baby pacifiers containing contaminated liquids. The second category of products is those which can be sold but must be labelled as hazardous. Standard symbols which denote poisonous, flammable, explosive, or corrosive properties must be attached to certain products.

Food and drug regulations are another important area. These regulations are designed to protect the public from possible risk to health, fraud, and deception in relation to food, drugs, cosmetics, and therapeutic devices. For example, the Food and Drug Act prohibits the sale of a food which contains any poisonous or harmful substances; is unfit for human consumption; consists in whole or in part of any rotten substances; is adulterated; or was manufactured under unsanitary conditions. The act also provides that no person can sell or advertise a food in a manner which is misleading or deceptive with respect to its value, quantity, composition, or safety.

The Consumer Packaging and Labelling Act has two main purposes. The first is to provide a comprehensive set of rules for packaging and labelling of consumer products. The second is to ensure that full and factual information is provided on labels by the manufacturer. All pre-packaged products must state the quantity enclosed in French and English, in metric as well as traditional units. The name and description of the product must also appear on the label in both French and English.

Regulations under the Weights and Measures Act complement the packaging and labelling regulations. The Weights and Measures Act sets standards of accuracy for weighing and measuring devices.

The Textile Labelling Act deals with the labelling, sale, importation, and advertising of consumer textile articles. The National Trade Mark and True Labelling Act provides that products authorized under the regulations can be designated by the term "Canada Standard". A familiar application is with children's garments which bear the Canada Standard trademark.

Environmental regulations. Most of the industrial sources of environmental pollution are subject to provincial regulation. The federal role is limited to areas where there are inter-provincial or international implications.

One of the major pieces of federal government environmental legislation is the Canada Water Act. Under it the federal government can control water quality in fresh and marine waters, when there is a formal federal-provincial agreement, when federal waters are involved, or when there is sufficient national urgency to warrant federal action.

Two other important environmental regulations are the Fisheries Act, which controls the discharge of any harmful substance into any water, and the Environmental Contaminants Act, which establishes regulations for airborne substances which are a danger to human health or the environment.

Government Action Which Changes Payoffs

Some of the regulations indicated above can affect the payoffs of certain options. This occurs if there are fines for certain actions, or if significant investment is required to comply with new regulations.

There are other government actions which affect the payoffs of certain options, making them either more or less attractive. These include taxes and tariffs, incentive programs (grants), low-cost or free governmental services, government demand for products and services, and macroeconomic policy management.

Taxes and tariffs. These charges, which governments levy on business firms, can have a big influence on the payoffs of business decisions. Taxes paid are a cost of doing business. Differences in tax levels in different countries or provinces can affect a company's decision about initial investment or increases in investment in those areas.

Canada imposes federal and provincial income taxes on profits of corporations. Profits of proprietorships and partnerships are taxed at the same rate as personal income. Corporate income taxes accounted for more than 13 per cent of total tax revenues in Canada in 1975.

Royalties

Local property taxes and provincial royalties are other forms of taxes. Companies pay property taxes directly on property they own and indirectly on property they rent. **Royalties are government charges levied for the use of Crown property.** Royalties on natural resources are levied by provincial governments because they control resources. The major source of royalties is from mining, oil, and timber properties.

The federal Excise Tax Act imposes a retail sales tax on most goods produced in Canada or imported into Canada. A tax of 12 per cent is imposed on the manufacturer's sales price of goods produced in Canada and on the duty-paid value of imports. Certain goods, such as production machinery and equipment, food products, and all exports, are exempt from the tax.

All provinces except Alberta impose a retail sales tax. The 1977 rates of sales tax varied from 10 per cent in Newfoundland to 5 per cent in Manitoba and Saskatchewan.

Canada has traditionally had high tariffs. These contributed to the centralized, small-scale, high-cost, and largely foreign-controlled manufacturing firms in Canada. Canada's tariff rates have been progressively lowered through seven post–World War II rounds of trade negotiations between the Western trading nations, known as the General Agreement on Tariffs and Trade (GATT).

Canada has also negotiated certain bilateral (two-country) agreements which have influenced business activity in Canada. These include the Auto Pact and the Defence Production Sharing Agreement with the United States. These are agreements for the reduction of duties on automobiles and defence products between the two countries.

Incentive programs

Incentive programs. Federal, provincial, and municipal governments offer incentive programs which business managers should take into account when they make decisions. Incentive programs can be very important in stimulating economic development. **Incentive programs are designed to encourage managers to make certain decisions and take certain actions desired by governments.** For example, they are designed to encourage managers to locate in one region rather than another, to invest in new product development, or to engage in export activities. Incentive programs improve the payoff from a certain option and hence encourage a manager to select that option.

An example of grant incentive programs is the one offered by the federal Department of Regional Economic Expansion (DREE). These grants are designed to stimulate increased manufacturing investment and employment in slow-growth areas. Cash grants and/or loan guarantees are used to encourage managers to locate in designated slow-growth areas. In 1978 these regions included all the Maritime provinces, Quebec (excluding Hull and its immediate area), Manitoba, Saskatchewan, the northern part of northern Ontario, Alberta, British Columbia, and all the Northwest Territories.

Incentive grants on approved projects are given as a percentage of approved capital cost and of approved annual wages and salaries. The incentive grant on a Category A project in the Atlantic region—one with approved capital cost of less than $200,000 and creating fewer than 40 direct jobs—is 25 per cent of approved capital cost and 30 per cent of approved wages and salaries in the second and third year of operation. In other regions the grant is 25 per cent of approved capital cost but only 15 per cent of approved wages and salaries.

An example of a grant program which encourages development of pollution control equipment is one administered by the federal Department of Fisheries and the Environment. This is the program for the Development and Demonstration of Pollution Abatement Technology. It is intended to assist in the development of new methods, procedures, processes, and equipment to prevent, eliminate, or reduce the release of pollutants into the environment. The program pays a percentage of the capital and operating costs incurred by firms and municipalities in such developments.

There are many other examples of grant programs. Some of these include municipal tax rebates for locating in certain areas, design assistance programs, and remission of tariffs on certain advanced technology production equipment.

Low-cost or free services. Governments offer many services of value to business firms. One example of such a program is the Trade Commissioner Service of the federal government. The Department of Industry, Trade and Commerce has approximately 300 trade commissioners in 70 countries. The Trade Commissioner Service responds to requests for assistance from Canadian exporters and assists foreign importers to locate Canadian sources of supply for products they wish to buy. Trade commissioners also participate in the development of programs to improve Canadian exports. This requires identification of market opportunities and the development of export programs.

Some provinces, such as British Columbia, Alberta, Ontario, and Quebec, also have very active programs to encourage exports.

The Export Development Corporation was designed to improve Canadian exports by offering export insurance for Canadian exporters against non-payment by foreign buyers; long-term loans to foreign purchasers of Canadian products or guarantees of private loans to purchasers; and insurance against loss of, or damage to, a Canadian firm's investment abroad arising from expropriation, revolution, or war.

There are many other grants and low- or no-cost incentive programs available to Canadian firms. Some of the additional federal government services are outlined in a publication entitled *Canadian Federal Government Services to Business* published by the Department of Industry, Trade and Commerce of the Government of Canada. Information on provincial incentive programs is available from the various provincial governments. Similarly, information on municipal incentive programs is available from the various municipal governments. CCH Canadian Limited publishes a book entitled *Industrial Assistance Programs in Canada* which is up-dated on a regular basis.

The federal Department of Energy, Mines and Resources provides geological maps of Canada's potential mineral-producing areas. This service gives companies interested in mineral exploration much better geological information about Canada than is available about most other countries. Provincial governments also provide geological services to the mining industry.

Statistics Canada is yet another valuable service to business firms. The prices charged are below the actual cost of providing the service.

WHAT DO YOU THINK?

Paper Burden

The following list indicates some of the different types of government programs which require companies or individuals to complete paperwork.

- federal sales tax collection
- excise duties
- customs clearance
- unemployment insurance commission deductions
- workmen's compensation
- hospitalization
- loan applications
- building permits
- equipment operating licences
- property taxes
- vehicles registration
- transport operating licences
- communication licences
- income tax forms
- detailed household surveys
- business licences
- drivers' licences
- restaurants and liquor taxes
- CMHC housing surveys
- safety inspections
- tax audits
- FIRA requirements
- elevator licences
- boiler licences
- subsidy applications
- development permits
- waste control
- minimum wage guidelines
- Statistics Canada Surveys
- manpower training programs
- employee hiring procedures
- government contracts—procurement
- grants and incentive programs
- census of population
- welfare and health benefits

This paper burden is one of the subtle ways which government changes the payoffs for certain options. The costs to all Canadians for completing all this paperwork have not been estimated. However, one U.S. estimate is that completion of government paperwork in that country costs the equivalent of $500 for each citizen. If Canadian costs are that high, can they be justified?

Government demand for goods and services. Government spending can also influence business decisions such as where to locate or what type of product to produce. Government purchases range from paper clips and pencils to warships, highways and high-rise office buildings. Many firms and industries are dependent upon government purchasing decisions, if not for their survival, at least for their level of prosperity. Examples include construction and architectural firms and companies in the aerospace industry.

In 1975-76 the federal Department of Supply and Services placed orders for over $1.8 billion of goods and services from a total of 21,000 Canadian firms. Although almost 80 per cent of these purchases have historically been made in Ontario and Quebec, attempts are being made to stimulate manufacturing activity in other regions of Canada by purchasing a greater proportion of supplies and services in them.

Macroeconomic management. Government is important to business in its role as manager of macroeconomic policy. This is done primarily through changes in monetary and fiscal policy. Macroeconomic conditions such as inflation, economic growth, exchange rates, and employment levels can, as we saw in the previous chapter, have a significant effect upon economic decisions.

The macroeconomic climate and the level of confidence which it generates will be a significant contributor to investment decisions made by company managers. It therefore affects the payoffs of decisions. The level of confidence is important because when business managers make a decision based on the expectation of some level of return, they want to feel reasonably sure that the return will be realized. If there are uncertainties about exchange rates, the rate of economic growth, or rates of income tax, it can result in deferral of investment decisions, or even decisions to not invest in Canada.

ADMINISTRATION OF GOVERNMENT PROGRAMS

The previous section on the role of government in the economic system illustrates the many government programs which are important to business decision makers. For these programs to be effective, they must be put into practice and managed properly. **Administration of government programs involves the day-to-day activities which are required in the implementation of government programs.** This creates the need for business decision makers and those responsible for implementation of government programs to relate to each other. This is important in the operation of Crown corporations and in understanding and administering government regulations, taxation, tariffs, incentive programs, and government purchasing programs.

Administration of government programs

For government representatives this implies a need to understand business strategy and how it is formulated and implemented. It also requires an emphasis on the design of efficient systems for administering government policies.

Business managers, on the other hand, must have a thorough knowledge of government programs and how they are implemented. This allows business to deal effectively with government and keep their costs for this activity as low as possible. Failure to understand government programs or regulations can be costly. One firm which developed a new product and spent money developing a media campaign for it found that under Food and Drug Act regulations they could not sell it. This firm could have avoided thousands of dollars of costs if it had taken the time and effort to check out the regulations.

Business firms working in a mixed economy must develop the capacity to deal with government. This includes the ability to make effective representations to regulatory tribunals such as the CRTC. The case presented by the firm may determine whether a price increase is granted or whether a firm will be able to keep its licence to conduct business.

Firms must also develop an ability to understand and work within a maze of government regulations. A part of this is the ability to deal with the tax laws of the country. Many large firms employ one or more full-time specialists who do nothing but work on tax problems. Similarly, firms involved in import and export operations must have specialists who understand tariffs and how to deal with the customs officials who administer them.

A thorough understanding of the available government incentive programs and how to utilize them to the best advantage of the firm is also necessary. With the many reporting requirements facing companies, an ability to deal efficiently with them is a capacity which must be developed by a firm.

Firms which sell products to the government must develop special skills. Some large firms such as IBM have sales groups which work only with government clients. For large multi-billion-dollar projects like the 1970 purchase of new fighter aircraft by the Canadian government, specialized sales efforts are required by airplane manufacturers.

CHANGE IN GOVERNMENT POLICY

Government policy formulation

Another type of interaction between business and government representatives is that which accompanies the process of changing government policy. **Government policy formulation is the process by which changes are made in current government policies.**

Some general characteristics of this process can be identified. First, there are a number of stages in the process. Secondly, there will typically be a number of people involved in the process; these people will change at different stages of the process. Thirdly, decisions will be made at each stage of the process, and these decisions will depend on the power of the people involved at each stage. (See Figure 19-3.)

No one person controls the process through all of the stages. The primary determiners of outcome vary from politicians at the "Decision to Proceed with Legislation" stage, to civil servants at the "Drafting of Legislation" stage. Similarly, the people and groups with access to the decision maker and, therefore, an opportunity to influence him, vary from stage to stage.

The primary determiners of outcomes at the various stages are not business managers. They are government representatives.

Understanding the process of change in government policy is important for business managers or others who wish to influence government policy. A major implication

Stage of the process	Outcome of stage	Primary determiner of outcome (decision makers)	Possible influencers
1. Societal need	The existence of a problem	Created by changes in technology or attitudes (economic or ecological)	
2. Perception of need	General awareness of a problem	Researchers, media, politicians	
3. Articulation of demand	Different groups with different ideas about what to do	Leaders of different groups; e.g., political parties, media, interest groups, businesses	Members of various groups
4. Decision to proceed with legislation	Government commitment to deal with the problem	Cabinet ministers	Political party members Civil servants Public Media Researchers Interest groups Etc.
5. Determination of nature of legislation	A decision about what type of legislation to introduce	Civil servant teams Cabinet	Same influences as for No. 4
6. Drafting of legislation	A draft of new legislation	Public service drafting expert	Cabinet minister Civil servant study team
7. Legislative consideration	Royal assent to legislation	Parliament or legislative assembly	Party loyalties Pressure groups Media
8. Formulation of regulations	The completed set of regulations	Civil servants	Industry representatives Politicians Interest groups
9. Implementation	The new policy in practice	Civil servants	Groups being affected

Figure 19-3. Stages of the process of government policy formulation and types of people involved at various stages

is that the would-be influencer should attempt to exert pressure at all stages of the process. He or she should not restrict efforts to one point in the process, such as the legislative consideration stage.

Business managers should also recognize that the decisions made at various stages

STAGES IN THE PROCESS OF GOVERNMENT POLICY FORMULATION*

a. Societal Need. Where for whatever reason, or combination of reasons, some problem of significance to society arises. For example, advances in technology which create a new form of environmental pollution.
b. Perception of Need. Where certain people come to be aware that there is a problem of significance to society, and that it requires attention.
c. Articulation of Demand. Where people begin to demand that the problem be addressed. At this stage different people will likely advocate different solutions to the problem. For example; some may advocate that the technology be banned; others will say that it really creates no problems and should be continued; others will advocate measures to reduce the level of pollution created by the new technology.
d. Decision to Proceed with Legislation. Will be made at some stage if the pressures from the first three stages are great enough. The first three stages can stretch over a long period of years. The decision to proceed with legislation may be precipitated by some event. Such an event could be a sudden worsening of a situation, a change in the elected government, or the appointment of a new elected or civil service official who is to be responsible for the problem.
e. Determination of the Nature of the Legislation. Will then be undertaken. This could involve an appointed task force or royal commission. Most likely some person or group will be appointed to study the problems and the solution options and to make recommendations to the Cabinet minister responsible. The Cabinet minister in consultation with his officials and other Cabinet ministers will arrive at the recommended course of action.
f. Drafting of Legislation. Will be the stage after it is decided what the characteristics of the program are to be.
g. Legislative Consideration. Involves the first, second, and third reading by the Canadian parliament or a provincial legislature. It also involves committee consideration between second and third reading. Federal legislation must also be considered by the Senate. Royal assent, or signing of the bill, by the Queen's representative (i.e., the Governor General or a Lieutenant Governor) is the final aspect of legislative consideration.
h. Formulation of Regulations. The next stage. The regulations may be necessary to put the legislation into effect. After regulations are formulated, the legislation will be proclaimed to be in effect.
i. Implementation. The final stage. The way in which the legislation and regulations are interpreted and administered by government departments, agencies, and the courts will ultimately determine exactly what effect the new policy has on economic decision making.

*The stages and process of the government policy formulation outlined here were presented in Brian E. Owen's article, "Business Managers Influence (or Lack of Influence) on Government," Autumn 1976, *The Business Quarterly*. Published by School of Business Administration, The University of Western Ontario.

of the process have a cumulative effect. Decisions at an early stage can have a profound effect on the ultimate outcome of the process.

The fact that different people are involved at different stages is also important. These people must be identified, and the appropriate method by which to approach them must be considered. Recognition of the stages of the process of government policy formulation and some thought about the best way to participate at each of the stages can result in an improved ability to influence the process.

Some motivation is required to move the process from stage to stage, for example, from "Societal Need", to "Perception of Need" to "Articulation of Demand". In the early stages, the movement from stage to stage is caused by a discrepancy between industry practice and the public perception of what proper industry practice should be.

SUMMARY AND LOOK AHEAD

This chapter has discussed the political environment of business. Initially, three ways of viewing the relationship between business and government were identified. These are: (1) the role of government in the structure of the economic system; (2) by the roles of government and business in the ongoing day-to-day administration of government programs, and (3) the interactions between business and government in the process of changing government policies.

Governments can participate in and influence economic decisions by actually making the decision (by using a Crown corporation or a regulatory tribunal), by limiting the options available to business managers, by specifying goals for the decision, or by influencing the payoffs associated with available options. The wide range of tools available to governments to influence economic decisions was indicated, including Crown corporations, administrative (regulatory) tribunals, regulations, taxes, tariffs, incentive programs, government purchases, information collection, and macroeconomic management.

Activities of government and business representatives involved in the administration of government policies were briefly indicated.

The formulation of new government policy was also discussed briefly. The stages involved in the process were identified, as were the types of people involved at various stages.

In the next chapter career opportunities are discussed. The choice of a career plan for you is also discussed.

KEY CONCEPTS

Administration of government programs The day-to-day activities which are required to implement the programs.

Competition policy Seeks to eliminate restrictive trade practices and thereby stimulate maximum production, distribution, and employment through open competition.

Co-operative federalism The techniques and arrangements for dealing with federal-provincial economic relations.

THE INTERNATIONAL POLITICAL ENVIRONMENT

When a government owns the means of production, foreign firms doing business in that country must deal with that government. A firm may have to enter into a "partnership" arrangement with a government before the firm can begin operations there.

A Canadian firm which deals with foreign governments and/or businesses gets involved in politics. It's hard to keep sharp dividing lines between politics and ethics. This often poses a dilemma for managers. What is illegal in one country may be acceptable activity in another country.

The governments of many of the emerging nations try to instil in their people a strong feeling of nationalism. This sometimes leads them to distrust "foreigners". This can hurt multinational business.

Business activity thrives under conditions of political stability. Management should study the past history of political stability and current trends before committing itself to operations in a given country.

Many laws restrict a parent firm's control of its overseas operations. These include laws which require a subsidiary to hire local nationals or restrict how much profit can flow out of the country. A multinational firm takes on a considerable risk, much of which is political.

A firm operating in different countries is subject to different legal and tax systems. Quite often an item that is tax deductible in one country is not in another. The firm is caught in the middle but must be careful not to violate the laws of any country in which it operates.

Crown corporation A corporation which is ultimately accountable, through a minister, to parliament for the conduct of its affairs.

Distribution of government powers Different levels of government have responsibilities over different matters.

Government policy formulation The process by which changes are made to the current framework of government policies.

Incentive programs Programs to encourage business managers to make certain decisions and take certain actions desired by governments.

Royalties Government charges levied for the use of Crown property.

QUESTIONS FOR DISCUSSION

1. What are the three different forms which business-government interaction can take?

2. What are the various levels of government in Canada and what are their characteristics?
3. What are the responsibilities of the various levels of government?
4. What are the elements of a business decision? How are goals an important element of an economic decision?
5. What are the tools government can use to influence business decisions?
6. How can these tools actually affect business decisions?
7. What is the difference between taxes and incentive programs? Can taxes ever be used to provide incentives to business decision makers?
8. Should the use of Crown corporations be limited? Why?
9. What different roles are business managers required to play as government programs are implemented on a day-to-day basis?
10. What are the stages in the process of government policy formulation?
11. Why is the political environment important to business managers?
12. How important to business is government as a manager of macroeconomic policy?
13. Do you regard the physical, ecological, technological, economic, social, or political environment as most important to business? Why?

INCIDENT:

Baumann's Ltd.

James Baumann, president of Baumann's Ltd., a manufacturer of bakery products, was considering what actions he should take about what he considered to be the maze of government regulations he operated within. There were relevant Food and Drug laws, packaging, environmental, consumer protection and competition regulations, and taxes to be paid and incentive programs to consider.

In recent months many changes in environmental and consumer protection legislation had been proposed by the provincial and federal governments. This had placed heavy demand upon Mr. Baumann as, in his small company, he could not afford to hire someone to deal exclusively with such matters. He was uncertain whether to fight the changes in regulation or to accept them and adapt his operation accordingly. He wondered whether his membership in the local chamber of commerce would help in this regard.

Questions:
1. What type of business-government interaction is Mr. Baumann participating in?
2. Should he concentrate on trying to prevent change in government policy, on adapting his operation to the proposed new regulations, or on looking for another option?

20
Your Career in Tomorrow's Business

20 YOUR CAREER IN TOMORROW'S BUSINESS

OBJECTIVES: After reading this chapter, you should be able to:

1. Point out at least two trends which will probably affect the business career you are thinking of.
2. Distinguish between positive and negative effects of each of these trends.
3. Summarize the general employment outlook for major classes of jobs (blue collar, white collar, etc.).
4. Name four industries which are growing.
5. Find two general and two specific sources of job information.
6. Draw up your personal job evaluation formula.
7. Write and evaluate your own personal resumé.
8. Fill out a job application form completely.
9. Write a letter of inquiry for a job.

KEY CONCEPTS: In reading this chapter, look for and understand these terms:

NATIONALISM
CAREER VALUES
LETTER OF INQUIRY

We have studied a great deal about today's business world. We have studied management and how managers plan, organize, staff, direct, and control profit-making businesses. We have seen that businesses take resources—materials, machines, money, and most important of all, men and women—and produce goods and services to sell. We have also studied the three functional areas of business—production, marketing, and finance. We have seen how they relate to each other in the profit-seeking process of business.

In all of this we tried to show how important the environment is. The social, ecological, physical, economic, technological, and political aspects of the environment all say something to decision makers in business. They affect what can be sold, how it can be produced, and how it can be financed. The manager must be sensitive to the environment in the present and must respond to change in it.

Institutions and people must adapt to new realities. You are no exception. In a few years you may be playing a major role in the business world as a manager. Your business world will be different from the one described in this book. It will contain many of the same things, but it will also have many new features which will come about during the course of your career. You will grow and change, too. You must grow and change so that you can meet the future on even terms.

But change is nothing new. Since its beginning, the human race has had to learn to adapt to change. Prehistoric people had to learn to live in communities. The small farmer who moves to the city to take a factory job has to learn a whole new way of life. The youth born and raised in another country who has immigrated to Canada has to adjust to the work environment of a national corporation. Whatever your background is, you will have to adapt to the accelerating rate of change in the era of computers and atomic energy.

It is not easy to predict change, but since your career is in the future, you should start doing some thinking about what the future business world will be like.

Although we can't really predict the future, there are three things this chapter will provide to help you prepare for it. First, we will review some of the most important environmental trends on the assumption that "the seeds of the future lie in the present". Secondly, we will examine some forecasts of industrial growth and how these affect your prospects for a job in various industries and lines of work. Finally, we'll get down to your encounter with the job market and how to prepare for it.

ENVIRONMENTAL TRENDS AFFECTING BUSINESS AND YOUR CAREER IN IT

You are training now to be a manager in the future. The types of occupations which will be important and the problems managers will deal with will be influenced by the environmental conditions which exist in the future. The seeds of future developments exist at present. In Chapter 17 we pointed out the difficulties of forecasting environmental trends. In this section, eight noticeable environmental trends are discussed in terms of implications they have for future job conditions and opportunities.

These trends are:

1. Greater leisure and technology.
2. Increasing assertion of women's rights.
3. Longer lives.
4. Continued emphasis on bilingualism.
5. Energy shifts.
6. New international realities.
7. Some shift of economic power to the West.
8. Continuation of the trend to more government involvement in economic decision making.

Leisure and Technology

Advancing technology is cutting down the amount of physical labour and increasing the amount of "brain" labour in the world of work. This is true in business, too. Computers provide the best example. Even a simple mechanical skill such as typing is changing as typewriters become more and more automated. This underlines the need for special training and education in tomorrow's job market. It also suggests that working hours will become shorter and vacations will become longer. Job opportunities in the recreation, hobby, entertainment, tourism, hotel, and restaurant industries will grow as a result. "Service" industries will grow faster than "product" industries.

Women's Rights

Both men and women will be affected by the struggle for women's rights. This struggle, together with strong government support of non-discrimination by sex, will mean greater opportunity for women. Higher-paying supervisory and technical jobs, previously thought of as strictly for men, are opening up to women. Law schools and business schools report a major increase in the percentage of female students. What represents greater opportunity for women may mean tougher competition for men in the job market. It will have a strong effect on family life styles and the ways in which some businesses will be conducted. Male secretaries and female production supervisors may become commonplace. This takes cultural adjustment, but it is happening.

Longer Lives

There is a clear trend towards longer lives in Canada. Not only are people living longer, but their later years are becoming healthier and more active because of higher incomes and better medical care. The elderly are organizing to put political and economic pressure on businesses. One of their main goals has been to combat the trend towards early retirement. They want to keep their jobs longer. In a sense, then, senior citizens are competing for jobs against you and other younger entrants into the job market. The anti-discrimination movement has already resulted in better treatment for elderly job seekers.

There are other by-products of the trend towards longer lives. For example, the construction industry will need to answer the demand for new types of housing to meet the needs of older people. Growing elderly markets are already resulting in the use of older persons in ads and more careful marketing treatment of this important group of the population. Anticipating longer retirement periods, workers and their unions will probably emphasize better retirement plans when they negotiate labour contracts.

Bilingualism

In recent years, there have been major efforts by the federal government to encourage bilingualism. These efforts will probably continue. People who are fluent in both official languages will have greater career opportunities than those who are unilingual. This will be so particularly (1) for people who work for the federal government, (2) for companies which deal extensively with the federal government, or (3) for companies which conduct a large amount of their business in Quebec.

Aside from the value of a second language within Canada, it is useful to be fluent in two or more languages if you aspire to a career in international business.

Energy Shifts

For the last several years our newspapers have reported the story of the petroleum shortage. You know of the efforts of Western nations to become independent of foreign sources of oil. The most obvious task facing industry is to find new sources of energy to satisfy our needs for industrial growth and for household use.

In the petroleum industry this means expansion of exploration for new oil fields and better ways of getting more oil from existing fields. In the coal mining industry this means a spectacular boom. Canada has a substantial supply of coal, so it's natural that we should turn to coal to meet our energy needs.

The principal problems with coal usage relate to ecology. Burning coal can release large amounts of poisonous wastes into the air. The demand for energy is so great that we are tempted to overlook these bad ecological side effects. We are also, of course, working to improve coal-burning systems to reduce their bad effect on the air.

The energy shortage is also creating vast new research programs in the fields of solar energy, nuclear energy, geothermal energy, and others. In the case of nuclear energy, possible harmful side effects of disposing of nuclear wastes is slowing down development. Despite this problem, it seems likely that the future will show greater dependence on coal and nuclear fuel and less on petroleum. Which way we turn for energy will affect your future in business.

New International Realities

The recent pressure put on the advanced industrial nations by the oil-producing nations of the Middle East has changed the relationship between the advanced and the under-developed countries of the world.

Some under-developed countries possess large amounts of the world's vital resources. After seeing the success of the Arab oil countries, it is possible that certain Latin American and African nations will threaten to withhold their vital raw materials. They would hope to gain greater political and economic power in the world and a higher standard of living for their people.

Such actions could lead to changes in the way we do business in the developed countries. In the first place, it could lead to efforts to produce substitutes for an increasing number of raw materials. It could eventually mean that many people in the less developed nations will begin to be able to afford to buy manufactured goods from us. This, in turn, could reduce world tension.

For still others of the under-developed nations there is little hope of economic improvement. Instead, there is a strong possibility of more hunger and starvation unless rapid progress is made in population control. This will mean the constant need to export agricultural surpluses from the United States and Canada to these starving nations without any permanent solution. Western businesses and farmers may benefit from exporting goods and medical supplies to these nations, but this may require a great amount of economic aid.

There is a growing spirit of nationalism in smaller, less developed countries. **Nationalism is a strong feeling of pride and independence among the people of a country.** This has also resulted in large-scale takeovers of the foreign-owned plants and distribution centres in these smaller nations. This movement has caused foreign firms to be much more cautious about foreign investments. Many such operations today are almost completely staffed by local people rather than by North Americans, and some are co-owned by local governments and investing firms.

Nationalism

Shift to the West

There are two main reasons why there will be some shift of economic power to the Western provinces—principally Alberta, British Columbia, and to some extent Saskatchewan. The first of these reasons is the natural resources of these provinces. The recent oil-price increase has greatly stimulated economic growth in Alberta. Potash and uranium are important in Saskatchewan.

The second reason is the continued development of trade opportunities with Pacific rim countries. This will stimulate related business activity in British Columbia. This does not mean that eastern Canada will decline in absolute importance, but that there will be an increase in activity in the Western provinces.

The implications for job opportunities are obvious. There will continue to be an increase in employment opportunities in the west. Students will perhaps have more opportunity to choose careers on a regional basis than they have had in the past.

Increased Government Involvement in Economic Decision Making

We expect a continuation of the increase in the importance of government in economic decision making in the future. The reasons for the trend in the past and for the expectation of its continuation were discussed in the chapter on the political environment.

The implication of this trend is that there will be more job opportunities with governments in the future. There will also be more opportunities for business jobs which involve liaison with government agencies or departments.

From this review of some likely areas of change for business, we turn to the role you will have in the changing business world. We hope the guidance we offer will help you make some important decisions.

CHOOSING A CAREER

As we have just seen, the business world of tomorrow will be different from that which exists today. The pace of change is expected to be rapid. In a matter of decades, whole new industries will arise because of new inventions and because of changes in resources and life styles. This can be a source of confusion for you, but it has its very definite positive aspect. A dynamic business world is a healthy business world. If nothing were to change, the future world would be a familiar one, but it would also provide very little opportunity. You should be optimistic and you should begin to plan for the business career which lies ahead of you.

What you will be doing in the business world depends on (1) job market conditions; (2) what you have to offer; and (3) how well you present your talents to potential employers. We will examine each of these factors.

The Job Marketplace

There are at least two ways to examine the marketplace for jobs. One way is to look at the market for individual occupations (regardless of the industry in which it is found). Another is to check out the projected growth of whole industries in which jobs may be available. Which way you choose should depend on how clear your ideas are about a career and on the "career values" you have established.

Career values

By career values we mean those things which you feel are important in selecting a career. You may, for example, value living in a certain place or dealing with the public

---

THINK ABOUT IT!

World Population Growth

The world's population is growing by 2.9 per cent annually. At the beginning of 1976 the world's population was 4.2 billion—90 million more people than at the beginning of 1975. If the present rate of population increase continues, the world's population will double in the next 30 years. THINK ABOUT IT!

BE INFORMED IN YOUR CAREER CHOICE

The publications listed below are available at your campus placement office, local Canada Manpower Centre, or address given:

- Manpower and Immigration, *Guide for the Job Hunter* (booklet), 1976.
- Manpower and Immigration, *Looking for a Job?* (folder), 1975.
- Manpower and Immigration, *Your Job Bank* (folder), 1976.
- Manpower and Immigration, *A Job Search Guide* (book), 1976.
    Publishing Centre, Mail Order Services, Supply and Services Canada, Ottawa, Ontario K1A 0S9 ($2.95).
- New York Life, *Making the Most of Your Job Interview* (booklet).
    New York Life Insurance Co., 51 Madison Ave., New York, N.Y. 10010.
- University and College Placement Assoc., *U.C.P.A. Employment Opportunities Handbook* (book containing employer lists and requirements plus job search information), published annually, U.C.P.A., Box 356, Markham, Ontario L3P 3J8.

The following references are available for consultation at your campus placement office, university library, or local Canada Manpower Centre (distribution is limited to counsellors and trained placement officers):

- Canada Pay Research Bureau, *Anticipated Recruiting Rates for University and Community College Graduates* (book containing salary information by kind of degree, diploma, or certificate), published annually, Canada Pay Research Bureau, Public Service Staff Relations Board.
- Manpower and Immigration, *Canadian Classification and Dictionary of Occupations,* volumes 1 & 2 (book).
- Manpower and Immigration, *Canadian Occupational Forecasting Program* (book containing forecasts of occupational demand to 1982 for some 500 occupations in Canada), 1975.
- Manpower and Immigration, *Directory of Employers of New Community College Graduates* (book), published biennially, 1975.
- Manpower and Immigration, *Directory of Employers of New University Graduates* (book), published biennially, 1975.
- Manpower and Immigration, *Forward Occupational Imbalance Listing* (book giving short-term detailed and localized forecasting of occupations in Canada by province), Economic Analysis and Forecasts Branch.
- Manpower and Immigration, *Supply, Demand and Salaries for New Graduates of Universities and Community Colleges* (book), published annually, Economic Analysis and Forecasts Branch, 1976.

Many of these publications are up-dated regularly. When using them, you should pay attention to their date of publication because they can become outdated.

or using your artistic talent or having the opportunity to rise rapidly in an organization. It may be that you are excited about working in a particular industry like the communications industry but haven't decided on the specific job you want in that industry. On the other hand, you may have a clear idea about the kind of work you want to do (e.g., accounting) but don't care much about the specific industry. In any case, you really ought to study both of these aspects of your job decision.

Canada Manpower is a good source of information about the job market. Analyses of the job outlook are available from your local or campus Canada Manpower office.

The publication entitled *Canadian Occupational Forecasting Program,* published by the federal Department of Manpower and Immigration gives an indication of job growth for selected occupations (Table 20-1). Although the classification does not include occupations generally requiring post-secondary education, it should be noted that among the largest growth areas in percentage and absolute terms are managerial, clerical, sales, and service occupations.

You may wish to get statistics on a particular job market from trade or professional associations such as the Investment Dealers Association or the Institute of Chartered Accountants. Some provincial departments of labour or departments of industry and commerce also have current job market statistics.

Of course, statistics don't present a complete picture. You will also want to know something about these jobs' expected earnings, working conditions, advancement, and so forth. A good place to start looking for such information is in a Canadian Government Manpower and Immigration series of publications entitled *Careers Canada.* The books in the series are intended to provide up-to-date, factual information about a variety of occupations. Figure 20-1 provides a sample of the kind of information available about a specific occupation—in this case accounting—in the book.

A U.S. government publication, the *Occupational Outlook Handbook* may also prove useful. The latest edition of this publication should be available in your Canada Manpower office. It, like *Careers Canada,* describes the nature of various careers.

Of particular interest to students of business are listings under the general headings of office occupations and sales occupations. These occupations are listed in Table 20-2.

You Are the Product

There are several ways of looking at the career selection process. Since we have been studying about markets and prices of products, why not look at yourself as a product which some business firm will "buy"? A good salesperson must know his or her product. You, then, must know yourself. You must know enough to know where your "product" will serve best. Ask yourself two questions: "What do I really want out of a job?" and "What are my skills?"

What Do You Really Want Out of a Job?

A job will bring you great unhappiness and little chance for success if it conflicts with what you really are. You must, therefore, explore your own values and interests in life. What are the things which really are important to you?

It may help you to examine the "Career Value Checklist". (See Table 20-3.) Look over each of these ten questions carefully. Can you give yourself a pretty clear answer

Table 20-1. Occupational demand in Canada*

CCDO CODE CCDP	Description	Demand 1974	Demand 1982	Net change 1974-1982	Net change In percentage of 1974	Withdrawals deaths 1974-1982	Req. supplies 1974-1982	Req. supplies In percentage of 1974
11	Managerial, Admin.	255,800	321,575	65,775	25.71	87,850	153,625	60.06
41	Clerical Occs.	1,611,175	2,049,450	438,275	27.20	323,200	761,475	47.26
51	Sales Occs.	980,450	1,185,400	204,950	20.90	251,450	456,400	46.55
61	Service Occs.	1,017,900	1,290,050	272,150	26.74	243,250	515,400	50.63
71	Farming Occs.	511,175	457,850	−53,325	−10.43	121,850	68,525	13.41
73	Fishing, Hunting	26,800	27,700	900	3.36	7,175	8,075	30.13
75	Forestry and Logging	73,500	80,750	7,250	9.86	17,200	24,450	33.27
77	Mining Occs.	56,825	66,250	9,425	16.59	15,325	24,750	43.55
81/82	Processing Occs.	384,250	418,200	33,950	8.84	92,025	125,975	32.78
83	Machining Occs.	291,200	324,650	33,450	11.49	71,900	105,350	36.18
85	Product Fab. Occs.	760,800	852,800	92,000	12.09	190,850	282,850	37.18
87	Construction Trades	642,800	716,850	74,050	11.52	169,400	243,450	37.87
91	Transport Eq. Operat.	400,850	477,575	76,725	19.14	112,050	188,775	47.09
93	Material-Handling	192,450	222,250	29,800	15.48	41,300	71,100	36.94
95	Other Crafts	117,450	140,875	23,425	19.94	29,175	52,600	44.79
99	Occs. not else. Class.	315,100	377,325	62,225	19.75	54,275	116,500	36.97

*Excluding occupations generally requiring post-secondary education.

Source: Department of Manpower and Immigration, *Canadian Occupational Forecasting Program, Forecasts of Occupational Demand to 1986 No. 5 Manitoba,* 1976, p. 19.

In all organizations, somebody has to take care of the finances to ensure that a profit is made. Money is paid out for wages, goods received, rent, taxes, and other expenses; money is received for goods or services sold. Someone has to keep records of all financial transactions so that the company knows its financial situation.

In a small business, the bookkeeping may be done by the owner or cashier who, from time to time, may hire an accountant to provide advice, set up an accounting system, or audit and check the bookkeeping records for accuracy.

In larger organizations, however, there is an accounting department staffed by many clerical workers and headed by an accountant, who, as you will see in later pages, may be known as an auditor, a comptroller, or a bursar, depending on the type of organization.

For a number of careers in finance, proof of qualification by membership in a professional association is an advantage and may be a requirement by some employers. There are several kinds of accountants, some of which are briefly described as follows.

Chartered Accountants (CAs) must first have a bachelor's degree. A degree in commerce or business administration is the most suitable, although degrees in other subjects are acceptable. While employed with an approved firm of accountants, they study part-time. Then they must pass a four-part examination for admission to the provincial institute of Chartered Accountants.

About half of all chartered accountants are employed in business and government where they audit and verify accounts, provide financial reports, and plan accounting systems to prevent errors or fraud. The rest, either in their own business or in partnership with other accountants, offer services for a fee. They may specialize in estate planning, taxation, auditing accounts, management planning, or other kinds of consulting work.

Certified General Accountants (GCAs) belong to a provincial Certified General Accountants Association. To qualify for membership, they must complete a program of correspondence studies extending over three terms of ten months each. They must provide proof of at least two years of practical experience obtained while following the course of studies. The usual minimum requirement for entering the program is graduation from Grade 12. Normally you will find CGAs doing general accounting work in industry and government.

Registered Industrial Accountants (RIAs) qualify for membership in the Society of Industrial Accountants through a combination of studies and practical experience.

The studies take about five or six years from high school graduation, the minimum educational requirement, but may be shorter if you have college or university credits. All courses can be completed by correspondence or through evening programs at most colleges and universities.

Examinations must be passed, and proof of practical experience is required. This can be done in as little as two years, or up to four years for students who have been required to complete the full program of correspondence studies.

RIAs are employed by many organizations, including industrial and commercial enterprises and government. They work in such areas as management, where they develop plans to help the company produce products at the best possible price, look for possible improvements in financial matters, and detect errors. They may also prepare financial reports to inform management on the control, planning, and evaluation of the company's operations.

Figure 20-1. Accountants—nature of the work. (Source: Department of Manpower and Immigration, *Careers Canada, Careers in Banking & Finance,* p. 23.)

Table 20-2. Business-related occupations

Clerical occupations	*Administrative & related occupations*
Bookkeepers	Accountants
Cashiers	Advertising workers
File clerks	City managers
Hotel front office clerks	College student personnel workers
Office machine operators	Credit officials
Postal clerks	Hotel managers and assistants
Shipping & receiving clerks	Industrial traffic managers
Statistical clerks	Lawyers
Stock clerks	Marketing research workers
Stenographers & secretaries	Personnel workers
Typists	Purchasing agents
Computer & related occupations	*Sales occupations*
Electronic computer operators	Auto parts counter workers
Programmers	Auto sales workers
Systems analysts	Auto service advisors
Banking occupations	Gasoline service station attendants
	Insurance agents and brokers
Bank clerks	Models
Bank tellers	Manufacturers sales workers
Insurance occupations	Real estate sales workers & brokers
Actuaries	Retail trade sales workers
Claims adjusters	Route drivers
Claim examiners	Securities sales workers
Underwriters	Wholesale sales workers

Table 20-3. Career value checklist

1. Do you like meeting people?
2. Do you welcome responsibility?
3. Do you want a flexible work schedule?
4. Do you want to be your own boss?
5. Do you want security in your job?
6. Do you expect high starting pay?
7. Is rapid promotion potential important?
8. Is staying in your hometown important?
9. Do you want public recognition?
10. Do you want to retire before age 60?

to each question? If so, you know yourself better than most college students do! You are in an excellent position to evaluate a specific job from the "interests" or "values" perspective. If you are unsure of some answers, this is typical. You can still use this checklist to clarify your value system for a job choice.

First, try to rank the ten questions in terms of their importance to you. You should be able to rank at least six or seven of them. Once you've done this, then pick out several jobs you know something about and rate them in terms of the important job values you have chosen. This process should help you to understand what a job means to your life in the ways which are important to you. It will also prepare you to make a final job choice. But first you must also evaluate your skills.

What Are Your Skills?

Your skills and interests are usually closely related. You are more likely to be interested in those things in which you excel. If you had trouble learning the multiplication tables, you would not be interested in a statistician's job.

While some skills are very specifically related to a job, others are required of most jobs. A skill which is needed in most jobs, particularly as you move up in a company, is human relations skill—the ability to communicate with others and to listen to and understand what they have to say. Another general skill is skill in organizing. If you have a knack for keeping things in order and putting them in perspective, there are many jobs in which this skill will prove valuable. More specific skills include skill in writing, skill in calculation, mechanical skills such as typing and shorthand, and skills relating to memory and precise observation.

Sometimes your physical characteristics, such as strength and appearance, are an asset or a handicap in getting a job. Consider your physical characteristics. Do they seem appropriate for the job you want? Sometimes a firm sets up physical specifications for a job applicant because of the demands of the position. Check to see if you meet these before you go too far in the job-seeking process.

Up to this point we have shown how you can survey the general job market to discover the occupations and industries which show promise for you. We have seen the importance of self-evaluation in terms of interest and competence. Now we turn to the process of finding a particular job.

Specific Job Search and Choice

You will probably have lots of advice—some you will ask for and some you will get without asking—about how to get a job. A bit more won't hurt, so we'd like to offer you a few tips about job hunting and the final decision.

Job Hunting

After making a general survey of opportunities and of yourself, you should start to narrow down the field.

Don't waste time with firms or jobs you are not actually interested in! To find out more about specific jobs, you can start by writing to the appropriate firm—either the personnel department or the department in which you want to work. You will be provided with general facts about the firm, job availability, wage scales, union status, how to apply for a job, and so on. You can supplement this information by talking to employees in different industries and firms. They can tell you what it's really like to work in a certain job or in a certain firm.

Next, you should start checking for specific job openings. Your school's Canada Manpower office or placement director will have good information about possible openings. They may have a schedule of employers coming to the campus for interviews. The classified ads are also a good source of information.

The best opportunity may sometimes result from a frank "letter of inquiry". This is a letter you send to a firm asking about the possibility of getting a job. It may be that you already know they are looking for someone with your qualifications. In this case your letter should start with that information. Such a letter is presented in the Appendix. It is also possible to send letters to firms without knowing if they are "looking". In that type of letter, you should stress your general interest in working for the firm and how you feel that your talents might best be used. Sometimes this brings surprising results.

Letter of inquiry

For a more detailed treatment of the job search process, please examine the Appendix carefully. We will concentrate now on making the final choice.

Making the Final Choice

If you are in a position to choose between or among job offers, it is wise to set up a practical system to reach a decision. To do this you need a standard, or set of values, to guide your choice. You will also need all the facts about each job. We have already discussed what you "want out of a job". These are the standards, or values. We assume that you have gained the facts about each job from the job search process.

A job selection formula could look like this. (See Table 20-4.) Assume that you have three jobs to choose from. These are labelled *A, B,* and *C* at the top of the three

Table 20-4. A job selection formula illustrated*

Directions: Rate each job by each value factor on a 1 to 10 scale. Multiply scores by indicated weights and then total the products for each job.

Value Factors and Weights	Job A	Job B	Job C
Interest × 5 =	3 15	5 25	8 40
Starting pay × 3 =	4 12	8 24	7 21
Promotion opportunity × 4 =	5 20	9 36	6 24
Prestige × 2 =	10 20	6 12	8 16
Location × 1 =	3 3	4 4	10 10
Total score	70	101	111

*This concept was developed from suggestions offered by Professor Robert S. Ristau of Eastern Michigan University.

columns of Table 20-4. Assume further that you have honestly searched your values and that only five factors are important to you in choosing: your level of interest in the job itself; the starting pay; the opportunity for promotion; the prestige of the job; and the location. You have assigned relative importance, or weights, of 5, 3, 4, 2, and 1, respectively, to each of these five factors. This means that, for example, level of interest is five times as important to you as location and that promotion opportunity is twice as important as prestige. Assigning these weights will require some careful thought.

Next, you start to "rate" each job on a 1 to 10 scale according to each of your value factors. Table 20-4 shows your interest level in Job *A* is rated only 3, while your interest in Job *B* is rated 5, and your interest in Job *C* is rated 8. Since the weight, or importance, of interest level is 5, you multiply the ratings times 5 and score them 15, 25, and 40 in that order. (These scores are underscored.) Next, you do the same in terms of the starting pay. In this case, Job *A* rates a 4, Job *B* an 8, and Job *C* a 7. Multiply these by 3—your "weight" for starting pay. The scores appear underscored below the ratings. They are 12, 24, and 21. You do the same for the other value factors and add up the scores for each job—all the underscored values in each column. The total scores in this illustration are 70, 101, and 111.

The results indicate that you should choose Job *C*. However, since Job *B* is a close second, you might want to go through the process once more for Jobs *B* and *C*. No formula is perfect, either. There may be something about one of these jobs which

> **YOUR CAREER AND CONTINUING YOUR EDUCATION**   CB
>
> In considering your career, do not forget the option of continuing your formal education. No one ever stops learning. There are always higher goals to reach, and further education is often the key. If your present plans are to take a full-time job after two years at a community college, give some thought to getting the bachelor's degree. The general rule is, the more formal education you have, the greater your career potential! Even more important is the fact that more education leaves you with more job alternatives to choose from. You are more flexible that way. Talk to your favourite instructor about programs available in nearby universities.
>
> If you are already in university, the chances are that your diploma is still a few years away, but it is never too early to think in terms of the master's degree or even a doctorate. There are universities in almost every province which offer the Master of Business Administration (MBA) degree. Thousands of young people and many people of all ages who have returned to school or who are studying evenings while they work are getting their MBAs.
>
> The MBA generally gives broad advanced training to students, usually with some quantitative or scientific management component. It usually takes one or two years of university work beyond the bachelor's degree. It leads people into responsible managerial or junior executive positions with excellent chances of advancement.
>
> The doctorate—either a Ph.D. (Doctor of Philosophy in Business) or a DBA (Doctor of Business Administration)—provides two or three additional years of study with some specialization. It generally leads to a teaching position or sometimes to a research job in industry. It is not unusual to move from the doctorate to a line executive position in industry.
>
> Even if you don't go for a degree beyond what you are working on now, you should think in terms of continuing your education in some way. Perhaps an occasional evening course on an interesting topic will be sufficient. Continuing to learn amounts to continuing your growth as a human being.

attracts you but you can't put your finger on it. It might be enough to make you select the second choice according to the formula.

When You Start Work

You may have noticed already that a successful career in school requires the ability to discipline yourself. You may sometimes "get by" because of a good memory or because of luck, but you will perform consistently well if you form good study habits, work at your studies on a regular schedule, and take every subject seriously. The payoff will be two-fold. You will earn the best grades your ability will allow. Also, you will be a much better potential employee because of what you have learned and, more importantly, because you have learned how to learn. The discipline, patience, and open-mindedness required to learn things well in school are exactly the same things you will need to learn and to grow in your job.

If you start demonstrating these qualities from the very start in your job (by working overtime if necessary, volunteering for assignments, and doing your work thoroughly), your efforts will be rewarded. Patience, too, will help, because any time you go to work for a specific boss there will be occasions when "the boss's way" and the way you learned previously will differ. Personality conflicts, too, will arise. They will require patience and tact. This does not mean that a new employee must be a doormat. It only means that you must present your views courteously. It is wise to assume that when you are new on a job, the "old hand" probably does know more about how things should be done. In the long run, your ability, good judgment, and creativity will become apparent to your employer.

Promotion and Changing Jobs

Sometimes the best way to get a promotion is not to seek it. This may sound funny, but it is often true. The worker or manager who concentrates on doing the best possible job with what he or she is assigned doesn't have the time to be looking for a promotion. The football team which looks ahead to next week's big game instead of this week's may get beaten by an underdog team. This is especially true if the other team takes "one game at a time".

Let's assume that you have been doing everything which is expected of you and perhaps a little bit more. If so, the word is likely to spread that you "more than carry your weight". The next time a better job in the organization comes up, the chances are that your name will come up. While obvious attention paid to superiors can backfire, making their acquaintance can help your career. The key ingredient, of course, is doing quality work!

What about changing jobs? Recent estimates predict widespread and frequent occupational shifts for young people now entering the job market. It seems likely that by the year 2000 many of the jobs done today will no longer even exist!

There are many reasons why you may need to change jobs during your lifetime. This can be caused by obsolescence, by government contract shifts, by the competition of imports, and by many other things. A wise person keeps a sharp eye for the kinds of events which can force a job change. He or she also keeps an eye out for opportunities. Otherwise, the time spent between jobs can be painful.

Two kinds of preparation for this can help. One is to stay as financially prepared as possible—to save something and to think ahead in terms of temporary "back-up" jobs you might take. The second, and more important, is to continue your education and training throughout your life. This has the effect of broadening your skills and increasing your ability to adapt to new and different jobs. In a world of dynamic change, the future belongs to those who remain flexible.

A FINAL WORD

You now have a taste of what the business world is like. It is such a huge and complicated set of industries and firms that it is almost impossible to describe adequately. Even if we could describe it well, you would still have to become a part of it to begin to really appreciate it. And even after you're in it for a while, you'll realize that because it's changing so rapidly, you'll never really learn all there is to know. You'll probably take other business courses in school which will concentrate on specific aspects of operating businesses to prepare you more completely for a specific job. They will also help you in choosing a career.

No matter what you choose as a career, the essential ingredients will be flexibility and willingness to learn. At some time in your career, you may have to be completely retrained if your job becomes obsolete. This is one of the bad features of technical advance. This specific kind of insecurity can be overcome if you have the right attitude. Think of your education as a process of learning how to learn, learning to keep an open mind, and learning about the relationships between yourself and your environment.

Such an attitude towards your education will sharpen your appreciation of it. How you approach the rest of your college experience will set a pattern for your lifetime. You should make sure that it is a pattern which includes the appreciation of learning as a continuing and unending process.

YOUR CAREER—COPING WITH CHANGE

In the first part of this chapter we reviewed some of the important changes we think the future will bring. They suggest strongly that you will probably have more than one kind of job during your life time. This could be caused by competition, by new technology, by resource shortages, or by changes in you and your value system. In any case, you must prepare for changing employment.

Job changes can be hard to face unless you convince yourself that you can handle them when they come. Self-confidence, of course, doesn't do it alone. Your self-confidence must be solidly based. You should have trained yourself all along to react positively to change—to view it as an opportunity rather than a threat. Change is an opportunity for the man or woman who has learned how to learn in school and who does not take a narrow view of life. You may have a second and even a third or fourth career in your life. Each one can be more rewarding than the last for the broadly educated, flexible person.

WHAT DO YOU THINK?

Leaving an Employer for Another Job

John Tatum took his first job with the Booker Company after graduating from college. The Booker Company is in the business of selling office machines.

John is happy with his job. Mr. Booker, the owner of the firm, taught John the business from the ground up. But a big office machine manufacturer recently approached John with a job offer. The opening is in the sales department. The job is very appealing to John, but he feels that leaving Mr. Tatum's firm would not be fair to Mr. Tatum. WHAT WOULD YOU DO?

KEY CONCEPTS

Career values Those factors which have an important effect upon the choice of a career or a job.

Letter of inquiry Letter sent to a possible employer to inquire about a job opportunity.

Nationalism A strong feeling of pride and independence within a nation. It is rising among many of the less developed nations of the world.

QUESTIONS FOR DISCUSSION

1. Identify five specific jobs which have become available to women for the first time in the last ten years.
2. What might happen to the economies of coal- and oil-producing provinces if a major shift to nuclear power occurred?
3. Demonstrate three effects of longer life spans on product design.
4. How can a resource shortage in North America give political power to an underdeveloped country?
5. Is it logical to think of yourself as a "product" in the job marketplace? Discuss.
6. List your major strengths and weaknesses as they might relate to becoming an accountant. Do the same for what you hope to be five years from now.
7. Prepare a complete resumé for yourself as of today. Do the same for what you hope to be five years from now.

INCIDENT:

Bill Mattoon

Bill Mattoon is the chief planner for Weigand Industries Ltd., a large firm with widespread operations in electronics, automation, and the aerospace industries. His forecasting geniuses have predicted (a) twenty years of peace for the major world powers, and (b) the invention of a revolutionary new communications system by a competitor. Weigland Industries Ltd. has a strong capital position.

Questions:
1. How could the firm react to each of the predictions?
2. What impact could each have on the employment of electronics technicians in the areas in which Weigand plants are located?

INCIDENT:

Myrna James

Myrna James is the daughter of a doctor. Her school record through high school is below average even though she has high intelligence. She has shown special interest in working with disturbed children. She just completed high school and is uncertain about going to university. Her study habits are poor.

Questions:
1. What advice would you give Myrna about university?
2. Examine your answer to Question 1. Does it reflect your own values or what you believe are her values? Discuss.
3. Write up a resumé for Myrna. Use your imagination.

ROCKETOY COMPANY IX

In recent months Rocketoy management began to feel that they were "losing touch" with the market. Even Terrence Phillips was now in his fifties, and he was the youngest of the present members of top management. The board of directors had a conference with Phillips and convinced him that it was time to find someone from the younger executive ranks to be promoted to a vice-presidency. Their idea was that, if the person chosen for this position were successful, he or she would succeed Phillips to the presidency of the company within a year or two.

After an extensive search of all the divisions of Rocketoy, a selection committee decided upon Paula Devine. Paula had come to Rocketoy from a large department store five years ago. She had been a toy buyer for seven years before coming to Rocketoy as chief of product design. In her short career at Rocketoy, Paula had made quite a name for herself as a clever analyst of public taste in toys. She had been particularly successful with an idea for a new line of dolls. Her vision and willingness to experiment had brought in some entirely new market segments, including the teen-age market.

Paula was named vice-president for development. She was given the charge to set up a "blue-skies committee," which was expected to study changes in society and technology and to translate such environmental change into new product policy for Rocketoy. Paula, now 33 years old, was ready to assume this challenge. Little did she think 10 years before, when she graduated from university, that she would succeed as well as she had.

Questions:
1. Give Paula some advice about choosing members of her "blue-skies" team.
2. What changes in the environment do you think should enter into the new committee's planning?
3. Assuming you are a recent graduate of the same university Paula attended and you wanted to start a career as a junior designer at Rocketoy, what kind of letter of inquiry would you write to her? Write such a letter.
4. What are the pros and cons of following a career in the toy design department at Rocketoy? Discuss this from the point of view of a 23-year-old male graduate of a western university engineering program and from the point of view of a 20-year-old female with artistic talent who is just completing a two-year community college program in art.

Appendix

Student Data Sheets, Resumes, Letters of Inquiry, and Job Interviews

In Chapter 20, we gave a general discussion of careers in business. The type of thinking you do about career opportunities is perhaps the most important type of analysis you will ever engage in. It affects your entire future life style. Needless to say, you should approach this with a great deal of careful thought.

But there is a big difference between firming up your ideas about career opportunities and landing a job which is related to your career plans. In this appendix we narrow our sights. We assume that you have some rather specific career choices in mind and are seeking employment.

Most of you will not be seeking a full-time job until after completing your studies. As a university student, however, you may be looking for a part-time job right now. Our discussion of student data sheets, resumes, letter of inquiry, and job interviews is helpful in getting both part-time and full-time jobs.

THE STUDENT DATA SHEET

Many schools have on-campus services (e.g., student testing and guidance counselling) which can be helpful to you in career preparation. They are sponsored by Canada Manpower and/or the university or community college which you are attending. This can be very helpful in your choice of a major area of study.

The student placement office is another service offered by many schools. Placement office personnel can help you find a job in the career field of your choice. Find out what services are offered. The personnel there are trained to help you make a smooth and productive transition from student to employee.

Your student placement office schedules on-campus interviews with prospective employers. These employers (business firms, government agencies, and other organizations) send campus recruiters to interview students who are about to complete their courses of study. Later in this appendix we will discuss the mechanics of interviewing in detail. At this point, let's look at the Canada Manpower Master Registration form.

A Master Registration form is a form filled out by the students who sign up for campus interviews at Canada Manpower offices. All students who establish placement files in the placement office fill out this form. A campus interviewer, therefore, can examine this form for each student who signs up for an interview. Figure A-1 shows a registration form.

THE RESUME

A resume is a biolgraphical summary of your education, experience, activities, interests, career goals, and so on. It contains much of the the same type of data which the student data sheet contains. The major difference is that it is tailor-made by the person who is seeking employment.

Your student data sheet is put into your placement file in your school's placement office for inspection by employers who send campus interviewers to your school. But if you want to apply for a job with employers who do not recruit on your campus, you must prepare a resume and mail it to them.

Preparing a good resume takes a lot of effort and care. Personnel departments receive many job application letters and resumes each day from persons seeking jobs. A well-prepared resume—one which is creative, neat, and complete—overshadows others which are poorly prepared, lacking in creativity, messy, and incomplete.

Usually, you will want to send your resume to several employers. Unless you are going to prepare a different resume for each employer or if you are going to type each employer's copy of the same resume separately, you will face the problem of reproducing your resume. Try to produce copies which are as clean and neat as the original.

Do not send carbon copies or poorly reproduced photocopies. Some methods of photo reproduction are acceptable, as are some offset-printed methods. You can get help here from a professional reproduction service either on campus or off campus.

Resume Style and Format

There is no one "best" style and format for a resume. If you are interested in several career options, you may want to prepare several different styles and formats for different employers. Each resume would be tailor-made to best appeal to each employer.

But there are several important guidelines for preparing your resume. One is the "KISS" rule—keep it short and simple! Unless you have had a great deal of prior work experience, your resume probably should not be longer than one typewritten page. To keep it short, do not use complete sentences. The reason is simple—personnel departments do not have unlimited time to devote to examining resumes.

Another good rule is to avoid creating a resume which appears crowded. Don't try to squeeze too much on the page. You've probably seen ads in the classified section of newspapers which have a lot of "white space". The purpose is to draw attention to them. They stand out against a background of ads which are practically all "black space". Skilful use of white space in a resume enables you to draw attention to important parts of your resume. It also increases the chance that your resume will be examined by the person to whom it is sent.

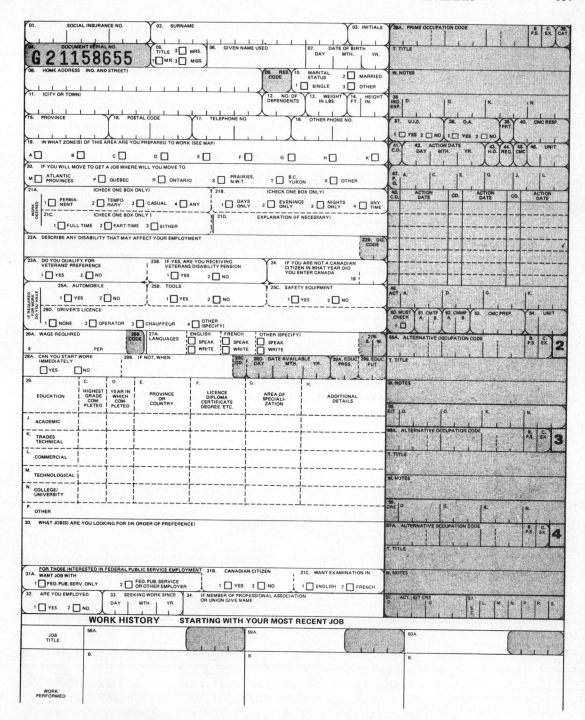

Figure A-1. Canada Manpower Master Registration Form. (Used with permission of Canada Manpower.)

Look at your resume as a type of promotional effort. Your task is to create interest in you as a prospective employee. Be honest, candid, and to the point.

Resume Contents

Remember, you are working with a limited amount of space. You cannot "tell all" you might want to tell. You must concentrate on or "highlight" your strongest points.

Work Experience

If you think that your major appeal is your previous work experience, then stress that experience. Of course, if you are a typical student, you won't have much to highlight here. But if you are an experienced worker, then highlight your experience.

In covering your work experience, list the last job you held first. Then work backwards to your earliest job. If you held summer jobs while in school, list them. If you show that you worked for the same employer for several summers, this "tells" the person who examines your resume that you performed well in your job. If you got a promotion, be sure to indicate that.

Extra-curricular Activities

By all means, list your extra-curricular activities. Membership in a collegiate chapter of a professional association, such as the American Marketing Association, shows that you are seriously interested in professional development. Membership in an honourary organization or society shows that you are a "notch above" the average. Membership in social organizations shows that you work well with others and are "people-oriented". If you held an office in any of these organizations, list that also. This shows leadership ability. The same is true of honours and other types of recognition you may have earned. In other words, try to give some insight into your interests and show how "well-rounded" you are.

Grades

While your resume is not the place to list each course you took and each grade you earned, you should give some indication regarding your overall grade point average. You could do this by stating "average grades in courses outside major; very good grades in courses in major"; or by stating your specific grade point average—"3.2 out of a possible 4.0".

It's hard to say how much importance a given employer will place on grades. Some employers consider grades to be the most objective thing they have to go on in the case of a young prospect with little work experience. Some others, particularly in recent years, "discount" grades due to their belief that "grade inflation" is widespread.

Your grade point average, of course, is not the only thing employers consider. An "average" grade point might look very impressive if you also were involved in on-campus activities and worked part-time. A "high" grade point average might be less impressive if you did nothing else but "kept your nose in a book" the whole time you were in school.

References

Generally, it's a good idea to include the names, titles, and addresses of persons who are in a position to evaluate your performance as a student and/or employee. Get their permission to list them on your resume. Do not list relatives or only friends. If you have reference letters in your file in your school's placement office, you might say that your references can be obtained from that source. Figure A-2 is a sample resume.

PREPARING YOUR LETTER OF INQUIRY

After preparing your resume, you should write a letter of inquiry. Both are mailed to employers with whom you are seeking employment.

Purposes of a Letter of Inquiry

The basic purpose of a letter of inquiry is communication. You are trying to communicate in writing to the employer your interest in working for that employer. You want it to lead to communication from the employer regarding your prospects for a job.

Your letter must first of all get the employer's attention. Secondly, it should spark the employer's interest in you as a prospective employee. Thirdly, it should motivate the employer to respond to your letter—hopefully, by sending a job application form and/or contacting you to set up a preliminary employment interview.

Situations Calling for a Letter of Inquiry

There are two situations in which you might write a letter of inquiry. The first is in response to an employer's ad in a newspaper (or any other medium) seeking applicants for a job in which you are interested. In your letter you would refer to the ad and express your desire to apply for that job. In a sense, this type of letter is solicited (or asked for) by the employer.

The second type of situation exists when you want to apply for a job with an employer even though you are unsure whether that employer is seeking job applicants. Your task of creating employer interest in you is tougher in this situation than in the case where an employer asks for such letters.

Contents of a Letter of Inquiry

In a letter of inquiry written in response to a job ad, you should refer to the ad and where you saw or heard it. This is important because the employer may be advertising

RESUME FOR ALFRED JOHN POWELL

(Date Prepared: April 15, 1979)

Address:	3060 Avenue A, Apt. 21-C Winnipeg, Manitoba R3T 4J7 (to June, 1979)
Telephone:	(204) 288-9543
Home Address:	2106 Myrtle Avenue St. Rose, Man.
Home Telephone:	(204) 721-0810
Personal Data:	Age—21; Weight—175 pounds; Height—6'1"; Health—excellent
Education:	B. Comm. (Honours), University of Manitoba (1979). Majored in marketing. Overall grade point average of 3.2 on 4.0 scale. In major, 3.9. Dean's list during my last two years. I received a scholarship which paid 100 per cent of my tuition during my last two years. I worked during summer vacations and held other part-time jobs to pay 75 per cent of my total expenses during my college career.
Career Goal:	Salesperson, with ultimate goal to move into sales management or marketing management.
Extra-curricular Activities:	Treasurer, Commerce Students Association Chairperson, Business Student Advisory Council (to the Dean, Faculty of Administrative Studies) Member, The University of Manitoba Marketing Association The Commerce Club
Work Experience:	June 1, 1977, to present, Zeppo Electronics Ltd., Winnipeg, Man. Part-time salesperson. Received outstanding part-time salesperson award for 1977. Summers, 1976 & 1975. Beta Retail Stereo Systems, Winnipeg, Man. Stock-clerk. Promoted to salesperson in June, 1976.
References:	Mr. Donald Sykes, Student Placement Officer University of Manitoba, Winnipeg, Man. R3T 2N2 (for references from my college professors). Mr. Anthony Pizzo, sales manager, Zeppo Electronics Ltd., 2001 Centennial St., Winnipeg, Man. R4K 3G4. Ms. Vera Callahan, owner, Beta Retail Stereo Systems, 7911 Portage Ave., Winnipeg, Man. R9A 2Y6.

Figure A-2. A sample resume

for several types of jobs. By referring to the specific ad, you leave no doubt about the job you want.

In an unsolicited letter of inquiry, your task is a little tougher. The task of specifying where you might "fit into" the employer's organization is entirely yours. Furthermore, the employer may be less interested in receiving such letters. You must overcome this barrier in order to avoid getting a polite rejection letter. You must do an extra-good job of motivating the employer to respond.

Regardless of the situation which prompts you to write a letter of inquiry, it must be brief and to the point. Put yourself in the employer's shoes. Write with the "you attitude"–the "you" being the employer. Show your interest and qualifications in a manner which motivates the employer to answer your letter favourably.

You indicate your interest by referring either to a specific job ad or to the employer in general. Again, it is easier to show interest when you are applying for a specific job. When sending an unsolicited letter, avoid being too narrow in stating your career goals. That narrows your appeal. But don't be too broad either. Leave some room for the employer to "fit" you into the organization.

Flexibility is good. It's much easier to "fit" a promising applicant who states his or her career interest as a "bookkeeper" than it is to "fit" the applicant who states his or her career interest as an "accounts payable clerk". If the employer doesn't need an accounts payable clerk, the applicant's letter may be answered with a "sorry, we can't use your talents now" letter. But the applicant for a "bookkeeping job" may be considered for an opening in the employer's accounts receivable department.

In general, try to communicate your career goals, education, experience, and personal qualifications as briefly as possible. Try to stimulate the employer to at least take the time to review your resumé. Figure A-3 is a sample letter of inquiry.

PREPARING FOR JOB INTERVIEWS

No matter what career you want, before you can actually land a job you will have to have a job interview. In fact, before landing a job, you will probably have several interviews.

The Preliminaries

A good way of looking at a job interview is to think of it as a "sales pitch". You are trying to sell yourself to the interviewer. As any good salesperson knows, you don't go into a selling situation without being prepared.

For campus interviews it's a good idea to start interviewing three or four months before you complete your program of study. Starting too early lessens the interviewer's interest in you, since you will not be available for employment in the short run. Few employers can accurately predict their employee needs beyond several months in the future. But don't wait to start interviewing until two weeks before you need a job. That will greatly limit your chances of finding a good job. In fact, you may be so desperate that you will accept a job which is below your skill level.

> 3060 Avenue A, Apt. 21-C
> Winnipeg, Manitoba
> R3T 4J7
> April 18, 1979
>
> Mr. Charles H. Browning
> Personnel Manager
> Nationwide Electronics Ltd.
> P.O. Box 1475
> Toronto, Ontario
> M5D 7P6
>
> Dear Mr. Browning:
>
> I read your advertisement for sales-trainee applicants in the *Winnipeg Tribune,* April 17, 1979.
>
> Your company came to my attention during the time I have been employed as a part-time salesperson with Zeppo Electronics Ltd. in Winnipeg, Manitoba. Ms. Barbara Robbins handles Zeppo's account with Nationwide. We have talked about sales career opportunities with your firm on several occasions. Those conversations, along with some reading I have done on Nationwide in business periodicals, have made me very interested in working for Nationwide.
>
> Although I have been very satisfied with my part-time position with Zeppo, I want to relocate with a larger firm after graduating from university next month. As you can see on my resumé, I majored in marketing. I will graduate in the top 15 per cent of those students receiving B. Comm. (Honours) degrees at the University of Manitoba.
>
> A career in selling has been my goal since my second year in university. Eventually, I want to be a sales manager or a marketing manager and am willing to work hard to reach that goal.
>
> Please consider my qualifications for employment with Nationwide. I am looking forward to hearing from you regarding an interview.
>
> Sincerely,
>
> Alfred J. Powell

Figure A-3. A sample letter of inquiry

A campus interview is an ideal opportunity for you. Prospective employers, in effect, come to you. After you leave school, you will have to find them on your own. In other words, the "search" process is much harder.

But how many interviews should you schedule? In general, it's a good idea to sign up only for those which you are really interested in. If you have never taken an interview, however, you should sign up for two or three to learn what it's all about. Experience is a good teacher of how to interview. But even these "practice interviews" should be with employers who are looking for people to fill jobs which you are interested in. There is always the chance that you could "pull it off" in your first interview. Your attitude is also better when you are really interested in the job.

Do Your Homework
Above all, an interview is an exercise in face-to-face communication. You are the major communicator. An interviewer is not there to sell you on the employer as much as you are there to sell the interviewer on your value to that employer. It's a good idea to look at it as a buyer's market. Remember, the interviewer is the buyer. You are the seller.

If you have an interviewing opportunity, make the most of it. Jot down the place and time where the interview is to be held. That is basic but often overlooked, forgotten, or incorrectly noted.

Gather some information on the prospective employer. If it's a company, go to the library and check reference works such as the *Dun and Bradstreet Reference Book*. If the firm is a corporation, try to locate and to read its recent annual reports to stockholders. Ask your instructors and any others who might be familiar with the firm about its operations. If it's a government agency or non-profit organization, check the *Canada Year Book* and other pertinent publications. You'd be surprised how many interviewees don't have the foggiest notion of the type of business or service the prospective employer is engaged in.

After you have a basic knowledge of the employer's operations (products sold, services performed, size, position in its industry, etc.), your next step is to learn something about the employer's representative—in this case, the interviewer. Don't overlook earlier interviewees as a source of information here. Knowing the interviewer's age, sex, hair style, style of dress, mannerisms, types of questions asked, and so on can help you relate better in the interview. Some interviewees will go into an interview and mispronounce the interviewer's name—and, even worse, the company's name. That does little for the interviewer's ego and even less for his or her first impression of you as a person.

Finally, try to anticipate and be prepared for one or more hard questions. Many interviewers ask questions such as, "Why do you want to work for _____?" "What can you do for _____?" "Why are you here?" If you walk in cold, chances are that you will have difficulty.

Get Your Head on Straight
Don't be nervous. Some people get so tense about interviews that they fall apart at the seams. Try your best to get the butterflies out of your stomach before you go into the interview.

But don't be so relaxed that you give the impression of not really caring whether or not you get the job. What is needed is a mild amount of anxiety—enough to keep you alert and responsive during the interview but not so much as to make you nervous.

Going to the Interview
Take a note pad and pencil and have questions prepared. An interviewer is always impressed when an interviewee has meaningful questions prepared in advance.

Be careful about your appearance. In recent years, many employers and their interviewers have adjusted to some of the newer clothing styles and have modified or done away with "dress codes". But you are wise not to appear in blue jeans and sandals. Cleanliness, neatness, and moderation are very desirable.

Most important, be on time!

The Interview

Your preparation for the interview will pay off once you go through the door to be interviewed. But that was only preparation. You did not rehearse for a part in a play. Interviewers are too diverse a group to permit you to walk in prepared for everything.

Introducing Yourself
By introducing yourself, you avoid the chance that the interviewer will mispronounce your name. Beyond that, it's a good idea to consider yourself a guest. Don't rush to sit down unless you are asked to. Don't walk in with your hand outstretched for a handshake. Let the interviewer make the invitation.

Interviewer-Interviewee Interaction
As we said earlier, an interview is an exercise in face-to-face communication. But who dominates the conversation depends on the interviewer's style of interviewing. Some interviewers like to do most of the talking. Others like the interviewee to do most of the talking. Regardless of how much you did or did not prepare, you will have to "play it by ear" once you are in the interview room.

If the interviewer asks simple, direct questions, that does not mean that "yes" or "no" answers are all that is expected of you. Elaborate, but not to the point of "running off at the mouth". The interviewer's reactions are your best guide here. They tell you how much to say if you are observant.

If there is a noticeable break in the conversation, it's a good clue that you are slipping. Pick up the ball and run with it! It's a good chance to stress your good points—to reinforce and to elaborate on items in your resumé. But don't "blow your horn" so loud that it forces the interviewer to "tune you out".

Above all, be honest and sincere in answering questions. Be creative in asking them. Don't give the impression that you are there simply because you had nothing better to do.

Things to Avoid

There are several interviewing "don'ts". Don't chew gum, smoke, squirm in your chair, stare at the ceiling, or crack your knuckles. Don't take notes as if you were in a lecture class. Don't try to "butter up" the interviewer. Don't be phoney. Don't give the impression that you believe that the business world has been waiting too long for you to become available. Don't try to pull a "snow job".

Don't try to be a comedian. On the other hand, don't be overly serious and pompous. Don't try to hide the "real you". Don't accept a job or schedule further interviews with higher-ups unless you really want the job. Don't "give up" if you get the feeling that you aren't "making it". That's instant death to your chances of landing the job. Finally, don't emphasize an interest in the firm's retirement plan. That tells a lot about your motivation and enthusiasm—all bad!

Getting to Specifics

There are certain important questions which you will want answered. A major one, of course, is the nature of the job. What does it involve? What about the chances for promotion to higher jobs? What is the starting salary? A discussion of the salary often is the most "ticklish" part of an interview. You may be asked what you would expect. You should have a "floor" salary in mind—what you need to get by on, given your personal circumstances. You should also have an idea of what the going rate for that particular job is in the area. But don't give the impression that you're interested only in the salary. Mention the job's challenge to you and your desire to see how you measure up to it. Emphasize that you want to learn.

If asked whether you are willing to relocate, give an honest answer. But you do limit your appeal when you say that you want to stay "close to home". If you want to relocate in a city which is very desirable to a lot of interviewees, be prepared for stiffer competition for jobs in that city.

A much-asked question is, "Where do you see yourself ten years from now?" How well you answer this question depends largely on how well you did your "homework" on the job and the company. The main thing to get across is that you want to move up in the firm. But be realistic about how far and how fast you expect to go.

When asked what kind of employees they want, many top-level executives answer in very general terms. "I want people who can think" is a typical response. In most cases, however, you are not interviewed by a top-level executive. An interviewer has more specific qualities and skills in mind when looking for prospective employees. This is why it's a good idea to try to learn as much as you can about the interviewer before and during the interview. He or she is a "gate-keeper". The "gate" will be closed to you if you do not make a good impression on the interviewer.

Closing the Interview

When you sign up for a campus interview, notice how far apart the interviews are spaced. That gives you an idea about how long a typical interview will last. Make sure you get all your questions answered in the allotted time.

Many people leave interviews in a total state of confusion. Very rarely is a person hired during the first interview. Few are even definitely offered a job. This, of course, is disappointing to many interviewees who expect to "sew it up" during the interview. They often leave feeling frustrated, rejected, or unsure of themselves.

Before leaving, however, you will get some indication as to your status. Usually, you will be told that the interviewer will be in touch with you in a couple of weeks. Thank the interviewer and leave.

After the Interview

Be patient! When the letter arrives, it will be either: (1) an outright rejection; (2) a notice that your skills are not needed now but that your resumé is being placed in the active file; or (3) a letter expressing a desire to arrange another interview with higher-ups.

If, however, you do not hear from the interviewer within a reasonable period after the interview, write a follow-up letter. If you get a letter saying that your resumé is being placed in the active file, write a follow-up after two or three months to keep your resumé active. If you get an outright rejection, do not become discouraged. If you get a letter or phone call for the purpose of setting up another interview, respond tactfully and promptly.

Remember, during your career, you will probably work for several employers. Your choice of a first job and employer is not a choice which will commit you to a lifetime with one employer. Approach that first job with a desire to learn all you can. That increases your mobility and promotability.

Index

Absolute advantage, 427–428, 450
Accountability, 84, 102
Accountants, 226–227, 247, 544–545
Accounting, 35, 226–248
    equations, 229, *231*, 247
    financial, 226, 228–231, 236–238, 247
    managerial, 226, 242–244, 248
    product cost, 243–244, 248
    responsibility, 243, 248
Accounts, 228–229, 247
    receivable, 262
Accrued expense, 233, 247
Activities, 78, 79, 80–81
Administration of government programs, 527–528, 531
Advantage, 427–428, 450
Advertising, 20, 178, 204–208, 458
    agencies, 205, 216
    co-operative, 414
    departments, 204
    media, 204, 205, 216
    regulation of, 207–208
After-tax profits, 38, 61
Age distribution, 498–499
Agency shop, 352, 364
Agent
    manufacturer's, 198, 217
    middleman, 198
    transfer, 261
AIDA process, 206–207, 216
Alberta, 462, 498, 499, 524, 525, 539
Amalgamation, 303, 304
Analysis
    cash flow, 254
    content, 475, 476
    value, 158–159, 162
    vendor, 159, 162
Arbitration, 348, 358–359, 362, 365, 368
Arithmetic mean, 390–391, 397
Articles of incorporation, 56
Assets, 229, 230, 232–233, 247, 267
Auditors, 227, 230
Authority, 83–84, 89, 103

Automation, 89, 149, 203, 362, 376, 397, 468, 469, 537

Background investigation, 318, 319, 333
Bad debts, 288
Balance
    of payments, 433–434, 450, 491
    of trade, 432–433, 450
Bank
    acceptance, 266, 435
    of Canada, 257, 259–261, 263
    chartered, 254, 255, 279, 295–296
    foreign, 435
    Industrial Development, 265, 280
    rate, 260, 279
    services, 256, 386, 418
Bankruptcy, 304, 305, 485
BASIC, 383
Belonging needs, 95, 100, 101
Bilingual/bicultural market, 178, 179, 499, 538
Bill of exchange, 435. *See also* Draft
Blacklist, 342, 365
Bond, 261, 268, *269,* 273, 275, 279, 418
Bondholders, 268, 292
Boom, 326, 489
Boycott, 359–360, 365
Brand, 194–196, 216
Break-even analysis, 392–394, 397
Breaking down, 140, 162, 170
British Columbia, 498, 499, 525, 539
British North America Act, 344, 365, 510, 512
Brokerage function, 269, 270, 271, 279
Business
    cycle, 488–490
    ethic, traditional, 500, 505
    firm, 34–44, 45
    insurance, 410
    policy, 41–44, 45

Caisses populaires, 261
Canada Labour Code, 342, 343, 346, 347, 365
Canada Savings Bonds, 253–254

Capacity
  plant, 144
  workers', 121
Capital, 7, 10, 22, 35
  account, 230
  budget, 291, 305
  formation, 15, 22
  investments, 146
  long-term, 286-287, 290-291, 297-299, 305
  working (short-term), 286-287, 290, 294-297, 305
Capital-intensive, 141, 162
Capitalism, 12, 13-19, 20, 22, 341
Careers in business, 21, 39, 43, 70, 99, 128, 153, 184, 207, 245, 277, 301, 331, 360, 389, 449, 540-547, 549
Career values, 540, 546, 552
Carrier
  common, 201, 216
  contract, 201, 216
  private, 201, 217
Cartel, 445, 450
Cash
  discounts, 213, 295
  payment, 435
  trading, 274, 278, 279
Cash-flow analysis, 254
Centralization, 86, 103
Central processing unit (CPU), 378, 379, 382, 397
Certificate
  of deposit, 287
  of occupancy, 409
Certified cheque, 266, 279
Certified Financial Analyst (CFA), 234
Certified General Accountant (CGA), 234, 244, 544
Chamber of Commerce, 409, 410, 422
Channel of distribution, 196-198, 216
Charter, corporate, 54, 57, 61
Chartered Accountant (CA), 234, 244, 544
Chartered bank, 254, 255, 279, 295-296
Chequing account, 252, 256, 257
Child labour, 342
Civil Service system, 328
Closed shop, 352, 365
Coal, 9, 464, 468, 538
COBOL, 383, 397
COLA clause, 357, 365
Collateral, 257
Collection, 161, 262
Collective bargaining, 341, 348, 358-362, 365
Collectivism, 13, *16*, 18, 19-20, 22
Combination, 140, 162
  export manager, 442, 450
Combines Investigation Act, 208, 522
Commercial letter of credit, 435
Commodity exchange, 278, 279
Common carrier, 201, 216

Common stock, 267, 279, 297
Communication, 121, 124, 129
  systems, 152
Comparative advantage, 428-429, 450
Compensation, employee, 325-328
Competition, 17, 175, 210, 217, 485
  and advertising, 204
  policy, 521, 531
Competitive bidding, 159
Compulsory arbitration, 359, 365
Compulsory overtime, 356
Computer, 89, 126, 152, 201, 203, 373-390, 469
  and accounting, 244, 246
  programs, 376, 383-384, 398
  service firms, 386
  systems analyst, 384
Conciliation, 348, 358, 365
Confiscation, 437, 450
Conflict
  labour-management, 355-358
  of personal goals, 77
Conglomerate, 64
Conspicuous consumption, 179-180, 186
Constitutional responsibilities of government, 511-512
Consumer
  finance company, 262, 280
  goods, 173, 174-177, 186
  protection, 180-183, 186, 196
  sovereignty, 15, 17, 22
Consumerism, 180-183, 186, 196
Consumers Association of Canada, 180, 182
Consumption, 170, 179-180, 186
Containerization, 203, 204, 216
Content analysis, 475
Continuous production, 140-141, 162
Contract
  carrier, 201, 216
  freedom of, 17
  labour, 148, 358-366
  manufacturing, 443, 450
  between old and new owners, 405
Control, 122-124, 129
  chart, 149, 162
  of corporations, 58-59, 292
  exchange, 436, 450
  and franchising, 199
  and sales effort, 209
  tax, 437, 452
Convenience goods, 175, 186
Co-operative, 51, 63, 65, 69
  advertising, 414
  federalism, 513, 531
Corporation, 51, 54, 56-62, 69
  bond sales, 261
  charter, 54, 57
  structure, *58*
Correlation, 474, 476
Cost accounting, 211

Costs of doing business, 35
Counselling Assistance to Small Enterprise (CASE), 417, 422
Countervailing duties, 432, 450
Countervailing power, 67, 69
Craft union, 343, 350, 365
Credibility in marketing, 180
Credit
    and banking, 257
    instruments, 264-266
    line of, 296
    ratios, key, 295
    record, 295
    report, 288, *289*
    revolving, 296, 305
    trade, 295, 305
    unions, 261, 280
Critical Path Method (CPM), 151-152
Crown corporation, 18-19, 519-520, 532
Cumulative quantity discount, 157
Current
    asset, 232, 247, 288
    liability, 233, 247, 293
    ratio, 239, 247, 293

Data processing, 244, 376, 398
Debenture bond, 298
Debt financing, 292, *293*, 305
Debt replacement, 297
Decentralization, 86, 103
Decisions, 124-128, 129
    marketing, 191-221
    non-routine, 127
    routine, 127
Deferred income, 328
Delayed consumption, 37
Delegation, 82, 84, 85, 91, 103
Delphi technique, 476
Demand, 29-32, 33, 45, 468
    approach to pricing, 211
    curve, *32*, 45
    deposit, 256, 280
    draft, 265
    law of, 29, *30*, 45
Demarketing, 180
Democratic socialism, 20
Demography, 178, 498-499, 504
Departmentation, 80-81, 103, 140
Department of Regional Economic Expansion (DREE), 263-264, 488, 525
Deposit
    demand, 256
    expansion, 258-259
    time, 256
Depreciation, 232, 247
Depression, 489
Devaluation, 434, 450
Development, 263, 485-487, 504
Direct transfer of funds, 253
Discharge, 330, 333

Discipline, 332-333
Discount, 157
    cash, 213, 295
    functional, 213, 216
    trade position, 213
Discretionary income, 168, *169*, 179, 186
Discrimination, laws against, 317-318
Dismissal, 330, 333
Disposable personal income, 6-7, *8*, 22
Dissolution of firm, 53, 54
Distribution, 172-173, 176, 186
    channel of, 196-198, 216, 217
    of government power, 510, 532
    of income, 487, 488, 504
Distributor's brands, 195
Diversion in transit, 203
Dividends, 236
    vs. retained earnings, 298
Documentation, 384, 398
Drafts, 265-266, 280, 435
Dun and Bradstreet Canada Ltd., 238, 239, *241*, 288, 419

Echelons of management, 108-109, 129
Ecology, 155, 195, 463-466, 476, 538
Economies of scale, 201, 412, 421
Economic
    development, 485-487, 504
    development council, 410, 422
    environment, 482-483, 504
    growth, 468, 487, 504
    objectives of unions, 352-354
    problem, 4-5, 22
    systems, 11-20
    theory of firm, 483, *484*, 504
Embargoes, 432, 450
Emerging nations, 437
Employee orientation, 320, 333
Entrepreneur, 404, 406, 407-408, 423
Entrepreneurship, 10-11, 23, 406
Environment, 455-458
    ecological, 463-466
    economic, 482-483, 504
    macroeconomic, 485-491, 504
    microeconomic, 483, 504
    physical, 461-463, 477
    political, 510-531
    social, 491-503, 505
    technological, 466-476
    trends, 537-546
Equal pay for equal work, 343
Equity, 229, 233-234, 247, 248. *See also* Capital
    financing, 292, *293*, 297, 298, 305
Ethic
    professional-managerial, 501-503, 505
    Protestant, 14, 23
    traditional business, 500, 505
Exchange, 5, 23
    control, 436, 450

Exchange (*continued*)
   rates, 491
Exit interview, 330, 333
Expense, 233, 247
   account, 229
Export Development Corporation, 263, 438, 520, 525
Exports, 421, 441–442
Expropriation, 437, 450, 525
Extrapolation, 474, 477
Extraterritoriality, 447, 450

Failures, business, 419
Fair employment practices, 346
Federal Business Development Bank, 263, 416, 423, 520
Finance companies, 262, 280, 281
Financial
   accounting, 226, 228–231, 236–238, 247
   institutions, 252–281
   intermediaries, 253, 254–264
   manager, 286, 288
   planning, 286–304
   statement, 231–236, 248
Financing
   debt, 292, *293*, 305
   equity, 292, *293*, 297, 298, 305
   of needs, asset, *294*
Firing, 330
Fixed asset, 232, 247, 290, 291, 419
Floor planning, 297, 305
Flowchart, 385
Flow of funds, 253, 264
Forecasting, 472–474, 477
Foreign
   assembly, 443, 450
   banks, 435
   direct investment, 447, 450
   exchange, 435
   ownership, 472
   sources of funds, 264
Foreign Investment Review Agency, 14, 448, 518, 520
Foremen, 83
Form utility, 170, 186
FORTRAN, 383, 398
Franchises, 199, 216, 411, 423
Free-market economy, 29
Frequency distribution, 392, 398
Fringe benefits, 328, 332
Front-end bonuses, 317
Functional authority, 89, 103
Functional discount, 213, 216
Futures markets, 278, 280

Generics, 195
Geological maps, 526
GNP. *See* Gross National Product
Gold, 433, 434–435
Goods, 173–177, 186, 187
   in process, 288

Goodwill, 233
Government
   bonds, 435
   contracts, 421
   financial institutions, 263–264
   insurance, 302
   loans, 418
   as market, 174, 211, 527
   ownership, 51
   policy formulation, 528–531, 532
   power, distribution of, 510
   program administration, 527–528, 531
   regulation, 13, 34, 61, 211, 346–347, 519, 521–523
   review boards, 520–521
   roles of, 16, 17–19, 516–519
   securities, 260, 287
   and small business, 415–418
   structure, 510–516
   subsidies, foreign, 439
Grants, 263, 525
Great Depression, 51, 490
Grievance procedures, 361–362, 365
Gross margin, 211, 215
Gross National Expenditure, 35
Gross National Product (GNP), 6–7, *8, 9*, 23, 429, 470, 487
Gross profit, 235–236, 247
Groups, 96–101, 494–495, 505
Growth
   economic, 487, 504
   population, 498
   small business, 420–422
   stock, 297
Guaranteed annual wage, 354, 365

Hardware (computer), 378–382, 398
Health, 331, 332
Hierarchy
   of human needs, 94–96, 103
   of organizational objectives, 79–80, 103
Holding company, 64, 303, 305
Human resource, 146, 311, 313–314, 333

Imports, 429, 430–433, 441
Incentive programs, 13, 34, 524–526, 532
Incentives, employee, 120, 326, 334
Income
   distribution of, 487, 488, 504
   personal, *8*
   statement, 235, 247
   tax, 20, 409, 487–488, 489, 505, 524
Incorporation, 56–57
Independent local union, 350, 365
Indirect ownership of stock, 67
Industrial
   goods, 173–174, 186
   Revolution, 341, 342, 463
   union, 350, 365
Inflation, 332

Informal groups, 97-101, 103
Information flow, 227-228
Informed judgment, 475-476, 477
Injunction, 361, 365
Input-output device, 381, 398
Inputs, 288
    to production, 138, 140
    productivity, 484-485
    supply and price, 483
Insolvency, 304
Institutions, 495-496, 498, 504
Insurance, 232, 299-302, 410
    companies, 298, 418
    premiums, 288
Interest
    bond, 272
    prime rate of, 257
    rates, 489
Intermediary, financial, 253
Intermittent production, 140, 162
International
    Monetary Fund, 438
    sources of funds, 264
    trade, 439-444, 451
    union, 350, 366
Interviews, personnel
    exit, 330, 333
    final selection, 319-320, 333
    in-depth, 319, 334
    preliminary, 317, 335
Inventory, 232, 288, 381-382
    control, 155
    turnover, 215, 217
Investment, 15, 275
    dealer, 271
    foreign direct, 447
    portfolio, 447, 451
Involuntary bankruptcy, 304
Isolationism, 436, 451

Japan, 439, 441, 487, 498
Job
    analysis, 314, 335
    application form, 317-318, 335
    description, 314, *315*, 335
    enlargement, 357
    enrichment, 357
    hunting, 547
    interviews, 561-566
    marketplace, 540
    point system for compensation, 327
    safety, 331
    security, 356, 362
    selection formula, 548
    specification, 314, 335
Job-skill training, 320-322
Joint venture, 64, 413, 443, 451

Key ratio, 239, 240-241, 247

Labour, 7, 9, 23, 35
    conflict with management, 355-358
    contract, 148, 358-366
    force, as individuals, 492-494
    legislation, 344-348
    turnover, 332, 421
    unions, 343-364, 366
Labour-intensive, 14, 162
Laissez faire, 12, 23
Land, 7, 8-9, 23, 35
Law. *See also* Legislation
    of demand, 29, *30, 36,* 45
    of large numbers, 302, 305
    of supply, 29, *30, 36,* 45
Layoff, 330, 361
Layout, plant, 145
Leader items, 200, 215
Leadership, 98, 118-119, 120, 129
Leasing, 291-292, 305
Legislation, 344-348, 519, 521-523
Letter of inquiry, 547, 552, 559, 561, 562
Letters patent, 57
Leverage, 292-293
Liability, 229, 233, 247, 267
    corporate, for employee wages, 57
    current, 233, 247
    joint, 53
    limited, 56-57, 60
    long-term, 233
    unlimited, 52, 70
Licences, 52, 409
Licensing, 276, 443, 451
Life insurance company, 261, 280
Limited liability, 56-57, 60
Limit order (securities), 271
Line
    authority, 86, 103
    conflict with staff, 87-89
    of credit, 296, 305
    functions, 87, 103
Linear programming, 396, 398
Liquidity, 287, 294, 305
Listed securities, 271
Loans, 254, 257, 261, 295-296
Local union, 350, 366
Location, 143-144, 170, 176, 185
Lockout, 348, 361, 366
Long-term capital, 286-287, 290-291, 297-299, 305

Macroeconomic environment, 485-491, 504, 527
Maintenance, 154, 162
Make-or-buy decision, 143, 162
Management, 42, 129
    and computers, 389-390
    development, 322, 335
    echelons of, 108-109, 129
    effectiveness, 117
    by exception, 82

INDEX

Management (*continued*)
  functions of, 113–124, 129
  goals, corporate, 57
  information system (MIS), 126
  inventory, 314, 335
  materials, 91
  matrix, 91
  by objectives (MBO), 324, 325, 335
  operations, 159–161, 162
  participative, 121–122, 130
  personnel, 311–333
  recruits, 316, 320
  of sales force, 208–209
  science, 395
  span of, 80, 81–86, 103
  training, franchisee, 412
  workers' view of, 112
Managers, 61, 83, 110–113, 122, 129, 132, 148, 328
Managerial
  accounting, 226, 248
  approach to marketing, 170–171, 186
  skills, 109–110, 129
Manitoba, 363, 499, 524, 525
Marginal utility, 170
Margin trading, 274, 280
Market
  economy, 29, 45
  penetration, 213, 217
  price, *31*
  research, and computers, 382
  segmentation, 177, 178, 186, 196
Marketing, 168–186
  boards, 521
  concept, 171, 186
  decisions, 191–221
  to government and non-profit institutions, 174
  managerial approach to, 170–171, 186
  mix, 171, 177, 185, 186
  research, 178–179, 186, 187
Markup, 211, 214–215, 217
Materials management, 156–159
Matrix management, 91
Maturity, bond, 292, 305
Maximum hours, 346
Mean, arithmetic, 390–391, 397
Measurement of performance, 122, 123
Mechanical processing (accounting), 244
Media advertising, 204, 205
Median, 391, 398
Mediation, 348, 358, 366
Memorandum of association, 57
Mercantilism, 11, 23
Merger, 302, 305, 421, 470
Merit rating, 323, 335
Microeconomic environment, 483, 504
Middlemen, 173, 187, 195, 198–200, 267, 386
Minicomputers, 381, 386–387
Minimum wage, 346

Minority hiring, 336
Mixed economic system, 12, 18–19, 23
Mode, 391, 398
Model building, 395–396, 398
Money, 252, 259, 280
Monopolistic competition, 210, 217
Monopoly, 33, 445
Morale, 208, 246, 316, 332, 456
Motivation, 36–37, 119–121, 129, 149, 414
Multinational company, 444–445, 451
Mutual company, 64
Mutual fund, 275, *276,* 280, 282

Nationalism, 539, 552
Nationalization, 437, 451
National union, 350, 366
"Near cash", 287
Needs, 5, 77, 94–96, 120
Negotiable promissory note, 264
Net profit before taxes, 236
New product development, 41
Non-profit institutions as market, 174
Non-routine decision, 127, 129
Non-tariff barrier, 432, 451
Norms, group, 98, 495, 504
Northwest Territories, 498, 525
Notes receivable, 232
Nuclear power, 155, 466, 538

Objectives, 76–80, 113–114
Obsolescence, 155, 162, 174, 192–193, 217, 291, 362, 467, 468, 550
Occupational licence, 409
Oil, 32, 441, 445, 461, 483, 524, 538
Oligopoly, 33, 216, 217
On-line system, 381, 398
Ontario, 463, 598, 499, 525
Open shop, 352, 366
Operational planning, 116, 128, 130
Operations management, 159–161, 162
Operations research, 394–395, 398
Opportunity, 39–40, 45, 288, 305
Order control, 149
Organization, 76, 103
  chart, 90–94, 103
  formal, 77, 86, *101*
  informal (group), 91, 97–101
Organizing, 116–117, 130
Output of production, 138, 141, 211
Overseas manufacturing, 443–444
Over-the-counter market, 272, 280
Overtime, compulsory, 356
Ownershop utility, 170, 187

Parkinson's Law, 90
Parliament, 57, 511, 519
Participative management, 121–122, 130
Partnership, 51, 52–55, 70
Patent, 196, 217, 467, 468, 470
Pension funds, 263, 280

Performance appraisal, 322-324, 335
Permits, 52, 409
Personal selling, 208-209, 217
Personnel, 311
    administration, 311-312, 335
    department, 312, 335
    management, 148, 311-333, 335
    services, 330, 332
PERT, 150, 151-152, 162
Peter Principle, 329
Petroleum. *See* Oil
Physical environment, 461-463, 477
Physiological needs, 95, 96, 101
Picketing, 359-360, 367
Piece rate, 326, 335
Piggyback exporting, 442, 451
Place utility, 170, 187
Planned obsolescence, 192-193, 217, 467
Planning, 115-116, 130, 422
    operational (tactical), 116, 128, 130
    strategic, 116, 128, 130
Plant
    capacity, 144, 162
    layout, 145, 162
    location, 143-144
Political
    borders, 440
    environment, 510-531
    objectives of unions, 351-352
Pollution control, 465, 466, 525
Portfolio, 275, 447, 451
Power, 67, 69, 496, 504
Preferential tariff, 436
Preferred stock, 267-268, 280
Prepaid expenses, 232, 288
Preventive maintenance, 154, 162
Price, 20, 29-32, *36*, 171, 172, 187, 210-215, 217, 272-273, 441, 468, 470, 483
Pricing, 210-215
    demand approach to, 211
    leader, 200
    market penetration, 213, 217
    model, 213, 217
    skimming, 214, 221
Primary demand advertising, 205
Primary research, 179, 187
Prime rate of interest, 65, 257, 280
Privacy, 61, 388, 394
Private
    brands, 195
    carriers, 201, 217
    enterprise, 23
    ownership, 51
Privy Council Order 1003, 344, 368
Product, 171-173, 187, 191
    as "bundle of utilities", 171, 191
    cost accounting, 243-244, 248
    demand, 483
    differentiation, 177, 187
    integrity, 183
    life cycle, 191-193, 217
    mix, 193, 194, 217
    price, 483
Production, 140-155, 162
    control, 149-155
    and ecology, 155
    factors of, 7, 11, 23
    inputs to, 140
    management, 141-155, 288
    mix, 143
    to order, 141
    organizing for, 146-148
    outputs of, 141
    planning, 142-145
    processes, 140-141, 362
    staffing for, 148
    for stock, 140-141
Productivity, 149, 327, 332, 358, 484-485, 504
Professional-managerial ethic, 501-503
Profit, 10, 15, 35-38, 45, 59, 76, 113, 142, 145, 161, 171, 195, 211, 213, 229, 230, 235-236, 242, 286, 287, 325, 328, 356, 396, 404, 406, 421, 489, 490, 500, 502, 524
Program, computer, 376, 398
Progressive income tax, 487-488, *489*, 505
Promissory note, 264, *265*, 280
Promotion
    of employees, 120, 328-329, 335, 421, 550
    sales, 172-173, 185, 187, 204-210, 221, 413, 414
Property, private, 13, 14-15, 23
Property taxes, 143, 524
Prospectus, 277, 280
Protecting the firm's investment, 286, 299-302, 332
Protective tariff, 431
Protestant ethic, 14, 23
Provincial
    export development agencies, 438
    liquor boards, 521
    Premier, 514
Proxy, 58-59, 70, 303
Psychology, 54, 94-97, 179-183, 325, 387-398, 407-408, 466, 489, 492-494
Publicity, 209-210, 217
Public market (investment), 266-267
Public relations, 210, 217, 502
Purchasing, 156-159
Pure risk, 302, 305

Quality control, 152-154, 162
Quantitative tools, 390-396, 399
Quantity
    demanded, 29, 33
    discount, 213
    supplied, 29, 33
Quebec, 179, 345, 499, 511, 538
Quota, 326

INDEX

Rate of exchange, 434
Ratio, 238–242
    current, 239
    key, 239, 240–241
    sales-to-inventory, 239
Recapitalization, 304, 305
Recession, 489
Reciprocity, 158, 162
Recognition
    company, by public, 412, 414
    personal, as worker goal, 121
Record-keeping, 35, 61
Recruiting, 316, 335, 495
Recycling, 155
Redemption price, bond, 273
Rediscount rate, 260
Regional trading bloc, 436, 451
Registered Industrial Accountant (RIA), 234, 544
Registrar for corporations, trust company as, 261
Registration cards, product, 179
Reinvested profit, 38
Reorganization, 304, 305
Research, marketing, 178–179, 186, 187
Research and development (R & D), 469–472, 477
Reserve requirement, 259, 281
Resignation, 330, 335
Resource, 4, 7, 12, 32–34, 78, 79
Responsibility, 84, 103
    accounting, 243, 248
Resumé, 556, 558, 560
Retained earnings, 231, 298
Retirement, 328–330
Revenue, 229, 235, 248
    tariff, 431
Right
    to manage, 83, 356
    to privacy, 388
    to work, 353
Risk, 10, 15, 40–41, 45, 115, 126, 199, 275, 282, 299, 301, 404
Routine decision, 127, 130
Royal prerogative, 57
Royalties
    franchiser, 414
    government, 524, 532

Safety, 95, 96, 101, 331, 332, 346
Salary, 326, 336
Sales
    finance company, 262, 281
    forecast, 242, 248
    promotion, 172–173, 185, 187, 204–210, 221, 413, 414
    tax, 524
Sales-to-inventory ratio, 239
Sample, statistical, 392, 399

Saskatchewan, 437, 524, 525, 539
Savings certificate, 257
Scalar chain, 82
Scenarios, 474, 477
Sealed-bid purchasing, 299
Search, personnel, 314, 316
Secondary research, 178, 187
Secured loans, 254, 296, 305
Securities, 232
    commission, 61
    exchanges, 270–271, 281
    regulation, 276–278
Security, job, 352–354, 356
Selection of employees, 316–321, 333, 336
Self-evaluation, 43, 542, 546–547
Self-insurance, 299
Selling agent, 198
Seniority, 324, 336, 356–357
Sensitivity training, 322
Services, 141, 174, 183, 185
Severance pay, 332
Sex discrimination, 320, 343
Shareholders. *See* Stockholders
Shopping goods, 175, 187
Short selling, 274, 281
Sight draft, 265
Simulation, 475, 477
Sinking-fund, 297, 305
Skimming pricing, 214, 221
Small business, 404–422
Small group, 494–495, 505
Social
    auditing, 502
    environment, 491–499, 505
    esteem, 95, 96, 101, 121
    insurance, 302
    objectives of unions, 354
    problems, 114, 117
    responsibility of business, 39, 500–503, 505
Socialization, 96, 103
Societal variables, 498–499
Software, 383–384, 399
Sole proprietorship, 51–52, 70
Soviet Union, 20, 440, 441, 462
Span of management, 81, 82, 83–86, 103
Specialization, 4, 23, 60, 421
Specialty goods, 175, 187
Specific duty, 432
Speculative risk, 302, 305
Speculative trading, 274, 281
Staff, 87, 103, 118, 130
Stagflation, 490, 505
Standard
    hours, 346
    of living, 6, 15, 23, 327, 352
    performance, 122, 123
Statement
    of financial position, 231–236, 248
    income, 235, 247
State trading company, 440, 451

Statistics, 390–392, 399, 466
Stock
    common, 267, 279
    exchanges, 60, 269. *See also* Securities exchanges
    preferred, 267–268, 280
    prices, 272–273
Stockbroker, 60, 271, 276
Stockholders, 57, 58, 70, 268, 292
Strategic planning, 116, 128, 130
Strategy, 42–44, 45
Stress, 111–113
Strike, 348, 359–360, 361, 366
Structure, 76, 77–80, 90–94, 116
Sub-optimized objectives, 77
Subordinates, 118. *See also* Span of management
Supply, 29–32, 33, *36*, 45
Supply-and-demand approach to compensation, 327
Survival of the small firm, 418–420, 422
Systems
    concept, 116–117, 130
    analysts, computer, 384

Tangible net worth, 239, 248
Target market, 171, 187
Tariff, 13, 430–432, 436, 451, 473
Tax
    bracket, 328
    Control, 437, 452
    employer, 409–410
    excise, 524
    exemptions, 143
    income, 20, 409, 487–488, 489, 505, 524
    policy, 488
    real property, 143, 524
    royalties, 524
    sales, 409, 524
    shelters, 328
    social insurance, 409, 410
    unemployment insurance, 410
Taxation
    of corporations, 61
    in democratic socialist countries, 20
    of partnerships, 53
    of sole proprietorships, 52
    in United States, 20
Technology, 375, 462, 466–477, 537
    and foreign trade, 445
    gap, 471
    and management, 83
    prediction of, 291
Terminating employees, 329–330
Theft, 229, 267
"Theory X" vs. "Theory Y" managers, 122, 132
Time
    deposit, 256, 281

    draft, 265
    sharing, 381–382, 386, 399
    utility, 170, 187
Total cost concept, 201, 221
Trade
    acceptance, 265, 281, 435
    barriers, 430, 437, 439, 445, 452
    credit, 295, 305
    position discount, 213, 221
Trademark, 196, 221
Trading
    cash, 274, 278
    margin, 274, 280
    speculative, 274
    stamps, 210
Traditional values, 499, 505
Training
    job-skill, 320–322, 335
    management, franchisee, 412
    performance appraisal, 322
    sensitivity, 322
    "vestibule", 321
Transaction, 226, 229, 248
Transfer agent, 261
Transportation, 144, 201–204
Treasury bills, 287
Treatment, 140, 162, 170
Trust company, 260–261, 281
Truth in advertising, 208
TSE index, 273
Turnover
    labour, 332, 421
    stock, 239

Underdeveloped countries, 538–539
Underdeveloped in Canada, 486
Underwriting, 269, 270, 281
Unemployment, 114, 356, 468, 469, 489
    insurance, 302
Unfair lists, 343, 368
Union, labour, 69, 327, 341–364
    in Canadian history, 341–345
    craft, 343, 350, 366
    dues, 362
    of future, 362–364
    independent local, 350, 365
    industrial, 350, 365
    international, 350, 366
    local, 350, 365
    motivation to join, 354–355
    national, 350, 365
    objectives, 350–354
    types of, 350
Union shop, 352, 368
United States, 12, 20, 441, 462, 498, 517
Unity of command, 82
Universe, statistical, 392
Unlimited liability, 52, 70
Usefulness. *See* Utility

INDEX

USSR. *See* Soviet Union
Utility, 5, 6, 23, 169–170, 186, 187

Vacations, 346
Value analysis, 158–159, 162
Value (in exchange), 6, 23
Values, sociological, 499, 504, 505
Vendor analysis, 159, 162
Venture capital firm, 263, 281
"Vestibule" training, 321
Voluntary arbitration, 358, 368
Voluntary bankruptcy, 304

Wage, 326, 336
    rates, 149, 470
    and salary administration, 325, 336
Wage-and-price controls, 19, 32
Wants, 4, 5, 25

Waste
    disposal, 463, 464
    materials, recycling of, 155
Wholesalers, 199–200
Women's rights, 537
"Workaholics", 113
Working
    capital, 286–287, 290, 294–297, 305, 418, 419
    conditions, 354
    mothers, 364
Workmen's compensation, 302, 410
World Bank, 438

Youth and unions, 363

Zoning, 409